**Statistics for Biology and Health**

*Series Editors:*
M. Gail
K. Krickeberg
J. Samet
A. Tsiatis
W. Wong

AF785788

# Statistics for Biology and Health

*Bacchieri/Cioppa:* Fundamentals of Clinical Research
*Borchers/Buckland/Zucchini:* Estimating Animal Abundance: Closed Populations
*Burzykowski/Molenberghs/Buyse:* The Evaluation of Surrogate Endpoints
*Duchateau/Janssen:* The Frailty Model
*Everitt/Rabe-Hesketh:* Analyzing Medical Data Using S-PLUS
*Ewens/Grant:* Statistical Methods in Bioinformatics: An Introduction, 2nd ed.
*Gentleman/Carey/Huber/Irizarry/Dudoit:* Bioinformatics and Computational Biology Solutions Using R and Bioconductor
*Hougaard:* Analysis of Multivariate Survival Data
*Keyfitz/Caswell:* Applied Mathematical Demography, 3rd ed.
*Klein/Moeschberger:* Survival Analysis: Techniques for Censored and Truncated Data, 2nd ed.
*Kleinbaum/Klein:* Survival AnalysisL A Self-Learning Text, 2nd ed.
*Kleinbaum/Klein:* Logistic Regression: A Self-Learning Text, 2nd ed.
*Lange:* Mathematical and Statistical Methods for Genetic Analysis, 2nd ed.
*Lazar:* The Statistical Analysis of Functional MRI Data
*Manton/Singer/Suzman:* Forecasting the Health of Elderly Populations
*Martinussen/Scheike:* Dynamic Regression Models for Survival Data
*Moyé:* Multiple Analyses in Clinical Trials: Fundamentals for Investigators
*Nielsen:* Statistical Methods in Molecular Evolution
*O'Quigley:* Proportional Hazards Regression
*Parmigiani/Garrett/Irizarry/Zeger:* The Analysis of Gene Expression Data: Methods and Software
*Proschan/LanWittes:* Statistical Monitoring of Clinical Trials: A Unified Approach
*Siegmund/Yakir:* The Statistics of Gene Mapping
*Simon/Korn/McShane/Radmacher/Wright/Zhao:* Design and Analysis of DNA Microarray Investigations
*Sorensen/Gianola:* Likelihood, Bayesian, and MCMC Methods in Quantitative Genetics
*Stallard/Manton/Cohen:* Forecasting Product Liability Claims: Epidemiology and Modeling in the Manville Asbestos Case
*Sun:* The Statistical Analysis of Interval-censored Failure Time Data
*Therneau/Grambsch:* Modeling Survival Data: Extending the Cox Model
*Ting:* Dose Finding in Drug Development
*Vittinghoff/Glidden/Shiboski/McCulloch:* Regression Methods in Biostatistics: Linear, Logistic, Survival, and Repeated Measures Models
*Wu/Ma/Casella:* Statistical Genetics of Quantitative Traits: Linkage, Maps, and QTL
*Zhang/Singer:* Recursive Partitioning in the Health Sciences
*Zuur/Ieno/Smith:* Analysing Ecological Data

K.G. Manton · Igor Akushevich ·
Julia Kravchenko

# Cancer Mortality and Morbidity Patterns in the U.S. Population

## An Interdisciplinary Approach

Foreword by Herbert Kim Lyerly

Springer

K.G. Manton
Duke University
Durham, NC, USA
kgmanton@duke.edu

Igor Akushevich
Duke University
Durham, NC, USA
igor.akushevich@duke.edu

Julia Kravchenko
Duke University
Durham, NC, USA
yk27@duke.edu

*Series Editors*
M. Gail
National Cancer Institute
Bethesda, MD 20892
USA

K. Krickeberg
Le Chatelet
F-63270 Manglieu
France

J. Sarnet
Department of
Epidemiology
School of Public Health
Johns Hopkins University
615 Wolfe Street
Baltimore, MD 21205-2103
USA

A. Tsiatis
Department of Statistics
North Carolina State
University
Raleigh, NC 27695
USA

W. Wong
Department of Statistics
Stanford University
Stanford, CA 94305-4065
USA

ISSN 1431-8776
ISBN 978-1-4419-2680-7 e-ISBN 978-0-387-78193-8
DOI 10.1007/978-0-387-78193-8

© Springer Science+Business Media, LLC 2009
Softcover reprint of the hardcover 1st edition 2009
All rights reserved. This work may not be translated or copied in whole or in part without the written permission of the publisher (Springer Science+Business Media, LLC, 233 Spring Street, New York, NY 10013, USA), except for brief excerpts in connection with reviews or scholarly analysis. Use in connection with any form of information storage and retrieval, electronic adaptation, computer software, or by similar or dissimilar methodology now known or hereafter developed is forbidden. The use in this publication of trade names, trademarks, service marks, and similar terms, even if they are not identified as such, is not to be taken as an expression of opinion as to whether or not they are subject to proprietary rights.
While the advice and information in this book are believed to be true and accurate at the date of going to press, neither the authors nor the editors nor the publisher can accept any legal responsibility for any errors or omissions that may be made. The publisher makes no warranty, express or implied, with respect to the material contained herein.

Printed on acid-free paper

springer.com

*This book is dedicated to our parents*

# Foreword

Over the past 50 years, a dramatic change has occurred in the incidence of cancer and cancer death. For example, in the early 1950s, a significant percentage of the population were cigarette smokers. In addition, therapies and screening were not nearly as effective as they are today. These have been proved to change cancer mortality in a large variety of cancers. For example, the death rate due to a childhood cancer dramatically decreased over the past 30 years, making this a much more treatable disease than it had been in the past. Finally, an emerging recognition is that the rate of death of cardiovascular diseases has dramatically decreased allowing for individuals to have longer lives but increased their exposure to additional diseases such as cancer.

Collectively, these series of events have dramatically altered the landscape of cancer and the role of cancer mortality in our society. For example, it is widely accepted that cancer is one of the leading causes of loss of productive life in U.S. society. The ability to develop insights into the distribution of these cancer deaths is a critical need. This volume is a significant work by the authors which will dramatically enhance our ability to look carefully at the changing role of cancer and its impact on our society. The depth of the information provided is complimented with significant insights into the emerging biology of cancer and the recognition that individual cancers in fact, may represent specific molecular pathway perturbations that may dramatically influence the cause and treatment of these diseases. Largely unexploited opportunity is the ability to determine the impact of cancer in geographically distributed populations that may lend insight into causality and treatment or treatment availability. For many reasons, we are grateful to the authors for providing this master work to allow us to begin to interrogate the landscape of America and try to strategically plan for the needs for our population. I am grateful for the extraordinary effort put forth by the authors in this work and know they will have significant impact on our decision making and plans for the future.

Herbert Kim Lyerly

# Preface

*Scientists differ from ordinary mortals by their ability to admire loquacious and complicated delusions*

*(Anatole France)*

*Everything should be made as simple as possible, but not simpler.*
*As far as the laws of mathematics refer to reality, they are not certain, and as far as they are certain, they do not refer to reality*

*(Albert Einstein)*

This book is the result of the joint efforts of three authors, each with different backgrounds: in sociology and mathematical demography, in theoretical and mathematical physics, and in medicine and biochemistry. Moving through the step-by-step understanding of each others' specialties and expertise, they teamed together to perform studies on each research fields' "edges". The process required "penetrations" of each others' areas, gaining and exchanging new knowledge and understanding of their field's aims, analyzing results and making conclusions – thereby forming an interdisciplinary research team.

The idea of an interdisciplinary approach to research has been around for half of a century only to become realistic in the recent decade. A huge amount of knowledge accumulated in various areas of science and art often needs a fresh look from other perspectives. Math and physics are two disciplines whose "law and order" may be fruitfully applied to other areas. Could anybody have guessed that it is possible to dance Newton's laws (i.e., the ballet "Newton: Three Laws of Motion" by Julia Adam, Lawrence Pech Dance Company) or take Einstein's theory as inspiration for a dance (i.e., the ballet "Constant Speed" by Mark Baldwin, Rambert Dance Company), or implement in music the mathematical progression of Pascal's triangle (i.e., "9 through 99" by Peter Adriaansz)? We can only guess what Newton and Einstein would think about these interpretations.

Medical sciences spent thousands of years on observation and experimentation studying human diseases, trying to discover nature's secrets to extend human life span by gaining control over deadly cancer, cardiovascular disease, and infections. This fight was successful in some cases, and failed in others.

Huge amounts of accumulated knowledge are obviously needed to be analyzed from different perspectives to produce effective strategies for improving human health. This requires trying new "tools", even those "borrowed" from other scientific areas.

The interdisciplinary approaches used in our book are not as unprecedented as those in ballet and music, but they are innovative enough to be studied intensively. Who are the potential readers we wrote the book for? The spectrum is wide: epidemiologists, physicians, oncologists, medical researchers, biologists, demographers, geneticists, specialists in mathematical modeling, as well as public health, and environmental specialists may find considerable useful information. It may also be of interest for medical, biology, and math undergraduates and postgraduates, as well as for researchers (beginners and experienced), especially for those interested to widen the spectrum of their studies and work in close contact with other specialists in interdisciplinary teams.

All chapters have been written with these readers' interests in mind. Chapters 1, 2, 3, 6, 7, 8, and 9 contain information about historical aspects of cancer studies, biological mechanisms of carcinogenesis, cancer risk factor interventions, cancer morbidity and mortality, age-specific patterns and their time trends, specific features of cancer histotypes, as well as approaches to cancer prevention. For somebody who is primarily interested in the description of various modeling approaches in cancer studies, Chapters 2, 4, 5, 7, and 8 would be useful reading. A more detailed description of how this book is structured, and which chapter is dedicated to what topic, may be found in Section 1.3. A glossary may be used as a useful source for quick searching for biomedical and mathematical terminology used in this book.

We are particularly grateful to Dr. Herbert Kim Lyerly, Director of Duke Comprehensive Cancer Center, George Barth Geller Professor for Research in Cancer in the Duke University School of Medicine, who kindly agreed to provide an independent review of the results in this book as a substantive specialist in oncology.

Durham, NC
July 2008

Kenneth Manton
Igor Akushevich
Julia Kravchenko

# Contents

**1 Cancer Contra Human: Cohabitation with Casualties?** ........... 1
1.1 Endless Assault: Prehistory and History of Human Cancer........................................... 1
1.2 Global and U.S. Cancer Morbidity and Mortality Trends: Historic Perspectives........................... 6
1.2.1 Global Cancer Morbidity and Mortality: At the Beginning of 21st Century....................... 6
1.2.2 U.S. Cancer Morbidity and Mortality: At the Beginning of 21st Century ................. 12
1.2.3 Cancer Mortality: U.S. Historical Trends........... 21
1.3 Interdisciplinary Approach to Population Health Studies ................................. 31
References............................................. 34

**2 Cancer Modeling: How Far Can We Go?**....................... 37
2.1 Cellular Aspects of Carcinogenesis ..................... 38
2.1.1 Nuclear DNA Mutation ........................ 38
2.1.2 Mitochondrial DNA Mutation.................. 39
2.1.3 Damage to the Protein Generation Machinery of the Endoplasmic Reticulum and Golgi Apparatus ............................... 41
2.1.4 Cell–Cell Communication....................... 42
2.1.5 Telomere Control of Cell Division ............... 43
2.1.6 Apoptosis.................................. 44
2.1.7 Angiogenesis ................................ 45
2.1.8 Immunomodulation........................... 45
2.1.9 Metalloproteinases ........................... 46
2.2 Theories of Carcinogenesis.......................... 46
2.2.1 Somatic Mutation Theory of Carcinogenesis................................ 48
2.2.2 The Stem Cell Theory of Carcinogenesis ........... 48
2.2.3 Mutation versus Epigenetic Theories of Carcinogenesis................................ 48

2.2.4 The Tissue Organization Field Theory of Carcinogenesis and Neoplasia . . . 49
2.2.5 Telomere Dysfunction Theory . . . 49
2.3 An Overview of Formal Quantitative Models of Carcinogenesis . . . 49
2.3.1 Nordling, and Armitage and Doll . . . 53
2.3.2 The Moolgavkar–Venzon–Knudson Two-Stage Model . . . 55
2.3.3 The Generalized MVK and Armitage–Doll Models . . . 58
2.3.4 The Multiple Pathway Models of Carcinogenesis . . . 59
2.3.5 Mixed Models of Carcinogenesis . . . 60
2.3.6 Cancer at Old Age and Approaches to Modeling: If the Cancer Incidence Rates Are Declining? . . . 61
2.3.7 Complexity and Chaos Theory . . . 64
2.3.8 Statistical/Empirical Cancer Models . . . 64
2.3.9 The Other Common Modeling Approaches for Carcinogenesis . . . 65
2.4 Modeling for Populations with Heterogeneous Mutational Events and Tumor Growth Rates . . . 66
2.4.1 Model Innovations, Fuzzy State Processes, Heterogeneous Tumor Risks, and Tumor Growth Rates . . . 66
2.4.2 Intracellular Processes: Interactions Complicate Modeling . . . 66
2.4.3 Tumor Growth and Growth Heterogeneity . . . 71
2.4.4 Stochastic Multivariate Models of Carcinogenesis . . . 73
2.5 Summary . . . 74
References . . . 76

**3 Cancer Risk Factors** . . . 89
3.1 Overview of Cancer Risk Factors . . . 89
3.1.1 Biomarkers . . . 89
3.1.2 Genotoxic and Nongenotoxic Mechanisms . . . 90
3.1.3 Controllable and Noncontrollable Risk Factors . . . 94
3.1.4 Advanced Age as Cancer Risk Factor . . . 115
3.1.5 Factors to Consider in Cancer Risk Analysis and Cancer Risk Prediction . . . 116
3.2 Environmental Cancer Risk Factors . . . 118
3.2.1 Radiation Exposure . . . 119
3.2.2 Nutrition as a Cancer Risk Factor . . . 134
3.3 Summary . . . 140
References . . . 140

**4 Standard and Innovative Statistical Methods for Empirically Analyzing Cancer Morbidity and Mortality** . . . . . . . . . . . . . . . . . . . . 151
4.1 Survival Analysis and Life Table Models . . . . . . . . . . . . . . . . 152
4.1.1 Useful Modification of Standard Life Table Computations . . . . . . . . . . . . . . . . . . . . . . . . . . . . . . . . 155
4.1.2 Standard Regression Procedures and Some of Their Limitations in Describing Longitudinal Data. . . . . . . . . 158
4.2 Multiple Disease Stochastic Compartment Models for Complex Cancer Population Mortality Curves: A Two-Disease Analysis of U.S. Female Breast Cancer . . . . . . . . . . . . . . . . . 161
4.3 Stochastic Process Models of Cancer Risk: Latent and Observed State Variables Dependence. . . . . . . . . . . . . . . . . . 164
4.3.1 Microsimulation Estimation of Stochastic Process Parameters. . . . . . . . . . . . . . . . . . . . . . . . . . . . . . . . 170
4.4 Evaluation of Characteristics of Individual and Grouped Data . . . . . . . . . . . . . . . . . . . . . . . . . . . . . . . . 175
4.4.1 Detection of Disease Onset . . . . . . . . . . . . . . . . . . . . . . 177
4.4.2 Cancer Incidence in U.S. Elderly. . . . . . . . . . . . . . . . . . 177
4.4.3 Sensitivity Analysis . . . . . . . . . . . . . . . . . . . . . . . . . . . 179
4.4.4 Uncertainty in Onset Calculation . . . . . . . . . . . . . . . . . 179
4.4.5 Medicare Coverage and Censoring Uncertainties . . . . . . . . . . . . . . . . . . . . . . . . . . . . . . . 179
4.4.6 Age Reporting Uncertainties. . . . . . . . . . . . . . . . . . . . . 180
4.5 Generalized Frailty Model . . . . . . . . . . . . . . . . . . . . . . . . . . . 182
4.6 Summary . . . . . . . . . . . . . . . . . . . . . . . . . . . . . . . . . . . . . . . 185
References . . . . . . . . . . . . . . . . . . . . . . . . . . . . . . . . . . . . . . . . 186

**5 Stochastic Methods of Analysis** . . . . . . . . . . . . . . . . . . . . . . . . . . . 191
5.1 Introduction . . . . . . . . . . . . . . . . . . . . . . . . . . . . . . . . . . . . . 191
5.2 Stochastic Process Models . . . . . . . . . . . . . . . . . . . . . . . . . . . 193
5.3 Microsimulation and Interventions . . . . . . . . . . . . . . . . . . . . 196
5.4 Grade of Membership and Other Latent-State Analysis Methods . . . . . . . . . . . . . . . . . . . . . . . . . . . . . . . . . . . . . . . . 199
5.4.1 GoM Model. . . . . . . . . . . . . . . . . . . . . . . . . . . . . . . . . . 202
5.4.2 LLS Model . . . . . . . . . . . . . . . . . . . . . . . . . . . . . . . . . . 203
5.5 A Geo-epidemiological/Mapping Study of the U.S. Cancer Mortality Rates and Trends Based on an Empirical Bayes Approach . . . . . . . . . . . . . . . . . . . . . . . . . . . . . . . . . . . . . . . 205
5.6 Summary . . . . . . . . . . . . . . . . . . . . . . . . . . . . . . . . . . . . . . . 212
References . . . . . . . . . . . . . . . . . . . . . . . . . . . . . . . . . . . . . . . . 213

**6 U.S. Cancer Morbidity and Mortality Trends** . . . . . . . . . . . . . . . . . 217
6.1 Introduction: Cancer Mortality and Morbidity Registration . . . . . . . . . . . . . . . . . . . . . . . . . . . . . . . . . . . . . . . 217

6.2 U.S. Cancer Mortality ... 218
6.3 U.S. Cancer Incidence ... 223
6.4 Morbidity and Mortality of Specific Cancer Sites ... 232
6.4.1 Lung Cancer ... 232
6.4.2 Breast Cancer ... 235
6.4.3 Prostate Cancer ... 235
6.4.4 Colorectal Cancer ... 236
6.4.5 Esophageal Cancer ... 237
6.4.6 Stomach Cancer ... 237
6.4.7 Cancer of Pancreas ... 238
6.4.8 Cancer of Liver and Intrahepatic Duct ... 238
6.4.9 Cancer of Corpus Uteri ... 239
6.4.10 Cervical Cancer ... 240
6.4.11 Kidney Cancer ... 240
6.4.12 Thyroid Cancer ... 241
6.5 Summary ... 241
References ... 242

**7 U.S. Cancer Morbidity: Modeling Age-Patterns of Cancer Histotypes** ... 247
7.1 Introduction ... 247
7.2 Analyses of Trends of Cancer Histotypes in the U.S. Population ... 248
7.2.1 Lung Cancer ... 248
7.2.2 Esophageal Cancer ... 256
7.2.3 Stomach Cancer ... 257
7.2.4 Colorectal Cancer ... 257
7.2.5 Cancer of Pancreas ... 258
7.2.6 Liver Cancer ... 258
7.2.7 Breast Cancer ... 259
7.2.8 Cancer of Uterus ... 259
7.2.9 Ovarian Cancer ... 260
7.2.10 Cancer of Cervix Uteri ... 260
7.2.11 Prostate Cancer ... 261
7.2.12 Cancer of Kidney ... 261
7.3 Analyses of Selected Cancer Histotypes for the Two-Disease Model ... 262
7.3.1 Breast Cancer: If Genetic Background May Result in Two Forms of Disease? ... 279
7.3.2 Cervical Cancer: Age Periods of Increased Susceptibility and Cancer Risk ... 283
7.3.3 Hepatocellular Carcinoma: Behavioral Risks and Viral Hepatitis Infections ... 290
7.3.4 Prostate Cancer: Screening Effects, Genetic Predisposition, or Something Else? ... 293

7.4 Squamous Cell Carcinomas and Adenocarcinomas: The Time Trends ........ 295
7.5 Summary ........ 298
References ........ 300

**8 Risk Factors Intervention** ........ 313
8.1 Environmental Risk Factor Contribution to Cancer and Noncancer Diseases ........ 313
8.1.1 Health-Based Population Forecasting Effects of Smoking on Mortality from Cancer and Noncancer Diseases ........ 315
8.2 Atherosclerosis and Cancer: Anything in Common? ........ 327
8.3 Making Projections of Cardiovascular Disease Risk ........ 333
8.4 Discussion and Comparison of the Results of the Two Intervention Studies ........ 338
8.5 Summary ........ 341
References ........ 343

**9 Cancer Prevention** ........ 351
9.1 Brief Overview of Prevention History, Strategies, Conquests, and Uncertainty ........ 351
9.2 Smoking ........ 359
9.3 Diet ........ 362
9.4 Obesity ........ 371
9.5 Physical Activity ........ 372
9.6 Alcohol ........ 373
9.7 Nonsteroidal Antiinflammatory Drugs ........ 373
9.8 Statistical and Modeling Approaches for Prognoses ........ 373
9.9 The Present and Future of Microsimulation Models in Tasks of Cancer Prevention ........ 380
9.10 Summary ........ 381
References ........ 383

**Conclusion and Outlook** ........ 393

**Keywords** ........ 395

**Abbreviations** ........ 397

**Glossary** ........ 403

**Index** ........ 445

# Chapter 1
# Cancer Contra Human: Cohabitation with Casualties?

## 1.1 Endless Assault: Prehistory and History of Human Cancer

When writing about cancer, it is difficult to avoid the military terminology, both in telling the story of single cancer patients and in studying the problem at a population level. The books' titles remind us of messages from battlefields or reports of an FBI agent (e.g., several most recent books: "The Secret History of the War on Cancer", "The Breast Cancer Wars...", "The War on Cancer: An Anatomy of Failure", and even "Cancer Wars: How Politics Shapes What We Know and Don't Know About Cancer" and "Cancer-Gate: How to Win the Losing Cancer War" (Davis, 2007; Lerner, 2003; Faquet, 2005; Proctor, 1995; Epstein, 2005)). That warfare started from the first steps of humankind. A popular opinion is that cancer is the price human pay for civilization. However, cancer is not a disease of our modern industrialized age. The origins of cancer go back far into the evolutionary time. Direct evidence shows how cancer was manifested in archeological records. A British/Kenyan archeologist Louis Leakey (1903–1972) in 1932 found what is likely the oldest hominid malignant tumor in the remains of a body of a *Homo erectus*, or possibly *Australopithecus*. This tumor had features suggestive of a Burkitt's lymphoma (although that nomenclature was not in use then). The mummified skeletal remains of Peruvian Incas who lived approximately 2400 years ago contained abnormalities suggestive of signs of melanoma (Morton and Moore, 1997). The tissue specimen of a human cancer was found in the remains of the skull of a female who lived in the Bronze Age (1900–1600 BC) – the tumor characteristics suggested a head and neck cancer. The oldest known "clinical" description of human cancer in written records is found in Egyptian papyri dating between 3000–1500 BC. Two of them, known as the "Edwin Smith" and "George Ebers" papyri, contain details of conditions that are consistent with modern descriptions of cancer. The "Edwin Smith surgical papyrus" (originally written in 3000 BC, it was purchased in 1862 by Edwin Smith in Luxor, Egypt) – undoubtedly one of the most significant medical texts ever discovered – describes 8 cases of breast tumors (or "ulcers) concluding that there is no treatment for this condition and recommending cauterization (it was performed with a tool called "the fire drill") as a

K.G. Manton et al., *Cancer Mortality and Morbidity Patterns in the U.S. Population*, Statistics for Biology and Health, DOI 10.1007/978-0-387-78193-8_1,
© Springer Science+Business Media, LLC 2009

palliative measure (Feldman and Goodrich, 1999; Atta, 1999; Morton and Moore, 1997; Udwadia, 2000). Hieroglyphic inscriptions and papyri manuscripts suggest that ancient physicians were able to distinguish between benign and malignant tumors and suggested that the surface tumors may be removed surgically. Compounds of barley, pig ears, and other indigenous materials were proposed as treatment for certain cancers, e.g., cancers of uterus and stomach. At about the same time a Chinese physician Huang Ti (born 2698 BC) wrote the "Nei Ching" – the oldest treatise of internal medicine, in which he gave the first detailed description of tumors (Lee, 2000).

Hippocrates (460–370 BC) is thought to be the first person to clearly recognize the difference between benign and malignant tumors. He noticed that blood vessels around a malignant tumor looked like the claws of a crab. He named this disease *karkinos* (the Greek name for crab). In English, this term translates to *carcinos* or *carcinoma.* Hippocrates believed that it was better to leave cancer alone because those who were treated did not survive as long as those untreated. He attributed cancer to an excess of black bile (thought to be one of the body's four fluids or humors: blood, phlegm, yellow bile, and black bile). This theory was passed through the period of Roman Empire and remained the "gold standard" through the Middle Ages – totally for more than 1300 years. During that period, autopsies were prohibited almost everywhere for religious reasons, thus limiting medical knowledge (Morton and Moore, 1997; Udwadia, 2000).

In 168 AD, a Roman physician and philosopher, Claudius Galen of Pergamum (129–216 AD), who is considered to be the first oncologist, believed that some cancer cases could be curable in the early stages, and that advanced tumors should be operated by cutting around the affected area or by cauterization. However, he considered that in most cases, a patient was incurable after a diagnosis of cancer. As did Hippocrates, Galen believed breast cancer to be a side effect of melancholia, directly connected to unhealthy diet and bad climate. Paul of Aegina (625–690), one of the famous Greek and Byzantine physicians, wrote in 657 AD the "Epitomae Medicae Libri Septem", where he described breast and uterus cancers as the most common tumors (Lee, 2000). For breast cancer, he recommended removal as opposed to cauterization, and he asserted that uterine cancer surgery was useless (Gurunluoglu and Gurunluoglu, 2003). Moses Maimonides (1135–1204), a physician, philosopher, and rabbi, treated cancer by "uprooting the entire tumor and its surroundings up to the point of healthy tissue, except if the tumor contains large vessels and/or the tumor happens to be situated in close proximity to any major organ, then excision is dangerous" (Rosner, 1998).

During the Renaissance, in the 15th and 16th centuries, the theory that cancer was caused by excess of black bile continued to prevail. While considered an incurable disease, temporary measures were used for symptom relief in cancer patients (e.g., arsenic-containing creams). An English physician, William Harvey (1578–1657), performed autopsies providing insights into the circulatory system. At approximately the same time, an Italian physician,

Gaspare Aselli (1581–1625), discovered the lymphatic system, which led to the end of the old theory of black bile as a cause of cancer: the new theory suggested that abnormalities in the lymphatic system caused cancer. Claude Deshaies-Gendron (1663 ca.–1750), a French physician, suggested that cancer was untreatable with drugs and must be surgically removed with all its "filaments". A Dutch surgeon Adrian Helvetius (1661–1727) performed lumpectomy and mastectomy for breast cancer, claiming surgical removal being a curative procedure.

In the 18th century, two French scientists – a physician, Jean Astruc (1684–1766) and a chemist, Bernard Peyrilhe (1735–1804) – made the first step in experimental oncology. That led to the consideration of oncology as a medical specialty, with the specialized study of cancer and establishing hospitals for cancer patients. Bernardino Ramazzini (1633–1714), an Italian physician, noticed in 1713 the virtual absence of cervical cancer and the high incidence of breast cancer in nuns; that led to the discovery of the importance of hormonal factors, such as pregnancy and infections related to sexual contact in modifying cancer risk. George Ernst Stahl (1660–1734) and Friedrich Hoffman (1660–1742), German physicians and chemists, theorized that cancer was composed of fermenting and degenerating lymph varying in density, acidity, and alkalinity. John Hunter (1728–1793), a Scottish surgeon, also supported that theory and suggested that some cancers might be cured by surgery. He described methods by which it was possible to distinguish the surgically removable tumors: if a tumor has not encroached to the nearby tissues and was still moveable, it should be removed. An Italian anatomist, Giovanni Morgagni of Padua (1682–1771), known as "the father of the modern anatomical pathology", was the first to perform in 1761 an autopsy to look for the pathological findings in a patient after death, which laid the foundation for scientific oncology. The same year Sir John Hill (1714–1775), a physician from London, was the first to recognize the potential carcinogeneity of tobacco: he published a book "Cautions Against the Immoderate Use of Snuff". In 1775, Percival Pott (1714–1788), a surgeon of Saint Bartholomew's Hospital in London, showed that chimney sweeps had an occupation-related risk of cancer of the scrotum caused by soot collecting under their scrotums (Greaves, 2002). This led to identification of various occupational carcinogenic exposures and public health measures to reduce cancer risk. In 1779, the first cancer hospital was founded in Reims, France. However, it was moved away from the city due to a widespread fear that cancer was contagious.

The 19th century saw the birth of scientific oncology with several very important inventions: the modern microscope, the discovery of anesthesia, and X-rays. In 1829, Joseph Claude Anthelme Recamier (1774–1852), a French gynecologist, in his treatise "Recherches sur le traitement du cancer" coined the term "metastasis" as a definition for the spread of cancer. A German physiologist and comparative anatomist, Johannes Muller (1801–1858), has suggested in 1838 that cancer is made up of cells and not lymph. He described that cancer cells came from "blastema" (the undifferentiated tissue from which it was believed that cells arose). The introduction of anesthesia to the mainstream by American dentists Horace Wells (1815–1848) in 1844 and William Thomas Green Morton (1819–1868) in 1846, allowed the performance of radical mastectomy to treat female breast cancer. By the end of the 19th century, with the development of better microscopes, it was discovered that cancer cells differed in appearance from the normal cells from which they originated. A German physician, public health activist, pathologist, and biologist Rudolf Virchow (1821–1902), a student of Johannes Muller, provided the scientific basis for the modern pathologic study of cancer and correlated the clinical course of illness with microscopic findings: tissues removed by surgeons were examined under the microscope to make a precise diagnosis of cancer. He suggested that chronic inflammation could lead to cancer development, and that malignant cells might spread through an identified "liquid". Virchow's method not only allowed a better understanding of the damages caused by cancer in human organs and tissues but also created a basis for the development of cancer surgery.

Since anesthesia became available, the next hundred years became known as "the century of the surgeon". Surgeons Christian Theodore Billroth (1829–1868) from Germany, W. Sampson Handley (1872–1962) from England, and William Stewart Halsted (1852–1922) from the United States developed the "cancer operations" designed to remove all of tumor together with the regional lymph nodes. Theodore Billroth performed the first esophagectomy (1871), the first laryngectomy (1873), and most famously, the first successful gastrectomy (1881) for gastric cancer, after many ill-fated attempts (legend has it that Billroth was nearly stoned to death in the streets of Vienna when his first patient died after the procedure). In 1890, William Stewart Halsted, a Professor of Surgery at Johns Hopkins, Harvard, and Yale, performed radical mastectomies to treat breast cancer: he removed breast with underlying muscles and axillary lymph nodes and managed to achieve an impressive 72% of 5-year cure rate for patients whose tumor had not spread to adjoining glands (Olson, 2002).

A German surgeon, Karl Thiersch (1822–1895), suggested that cancer metastasized through the spread of malignant cells and not through an unidentified fluid. In 1889, Steven Paget (1855–1926), an English surgeon and the son of a famous British surgeon and pathologist Sir James Paget (1814–1899) (he is best remembered for Paget's disease), proposed the "seed and soil" theory of cancer (even though in his paper "The distribution of secondary growths in cancer of the breast", The Lancet 1889, he clearly stated that "...the chief advocate of this theory of the relation between the embolus and the tissues which receive it is Fuchs..." [Ernst Fuchs (1851–1930), an Austrian ophthalmologist]). He reanalyzed approximately 1000 autopsies of women with breast cancer and noticed that the patterns of metastasis were not random; he proposed that tumor cells (the "seeds") have a specific affinity for specific organs (the "soil"), and metastasis would only result if the "seed" and "soil" were compatible (Fidler et al., 2002; Pantel et al., 1999).

In the late 1800s, Sir George Thomas Beatson (1848–1933), an England surgeon and scientist who has been called later "the father of endocrine ablation in cancer management", had discovered the stimulating effect of estrogen on breast cancer. He found that oophorectomy often resulted in improvement of breast cancer patients, proving the stimulating effect of the female ovarian hormone – estrogen, on breast cancer far before that hormone itself was discovered. His findings initiated the development of hormone therapy for breast cancer patients which is widely used now.

The German physicist Wilhelm Conrad Roentgen (1845–1923), who is considered "the father of diagnostic radiology", at his lecture "Concerning a new kind of ray" in 1896 presented a term "X-rays" (X was the algebraic symbol of an unknown quantity), and later that year the X-rays apparatus was devised for diagnosis (he was awarded the first Nobel Prize in physics in 1901 for the work) (Morton and Moore, 1997). Roentgen died in 1923 of carcinoma of the intestine. It is still unclear whether this tumor was a consequence of his repeating exposure to X-rays, or it was spontaneous (Roentgen was one of the few pioneers in the field who used protective lead shields routinely). Within 3 years of X-rays discovery, radiation was used in cancer treatment: radiation therapy began with radium and with relatively low-voltage equipment. It was later discovered that radiation could cause cancer, as well as cure it.

The 20th century has brought further discoveries in cancer. In 1911, Francis Peyton Rous (1879–1970), an American pathologist from the Rockefeller Institute in New York, provided the scientific background for a viral theory of cancer by injecting chickens with cell-free liquids obtained from chicken's sarcomas (that was later known as the *Rous sarcoma virus*), thus observing the formation of sarcomas on the injected hens (he was awarded the Nobel Prize for that work in 1968). In 1915, cancer was induced in laboratory animals at the Tokyo University by applying coal tar onto the skin of rabbits. In 1926, the Nobel Prize was awarded to a Danish scientist Johannes Andreas Grib Fibiger (1867–1928) for his work that showed nematodes caused cancer in mice and rats. However, his findings were unable to be replicated by other scientists and were later discredited (nevertheless, some credit Fibiger for showing that external stimuli can induce cancer).

During the 20th century, many chemical and physical carcinogens were identified. In 1939, based on his research of androgen levels and prostate cancer in dogs Charles Brandon Huggins (1901–1997), an urologist at the University of Chicago, has reported a regression of metastatic prostate cancer following removal of the testis. That was an important observation provided a background for the use of hormone therapy for certain cancers. After World War II, a compound called nitrogen mustard was studied and found to have substantial cytotoxic activity against lymphoma. This substance (not by breathing the irritating gas, but by injecting the prescribed drug intraveinously) was used to treat patients with advanced lymphomas. This agent served as a model for a group of more effective alkylating agents. Sidney Farber (1903-1973), a pediatric pathologist at Harvard Medical School has demonstrated that aminopterin, a compound related to folic acid, produces remission in acute leukemia in children, thus becoming the predecessor of methotrexate. In 1956, methotrexate was used to treat a metastatic form of rare tumor – choriocarcinoma. Acute childhood leukemia, testicular cancer, and Hodgkin's disease now can be cured with chemotherapy, and many other cancers can stay under control for a fairly long period, even if not actually cured. Different approaches are being studied now to improve the efficacy of chemotherapy and to reduce its adverse effects – e.g., liposomal therapy, monoclonal antibody therapy, adjuvant therapy, combination chemotherapy, colony-stimulating factors, chemopreventive agents, hematopoietic stem cell therapy, agents overcoming multidrug resistance.

Advances in technology now make it possible to improve the methods of radiation therapy. Various techniques were implemented since Roentgen's discovery – e.g., three-dimensional conformal radiation therapy, conformal proton beam radiation therapy, stereotactic radiosurgery and stereotactic radiotherapy, and intraoperative radiation therapy. Other methods such as hyperthermia, chemical modifiers/radiosensitizers, and boron neutron capture therapy might enhance the therapeutic effect of ionizing radiation.

In 1960, Howard Temin (1934–1994), a U.S. geneticist, showed that certain RNA viruses are capable of inserting their genetic material in the host cells' DNA, and thus may contribute to cancer formation. In 1976, Harold E. Varmus (born 1939) and J. Michael Bishop (born 1936), American scientists, "for their discovery of the cellular origin of retroviral oncogenes," were awarded the 1989 Nobel Prize in physiology or medicine. In 1986, Stephen H. Friend, with coauthors, isolated the first tumor suppressor gene – the retinoblastoma gene. In 1995, the first DNA microarray chip was constructed: initially created to measure gene expression in plants, this technology is now used to study human cancer. In 1999, human epithelial and fibroblast cells were transformed for the first time into tumor cells in a laboratory. Studies of gene mapping and gene expression, angiogenesis, signal transduction, and other sophisticated methods can make a deeper insight into mechanisms of carcinogenesis, thus helping to individualize preventive and therapeutic strategies in oncology.

## 1.2 Global and U.S. Cancer Morbidity and Mortality Trends: Historic Perspectives

### *1.2.1 Global Cancer Morbidity and Mortality: At the Beginning of 21st Century*

More than 10 million new cancer cases occurred worldwide in 2000, accounting for over 7 million deaths worldwide (that was 13% of total mortality and 22% of noncommunicable disease mortality), being exceeded only by cardiovascular disease (CVD) (29% of total mortality), and infectious and parasitic diseases (19% of total mortality) (Shibuya et al., 2002). The number of deaths caused by cancer is expected to rise worldwide to 10.3 million in 2020 (Cancer Facts & Figures, 2006).

Various attempts have been made to quantify the global burden of cancer and to estimate site-specific cancer mortality and morbidity (Murray and Lopez, 1997; Pisani et al, 1999; Parkin et al., 1999; Ferlay et al., 2001). The efforts made by the International Agency for Research on Cancer (IARC) have led to the CLOBOCAN estimates, which has also used information on incidence and survival to estimate cancer deaths from various sources, including tumor registries. The GLOBOCAN database has been built up using the huge amount of data available in the Descriptive Epidemiology Group of IARC. Incidence data are available from cancer registries and cover entire national populations, or samples of such populations from selected regions. Cause-specific mortality data are available for many countries, however their precisions vary considerably depending on the country. Because the sources of data are continuously improving in quality and extent, estimates may not be truly comparable over time, and care should be taken when comparing these estimates with those published earlier (the observed differences may be the result of a change in the methodology and should not be interpreted as a time trend effect) (GLOBOCAN, 2002). Compared to the estimates of the GLOBOCAN, the Global Burden of Disease (GBD) study estimates for global cancer mortality and incidence were higher (for 2000 year – 11 and 3% higher, respectively). These differences were predominantly due to a substantially large difference in the African countries, EMRO[1] and SEARO[2] regions. According to GBD, cancer mortality as a proportion of total mortality differed substantially by region – from less than 5% in African regions to approximately 30% in Australia, Japan, and New Zealand. Lung cancer, accounting for 17% of total cancer mortality, holds the leading position in causing cancer death, followed by cancers of the stomach (12%), liver (9%), colorectal (9%), and breast (7%) (Shibuya et al., 2002). Lung, stomach, and liver cancers are the most common cause of cancer deaths in males, whereas breast and lung cancers in females. In African regions, the leading causes of cancer mortality are liver and cervical cancers, both of which are primarily due to viral infections. In European countries, the pattern of cancer mortality is typical for industrialized countries: lung, colorectal, and prostate cancers for males and breast, colorectal, and lung for females.

The worldwide cancer death rates for 2002 are presented in Table 1.1. For all sites (in total), the top three positions are held by Hungary, Slovakia, and Kazakhstan for males and Zimbabwe, Denmark, and Hungary for females (in 1988–1992, the leading position was held by Hungary, Czechoslovakia, and New Zealand for males and Hungary, Israel, and Czechoslovakia for females [Tominaga and Oshima, 2000]). Analysis of cancer sites showed that in 2002 the highest death rates among 50 countries were as follows – for colorectal cancer:

[1] EMRO – Bahrain, Cyprus, Iran, Jordan, Kuwait, Lebanon, Libya, Oman, Qatar, Saudi Arabia, Syria, Tunisia, United Arab Emirates, Egypt, Iraq, Morocco, and Yemen.

[2] SEARO – Indonesia, Sri Lanka, Thailand, Malaysia, Phillippines, Brunei, Singapore, Bangladesh, Bhutan, India, Maldives, Nepal, Afganistan, and Pakistan.

**Table 1.1** Cancer around the world, 2002. Death rates* for 50 countries (from Cancer Facts & Figures, 2006)

| Country | All sites | | Colorectal | | Liver | | Lung and bronchus | | Breast | | Uterus | | Esophagus | | Stomach | |
|---|---|---|---|---|---|---|---|---|---|---|---|---|---|---|---|---|
| | Male | Female | Male | Female | Male | Female | Male | Female | Male | Prostate | Cervix | Corpus | Male | Female | Male | Female |
| United States | 152.6 | 111.9 | 15.2 | 11.6 | 4.4 | 2.0 | 48.7 | 26.8 | 19.0 | 15.8 | 2.3 | 2.6 | 5.1 | 1.2 | 4.0 | 2.2 |
| | (30) | (16) | (29) | (24) | (31) | (37) | (16) | (2) | (18) | (28) | (44) | (18) | (21) | (23) | (50) | (50) |
| Australia | 147.1 | 99.0 | 18.7 | 13.3 | 3.4 | 1.5 | 34.7 | 13.8 | 18.4 | 17.7 | 1.7 | 1.6 | 4.9 | 1.8 | 5.7 | 2.8 |
| | (33) | (31) | (16) | (15) | (42) | (45) | (30) | (12) | (22) | (22) | (50) | (40) | (24) | (14) | (47) | (48) |
| Austria | 156.0 | 106.7 | 20.1 | 13.9 | 7.1 | 2.5 | 37.7 | 12.1 | 20.6 | 18.4 | 4.1 | 2.5 | 3.8 | 0.7 | 10.3 | 6.5 |
| | (28) | (22) | (9) | (11) | (19) | (27) | (28) | (17) | (12) | (17) | (29) | (19) | (34) | (37) | (28) | (23) |
| Azerbaijan | 132.7 | 80.2 | 3.8 | 2.8 | 3.3 | 2.0 | 28.1 | 5.1 | 13.7 | 4.5 | 2.8 | 6.0 | 10.1 | 6.1 | 30.0 | 13.1 |
| | (41) | (47) | (50) | (50) | (45) | (37) | (38) | (44) | (39) | (49) | (38) | (2) | (6) | (6) | (5) | (7) |
| Bulgaria | 139.5 | 86.3 | 17.1 | 11.4 | 7.3 | 3.2 | 39.1 | 6.9 | 16.0 | 8.9 | 8.0 | 2.8 | 2.4 | 0.5 | 15.0 | 7.6 |
| | (39) | (41) | (25) | (25) | (17) | (20) | (25) | (35) | (33) | (39) | (13) | (14) | (44) | (44) | (21) | (19) |
| Canada | 156.6 | 114.3 | 16.1 | 11.7 | 3.8 | 1.7 | 48.5 | 25.6 | 21.1 | 16.6 | 2.5 | 1.9 | 4.7 | 1.3 | 5.9 | 2.8 |
| | (27) | (15) | (27) | (23) | (38) | (42) | (17) | (3) | (11) | (25) | (40) | (32) | (26) | (21) | (46) | (48) |
| Chile | 148.9 | 114.4 | 7.7 | 7.8 | 6.6 | 4.1 | 21.0 | 7.6 | 13.1 | 20.8 | 10.9 | 1.3 | 7.4 | 3.4 | 32.5 | 13.2 |
| | (31) | (14) | (39) | (37) | (21) | (13) | (42) | (31) | (42) | (10) | (9) | (44) | (13) | (9) | (3) | (6) |
| China | 159.8 | 86.7 | 7.9 | 5.3 | 35.3 | 13.3 | 36.7 | 163 | 5.5 | 1.0 | 3.8 | 0.4 | 21.6 | 9.6 | 32.7 | 15.1 |
| | (23) | (40) | (37) | (45) | (1) | (1) | (29) | (9) | (50) | (50) | (30) | (49) | (1) | (3) | (2) | (4) |
| Colombia | 141.1 | 122.5 | 7.3 | 7.6 | 7.6 | 7.1 | 19.9 | 10.0 | 12.5 | 21.6 | 18.2 | 1.5 | 4.7 | 2.1 | 27.8 | 15.7 |
| | (36) | (8) | (40) | (38) | (15) | (4) | (43) | (21) | (44) | 99) | (5) | (42) | (26) | (12) | (7) | (2) |
| Croatia | 212.6 | 104.6 | 23.4 | 13.0 | 7.3 | 3.2 | 65.3 | 9.7 | 20.0 | 13.5 | 5.0 | 2.5 | 5.8 | 0.8 | 19.4 | 8.0 |
| | (5) | (25) | (6) | (17) | (17) | (20) | (4) | (23) | (14) | (32) | (24) | (19) | (18) | (33) | (13) | (18) |
| Cuba | 139.8 | 100.2 | 10.7 | 13.5 | 4.2 | 3.8 | 38.0 | 16.2 | 14.6 | 26.4 | 8.3 | 5.8 | 4.4 | 1.4 | 6.9 | 3.6 |
| | (38) | (29) | (35) | (14) | (35) | (16) | (26) | (10) | (37) | (4) | (12) | (3) | (32) | (18) | (43) | (42) |
| Czech Republic | 216.4 | 126.6 | 34.0 | 18.0 | 7.7 | 3.6 | 61.8 | 12.8 | 20.0 | 17.2 | 5.5 | 4.6 | 4.7 | 0.7 | 12.1 | 6.4 |
| | (4) | (5) | (2) | (4) | (14) | (18) | (7) | (15) | (14) | (24) | (23) | (5) | (26) | (37) | (24) | (25) |
| Denmark | 179.2 | 148.1 | 23.3 | 19.2 | 3.4 | 2.3 | 45.2 | 27.8 | 27.8 | 22.6 | 5.0 | 2.9 | 7.0 | 1.9 | 5.4 | 3.3 |
| | (15) | (2) | (7) | (2) | (42) | (32) | (20) | (1) | (1) | (7) | (24) | (12) | (14) | (13) | (48) | (45) |
| Estonia | 201.7 | 106.3 | 17.9 | 12.6 | 3.6 | 1.6 | 62.2 | 7.3 | 20.4 | 17.6 | 6.6 | 3.6 | 4.6 | 0.4 | 24.1 | 11.4 |
| | (8) | (23) | (22) | (18) | (41) | (44) | (6) | (34) | (13) | (23) | (20) | (7) | (31) | (48) | (8) | (9) |

**Table 1.1** (continued)

| Country | All sites | | Colorectal | | Liver | | Lung and bronchus | | Breast | | Uterus | | Esophagus | | Stomach | |
|---|---|---|---|---|---|---|---|---|---|---|---|---|---|---|---|---|
| | Male | Female | Male | Female | Male | Female | Male | Female | Male | Prostate | Cervix | Corpus | Male | Female | Male | Female |
| Finland | 130.2 | 93.0 | 11.5 | 9.8 | 4.2 | 3.0 | 34.4 | 8.2 | 17.4 | 18.0 | 1.8 | 2.7 | 2.5 | 1.2 | 7.9 | 4.5 |
| | (43) | (37) | (34) | (33) | (35) | (24) | (32) | (27) | (28) | (20) | (49) | (17) | (43) | (23) | (40) | (36) |
| France | 191.7 | 96.3 | 18.2 | 11.8 | 11.4 | 2.5 | 47.5 | 8.0 | 21.5 | 18.2 | 3.1 | 2.2 | 8.6 | 1.2 | 7.0 | 3.1 |
| | (12) | (33) | (18) | (22) | (8) | (27) | (18) | (30) | (10) | (19) | (35) | (26) | (9) | (23) | (42) | (46) |
| Germany | 161.8 | 110.4 | 19.9 | 15.7 | 4.9 | 2.1 | 42.4 | 10.8 | 21.6 | 15.8 | 3.8 | 1.9 | 5.0 | 1.0 | 10.3 | 6.4 |
| | (21) | (18) | (12) | (7) | (28) | (36) | (23) | (19) | (9) | (28) | (30) | (32) | (23) | (28) | (28) | (25) |
| Greece | 148.2 | 81.9 | 9.7 | 8.0 | 11.3 | 5.1 | 49.8 | 7.6 | 15.4 | 11.2 | 2.5 | 1.3 | 1.3 | 0.4 | 8.9 | 4.3 |
| | (32) | (45) | (36) | (36) | (9) | (8) | (14) | (31) | (36) | (37) | (40) | (44) | (50) | (48) | (35) | (37) |
| Hungary | 271.4 | 145.1 | 35.6 | 21.2 | 7.8 | 3.8 | 83.9 | 22.3 | 24.6 | 18.4 | 6.7 | 4.1 | 9.1 | 1.3 | 18.2 | 8.5 |
| | (1) | (3) | (1) | (1) | (13) | (16) | (1) | (5) | (4) | (17) | (19) | (6) | (7) | (21) | (14) | (16) |
| Iceland | 145.8 | 118.6 | 12.8 | 13.2 | 43 | 2.2 | 33.1 | 25.2 | 19.6 | 23.0 | 4.7 | 1.9 | 4.7 | 1.6 | 9.0 | 3.5 |
| | (34) | (11) | (32) | (16) | (33) | (33) | (34) | (4) | (16) | (6) | (26) | (32) | (26) | (16) | (34) | (43) |
| Ireland | 168.4 | 123.7 | 23.6 | 13.7 | 3.4 | 1.7 | 37.9 | 18.1 | 25.5 | 19.7 | 3.5 | 1.6 | 7.9 | 4.0 | 8.5 | 4.8 |
| | (18) | (6) | (5) | (12) | (42) | (42) | (27) | (8) | (3) | (14) | (32) | (40) | (11) | (8) | (38) | (34) |
| Israel | 132.6 | 105.0 | 18.8 | 14.6 | 3.0 | 2.2 | 26.9 | 8.6 | 24.0 | 13.4 | 2.3 | 2.2 | 1.6 | 0.8 | 8.9 | 4.7 |
| | (42) | (24) | (15) | (8) | (46) | (33) | (39) | (25) | (7) | (33) | (44) | (26) | (48) | (33) | (35) | (35) |
| Italy | 170.9 | 95.2 | 16.5 | 10.9 | 12.6 | 4.8 | 50.1 | 8.5 | 18.9 | 12.2 | 2.2 | 2.2 | 3.4 | 0.7 | 12.6 | 6.5 |
| | (17) | (34) | (26) | (31) | (6) | (10) | (13) | (26) | (19) | (36) | (47) | (26) | (35) | (37) | (23) | (23) |
| Japan | 154.3 | 82.2 | 17.3 | 11.1 | 21.0 | 6.7 | 32.4 | 9.6 | 8.3 | 5.7 | 2.8 | 1.3 | 7.5 | 1.1 | 28.7 | 12.7 |
| | (29) | (44) | (24) | (29) | (4) | (6) | (36) | (24) | (49) | (45) | (38) | (44) | (12) | (27) | (6) | (8) |
| Kazakhstan | 221.2 | 120.1 | 6.2 | 5.1 | 12.5 | 4.8 | 66.8 | 10.0 | 18.7 | 6.0 | 7.9 | 7.4 | 19.1 | 10.0 | 34.7 | 15.4 |
| | (3) | (9) | (44) | (46) | (7) | (10) | (3) | (21) | (20) | (43) | (14) | (1) | (3) | (2) | (1) | (3) |
| Latvia | 196.6 | 101.4 | 18.0 | 12.3 | 4.4 | 2.0 | 58.9 | 6.3 | 18.5 | 13.4 | 7.4 | 3.2 | 5.6 | 0.6 | 22.2 | 10.4 |
| | (10) | (28) | (20) | (20) | (31) | (37) | (9) | (38) | (21) | (33) | (18) | (10) | (19) | (41) | (10) | (10) |
| Lithuania | 194.4 | 100.1 | 18.0 | 11.3 | 3.8 | 1.8 | 55.9 | 5.3 | 17.6 | 16.6 | 9.0 | 3.6 | 6.0 | 0.6 | 22.4 | 9.7 |
| | (11) | (30) | (20) | (27) | (38) | (41) | (11) | (42) | (27) | (25) | (911) | (7) | (17) | (41) | (9) | (12) |
| Macedonia | 145.6 | 89.6 | 12.3 | 8.4 | 7.4 | 3.9 | 41.5 | 7.5 | 17.7 | 8.7 | 7.6 | 2.1 | 1.4 | 0.4 | 20.3 | 8.7 |
| | (35) | (38) | (33) | (35) | (16) | (14) | (24) | (33) | (25) | (40) | (17) | (29) | (49) | (48) | (11) | (15) |
| Mali | | | | | | | | | | | | | | | | |

**Table 1.1** (continued)

| | All sites | | Colorectal | | Liver | | Lung and bronchus | | Breast | | Uterus | | Esophagus | | Stomach | |
|---|---|---|---|---|---|---|---|---|---|---|---|---|---|---|---|---|
| Country | Male | Female | Male | Female | Male | Female | Male | Female | Male | Prostate | Cervix | Corpus | Male | Female | Male | Female |
| | 86.0 | 98.8 | 4.7 | 4.3 | 29.3 | 13.2 | 2.8 | 0.1 | 13.1 | 6.0 | 28.4 | 0.6 | 2.8 | 1.4 | 16.1 | 18.3 |
| | (49) | (32) | (48) | (47) | (2) | (2) | (50) | (50) | (42) | (43) | (3) | (48) | (940) | (18) | (200) | (1) |
| Mauritius | 83.3 | 60.6 | 6.0 | 4.0 | 4.6 | 2.4 | 16.1 | 4.3 | 9.3 | 7.5 | 10.2 | 0.1 | 3.4 | 1.5 | 10.1 | 5.1 |
| | (50) | (49) | (45) | (49) | (30) | (29) | (46) | (47) | (48) | (42) | (10) | (50) | (35) | (17) | (30) | (932) |
| Mexico | 92.3 | 86.0 | 4.5 | 4.1 | 7.1 | 7.0 | 16.6 | 6.6 | 10.5 | 14.8 | 14.1 | 1.9 | 1.9 | 0.7 | 9.9 | 7.2 |
| | (48) | (42) | (49) | (48) | (19) | (5) | (45) | (37) | (46) | (931) | (7) | (932) | (47) | (37) | (31) | (20) |
| Moldova | 141.1 | 84.0 | 16.1 | 10.5 | 8.4 | 3.1 | 33.3 | 6.0 | 17.7 | 4.7 | 7.8 | 2.9 | 2.7 | 0.6 | 17.8 | 7.1 |
| | (36) | (43) | (27) | (37) | (11) | (23) | (33) | (40) | (25) | (48) | (15) | (12) | (42) | (41) | (15) | (21) |
| New Zealand | 159.7 | 127.0 | 23.2 | 18.5 | 3.8 | 1.3 | 34.7 | 19.0 | 24.5 | 20.3 | 3.2 | 2.5 | 4.4 | 1.8 | 8.0 | 4.1 |
| | (24) | (4) | (8) | (3) | (38) | (48) | (30) | (7) | (5) | (11) | (34) | (919) | (32) | (14) | (39) | (38) |
| Norway | 156.7 | 109.1 | 20.1 | 16.8 | 2.0 | 1.3 | 32.7 | 13.5 | 17.9 | 28.4 | 3.5 | 2.3 | 3.3 | 0.9 | 9.4 | 5.0 |
| | (26) | (20) | (9) | (5) | (50) | (48) | (35) | (13) | (24) | (2) | (32) | (24) | (38) | (29) | (32) | (33) |
| Poland | 203.5 | 110.6 | 18.2 | 11.4 | 4.3 | 3.2 | 68.4 | 12.3 | 15.5. | 12.4 | 7.8 | 2.8 | 4.7 | 0.8 | 16.6 | 6.2 |
| | (7) | (17) | (18) | (25) | (33) | (20) | (2) | (16) | (35) | (35) | (15) | (14) | (26) | (33) | (18) | (28) |
| Portugal | 160.2 | 87.3 | 20.0 | 11.9 | 5.5 | 1.9 | 29.9 | 5.3 | 17.0 | 19.9 | 4.5 | 1.9 | 5.6 | 0.9 | 20.3 | 10.1 |
| | (22) | (39) | (11) | (21) | (27) | (40) | (37) | (42) | (30) | (12) | (28) | (32) | (19) | (29) | (11) | (11) |
| Romania | 159.4 | 93.7 | 13.6 | 9.0 | 8.8 | 3.9 | 47.1 | 8.1 | 16.7 | 9.0 | 13.0 | 2.0 | 2.8 | 0.5 | 17.0 | 6.6 |
| | (25) | (36) | (31) | (34) | (10) | (14) | (19) | (29) | (31) | (38) | (8) | (30) | (40) | (44) | (16) | (22) |
| Russian Federation | 205.0 | 101.6 | 18.9 | 13.6 | 5.8 | 2.6 | 63.0 | 6.2 | 18.0 | 8.2 | 6.5 | 3.6 | 6.9 | 1.2 | 31.8 | 13.5 |
| | (6) | (27) | (13) | (13) | (25) | (26) | (5) | (39) | (23) | (41) | (21) | (7) | (15) | (23) | (4) | (5) |
| Saudi Arabia | 92.5 | 742 | 6.0 | 5.5 | 13.7 | 5.3 | 9.6 | 2.6 | 10.9 | 5.3 | 2.5 | 1.8 | 3.4 | 2.9 | 4.9 | 3.0 |
| | (47) | (48) | (45) | (43) | (5) | (7) | (48) | (48) | (45) | (46) | (40) | (38) | (35) | (10) | (49) | (47) |
| Slovakia | 224.5 | 110.3 | 33.2 | 16.0 | 6.6 | 2.9 | 59.9 | 8.2 | 19.3 | 16.5 | 6.1 | 5.1 | 8.2 | 0.5 | 16.6 | 6.4 |
| | (2) | (19) | (3) | (6) | (21) | (25) | (8) | (27) | (17) | (27) | (22) | (4) | (10) | (44) | (18) | (25) |
| Slovenia | 200.6 | 117.1 | 24.1 | 14.0 | 6.6 | 2.4 | 54.0 | 11.9 | 22.1 | 18.8 | 4.7 | 3.0 | 4.8 | 0.9 | 17.0 | 8.2 |
| | (9) | (13) | (4) | (10) | (21) | (29) | (12) | (18) | (8) | (16) | (26) | (11) | (25) | (29) | (16) | (17) |

**Table 1.1** (continued)

| | All sites | | Colorectal | | Liver | | Lung and bronchus | | Breast | | Uterus | | Esophagus | | Stomach | |
|---|---|---|---|---|---|---|---|---|---|---|---|---|---|---|---|---|
| Country | Male | Female | Male | Female | Male | Female | Male | Female | Male | Prostate | Cervix | Corpus | Male | Female | Male | Female |
| South African | 163.6 | 107.6 | 7.9 | 6.4 | 5.8 | 2.2 | 23.0 | 6.9 | 16.4 | 22.6 | 21.0 | 1.5 | 19.2 | 6.9 | 7.6 | 3.4 |
| Republic | (19) | (21) | (37) | (40) | (25) | (33) | (40) | (35) | (32) | (7) | (4) | (42) | (2) | (5) | (41) | (44) |
| Spain | 173.6 | 81.9 | 18.5 | 11.3 | 8.4 | 3.3 | 49.2 | 4.7 | 15.9 | 14.9 | 2.2 | 2.4 | 5.1 | 0.5 | 11.4 | 5.4 |
| | (16) | (45) | (17) | (27) | (11) | (19) | (15) | (46) | (34) | (30) | (47) | (22) | (21) | (44) | (25) | 929) |
| Sweden | 135.1 | 102.8 | 14.9 | 11.1 | 4.2 | 2.4 | 22.6 | 12.9 | 17.3 | 27.7 | 3.1 | 2.3 | 3.3 | 0.9 | 6.8 | 3.8 |
| | (40) | (26) | (30) | (29) | (35) | (29) | (41) | (14) | (29) | (3) | (35) | (24) | (38) | (29) | (44) | (41) |
| The | 181.6 | 119.8 | 18.9 | 14.4 | 2.5 | 1.3 | 57.6 | 15.6 | 27.5 | 19.7 | 2.3 | 2.4 | 6.8 | 2.2 | 9.1 | 4.1 |
| Netherlands | (14) | (10) | (13) | (9) | (48) | (48) | (10) | (11) | (2) | (14) | (44) | (922) | (916) | (11) | (33) | (38) |
| Turkey | 107.8 | 58.7 | 5.8 | 5.4 | 2.5 | 1.4 | 44.1 | 4.9 | 9.7 | 5.0 | 2.4 | 2.0 | 2.0 | 1.4 | 10.4 | 5.4 |
| | (45) | (50) | (47) | (44) | (48) | (47) | (21) | (45) | (47) | (947) | (43) | (30) | (45) | (18) | (26) | (29) |
| Uganda | 123.6 | 118.5 | 7.0 | 6.2 | 6.1 | 5.9 | 3.3 | 2.1 | 13.4 | 32.5 | 29.2 | 1.2 | 12.5 | 11.3 | 6.6 | 5.2 |
| | (44) | (12) | (41) | (41) | (24) | (9) | (49) | (49) | (40) | (1) | (2) | (47) | (5) | (1) | (45) | (31) |
| United Kingdom | 162.3 | 122.7 | 17.5 | 12.4 | 2.8 | 1.5 | 42.9 | 21.1 | 24.3 | 17.9 | 3.1 | 1.8 | 9.0 | 4.1 | 8.7 | 4.0 |
| | (20) | (7) | (23) | (19) | (47) | (45) | (22) | (6) | (5) | (21) | (35) | (38) | (8) | (7) | (37) | (40) |
| Venezuela | 101.5 | 95.1 | 6.4 | 6.7 | 4.8 | 4.3 | 18.1 | 10.2 | 13.4 | 19.8 | 16.8 | 1.9 | 2.4 | 0.8 | 14.5 | 9.3 |
| | (46) | (35) | (43) | (39) | (29) | (12) | (44) | (20) | (40) | (13) | (6) | (32) | (44) | (33) | (22) | (13) |
| Zimbabwe | 183.6 | 165.4 | 6.5 | 6.2 | 25.4 | 10.5 | 12.0 | 5.8 | 14.1 | 23.5 | 43.1 | 2.8 | 17.6 | 8.4 | 10.4 | 9.1 |
| | (13) | (1) | (42) | (41) | (3) | (3) | (47) | (41) | (38) | (5) | (1) | (14) | (4) | (4) | (26) | (14) |

Death rates are per 100,000 population, age adjusted to the WHO world standard population; figures in parentheses are in order of rank within site and gender group.
Source: Ferlay et al., 2006.

Hungary, Czech Republic, and Slovakia for males and Hungary, Denmark, and New Zealand for females; for lung and bronchus cancer: Hungary, Poland, and Kazakstan for males and Denmark, United States, and Canada for females; for breast cancer: Denmark, Netherlands, and Ireland; for prostate cancer: Uganda, Norway, and Sweden; for Cervical cancer: Zimbabwe, Uganda, and Mali; for stomach cancer: Kazakhstan, China, and Chile for males and Mali, Colombia, and Kazakhstan in females. The death rate for female lung and bronchus cancer in the United States in 2002 held the second position among 50 countries (Denmark held the first position with the highest lung cancer death rates). In the United States, 2002 death rates for stomach cancer in both males and females were the lowest among 50 countries (holding the 50th position).

### *1.2.2 U.S. Cancer Morbidity and Mortality: At the Beginning of 21st Century*

Cancer as a cause of death in the United States in 2001 was ranked in the second position (22.9% of all deaths), being 6.1% lower than the number one ranked cause, heart disease (29% of all deaths), and being far ahead of the number three cause of death, cerebrovascular disease (6.8% of all deaths) (US Mortality Public Data Tape, 2001). In 2005, these percentages were 22.8% vs. 26.6% and 5.9% for cancer, heart disease and cerebrovascular disease, respectively (US Mortality Files, 2005).

Malignant neoplasms keep the 7th position in average years of life lost per person – 5.7 years, while the "leaders" are homicide (45.8 years), suicide and self-inflicted injury (34.2 years), HIV (33.9 years), and accidents (31.8 years). However, cancer has a leading position in person-years of life lost – 8.8 million years (compare to cause number two "all other causes" – 7.9 million years, and cause number three "heart disease" – 7.7 million years). Among the cancer sites, the highest person-years of life lost has lung and bronchus cancer (2437.8 thousand years), followed by female breast (794.8 thousand years), and colorectal (759.7 thousand years) cancers (see Fig. 1.1a and b).

In 2001–2005, the median age at diagnosis for cancer of all sites in the United States was 67 years; the median age at death from cancer was 73 years (SEER, Table I-11; SEER, Table I-13). Approximately 26% of patients with cancer died at age 65–74, 30% at age 75–84, and 14% at ages 85+. NCI estimated that approximately 10.1 million Americans with a history of cancer were alive in January 2002. These included cancer-free individuals, as well as patients with cancer who were undergoing anti-cancer therapy (Cancer Facts & Figures, 2006). About 1,437,180 new cancer cases are expected to be diagnosed in the United States in 2008 (this estimate does not include carcinoma *in situ* of any site except urinary bladder, and it does not include basal and squamous cell skin cancers). In 2008, about 565,650 Americans are expected to die from cancer, approximately 1550 people a day (Cancer Facts & Figures, 2008). The

NIH estimated the overall costs for cancer in 2005 at $209.9 billion: $74.0 billion for direct medical costs (total of all health care expenditures), $17.5 billion for indirect morbidity costs (cost of lost productivity due to illness), and $118.4 billion for indirect mortality costs (cost of lost productivity due to premature death).

The probability of developing the majority of invasive cancers tends to increase with age (see Table 1.2), thus allowing one to characterize cancer as an aging-associated disease (cancer incidence in populations older than 85 years will be discussed in Chapter 2).

One of the ways to estimate the success of "the war against cancer" over time is to examine the changes/improvements in cancer survival rates. The 5-year survival rate is perhaps the most common statistics used to report progress in "the war against cancer". Improvements in 5-year survival rates are held up as an unambiguous sign of success: if cancer patients live longer now than compared to the past, it is arguable that society's enormous investment in cancer research must be paying off. The 5-year relative survival rate in the United States for all cancers diagnosed between 1995 and 2001 was 65%, compared to 50% in 1974–1976 (see Table 1.3). This improvement in survival, however,

(a)

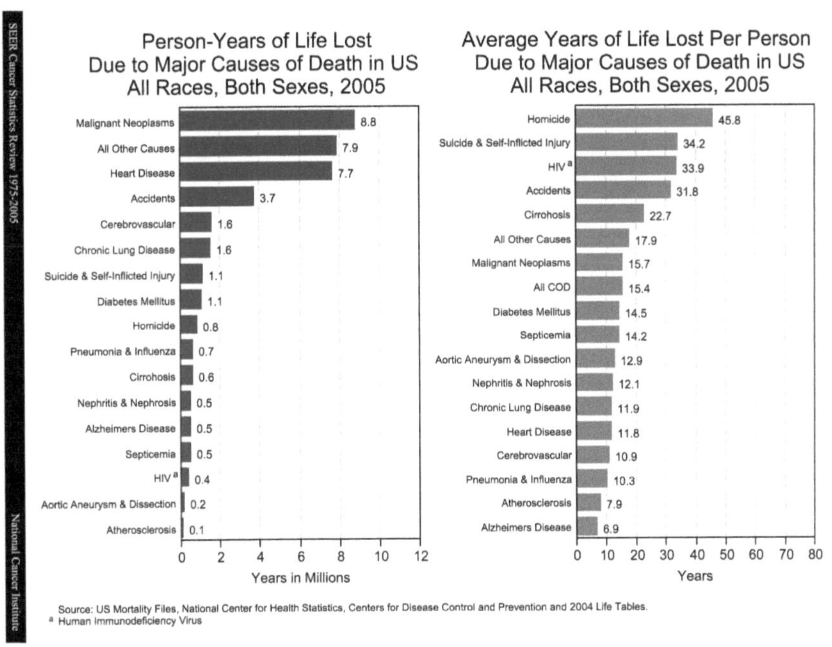


**Fig. 1.1** (**a**) Person-years of life lost and average years of life lost per person due to major causes of death in the United States, all races, both sexes, 2005. (**b**) Person-years of life lost due to cancer and average years of life lost per person dying of cancer in United States, all races, both sexes, 2005

(b)

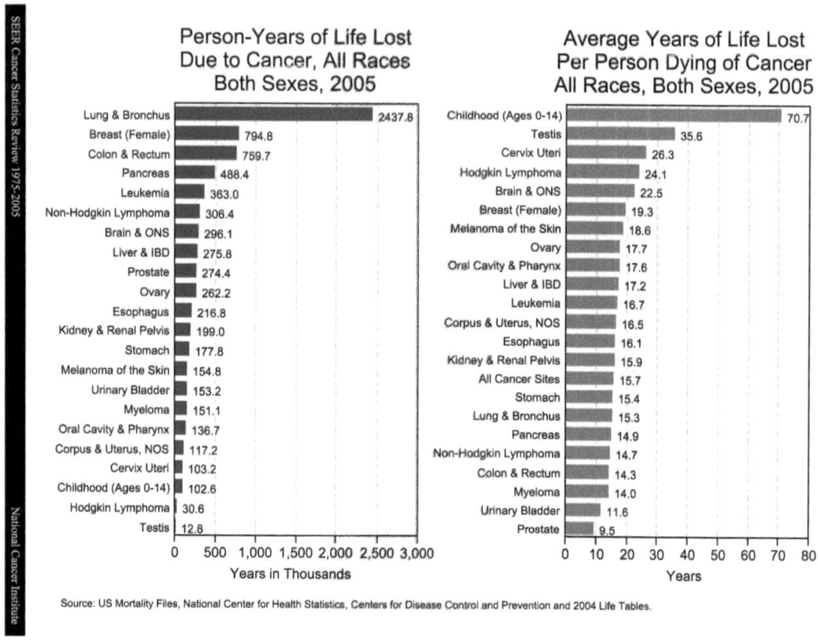


**Fig. 1.1** (continued)

**Table 1.2** Probability of developing selected invasive cancers over selected age intervals by sex in the United States from 2000 to 2002 (from Cancer Facts & Figures, 2006)

| Cancer site | Sex | Birth to 39 years old, % | 40–59 years old, % | 60–69 years old, % | > 70 years old, % |
|---|---|---|---|---|---|
| All sites[1] | Male | 1.43 (1 in 70) | 8.57 (1 in 12) | 16.46 (1 in 6) | 39.61 (1 in 3) |
| | Female | 1.99 (1 in 50) | 9.06 (1 in 11) | 10.54 (1 in 9) | 26.72 (1 in 4) |
| Breast | Female | 0.48 (1 in 209) | 4.11 (1 in 24) | 3.82 (1 in 26) | 7.13 (1 in 14) |
| Colorectal | Male | 0.07 (1 in 1399) | 0.90 (1 in 111) | 1.66 (1 in 60) | 4.94 (1 in 20) |
| | Female | 0.06 (1 in 1567) | 0.70 (1 in 143) | 1.16 (1 in 86) | 4.61 (1 in 22) |
| Lung and bronchus | Male | 0.03 (1 in 3244) | 1.00 (1 in 100) | 2.45 (1 in 41) | 6.33 (1 in 16) |
| | Female | 0.03 (1 in 3103) | 0.80 (1 in 125) | 1.68 (1 in 60) | 4.17 (1 in 24) |
| Prostate | Male | 0.01 (1 in 10149) | 2.66 (1 in 38) | 7.19 (1 in 14) | 14.51 (1 in 7) |

[1]All sites excluding basal and squamous cell skin cancers and in situ cancers excluding urinary bladder.
Source: DevCan, 2005.

reflects both the progress in diagnosing certain cancers at an earlier stage and the widespread use of improved treatments. Over the last decade, noticeable increases in the 5-year survival rate were found for prostate, colorectal, brain, stomach cancers, and leukemia. The highest 5-year relative survival rates in 1995–2002 were for prostate, testis, female breast, urinary bladder, corpus uteri,

**Table 1.3** Trends in 5-year relative survival rates[1] (%) in the United States, from 1960 to 2001

| Cancer site | 1960–1963 White/ African American | 1974–1976 | 1983–1985 | Percent of changes (1983–1985)/(1974–1976) | 1995–2001 | Percent of changes (1995–2001)/(1983–1985) |
|---|---|---|---|---|---|---|
| All cancers | 39/27 | 50 | 53 | +6% | 65* | + 22.6% |
| Oral cavity | 45/N/A | 54 | 54 | 0% | 59* | + 9.3% |
| Esophagus | 4/1 | 5 | 8 | +60% | 15* | + 87.5% |
| Stomach | 11/8 | 15 | 17 | +13.3% | 23* | + 35.3% |
| Colon | N/A/N/A | 50 | 58 | +16% | 64* | + 10.3% |
| Rectum | N/A/N/A | 49 | 55 | +12.2% | 65* | + 18.2% |
| Liver[2] | N/A/N/A | 4 | 6 | +50% | 9* | + 50% |
| Pancreas | 1/1 | 3 | 3 | 0% | 4* | + 33.3% |
| Larynx | 53/N/A | 66 | 67 | +1.5% | 66 | −1.5% |
| Lung and bronchus | 8/5 | 12 | 14 | +16.7% | 15* | + 7.14% |
| Melanoma of skin | 60/N/A | 80 | 85 | +6.3% | 92* | + 8.2% |
| Breast (female) | 63/46 | 75 | 78 | +4% | 88* | + 12.8% |
| Cervix uteri | 58/47 | 69 | 69 | 0% | 73* | + 5.8% |
| Corpus uteri | 73/31 | 88 | 83 | −5.7% | 84* | 1.2% |
| Ovary[3] | 32/32 | 37 | 41 | +10.8% | 45* | + 9.8% |
| Prostate | 60/35 | 67 | 75 | +11.9% | 100* | + 33.3% |
| Testis | 63/N/A | 79 | 91 | +15.2% | 96* | + 5.5% |
| Urinary bladder | 53/24 | 73 | 78 | +6.8% | 82* | + 5.1% |
| Kidney and renal pelvis | 37/38 | 52 | 56 | +7.7% | 65* | + 16.1% |
| Brain and other nervous | 18/19 | 22 | 27 | +22.7% | 33* | + 22.2% |
| Thyroid | 83/N/A | 92 | 93 | +1.2% | 97* | + 4.3% |
| Hodgkin's lymphoma | 40/N/A | 71 | 79 | +11.3% | 85* | + 7.6% |

**Table 1.3** (continued)

| Cancer site | 1960–1963 White/ African American | 1974–1976 | 1983–1985 | Percent of changes (1983–1985)/(1974–1976) | 1995–2001 | Percent of changes (1995–2001)/(1983–1985) |
|---|---|---|---|---|---|---|
| Non-Hodgkin's lymphoma | 31/N/A | 47 | 54 | +14.9% | 60* | + 11.1% |
| Leukemia | 14/N/A | 34 | 41 | +20.6% | 48* | + 17.2% |
| Multiple myeloma | 12/N/A | 25 | 28 | +12% | 32* | + 14.3% |

[1]Survival rates are adjusted for normal life expectancy.
[2]Includes intrahepatic bile ducts.
[3]Recent changes in classification of ovarian cancer, namely excluding borderline tumors, have affected 1995–2001 survival rates.
*The difference in rates between 1974–1976 and 1995–2001 is statistically significant ($p<0.05$). N/A = not available.
Sources: Surveillance, Epidemiology, and End Result Program, 1975–2002; End Results Group, 1960–1973.

and thyroid cancers, as well as for melanoma of skin and for Hodgkin's lymphoma. At present, the worst prognoses are for liver, pancreas, esophagus, and lung and bronchus cancers (even taking into account that compared to initial survival rates, "relative progress" in liver and esophagus cancers was impressive over a 25-year period). But while the 5-year relative survival rate is useful in monitoring progress in the early detection and treatment of cancer, it does not accurately represent the proportion of people who are cured permanently, since cancer can affect survival more than 5 years after diagnosis.

Even when the 5-year survival is a valid and widely used characteristic in evaluating the results of cancer therapies (e.g., in randomized trials), using this measure throughout time cannot provide a doubtless criteria for comparison (Welch et al., 2000). The 5-year survival may increase, when the treatment of established cancers improves, allowing patients to live to an older age, or when more patients are diagnosed with cancer earlier in their disease course, and if early treatment is effective, there will be a further increase in 5-year survival (mortality will also decrease). However, if cancer is diagnosed at early stages, but early treatment will be ineffective, mortality will be unchanged. If cancer patients in the past always had palpable tumors at the time of diagnosis, the current cancer patients include those diagnosed with microscopic abnormalities, and 5-year survival rates would be expected to increase over time due to earlier ages at detection. The expectation might be that a large increase in 5-year survival would be associated with mortality declines. However, no obvious relationship is evident (see Table 1.4): no correlation have been found between the increase in 5-year survival and the change in tumor-related mortality. A positive correlation exists between the increase in 5-year survival for a specific tumor and the change in the tumor incidence rate, as well as between mortality changes and incidence changes.

Table 1.5 presents the 5-year survival rates by stage at diagnosis for a 15-year interval: there are no significant changes in survival rates for this period, except for the positive trend in survival rate for ovary cancer with regional metastases, thyroid cancer with distant metastases, and oropharyngeal cancer with regional and distant metastasis. Cancers of pancreas, liver and intrahepatic ducts, lung and bronchus, and esophagus have the poorest prognosis for 5-year survival: approximately 5, 12, 15, and 16%, respectively (for regional cancer stage – about 8, 8, 21, and 17%, respectively). The best survival prognoses have cancers of breast (female) (89%), melanoma of the skin (91%), prostate (99%), testis (96%), and corpus uteri (83%), especially when diagnosed at local stages – 98, 99, 100, 99, and 96%, respectively.

When the NCI asked experts to assess the various measures of progress in "the war against cancer", the committee was clear that mortality rate is the most important (Extramural Committee, 1990). Mortality rates would be expected to decrease with any improvement in cancer control: be it risk factor reduction (primary prevention), successful early cancer detection (screening), or improved cancer treatment.

**Table 1.4** Changes in 5-year survival rates (1950–2004), and cancer mortality and incidence (1950–2005), for 20 solid tumors, for white population, males and females, by primary cancer site

| | 5-year survival, % | | | Absolute increase in 5-year survival, % | | Mortality, percent change | | | |
|---|---|---|---|---|---|---|---|---|---|
| Primary site | 1950–1954 | 1989–1995 | 1996–2004 | 1995/ 1950 | 2004/ 1950 | 1995/ 1950 | 2005/ 1950 | APC* | Incidence, percent change, 1995/50 |
| Oral cavity and pharynx | 46 | 56 | 62.4 | 10 | 16.4 | –37 | –52.1 | –1.3 | –38 |
| Esophagus | 4 | 13 | 18.1 | 9 | 14.1 | 22 | 32.5 | 0.7 | –8 |
| Stomach | 12 | 19 | 23.1 | 7 | 9.1 | –80 | –86.2 | –3.5 | –78 |
| Colon | 41 | 62 | 66.0 | 21 | 25 | –21 | –39.4 | –0.7 | 12 |
| Rectum | 40 | 60 | 67.2 | 20 | 27.2 | –67 | –70.5 | –2.5 | –27 |
| Liver and intrahepatic ducts | 1 | 6 | 10.8 | 5 | 9.8 | 34 | 35.6 | 0.7 | 140 |
| Pancreas | 1 | 4 | 4.9 | 3 | 3.9 | 16 | 24 | 0.1 | 9 |
| Larynx | 52 | 66 | 65.8 | 14 | 13.8 | –14 | –30 | –0.6 | 38 |
| Lung and bronchus | 6<br>M[1]–5<br>F[2]–9 | 14<br>M–15.7<br>F–18.4 | 15.9 | 8 | 9.9<br>M–10.7<br>F–9.4 | 259 | 250.6<br>M–176.9<br>F–600.9 | 1.9<br>M–1.2<br>F–3.7 | 249 |
| Melanoma | 49 | 88 | 92.1 | 39 | 43.1 | 161 | 174.6 | 1.5 | 453 |
| Breast (female) | 60 | 86 | 90.5 | 26 | 30.5 | –8 | –28.4 | –0.4 | 55 |
| Cervix uteri | 59 | 71 | 73.7 | 12 | 14.7 | –76 | –81.4 | –3.4 | –79 |
| Corpus and uterus, NOS | 72 | 86 | 86.2 | 14 | 14.2 | –67 | –68.8 | –1.9 | 0 |
| Ovary | 30 | 50 | 45.3 | 20 | 15.3 | –2 | –1 | –0.2 | 3 |
| Prostate | 43 | 93 | 99.4 | 50 | 56.4 | 10 | –22.1 | 0.0 | 190 |
| Testis | 57 | 96 | 96.4 | 39 | 39.4 | –73 | –73.4 | –3.0 | 106 |

**Table 1.4** (continued)

| Primary site | 5-year survival, % | | | Absolute increase in 5-year survival, % | | Mortality, percent change | | | Incidence, percent change, 1995/50 |
|---|---|---|---|---|---|---|---|---|---|
| | 1950–1954 | 1989–1995 | 1996–2004 | 1995/1950 | 2004/1950 | 1995/1950 | 2005/1950 | APC* | |
| Urinary bladder | 53 | 82 | 81.6 | 29 | 28.6 | −35 | −30.9 | −0.9 | 51 |
| Kidney and renal pelvis | 34 | 61 | 66.6 | 27 | 32.6 | 37 | 38.8 | 0.6 | 126 |
| Brain and other nervous | 21 | 30 | 33.7 | 9 | 12.7 | 45 | 51.8 | 0.6 | 68 |
| Thyroid | 80 | 95 | 97.1 | 15 | 17.1 | −48 | −44.1 | −1.3 | 142 |

[1]Males.
[2]Females.
*APC – the Annual Percent Change over the time interval. Rates used in the calculation of the APC are age adjusted to the 2000 U.S. standard population (18 age groups – Census P25-1130).
Sources: SEER Cancer Statistics Review 1975–2005 (Summary); Welch et al., 2000.

**Table 1.5** Five-year relative survival rates by stage at diagnosis, 1986–1992 and 1996–2004

| | (1986–1992) / (1996–2004) | | | |
|---|---|---|---|---|
| Cancer site | All stages, % | Local*, % | Regional**, % | Distant***, % |
| Oral / oral cavity and pharynx, invasive | 52.8 / 59.7 | 81.2 / 82.2 | 42 / 52.7 | 18.1 / 28.4 |
| Esophagus, invasive | 10.7 / 15.8 | 22.3 / 34.4 | 10.5 / 17.1 | 1.9 / 2.8 |
| Stomach, invasive | 20.5 / 24.7 | 60.6 / 60.7 | 22.6 / 24.8 | 2.2 / 3.7 |
| Colon and rectum, invasive | 61.4 / 64.4 | 91.4 / 89.7 | 63.4 / 68.4 | 7.1 / 10.9 |
| Liver and intrahepatic ducts, invasive | 5.9 / 11.7 | 13.4 / 23.6 | 7.5 / 7.7 | 1.8 / 2.9 |
| Pancreas, invasive | 3.8 / 5.1 | 12.7 / 20.0 | 5 / 8.2 | |
| Larynx[1], invasive | 66.2 / 62.9 | 84.1 / 81.1 | 53.8 / 50.0 | 39.8 / 23.9 |
| Lung and bronchus, invasive | 13.7 / 15.2 | 48 / 49.5 | 17.8 / 20.6 | 1.9 / 2.8 |
| Melanoma of the skin, invasive | 87.4 / 91.2 | 94.6 / 98.7 | 60.8 / 65.1 | 16 / 15.5 |
| Breast (female), invasive | 83.8 / 88.7 | 96.5 / 98.1 | 75.6 / 83.8 | 20.2 / 27.1 |
| Cervix uteri, invasive | 68.7 / 71.2 | 91.3 / 91.7 | 49.9 / 55.9 | 8.7 / 16.6 |
| Corpus and uterus, NOS, invasive | 83.7 / 82.9 | 95.3 / 95.5 | 66 / 67.5 | 26.1 / 23.6 |
| Ovary[2], invasive | 45.5 / 45.5 | 91.5 / 92.7 | 51.1 / 71.1 | 24.5 / 30.6 |
| Prostate[3], invasive | 86.7 / 98.9 | 99.1 / 100 | 92.7 /100 | 30.3 / 31.7 |
| Testis, invasive | 95 / 95.5 | 98.5 / 99.3 | 97.3 / 95.7 | 72.2 / 71.7 |
| Kidney and renal pelvis, invasive | 58.8 / 66.5 | 88.1 / 89.9 | 59.8 / 61.3 | 9.3 / 9.9 |
| Urinary bladder, invasive + in situ | 81.1 / 79.8 | 93.4 / 92.5 | 48.5 / 44.7 | 6.1 / 6.1 |
| Thyroid, invasive | 95.2 / 96.9 | 99.8 / 99.7 | 93.7 / 96.9 | 46.8 / 57.8 |

[1]Data for larynx cancer obtained for 1975–2004.

[2]Recent changes in classification of ovarian cancer, specifically excluding borderline tumors, have affected the 1995–2001 survival rates.

[3]The rate for local stage represents local and regional stages combined.

Rates are adjusted for normal life expectancy, and based on cases diagnosed from 1986 to 1992, followed through 1993, and from 1995 to 2001, followed through 2002.

* Local – an invasive malignant cancer confined entirely to the organ of origin.

** Regional – a malignant cancer that (1) has extended beyond the limits of the organ of origin directly into surrounding organs or tissues; (2) involves regional lymph nodes by way of lymphatic system; or (3) has both regional extension and involvement of regional lymphatic nodes.

*** Distant – a malignant cancer that has spread to parts of the body remote from the primary tumor either by direct extension or by discontinuous metastasis to distant organs, tissues, or via the lymphatic system to distant lymph nodes.

Data sources: SEER Cancer Statistics Review 1975–2005; SEER Cancer Statistics Review 1975–2004; SEER Cancer Statistics review 1973–1993.

### 1.2.3 *Cancer Mortality: U.S. Historical Trends*

Looking at cancer death rates in the United States from a historical perspective, differences exist between male and female patterns (see Fig. 1.2a and b). In males, the most obvious changes over the last 70 + years were decrease in stomach cancer and dramatic increase in lung and bronchus cancer death rates (in 1985–1992, with its recent decrease since 1993). In females, the most significant changes were decreases in stomach and uterine cancer death rates, and increase in lung and bronchus cancer death rate, beginning from 1965 to 1970. It is supposed that lung cancer mortality in females reached its peak in 1995–2000 (not as dramatically high as for males, but still very significant), making lung and bronchus cancer the current leading cancer cause of death for females in the United States.

Figure 1.3 presents the cancer death rates for malignant neoplasms in the United States from 1900 to 1960, which shows gradual increase in cancer death rates.

Historical cancer mortality trends differ by cancer sites and sex (see Table 1.6).

The positive historical trends exist for cancer death rates from 1950 to 1998 for oropharyngeal, digestive, male respiratory (based on dynamics since 1993), female breast, and genital cancers (see Table 1.7).

(a)

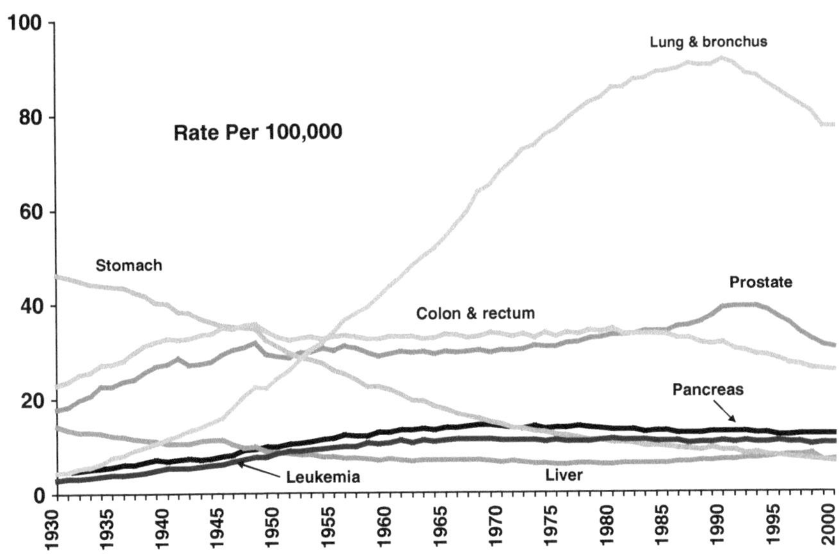


*Age-adjusted to the 2000 US standard population.
Source: US Mortality Public Use Data Tapes 1960-2000, US Mortality Volumes 1930-1959, National Center for Health Statistics, Centers for Disease Control and Prevention, 2003. From Cancer Statistics 2004, a presentation from the American Cancer Society.

**Fig. 1.2** (**a**) Cancer death rates for men, United States, 1930–2000. (**b**) Cancer death rates for women, United States, 1930–2000

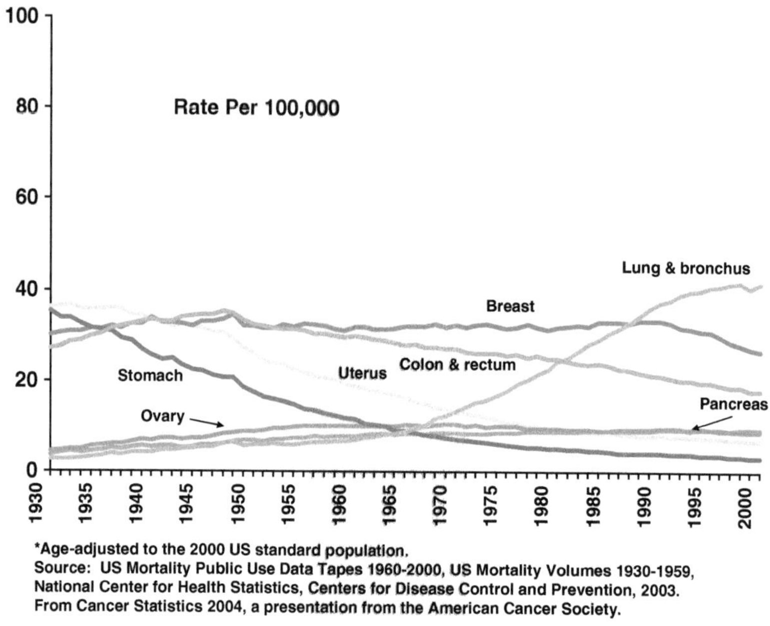


**Fig. 1.2** (continued)

More detailed analysis of the changes (decreases) of the 30-year trends demonstrated that the decline of death rates began from the early 1990s. The most dramatic changes were observed from the 1990s for female breast, prostate, colorectal, and male lung cancers, and for melanoma and for non-Hodgkin's lymphoma (see Table 1.8). Death rates for male esophagus and female lung cancers are still increasing, but in the last 15 years, these increases are not as rapid as in 1971–1990. These changes are probably resulting from the efforts of primary cancer prevention made in early 1970s in the United States, so it takes several decades for cancers to respond by showing declines in their death rate. For cancer of liver and intrahepatic bile ducts, there is not a noticeable "slowing" of increase of death rates: the efforts been made to decrease alcohol consumption, and hepatitis B and C primary and secondary prevention are huge, but with variable degrees of success, and it appears that it is too early to expect this effort to produce liver cancer decreases now.

Comparing the historical changes in cancer death rates with deaths caused by other diseases gives us additional information about the current situation "in the fields of cancer war" (see Table 1.9): while there were gradual declines in death caused by cerebrovascular disease and pneumonia/influenza starting in the 1900s–1920s, and decline in heart disease death starting from 1940s to 1960s, cancer mortality only began to decrease in the 1990s (for both males and females).

The historical trends of the contributions of cancer and non-cancer diseases to human mortality differ by age. At ages younger than 65, during the last

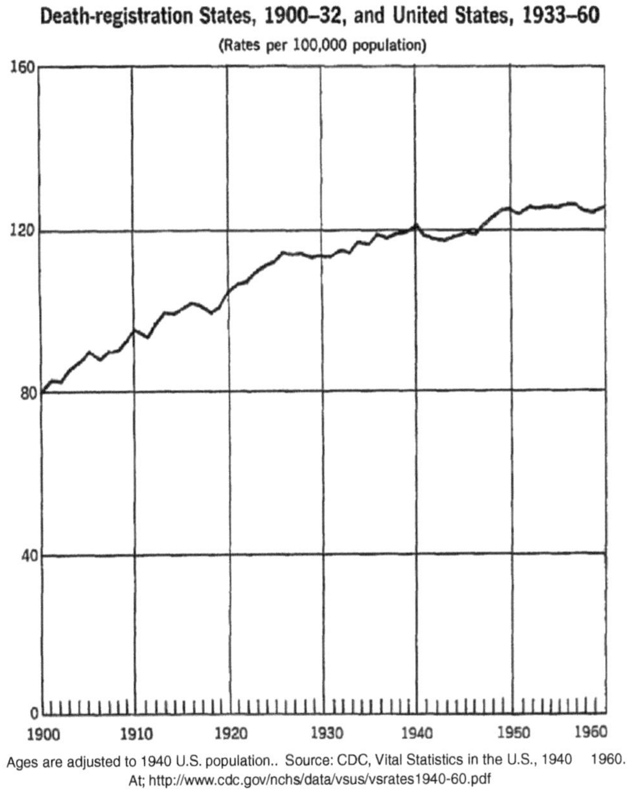


**Fig. 1.3** Age-adjusted death rates for malignant neoplasms: death-registration states, 1900–1932, and United States, 1933–1960

30 years, cancer has changed from the second-ranked cause of death to become the leading cause (see Fig. 1.4). At older ages, heart disease is still the number one cause of death, however, the distance between heart disease and cancer contribution to mortality declined from 26% in 1975 to 9% in 2003, predominantly due to decreases in heart disease mortality. In order to understand the magnitude of the progress in fighting with cancer, it is important to recognize that most of the discussed parameters are relative to normal life expectancy, which is increased during the century for both males and females (some of these changes are discussed more detailed in Chapter 4).

It is obvious that anticancer medicine has made progress, but results in most areas are still by a long way from very impressive. Part of this may be due to the fact that the national research emphasis on circulatory disease increased started in 1950 with concern expressed about increases in CVD mortality (e.g., from 1954 to 1968). Actually, the real "war on cancer" in the United States was not started until 1972.

**Table 1.6** Historical trends in cancer mortality, 1979–2003: the annual percent change (APC) in the rate, United States, for males and females, all races (calculated using web source at NCI, CDC, State Cancer Profiles)

| Cancer site | | 1979 | 1980 | 1981 | 1982 | 1983 | 1984 | 1985 | 1986 | 1987 | 1988 | 1989 | 1990 | 1991 | 1992 | 1993 | 1994 | 1995 | 1996 | 1997 | 1998 | 1999 | 2000 | 2001 | 2002 | 2003 |
|---|---|---|---|---|---|---|---|---|---|---|---|---|---|---|---|---|---|---|---|---|---|---|---|---|---|---|
| All cancers | Male | 0.3 (0.2–0.3)* | | | | | | | | | | | | | | −1.5 (−1.6 to −1.5) | | | | | | | | | | |
| | Female | 0.6 (0.5–0.7) | | | | | | | | | | | | −0.1 (−0.8 to 0.5) | | | | −0.9 (−0.1 to −0.7) | | | | | | | | |
| Lung and bronchus | Male | 1.2 (0.8–1.6) | | | | | | 0.4 (0.0–0.8) | | | | | | −1.1 (−2.6 to 0.4) | | | −1.9 (−2.0 to −1.8) | | | | | | | | | |
| | Female | 5.5 (4.4–6.7) | | | | | 4.1 (3.6–4.6) | | | | | | | 1.7 (1.0–2.5) | | | | | 0.3 (0.1–0.5) | | | | | | | |
| Colon and rectum | Male | −0.4 (−0.8 to 0.2) | | | | | | −1.3 (−1.8 to −0.8) | | | | | | −2.1 (−2.2 to −2.0) | | | | | | | | | | | | |
| | Female | −1.1 (−1.6 to −0.5) | | | | | | −1.8 (−1.9 to −1.8) | | | | | | | | | | | | | | | | | | |
| Breast | Female | 0.6 (0.4–0.8) | | | | | | | | | | | −1.5 (−2.0 to −1.0) | | | | | | −3.6 (−5.7 to −1.4) | | | −1.6 (−2.1 to −1.2) | | | | |
| Liver and bile ducts | Male | 2.3 (1.8–2.7) | | | | | | | | | 3.9 (3.5–4.3) | | | | | | | | | 0.3 (−2.3 to 3.0) | | | 2.8 (2.0–3.6) | | | |
| | Female | 1.3 (0.8–1.9) | | | | | | | | | | 3.8 (2.9–4.8) | | | | | | | 0.6 (0.1–1.1) | | | | | | | |
| Stomach | Male | −2.0 (−2.2 to −1.7) | | | | | | | | | | | | | | | −3.6 (−4.0 to −3.2) | | | | | | | | | |
| | Female | −2.9 (−3.2 to −2.5) | | | | | | | | | −0.4 (−4.0 to 3.3) | | | −2.6 (−2.8 to −2.4) | | | | | | | | | | | | |

*95% confident interval; 1 – The APC is the Annual Percent Change over the time interval. Rates used in the calculation of the APC are age adjusted to the 2000 U.S. standard population. Explanation of the calculation of the trend: if the APC is less than −1.5, the trend is *falling*. If the APC is between −1.5 and −0.5, the trend is *slightly falling*. If the APC is between −0.5 and 0.5, the trend is *statistically stable*. If the APC is between 0.5 and 1.5, the trend is *slightly rising*. If the APC is greater than 1.5, the trend is *rising*.

**Table 1.7** Age-adjusted death rates for total cancer from 1900 to 1998, and for selected cancer groups from 1950 to 1998, by sex in the United States

| Cancer site | Sex | 1900 | 1920 | 1940 | 1950 | 1960 | 1967 | 1980 | 1990 | Percent of changes, 1990/1950 | 1995 | 1998 | Percent of changes, 1998/1990 |
|---|---|---|---|---|---|---|---|---|---|---|---|---|---|
| All sites, including lymphatic and hematopoietic tissues | Male | 90.8 | 139.7 | 187.2 | 208.1 | 225.1 | 242.6 | 271.2 | 280.4 | +34.7% | 268.8 | 252.4 | −10% |
| | Female | 138.6 | 180.0 | 189.6 | 182.3 | 168.7 | 163.9 | 166.7 | 175.7 | −3.6% | 175.4 | 169.2 | −3.7% |
| Lip, oral cavity, and pharynx | Male | N/A | N/A | N/A | 8.2[1] | 7.3[1] | 6.9[1] | 6.7 | 5.6 | N/A | 4.9 | 4.5 | −19.6% |
| | Female | N/A | N/A | N/A | 2.0[1] | 2.1[1] | 2.0[1] | 2.3 | 2.0 | N/A | 1.9 | 1.7 | −15% |
| Digestive organs and peritoneum | Male | N/A | N/A | N/A | 89.7 | 80.8 | 76.2 | 71.1 | 67.5 | −24.8% | 63.9 | 61.2 | −9.3% |
| | Female | N/A | N/A | N/A | 69.6 | 59.0 | 52.6 | 45.9 | 41.3 | −40.7% | 39.8 | 38.3 | −7.3% |
| Respiratory and intrathoracic organs | Male | N/A | N/A | N/A | 28.9[2] | 47.7[2] | 63.6[2] | 89.3 | 95.0 | N/A | 88.3 | 83.2 | −12.4% |
| | Female | N/A | N/A | N/A | 6.9[2] | 7.6[2] | 10.9[2] | 25.3 | 38.1 | N/A | 41.6 | 42.4 | +11.3% |
| Breast | Male | N/A | N/A | N/A | 0.5 | 0.3 | 0.3 | 0.3 | 0.3 | −40% | 0.3 | 0.3 | 0% |
| | Female | N/A | N/A | N/A | 31.9 | 31.7 | 32.1 | 31.9 | 33.3 | +4.4% | 30.8 | 27.9 | −16.2% |
| Genital organs | Male | N/A | N/A | N/A | 30.3 | 30.0 | 30.6 | 33.6 | 38.9 | +28.4% | 37.6 | 32.4 | −16.7% |
| | Female | N/A | N/A | N/A | 36.5 | 30.3 | 26.1 | 20.0 | 18.4 | −49.6% | 17.6 | 16.9 | −8.2% |
| Urinary organs | Male | N/A | N/A | N/A | 13.0 | 13.9 | 14.5 | 14.8 | 14.4 | +10.8% | 14.2 | 14.1 | −2.1% |
| | Female | N/A | N/A | N/A | 6.3 | 6.1 | 5.9 | 5.4 | 5.4 | −14.3% | 5.5 | 5.4 | 0% |
| Malignant neoplasms of all other and unspecified sites | Male | N/A | N/A | N/A | 21.6 | 23.7 | 26.9 | 31.3 | 32.3 | +49.5% | 32.3 | 30.4 | −5.9% |
| | Female | N/A | N/A | N/A | 18.4 | 18.4 | 19.4 | 20.7 | 21.0 | +14.1% | 21.1 | 19.9 | −5.2% |

**Table 1.7** (continued)

| Cancer site | Sex | 1900 | 1920 | 1940 | 1950 | 1960 | 1967 | 1980 | 1990 | Percent of changes, 1990/1950 | 1995 | 1998 | Percent of changes, 1998/1990 |
|---|---|---|---|---|---|---|---|---|---|---|---|---|---|
| Leukemia | Male | N/A | N/A | N/A | 8.2[3] | 10.4[3] | 11.3[3] | 11.0 | 10.6 | N/A | 10.7 | 10.0 | −5.7% |
| | Female | N/A | N/A | N/A | 5.7[3] | 6.7[3] | 6.8[3] | 6.4 | 6.2 | N/A | 6.0 | 5.9 | −4.8% |
| Other malignant neoplasms of lymphatic and hematopoietic tissues | Male | N/A | N/A | N/A | 7.6[4] | 10.7[4] | 12.3[4] | 13.0 | 15.7 | N/A | 16.6 | 16.3 | +3.8% |
| | Female | N/A | N/A | N/A | 4.8[4] | 6.9[4] | 8.1[4] | 8.9 | 10.1 | N/A | 11.1 | 10.9 | +7.9% |

[1]Malignant neoplasms of buccal cavity and pharynx.
[2]Malignant neoplasms of respiratory system, not specified as secondary.
[3]Leukemia and aleukemia (code 204 of International Classification of diseases, 7th revision, 1955).
All rates per 100,000 are age adjusted using year 2000 standard population. N/A – not available.
Data sources: CDC/NCHS, National Vital Statistics System, Mortality, 2000d.
[4]Lymphosarcoma and other neoplasms of lymphatic and hematopoietic tissues (codes 200-203, 205, ICD-7).

**Table 1.8** The 30-year trends in cancer death rates per 1000,000 of population, 1971–2003

| Cancer site | Sex | 1971–1973 | 1990 | Percent of changes, 1990/(1971–1973) | 1995 | Percent of changes, 1995/1990 | 2003 | Percent of changes, 2003/1990 |
|---|---|---|---|---|---|---|---|---|
| All sites | Male | 204.5 | 279.8 | +36.8% | 268.5 | –4.0% | 234.1 | –16.3% |
| | Female | 132.0 | 174.7 | +32.3% | 173.4 | –0.7% | 160.5 | –8.1% |
| Oral cavity and pharynx | Male | 5.9 | 5.6 | –5.1% | 4.9 | –12.5% | 4.1 | –26.8% |
| | Female | 1.9 | 2.0 | +5.3% | 1.8 | –10.0% | 1.5 | –25.0% |
| Esophagus | Male | 5.0 | 7.2 | +44% | 7.6 | +5.6% | 7.7 | +6.9% |
| | Female | 1.4 | 1.8 | +28.6% | 1.8 | 0% | 1.7 | –5.6% |
| Stomach | Male | 10.4 | 8.9 | –14.4% | 7.7 | –13.5% | 5.7 | –36% |
| | Female | 5.0 | 4.2 | –16.0% | 3.7 | –11.9% | 3.0 | –28.6% |
| Colon and rectum | Male | 25.3 | 30.8 | +21.7% | 27.7 | –10.1% | 23.0 | –25.3% |
| | Female | 20.0 | 20.6 | +3% | 19.1 | –7.3% | 16.1 | –21.8% |
| Liver and intrahepatic bile ducts | Male | 3.4 | 5.3 | +55.9% | 6.3 | +18.9% | 7.4 | +39.6% |
| | Female | 1.8 | 2.4 | +33.3% | 2.9 | +20.8% | 3.1 | +29.2% |
| Pancreas | Male | 11.1 | 12.6 | +13.5% | 12.1 | –4.0% | 12.0 | –4.8% |
| | Female | 6.7 | 9.3 | +38.8% | 9.2 | –1.1% | 9.2 | –1.1% |
| Larynx | Male | 2.8 | 3.0 | +7.1% | 2.8 | –6.7% | 2.4 | –20.0% |
| | Female | 0.3 | 0.5 | +66.7% | 0.5 | 0% | 0.5 | 0% |
| Lung and bronchus | Male | 61.3 | 90.6 | +47.8% | 84.4 | –6.8% | 71.9 | –20.6% |
| | Female | 12.7 | 36.8 | +189.8% | 40.3 | +9.5% | 41.2 | +12.0% |
| Melanoma | Male | 2.0 | 3.8 | +90.0% | 3.9 | +2.6% | 3.9 | +2.6% |
| | Female | 1.3 | 2.0 | +53.8% | 1.8 | –10.0% | 1.7 | –15.0% |
| Breast | Male | 0.3 | 0.3 | 0% | 0.4 | +33.3% | 0.3 | 0% |
| | Female | 26.6 | 33.1 | +25.9% | 30.6 | –7.6% | 25.2 | –23.9% |
| Cervix uteri | Female | 5.6 | 3.7 | –33.9% | 3.2 | –13.5% | 2.5 | –32.4% |
| Corpus uteri | Female | 4.7 | 4.3 | –8.5% | 4.1 | –4.7% | 4.1 | –4.7% |
| Ovary | Female | 8.6 | 9.3 | +8.1% | 9.1 | –2.2% | 8.8 | –5.4% |
| Prostate | Male | 21.4 | 38.6 | +80.4% | 37.3 | –3.4% | 26.6 | –31.1% |
| Testis | Male | 0.7 | 0.3 | –57.1% | 0.2 | –33.3% | 0.2 | –33.3% |

**Table 1.8** (continued)

| Cancer site | Sex | 1971–1973 | 1990 | Percent of changes, 1990/(1971–1973) | 1995 | Percent of changes, 1995/1990 | 2003 | Percent of changes, 2003/1990 |
|---|---|---|---|---|---|---|---|---|
| Urinary bladder | Male | 7.2 | 8.0 | +11.1% | 7.8 | –2.5% | 7.4 | –7.5% |
| | Female | 2.2 | 2.4 | +9.1% | 2.3 | –4.2% | 2.2 | –8.3% |
| Kidney and renal pelvis | Male | 4.3 | 6.2 | +44.2% | 6.2 | 0% | 6.1 | –1.6% |
| | Female | 1.9 | 2.8 | +47.4% | 3.0 | +7.1% | 2.7 | –3.6% |
| Brain and other nervous system | Male | 4.7 | 6.0 | +27.7% | 5.7 | –5.0% | 5.4 | –10.0% |
| | Female | 3.2 | 4.0 | +25.0% | 3.9 | –2.5% | 3.6 | –10.0% |
| Thyroid | Male | 0.4 | 0.4 | 0% | 0.4 | 0% | 0.4 | 0% |
| | Female | 0.5 | 0.5 | 0% | 0.4 | –20.0% | 0.5 | 0% |
| Hodgkin lymphoma | Male | 1.9 | 0.9 | –52.6% | 0.7 | –22.2% | 0.5 | –44.4% |
| | Female | 1.1 | 0.5 | –54.5% | 0.4 | –20.0% | 0.4 | –20.0% |
| Non-Hodgkin's lymphoma | Male | 5.8 | 10.0 | +72.4% | 10.8 | +8.0% | 9.3 | –7.0% |
| | Female | 3.9 | 6.3 | +61.5% | 7.2 | +14.3% | 5.9 | –6.3% |
| Leukemia | Male | 8.9 | 10.7 | +20.2% | 10.8 | +0.9% | 9.9 | –7.5% |
| | Female | 5.3 | 6.2 | +17.0% | 6.0 | –3.2% | 5.6 | –9.7% |
| Multiple myeloma | Male | 2.8 | 4.8 | +71.4% | 4.9 | +2.1% | 4.6 | –4.2% |
| | Female | 1.9 | 3.1 | +63.2% | 3.4 | +9.7% | 3.1 | 0% |

Rates are adjusted to the age distribution of the 1970 U.S. census population for 1971–1993 trends, age adjusted to the 2000 U.S. standard population (19 age groups – Census P25-1130).
Sources: Vital Statistics of the United States, 1993.

**Table 1.9** Age-adjusted death rates for leading human death causes in the United States from 1900 to 2001

| Disease | Sex | 1900 | 1920 | 1940 | 1960 | 1980 | 1990 | 2003 |
|---|---|---|---|---|---|---|---|---|
| Heart disease | Male | 285.6 | 383.3 | 631.9 | 687.6 | 538.9 | 412.4 | 286.6 |
| | Female | 247.6 | 367.0 | 486.4 | 447.0 | 320.8 | 257.0 | 190.3 |
| Cancer | Male | 90.8 | 139.7 | 187.2 | 225.1 | 271.2 | 280.4 | 234.1 |
| | Female | 138.6 | 180.0 | 189.6 | 168.7 | 166.7 | 175.7 | 160.5 |
| Cerebrovascular disease | Male | 248.0 | 229.5 | 179.8 | 186.1 | 102.4 | 68.7 | 54.1 |
| | Female | 240.4 | 246.9 | 174.9 | 170.7 | 91.9 | 62.7 | 52.3 |
| Pneumonia/influenza | Male | 287.2 | 280.8 | 127.8 | 65.8 | 42.1 | 47.8 | 26.1 |
| | Female | 304.7 | 276.1 | 107.6 | 43.8 | 36.0 | 30.5 | 19.4 |

Rates are per 100,000 population and age adjusted to 2000 U.S. standard population.
Sources: CDC/NCHS, 2000a, b, c; CDC, 2006.

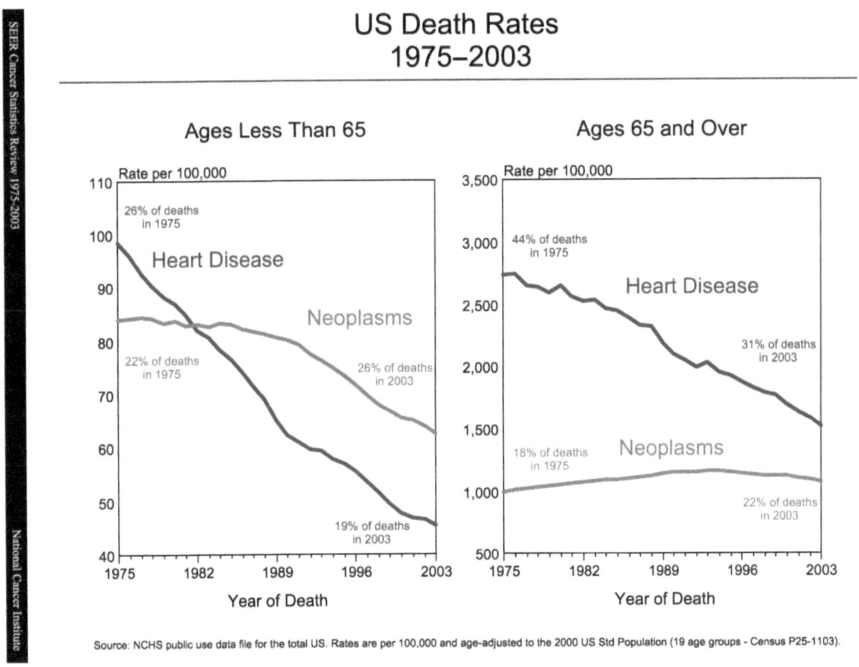


**Fig. 1.4** Trends of the U.S. death rates for neoplasms and heart disease, for ages less than 65, and 65 and over, 1975–2003

To move forward, it is important to pay attention not only to the diagnosis of cancer at early stages and treatment, which is obviously of great importance, but also to intervention in potentially controllable cancer risk factors. According to some forecasts, it might be 23% fewer cases of cancer in the developing world in 1990, if infections such as hepatitis B and C viruses, and human papillomavirus (HPV) had been prevented (Parkin, 1999). Another estimate suggests that 230,000 deaths (more than 4% of the worldwide cancer deaths total) from liver cancer could have been avoided with only immunization against hepatitis B (Pisani et al., 1999). Smoking was estimated to be responsible for another 20% of all cancer deaths, and most of them were preventable (Pisani et al., 1999). Because of a still high incidence of several potentially preventable cancers (in both developed and developing countries) it is very important to precisely evaluate the role of primary prevention, early cancer detection (screening), and methods of cancer treatment (such as cost-effectiveness analysis, and others) which can be generalized and are comparable across both existing and potentially feasible interventions (Murray et al., 2000) (the life table methods for these evaluations are discussed in Chapter 4).

In the United States, one-third of the 564,830 cancer deaths (expected to occur in 2006) were related to nutrition, physical inactivity, and overweight or obesity, and could potentially be prevented (see Table 1.10). Certain cancers are related to infectious agents, such as hepatitis B and C viruses, HPV, human immunodeficiency virus (HIV), Helicobacter pylori (Hp) and are potentially preventable through behavioral changes, vaccines, or antibiotics. Opportune protection from exposure to the solar UV radiation can prevent many of the more than one million skin cancers that are expected to be diagnosed in the United States per year (Cancer Facts & Figures, 2006). However, interventions in cancer risk start to be paid off (by observable decrease in associated with risk cancer death rate) in several decades: i.e., the 30-year period was registered between decrease of cigarette consumption in the United States and beginning of the decrease of lung cancer death rate (see Fig. 1.5).

**Table 1.10** Prevalence of major cancer risk factors by race/ethnicity in adults 18 and older in the United States, 2000 (from Cancer Facts & Figures, 2004)

| | Current smokers (%) | | Reporting no leisure-time physical activity (%) | | Obese (%)[1] | |
|---|---|---|---|---|---|---|
| Race/ethnicity | Male | Female | Male | Female | Male | Female |
| White (non-Hispanic) | 25.7 | 23.0 | 33.1 | 36.8 | 21.3 | 19.6 |
| African American (non-Hispanic) | 25.5 | 20.4 | 47.3 | 55.7 | 24.4 | 35.9 |
| Hispanic/Latino | 23.2 | 12.8 | 51.9 | 56.5 | 23.0 | 26.1 |
| American Indian and Alaska Native[2] | 27.4 | 38.6 | 46.5 | 52.1 | 38.1 | 43.2 |
| Asian American[3] | 19.6 | 7.9 | 29.1 | 42.1 | 6.0 | 8.3 |

[1]BMI ≥30 kg/m2.
[2]Estimates should be interpreted with caution because of the small sample sizes.
[3]Does not include Native Hawaiians and other Pacific Islanders.
Percentages are age adjusted to the 2000 U.S. standard population.
Source: National Health Interview Survey 2000, National Center for Health Statistics, Centers for Disease Control and Prevention.

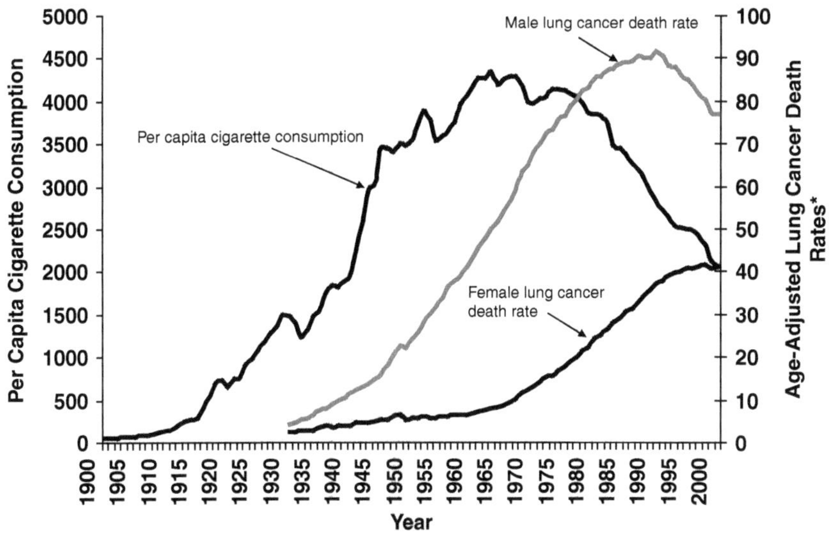


*Age-adjusted to 2000 US standard population.
Source: Death rates: US Mortality Public Use Tapes, 1960–2000, US Mortality Volumes, 1930–1959, National Center for Health Statistics, Centers for Disease Control and Prevention, 2002. Cigarette consumption: US Department of Agriculture, 1900–2000.

**Fig. 1.5** Historic trends of tobacco use and lung cancer death rates in the United States, 1900–2000, for male and female

## 1.3 Interdisciplinary Approach to Population Health Studies

Frequently reference is made to multidisciplinary and interdisciplinary studies of health without a clear distinction between the two types of efforts. One way to clearly distinguish these efforts is to conceive of multidisciplinary studies as involving the effort of experts from a number of distinct disciplines without integration. In that case, each discipline yields specific results while any integration would be left to a third party observer. In interdisciplinary studies, there is a greater reliance on the cross-training of individuals, so that medical specialists will have a clear understanding of mathematical modeling efforts, so that they can fully participate in the development of the models to be formulated by mathematicians, biophysicists, epidemiologist, statisticians, and computer scientists. Thus the interdisciplinary approach involves different specialists with the goals of connecting and integrating several academic disciplines, professions, or technologies, along with their specific perspectives, in the pursuit of a common task. An interdisciplinary team attacks a subject from various angles and methods, eventually cutting across disciplines and forming a new method for understanding the subject. This "penetration" of expertise and methods may lead to development of new academic disciplines.

An interdisciplinary approach might be used in a wide spectrum of translational studies, making it possible to model disease process across the biological spectrum from human population to inter- and intracellular levels. One of the important tools related to this approach is the use of theoretical and computational mathematical modeling. This complex approach allows one to analyze biological processes from the prospective of mechanisms defining the basic characteristics of processes, such as mechanisms of carcinogenesis. Thus, the modeling approach may not only be used for traditional analyses of cost effectiveness of preventive or therapeutic strategies (e.g., screening or vaccination) but provide a deeper insight in cancer mechanisms, a better understanding of which will likely prove valuable for improving therapies.

These methods use new opportunities, such as construction of models for specific tasks/purposes, which appear more complicated, say, compared to the classical Armitage-Doll, may be developed with participation of clinicians and basic laboratory researchers and use data from human populations. This approach allows one to combine populations with different characteristics to work with large masses of data, to combine data on various cancers to study certain common mechanisms of carcinogenesis, to create linkages between case–control studies and large population data (e.g., SEER register) and conveyed results obtained from case–control studies to large populations to check hypotheses in "real-life" human data.

In the analyses of cancer and carcinogenesis presented in this monograph, we attempted to apply truly interdisciplinary methods and approaches because they address disease initiation and expression all the way from mechanisms at the cellular/molecular level up to its impact on the health and longevity of national populations. This is an ambitious effort involving the integration, not only of cross-trained experts but also the utilization of data generated at multiple levels of biological complexity, and produced using different measurement systems and observational plans. Thus, these interdisciplinary studies involve the complications of multidimensionality, extension over biological scale and complexity and emergent self-organization in conducting both statistical analyses and in making forecasts of disease effects. Generally this involves assessing how nonlinear effects are expressed in interactions and how remote effects operate in complex biological systems.

We approach our assessment of this problem by beginning to describe mathematical and statistical models at simple and basic levels and extend analyses in complexity in an incremental fashion. In this way, we hope to produce an interdisciplinary analysis of cancer morbidity and mortality in a comprehensive fashion with the interactions and nonlinear effects emerging naturally. The spectrum of specific tasks which are solved by the interdisciplinary team is much broader than what is possible to investigate by a homogenous group of specialists. The properties of the U.S. population are analyzed using demographic methods specialized to deal with the structure of the available datasets coming from demographic and epidemiologic studies. Such methods might be generalized to use quite advanced modeling technologies developed in different branches of mathematics and/or theoretical physics. Results of such analyses allow us to make the conclusions about risk factors of specific diseases and to determine the efficacy of different medical strategies/risk factor interventions over time.

For such investigations, the expertise of scientists in medical sciences is irreplaceable. Detailed analysis of demographic datasets by combined groups of specialists often allows the formulation and testing of new hypotheses about the biomedical mechanisms of the underlying biological processes responsible for generating observable data. One example from Chapter 4 is the quadratic hazard model, where dynamics of covariate/risk factors are described by generalized equations like those due to Fokker–Planck (as in theoretical physics) or Kolmogorov (as in stochastic mathematics) with mortality selection modeled by a multidimentional quadratic form, reflecting the fact that the probability of death or disease onset increases when an individual has physiologically non-optimal values of risk factors.

What is of interest is that the recent rapid progress in theoretical physics related to the formulation of new statistical mechanics based on Tsallis entropy and the notion of abnormal diffusion admits natural generalization of the stochastic process modeling approach in biodemography. In demographic models, using this generalization, one can describe cohorts of individuals related in different senses (e.g., genetically) or alternately of the population of interrelated and interacting cells in an organ or specific tissue.

Another example comes from the so-called two-disease model, where the observed age pattern of disease onsets is modeled as a mixture of two population groups with different susceptibilities for a given disease (see Chapter 7). Detailed analysis of the properties discovered for each group allows us to formulate hypotheses on biomedical grounds of these susceptibilities (e.g., genetic and/or environmental risk factors, etc.)

The next example is the use of microsimulation as an estimation technique, which came from experimental physics, where each experiment is simulated by the methods of stochastic mathematics and informatics. In social and biomedical sciences, these approaches allow one not only to carefully plan new studies where data collection is costly but to provide estimation strategies, where standard, analytical methods cease to work. In Chapter 5, a reader can find (i) analytic approaches based on fuzzy set logic and latent structure analyses which are irreplaceable for analysis of highly multivariate categorical data, (ii) methods known as stochastic process models which generalizes the quadratic hazard model by allowing it to be very flexible for application to different kind of longitudinal data (e.g., to adjust for different types of concerns), (iii) empirical Bayesian approaches for analysis of the variation of cancer incidence and mortality in various geographic regions. The combined application of these approaches in Chapters 6 and 7 allows us to comprehensively investigate U.S. mortality and morbidity age patterns, to perform sensitivity analysis of the obtained results and to relate these patterns with risk factors measured in studies of different kinds of research design.

## References


Atta H.M., 1999. Edwin Smith surgical papyrus: the oldest known surgical treatise. Am Surg 15(12):1190–1192.

Cancer Facts & Figures, 2004. At: http://www.cancer.org/downloads/STT/CAFF_finalPWSecured.pdf

Cancer Facts & Figures, 2006. At: http://www.cancer.org/downloads/STT/CAFF_finalPWSecured.pdf

Cancer Facts and Figures, 2008. At: http://www.cancer.org/downloads/STT/CAFF_finalPWSecured.pdf

CDC, 2006. National Vital Statistics Reports, Vol. 54, Number 13 "Death: Final data for 2003" (by Hoyert D.L., Heron M.P., Murphy S.L., Kung H-C.), at: http://www.cdc.gov/nchs/data/nvsr/nvsr54/nvsr54_13.pdf

CDC/NCHS, 2000a. Vital Statistics of the United States. HIST293. Table: Age-adjusted rates for selected causes by race and sex using year 2000 standard population: death registration states, 1900–32 and United States, 1933–49. At: http://www.cdc.gov/nchs/data/dvs/hist293-1900-49.pdf

CDC/NCHS, 2000b. Vital Statistics System, HIST293. Table: Age-adjusted rates for 60 selected causes by race and sex using year 2000 standard population: United States, 1960–67. At: http://www.cdc.gov/nchs/data/mortab/aadr6067.pdf

CDC/NCHS, 2000c. Vital Statistics System, HIST293. Table: Age-adjusted rates for 72 selected causes by race and sex using year 2000 standard population: United States, 1979–98. At: http://www.cdc.gov/nchs/data/mortab/aadr7998s.pdf

CDC/NCHS, National Vital Statistics System, Mortality, 2000d. HIST293. Table: Age-adjusted death rates for 72 selected causes by race and sex using year 2000 standard population: United States, 1978–98. At: http://www.cdc.gov/nchs/data/mortab/aadr7998s.pdf HIST293. Table: Age-adjusted death rates for 60 selected causes by race and sex using year 2000 standard population: United States, 1960–67. At: http://www.cdc.gov/nchs/data/mortab/aadr6067.pdf HIST293. Table: Age-adjusted death rates for approximately 64 selected causes, by race and sex: United States, 1950–59. At: http://www.cdc.gov/nchs/data/dvs/hist293_1950_59.pdf HIST293. Table: Age-adjusted death rates for selected causes by race and sex using year 2000 standard population: death registration states, 1900–32 and United States, 1933–49. At: http://www.cdc.gov/nchs/data/dvs/hist293_1900_49.pdf

Davis D., 2007. The Secret History of the War on Cancer. New York, NY: Basic Books, 528 pages.

DevCan, 2005. Probability of Developing or Dying of Cancer Software, Version 6.0. Statistical Research Applications Branch, NCI, 2005. http:// srab.cancer.gov/devcan American Cancer Society, Surveillance Research, 2006.

Epstein S.S., 2005. Cancer-Gate: How to Win the Losing Cancer War (Policy, Politics, Health and Medicine). Policy, Politics, Health and Medicine Series, Vicente Navarro, Series. Amityville, NY: Baywood Publishing Company. 377 pages.

Extramural Committee to Assess Measures of Progress against Cancer, 1990. Measurement of progress against cancer. J Natl Cancer Inst 82:825–835.

Faquet G.B., 2005. The War on Cancer: An Anatomy of Failure. A Blueprint for the Future. New York: Springer. 227 pages.

Feldman R.P., Goodrich J.T., 1999. The Edwin Smith surgical papyrus. Childs Nerv Syst 15(6–7):281–284.

Ferlay J., Bray F., Pisani P., 2001. Cancer Incidence, Mortality and Prevalence Worldwide, Version 1.0. IARC CancerBase No.%, Lyon: IARCPress.

Ferlay J., Bray F., Pisani P., Parkin D.M., 2006. GLOBOCAN 2002: Cancer Incidence, Mortality and Prevalence Worldwide IARC Cancer Base No. 5, Version 2.0, Lyon: IARC Press, 2004. American Cancer Society, Surveillance Research, 2006.

Fidler I.J., Yano S., Zhang R.D. et al., 2002. The seed and soil hypothesis: vascularization and brain metastasis. Lancet Oncol. 3:53–57.

GLOBOCAN 2002. At: http://www-dep.iarc.fr/globocan/database.htm

Greaves M., 2002. Cancer: The Evolutionary Legacy. New York: Oxford University Press. 288 pages.

Gurunluoglu R., Gurunluoglu A., 2003. Paul of Aegina: landmark in surgical progress. J Sur 27:18–25.

Lee H.S. (ed.), 2000. Dates in Oncology: A Chronological Record of Progress in Oncology over the Last Millenium (Landmarks in Medicine). Lancaster: Parthenon Publishing Group. 124 pages.

Lerner B.H., 2003. The Breast Cancer Wars: Hope, Fear, and the Persuit of a Cure in Twentieth-Century America. New York: Oxford University Press. 416 pages.

Morton L.T., Moore R.J., 1997. A Chronology of Medicine and Related Sciences. Aldershot, England: Scholar Press.

Murray C.J., Evans D.B., Acharya A., 2000. Development of WHO guidelines on generalized cost-effectiveness analysis. Health Econ 9:235–251.

Murray C.J.L., Lopez A.D., 1997. Mortality by cause for eight regions of the world: global burden of disease study. Lancet 349:1269–1276.

NCI, CDC, State Cancer Profiles. At: http://statecancerprofiles.cancer.gov/historicaltrend/joinpoint.noimage.html

Olson J.S., 2002. Bathsheba's Breast: Women, Cancer, and History. Baltimore, MD: The John Hopkins University Press. 320 pages.

Pantel K., Cote R.J., Fodsyad O., 1999. Detection and clinical importance of micrometastatic disease. J Natl Cancer Inst 91:1113–1124.

Parkin D.M., 1999. The global health burden of infection associated cancers. Cancer Surv 33:5–33.

Parkin D.M., Pisani P., Ferlay J., 1999. Estimates of the worldwide incidence of 25 major cancers in 1990. Int J Cancer 80(6):827–841.

Pisani P., Parkin D.M., Bray F., 1999. Estimates of the worldwide mortality from 25 cancers in 1990. Int J Cancer 83(1):18–29.

Proctor R.N., 1995. Cancer Wars: How Politics Shapes What We Know and Don't Know About Cancer. New York, NY: Basic Books. 368 pages.

Rosner F., 1998. The Medical Legacy of Moses Maimonides. Hoboken, NJ: KTAV Publishing House Inc.

SEER Cancer Statistics Review 1973–1993. Survival rates, by race, diagnosis year, stage and age. Tables IV-5, V-4, VI-7, VII-4, VIII-4, XI-4, XII-4, XIV-4, XV-7, XVI-4, XIX-4, XX-4, XXI-4, XXII-4, XXIII-4, XXIV-4, XXV-4, XXVI-4. At: http://seer.cancer.gov/csr/1973_1993

SEER Cancer Statistics Review 1975–2004. Survival rates, by race, sex, diagnosis year, stage and age. Table XII-5. Larynx. At: http:// seer.cancer.gov/csr/1975_2004/results_-merged/sect_12_larynx.pdf

SEER Cancer Statistics Review 1975–2005 (Summary). Summary of changes in cancer mortality, 1950–2005, and 5-year relative survival rates, 1950–2004. Males and females, by primary cancer site. Table I-3. At: http://seer.cancer.gov/csr/1975_2005/results_single/sect_01_table.03.pdf

SEER Cancer Statistics Review 1975–2005. Survival rates, by race, sex, diagnosis year, stage and age, 1996–2004. Tables IV-10, V-5, VII-5, VIII-5, XI-5, XII-5, XIV-5, XV-9, XVI-5, XX-5a, XXI-5, XXII-5, XXIII-5, XXIV-5, XXV-5, XXVI-5, XXVII-5. At: http://seer.cancer.gov/csr/1975_2005/results_merged/topic_survival.pdf

SEER End Results Group, 1960–1973. NCI Surveillance, Epidemiology and End Results Program, 1996.

SEER, Table I-11. Median age of cancer patients at diagnosis, 2000–2003. At: http://seer.cancer.gov/csr/1975_2003/results_single/sect_01_table.11_2pgs.pdf

SEER, Table I-13. Median age of cancer patients at death, 2000–2003. At: http://seer.cancer.gov/crs/1975_2003/results_single/sect_01_table.13_2pgs.pdf
Shibuya K., Mathers C.D., Boschi-Pinto C., Lopez A.D., Murray C.J.L., 2002. Global and regional estimates of cancer mortality and incidence by site: II. Results for the global burden of diseases 2000. BMC Cancer 2:37.
SEEResult Program, 1975–2002. Division of Cancer Control and Population Sciences, Bethesda, MD: NCI, 2005. American Cancer Society Surveillance Research, 2006. At: http://seer.cancer.gov/csr/1975_2003/results_merged/topic_survival.pdf
Tominaga S., Oshima A. (eds.), 2000.http://Cancer mortality and morbidity statistics. Japan and the world – 1999. Japanese Cancer Association, Gann Monograph on Cancer Research No.47. Tokyo: Japan Scientific Societies Press.
U.S. Mortality Public Use Data Tape, 2001. National Center for Health Statistics, Centers for Disease Control and Prevention, 2003.
Udwadia F.E., 2000. Man and Medicine: A History. Oxford: Oxford University Press.
US Mortality Data Files, 2005. Leading causes of death in US. Percent of all causes of death. 1975 vs. 2005. National Center for Health Statistics, Centers for Disease Control and Prevention. At: http://seer.cancer.gov/csr/1975_2005/results_merged/topic_lead_cod.pdf
Vital Statistics of the United States, 1993. United States Cancer Statistics: 2002 Incidence and Mortality, http://www.cdc.gov/cancer/npcr/uscs
Welch H.G., Schwartz L.M., Woloshin S., 2000. Are increasing 5-year survival rates evidence of success against cancer? JAMA 283(22):2975–2978.

# Chapter 2
# Cancer Modeling: How Far Can We Go?

Cancer, a disease which occurs in complex multicellular organisms, appears to reflect a “throw-back” in the evolutionary process (Trosko and Ruch, 1998): cancer cells resemble primitive bacterial cells that survive in relatively unstructured cell colonies, characterized by uncontrolled proliferation, which do not functionally differentiate to support colony survival. In contrast, normal cells in higher multicellular organisms have genes coding connexins – proteins that determine the structural and functional relation of specific cells in a tissue by the alignment of cell pores and ion channels for intercellular communication. Cells that are cancerous appear not to respond to “contact inhibition,” fail to terminally differentiate, appear to be clonally derived from a stem-like cell (i.e., one reverting to a less functionally differentiated state) and continue to change geno- and phenotypically with tumor progression and growth. The biological processes of signal transduction and apoptosis (programmed cell death) appears also to be often altered in cancer cells compared to normal parent cells. Some of the major observations that should be explained by a comprehensive theory of carcinogenesis include (a) normal cell growth is inhibitable, while the reproduction of cancer cells are not; (b) normal cells derived from stem/progenitor cells are capable of terminal differentiation, cancer cells under “normal” conditions are not (the teratomas represent a special case); (c) most, if not all, tumors appear to be derived from a single cell; and (d) during the long, frequently chaotic process of carcinogenesis, the tumor cell acquires multiple genotypic and phenotypic change, often including drug resistance traits (Trosko and Ruch, 1998).

To seek answers as to how cancer initiates and progresses, scientists around the world have approached the problem from two fundamental perspectives. One group comprised of molecular biologists, biochemists, geneticists, and oncologists, who have done extensive laboratory work on *in vitro* cellular and molecular test systems, *in vivo* studies of animal models, and clinical trial assessments of tumor response to radiation and chemotherapy to identify the basic components and fundamental dynamics of carcinogenesis. Another group involved epidemiologists and demographers who have been trying to develop mathematical human population models for carcinogenesis to integrate animal

K.G. Manton et al., *Cancer Mortality and Morbidity Patterns in the U.S. Population*, Statistics for Biology and Health, DOI 10.1007/978-0-387-78193-8_2, 
© Springer Science+Business Media, LLC 2009

experiments, human tumor registries, and human epidemiological data. While making some progress in understanding certain mechanisms of human carcinogenesis, however, it appears that compared to the rapid progress in clinical and laboratory findings, epidemiologists, demographers, and biostatisticians have not advanced as rapidly in developing biologically meaningful models of carcinogenesis and tumor growth and progression in large human populations which are exposed to various risk factors. Indeed, there is debate over how much information may be extracted from the applications of such models to human population data (Hazelton et al., 2005) and of what kind of practical use for oncologists and public health specialists might be the results of such models.

Studying carcinogenesis over the decades, many theories included elements that explain certain observations, but, presently, none of them provide a general framework for a fully integrated explanation of cancer. The following two quotations set the stage for understanding "reductionalistic" versus "holistic" views of the cancer problem: (a) "The understanding of the cellular basis of cancer means being able to describe the biochemistry of the regulated pathways between the cell surface and the nucleus that control cell growth" (Hunter, 1986), and (b) "The cancer problem is not merely a cell problem, it is a problem of cell interaction not only within tissues, but with distant cells in other tissues" (Potter, 1973).

Below we examine the status of a number of different conceptual models of carcinogenesis and the various quantitative implications of those models.

## 2.1 Cellular Aspects of Carcinogenesis

Carcinogenesis is a complex process involving a number of cellular mechanisms. Among the key aspects of cancer biology are the cell proliferation, chromosome instability, gene aberration, telomere length and telomerase activity, cell senescence, apoptosis, anchorage-independent growth, and formation of cancer (Reddel, 2000; Heselmeyer et al., 1998; Rhim, 2001; Yang et al., 2000; Kim et al., 1999; Cifone and Fidler, 1980; Harris, 1987; Shen et al., 2001). It has been suggested an existence of six hallmarks in virtually all cancers (these hallmarks may vary both mechanistically and chronologically, depending on cancer type, tissue type, age at onset, etc.), such as (1) self-sufficiency in growth signals, (2) insensitivity to anti-growth signals, (3) evading apoptosis, (4) limitless replicative potential, (5) sustained angiogenesis, and (6) tissue invasion and metastasis (with genome instability placed apart from these characteristics as an "enabling characteristic" facilitating the acquisition of other mutations caused by DNA repair defects) (Hanahan and Weinberg, 2000; Spencer et al., 2006).

### *2.1.1 Nuclear DNA Mutation*

Nuclear DNA mutation has traditionally been viewed as the primary mechanism of carcinogenesis by which environmental stresses are assumed to leave their imprint on those surviving cells that remain capable of replication after injury. Unresolved DNA damages caused by exogenous factors, replication errors and "by-products" of cellular respiration, lead to permanent mutations in genetic code that may be oncogenic. Pathways involved in reparation and controlling DNA damage are important for anticancer defense. There are various pathways that repair DNA damage, as well as those which prevent cellular replication or induce cell death. Results of several studies on mice suggested that a potential consequence of DNA damage and responses to DNA damage was aging, supporting the hypothesis that at least some aspects of normal aging are the consequences of anticancer mechanisms designed to deal with damaged DNA (Hasty, 2005).

However, these mutations may occur not only in the nucleus but also in other DNA-containing organelles. There are several other mechanisms that can affect core cellular and molecular processes that should be identified for the modeling effort. One of the necessary innovations in modeling carcinogenesis is determining how to represent the influence of these subsystems on nuclear DNA mutations processes. Below we briefly describe some of those subsystems.

### 2.1.2 Mitochondrial DNA Mutation

It is assumed that more than 1.5 billion years ago the energy-converting organelles of eukaryotes evolved from procariotes that were "swallowed" by primitive eucaryotic cells, thus developing a symbiotic relationship. That might explain why mitochondria have their own DNA (mtDNA). The mitochondrial genome is much simpler than the nuclear genome, involving only 37 genes and lacking much of the cell error detection and repair machinery of nuclear DNA. One of the two mitochondrial DNA strands – the heavy strand – contains 12 of the 13 polypeptide-encoding genes, 14 of the 22 tRNA-encoding genes, and both rRNA-encoding genes (Wallace et al., 1992; Wallace, 1995; Zeviani et al., 1998). While damaged by free radicals, mtDNA lacking the protective action by histones (protective proteins) and has the limited capacity of an efficient DNA repair system (Bogenhagen, 1999; Pettepher et al., 1991). That may accelerate the rate of mtDNA mutation (Lightowlers et al., 1997). This is probably why the accumulation of polymorphisms in mtDNA is approximately 10–17 times higher than in nuclear DNA (Neckelmann et al., 1987; Wallace et al., 1997). The phenotype is normal until a critical proportion of mutant mtDNA is present within the tissue and the genotype expression threshold is exceeded, and then substantial changes in phenotype happen with minor increases in mutant mtDNA proportion (Wallace et al., 1997).

Most of inherited mutations are not enough for suppression of mitochondrial oxidative phosphorylation (OXPHOS) below the expression threshold, and the accumulation of somatic mutations in postmitotic tissues is needed to exacerbate the inherited OXPHOS defect, leading to phenotypic expression (Wallace et al., 1992; Wallace, 1995). Particularly vulnerable to oxidative stress is the lipid bi-layer membrane of the mitochondria, which allows leakage of electrons from the mitochondria to the rest of the cell extending the range and intensity of oxidative damage. This process can induce an enhancement of damage to nuclear DNA, thereby further increasing gene mutation and cancer risks. This effect is of greatest importance in tissues with the high density of mitochondria and endogenous oxidative stress, due to high levels of metabolism and energy production, e.g., in the central nervous system (CNS) (Manton et al., 2004) and in tissues with a high mitotic index. The structure and number of mitochondria are also strongly influenced by the neuroendocrine system, especially by thyroid hormones (Wrutniak-Cabello et al., 2001).

A German biochemist and physiologist, the Nobel Laureate Otto Warburg, (1883–1970) proposed in 1956 that cancer was caused by altered metabolism and by deranged energy processing in mitochondria (Warburg, 1956). Presently, it has been shown in numerous studies that mtDNA mutations were associated with severe neurodegenerative disorders, primary hereditary neoplasias, such as inherited pheochromocytomas and paragangliomas, cutaneous and uterne leiomyomas, sporadic kidney oncocytoma, etc., as well as mtDNA mutations, were linked to nonhereditary tumors (DiMauro and Schon, 2003; Eng et al., 2003; Zanssen et al., 2004). High-incidence mtDNA alterations has been recently described in various cancers, such as oesophageal (Hibi et al., 2001b; Kumimoto et al., 2004), gastric (Maximo et al., 2001; Wu et al., 2005), colorectal (Polyak et al., 1998; Hibi et al., 2001a; Lievre et al., 2005), pancreatic (Jones et al., 2001), hepatocellular (Nishikawa et al., 2001), breast (Richard et al., 2000; Tan et al., 2002; Zhu et al., 2005), uterine (Pejovic et al., 2004), prostate (Jeronimo et al., 2001; Petros et al., 2005), renal (Nagy et al., 2002), urinary bladder (Fliss et al., 2000), thyroid (Yeh et al., 2000), and skin (Girald-Rosa et al., 2005) [some studies suggested that somatic mtDNA mutations in renal carcinoma does not indicate a major contribution of these alterations in tumor development (Meierhofer et al., 2006)].

While the functional significance of somatic mutations in the mtDNA has been debated vis-à-vis their "cause and effect" relationship in cancer cells, there is little doubt that these mutations can play an important role as a biomarker for human cancers (Jakupciak et al., 2005; Kagan and Srivastava, 2005; Parr et al., 2006). Mutations of mtDNA, even driven by random process during malignant transformation, present an excellent possibility for early tumor detection, e.g., using D-loop analysis of bodily fluids from patients with tumors (Fliss et al., 2000). Findings of somatic mtDNA alterations in precancer lesions (even in the absence of histopathologically identified dysplasia) of gastrointestinal tract let speculate about susceptibility of mitochondrial genome at early stages of tumorogenesis (Sui et al., 2006). That might make it reasonable to explore the mitochondrial genome as a biomarker for the early diagnosis of cancer.

This nexus of cancer and cellular energetics may provide a basis for explaining the linkage of cancer and aging. It may also provide the basis for the explanation for mechanisms underlying certain degenerative diseases, and likely human senescence, and limits to human longevity (Economos, 1982).

The proposed hypothesis that mtDNA mutations and respiratory dysfunction might be linked directly to carcinogenesis via apoptotic or reactive oxygen species-mediated pathway needs urgent experimental proofs; it should be also clarified whether the hypoxia-inducible factor (HIF)-mediated pathway is also initiated in hypoxia and mitochondrial deficiency, which both are tumor characteristics (Doege et al., 2005). If these pathways are confirmed as being involved in tumorigenesis, metabolic targeting of mitochondrias in cancer, such as blocking the HIF pathway by administration of $\alpha$-ketoglutarate, may be used in developing new approaches to anticancer therapies and cancer prevention (Zanssen and Schon, 2005).

### 2.1.3 Damage to the Protein Generation Machinery of the Endoplasmic Reticulum and Golgi Apparatus

The endoplasmic reticulum and the Golgi apparatus are the primary structures within which RNA is translated into functional proteins by the encoding of specific amino acid sequences. Any damage in these cell structures could lead to miscoding in protein sequences which could lead to the loss of cell function and to functional dedifferentiation of the cell. This mechanism also includes the cytoplasmic operation of heat shock proteins (e.g., HSP70 and HSP90) and other chaperone molecules to help determine and maintain the spatial–organizational structure of the protein, which largely determines its function under various types of stress.

The endoplasmic reticulum is the major organelle for protein synthesis and maturation as well as regulation of intracellular calcium ($Ca^{2+}$) homeostasis. The accumulation of unfolded proteins in the endoplasmic reticulum lumen or depletion of $Ca^{2+}$ from the lumen of endoplasmic reticulum leads to the stress response (Harding et al., 2002; Rutkowski and Kaufman, 2004), when the reticulum resident chaperons (proteins such as the nucleoplasmins, the chaperonins, the heat-shock proteins 70, and the heat-shock proteins 90, mediating correction of assembly or disassembly of other polypeptides, but which are not components of final oligomeric structures) such as Bip are induced, thus protecting the cell from reticular stress and improving the protein-folding abilities. However, the mechanism by which the excess stress from which the cell cannot recover triggers apoptosis remains unclear (Breckenridge et al., 2003; Oyadomari and Mori, 2004; Rao et al., 2004). Recently it is has been shown that transcriptional regulator CHOP, which is upregulated by the stress of endoplasmic reticulum, regulates stress-induced apoptosis, at least in part, through enhancing DR5 expression [death receptor 5 (DR5), also called TRAIL-R2, Apo2 or KILLER] which is the members of the cytokine tumor necrosis factor (TNF)-receptor family (Yamaguchi and Wang, 2004). Endoplasmic stress-mediated DR5 induction is p53 independent, what is important for advanced therapies developing, because agents that cause endoplasmic stress could be used for therapy of human cancers with p53 mutations. Recently several agents have been developed based on these findings: e.g., prodrug that couples with PSA for prostate cancer treatment, and NSAIDs and proteasome inhibitors that trigger the stress in endoplasmic reticulum and induce expression of DR5 (Nishitoh et al., 2002; Tsutsumi et al., 2004; He et al., 2002a, b, 2004). Also the recent studies demonstrated that endoplasmic reticulum homeostasis may be used as biomarker of precancer and early-stage cancer: e.g., reticulum homeostasis of colon tumors becomes abnormal at a premalignant stage – adenomas, with the deepening of this defect during the progression to carcinoma (Brouland et al., 2005).

### *2.1.4 Cell–Cell Communication*

In the multicellular organism, homeostasis is regulated by three communication processes: (1) extracellular communication via hormones, growth factors, neurotransmitters and cytokines, which trigger; (2) intracellular communication via alterations in second messengers (e.g., $Ca^{2+}$, pH, ceramides, NO, c-AMP, reactive oxygen species, etc.); and (3) signal transduction systems to modulate intercellular communication, mediated by gap junction channels. This set of communication processes must control a cell's ability to proliferate, to functionally differentiate, to apoptose, and to respond adaptively to changing environmental conditions. Disruption of any of these ways of communication could affect cell proliferation, apoptosis, cause abnormal cell differentiation, and lead to abnormal adaptive responses of these de-differentiated cells to stress (Trosko and Ruch, 1998).

Functional gap junctional intercellular communications (GJICs) exist in most solid tissues. Free-standing cells (e.g., neutrophils, red blood cells, etc.) and most stem cells have no GJICs. One of the most significant physiological implications for GJIC is that gap junction "coupled" cells within a given tissue are highly functionally integrated with their "neighbors" into tissue systems or functional fields, whose structure is defined and maintained by various autocrine and paracrine factors. That facilitates tissue homeostasis and also permits the rapid, direct transfer of second messengers between cells to coordinate cellular responses within the tissue (e.g., islet cell production of insulin in the pancreas).

Among the many differences between a cancer cells and their "normal" parental cells, there is one that involves the transition from a normal, GJIC-competent cell to one that is defective in GJIC. Cancer cells have fewer gap junctions, while growth stimuli inhibit GJIC and growth inhibitors stimulate GJIC, and GJIC has cell cycle-related changes – these are the possible ways for the gap junction to be involved in carcinogenesis. Modeling of these mechanisms may identify new possibilities for the development of innovative approaches to cancer treatment and prevention. One possible mechanism to increase drug penetration and dispersal in tumors would be to increase GJIC by increasing tumor cell connexin expression pharmacologically (e.g., with steroids, retinoids, etc.), or by introducing active connexin genes (gene therapy) (Trosko and Ruch, 1998). GJIC may improve cancer therapy involving a lethal gene introduction, such as *Herpes simplex virus* thymidine kinase (HSV-TK) gene introduced into tumor cells: only a small percentage of the tumor cells take up and express the HSV-TK gene, but a much higher percentage of these cells are killed following ganciclovir treatment, suggesting the presence of a significant "bystander effect" (Denning and Pitts, 1997; Paillard, 1997). The inhibition of GJIC might be one of the mechanisms by which an inflammation affects cancer, therefore, an intervention with antiinflammatory drugs (e.g., NSAID) during tumor promotion might be a highly efficacious anticancer strategy for certain tumor types (Trosko and Tai, 2006; Khuder et al., 2005).

### 2.1.5 Telomere Control of Cell Division

Another important type of cell regulatory structure relevant to carcinogenesis is that of the telomere – the repeating sequence of nucleic acids at the end of the human chromosome which is thought to determine how many times a cell can successfully replicate. The telomeric cap on the chromosome end can become shortened to the point where the fidelity of gene replication is threatened and a danger signal is generated, arresting the cell cycle. This signal is similar in function to the one that arrests the cycle when an uncapped DNA end is created by an accidental double-strand chromosome break. In a cell with a chromosome break the prevention of cell division for as long as the cell contains broken or inadequately capped DNA, allows time for DNA repair.

The telomere, however, can be lengthened due to the operation of an enzyme, telomerase, which can allow extension (reconstruction) of the telomeric sequence. The genetic capability to express telomerase is present in most types of human cells (as evidenced by its presence in most tumor types) with its expression in a tissue regulated by a specific hormonal (e.g., testosterone) and stress (e.g., in response to cellular injury in lung) signals.

In cancer cell the telomere is rapidly consumed during unregulated cell division and, as a consequence, cell function is further altered and degraded as additional damage is accumulated at the end of the chromosome sequence. The maintenance of telomere length is assumed to be an obligatory step in the progression and immortalization of most human cells (Shen et al., 2001; von Zglinicki, 1996; Rudolph et al., 1999). Alterations in the length of telomeric DNA have been documented in a variety of human neoplasms, where they were shorter than normal (Schwartz et al., 1995; Sommerfield et al., 1996; Yamada et al., 1993) or showed a broader range of length than in normal tissues (Schmitt et al., 1994; Hiyama et al., 1995a, b). It has been suggested that telomere dysfunction induces chromosomal instability as an early initiating event in most of human epithelial cancers (Meeker et al., 2004). It has been demonstrated that the persistent cell proliferation or rapid cell turnover through damage of hepatic cells resulted in a process of multistep hepatocellular carcinogenesis, thus progressive telomeres shortening and telomerase activation may be used as biomarkers for the early detection of liver carcinoma (Miura et al., 1997). Among others it has been assumed that induction of telomerase activity is an early event in gastric carcinogenesis (from intestinal metaplasia through adenoma) (Maruyama et al., 1997) and breast cancer (Artandi, 2003). Also telomeres are thought to play an important role in radiation carcinogenesis: recently the telomere dysfunction theory has been proposed for radiation carcinogenesis (see more in Section 2.2) (Kodama et al., 2006).

### 2.1.6 Apoptosis

Apoptosis plays a crucial role in many normal processes in the human organism, such as embryogenesis, cell maturation and differentiation, and development of the immune system. Apoptosis also is involved in immunodeficiency, developing of drug resistance, and carcinogenesis. It has been first described in the early 1970s as a basic biological phenomenon with wide-ranging implications in tissue kinetics, which is important as for the spontaneous elimination of potentially malignant cells and therapeutically induced tumor regression, as for tumor progression (Kerr et al., 1972; Zhivotovsky and Orrenius, 2006). It is a complex cell death mechanism triggered by various signals when certain aspects of cell functions are sufficiently degraded. Apoptosis has been suggested to have a barrier function against cancer; it has been supposed that in cell death cells might be antitumorigenic, while genes involved in negative regulation of cell death might act as oncogenes. Several genes [e.g., p21; p53, Bcl-2, Bax, c-myc, caspases, inhibitors of apoptosis proteins (IAPs), fas-APO-1, mdr-1, etc.] are related to the initiation of apoptosis with the most important being p53 with its mutations found in half or more of solid tumors (Minna and Gazdar, 1996; Delfino et al., 1997). Initiation of apoptosis is associated with activation of an upstream cascade, including the release of cytochrome *c* from an intermembrane space in mitochondria to cytoplasm, and the processing of proteolytic caspases (Jagat et al., 2000). Although recently the role of apoptosis in cancer treatment has been discussed in many publications (Fesik, 2005; Klein et al., 2005; Andersen et al., 2005; Reed and Pellecchia, 2005; Gerl and Vaux, 2005), the role cell death plays in carcinogenesis is still unclear.

### 2.1.7 Angiogenesis

Angiogenesis plays a critical role in providing malignant cells with nutrients for rapid growth (Bergers et al., 1999; Folkman, 1971). The endogenous negative regulators of angiogenesis (e.g., thrombospondin, angiostatin and glioma-derived angiogenesis inhibitory factor) are all associated with neovascularized tumors. The extent to which these regulators are decreased due to the angiogenic phenotype, dictates whether a primary tumor growth is rapid or slow, and whether distant metastases can grow successfully (Folkman, 1995). Angiogenesis receives an increasing attention as one of the candidate mechanisms which may be used for cancer prevention (Bisacchi et al., 2003; Pfeffer et al., 2003). Presently, studies of angiogenesis in early stages of precancers progression to cancer started receiving more attention. It has been recently shown that breast carcinomas induced new blood vessel formation to make a transition from intraductal proliferation (IDP) and ductal carcinoma *in situ* (DCIS) to invasive adenocarcinoma (AC), and this process was accompanied by the loss of basement membrane integrity. The anti-angiogenetic compounds might inhibit the progression of pre-malignant breast lesions to carcinomas and slow tumor growth by reducing the density of blood vessels both within a carcinoma and also in tissues immediately adjacent to the tumor (Thompson et al., 2004). Role of inhibition of angiogenesis as method of cancer chemoprevention (e.g., breast cancer) and its effectiveness depending on stage of tumorogenesis (i.e., at early stages) require further studies (Heffelfinger et al., 2003).

### 2.1.8 Immunomodulation

In the 1890s a New York surgeon William Coley had noticed that some of his cancer patients who developed systemic infections had a regression of their tumors. He hypothesized that these systemic infections activated nonspecific immune response in cancer patients, which somehow improved patients' conditions. Skepticism still characterizes some studies in cancer immunotherapy, but certain features of the immune system make them very promising (Pardoll, 2004). Many abnormal cells are removed by macrophages and other humorally mediated immune activities. A crucial part of this activity is the ability to identify cells that have changed phenotypically sufficiently to be targeted by cytokines as interleukins, lymphokines, and cell signal molecules, such as tumor necrosis factor and the interferons. Some tumor cells successfully escape immune surveillance, depending upon how the tumor cells express specific sequences of mutations. There were reported associations linking aggressive non-Hodgkin's lymphoma to TNF-$\alpha$, and gastric cancer to proinflammatory cytokine polymorphisms (Jillella et al., 2000; El-Omar et al., 2003). It has been supposed that tumor growth, paradoxically, may be decreased at later ages due to immune senescence, which may downregulate the production of certain types of inflammatory and growth factor responses (Huang et al., 2005).

It is likely that mechanisms by which nonvirus-associated cancers operate and mechanisms by which chronic viral infections avoid their immune elimination might have a common features. Studies of cancer immunotherapy such as targeting of different regulatory points of the immune response, from priming to amplification and to effector's function, as well as vaccination (in the context of bone marrow transplantation) may be very promising (Pardoll, 2004). The combinatorial immunotherapy becomes an important research area in various cancer types.

### 2.1.9 *Metalloproteinases*

Metalloproteinases, also known as matrixins, are a family of zinc-dependent enzymes – endopeptidases, that are involved in the degradation (proteolysis) of the extracellular matrix due to their capability to break down the proteins (e.g., collagen) that are normally found in the spaces between cells in tissues and participating in normal tissue remodeling events, such as embryogenesis, angiogenesis, ovulation, mammary gland involution, and wound healing. Abnormal expression of metalloproteinases contributes to such diseases as rheumatoid arthritis, osteoarthritis, pulmonary emphysema, and tumor growth, invasion, and metastasis (Chambers and Matrisian, 1997). They are crucial to tumor cells being able to enter the general circulation and lymphatic vessels by degrading basement membranes, penetrating vessel walls, and then repenetrating vessel walls to being seeded as metastases in distant tissues. That allows tumor cells to exist in vascular spaces and to form cell clones in distant tissues by remodeling of specific tissue membranes (Michor et al., 2006; Alber et al., 2006). Metalloproteinases have been linked to breast, ovarian, colorectal and lung cancers, and others (Duffy et al., 2000; Heslin et al., 2001; Yu et al., 2002; Kamat et al., 2006). Because matrix metalloproteinses are involved in certain cancers initiation and dissemination, inhibition of these proteinases may be important in cancer prevention and decreasing risk of cancer metastasis.

## 2.2 Theories of Carcinogenesis

**History**. Chemicals and cancer were first linked epidemiologically in the 1700s, while a direct link was proven experimentally in 1930s. In the mid-nineteenth century, the similarity between embryonic tissue and cancer was noticed, which suggested that tumors might arise from embryo-like cells. Rudolf Virchow hypothesized in 1858 that cancer arises from embryo-like cells. The concept that adult tissue contains embryonic remnants which generally lie dormant, but that could be activated to become cancer, was later formalized by Julius

Cohnheim (1839–1884), a German pathologist and a pupil of Rudolf Virchow, in 1875, and by an Italian surgeon and physiologist Durante in 1874 as the "embryonal rest" theory of cancer (Sell, 2003). Later John Beard (1857–1924), a Scottish Professor of embryology, proposed the trophoblast theory of cancer, where cancer represented primarily trophoblastic tissue derived either from an aberrant germ cell or from a somatic cell, when normally repressed "asexual generation" genes were abnormally reactivated.

In 1914, a German pathologist Theodor Boveri (1862–1915) asked a fundamental question: "If normal cells beget normal cells, and neoplastic cells beget neoplastic cells, what causes normal cells to become neoplastic?" Boveri advanced the notion that a neoplastic cell arose from "abnormal mitosis" that caused an uneven distribution of genetic material in their daughter cells (Boveri, 1929). For almost 100 years, it was assumed that phenotypic changes were due to mutations in nuclear genetic material. Initially this simply implied a change within the chromatin in the cell nucleus (genetic theory); proteins were then thought to be the genetic material (epigenetic theory). In 1976, Nowell published a paper in which he proposed that "most neoplasms arise from a single cell of origin, and tumor progression results from acquired genetic variability within the original clone allowing sequential selection of more aggressive sublines", and that "each patient's cancer may require individual specific therapy, and even this may be thwarted by emergence of a genetically variant subline resistant to the treatment" (Nowell, 1976). Several other major theories of cancer have stimulated a further research: (a) cancer as a "disease of differentiation" (Markert, 1968); (b) the "stem cell" theory of cancer (Till, 1982; Kondo, 1983), which has been pitted against; (c) the "dedifferentiation" theory of cancer (Sell, 1993); (d) the idea that combines these former two theories is found in the "oncogeny as partially blocked ontogeny" theory (Potter, 1978); (e) the "initiation/promotion/progression" concept of carcinogenesis, which was conceived as an operational description to explain distinct steps during the multistep process of carcinogenesis (Pitot et al., 1981); (f) the "nature versus nurture" theory (Trosko and Chang, 1979), which has been argued to explain whether genetics or the environment was the prime determinant in causing cancer; (g) classic disagreements have appeared as to whether mutagenic versus epigenetic mechanisms are most responsible for carcinogenesis (Trosko et al., 1983); (h) the "oncogene and tumor suppressor gene" theory has also been a driving force in cancer research (Brissette et al., 1991; Land et al., 1983); (i) the hypothesis that "cancer was the result of dysfunctional gap junctional intercellular communication" (Loewenstein, 1966; Trosko et al., 1993).

In general, theories of carcinogenesis represent two directions of studies: genetic (when carcinogens affect DNA) and epigenetic (when carcinogens affect proteins, enzymes, membranes, metabolism, etc.). Below, we briefly describe some of the current cornerstone theories of carcinogenesis.

### 2.2.1 Somatic Mutation Theory of Carcinogenesis

Somatic mutation theory of carcinogenesis was the most dominant theory during most of the twentieth century that inspired cancer researchers. It supposes the monoclonality of tumorigenesis – i.e., that cancer is caused by successive DNA mutations in a single cell. Its mechanisms include an altered growth factor signaling pathways, altered cell cycle effectors (cellular oncogenes, cyclins, etc.), altered inhibitory factors and suppressor genes, regulation of cell death, and differentiation pathways. The somatic mutation theory identifies carcinogenesis as operating at both the cellular and the subcellular-molecular levels (Sonnenschein and Soto, 2000).

### 2.2.2 The Stem Cell Theory of Carcinogenesis

The stem cell theory of carcinogenesis suggests that cancer develops from a single normal stem cell which has undergone a series of discrete genetic changes. All tissues consist of two types of cells: (i) differentiated cells which are the main component of most tissues and (ii) stem cells from which the various differentiated cells arise, and from which malignant tumors may also develop. The stem cell theory of carcinogenesis has recently been given a revival in that isolated human adult stem cells have been shown to be "targets" for neoplastic transformation, e.g., the oct4 (octamer-4, a homeodomain transcription factor, that is critically involved in the self-renewal of undifferentiated embryonic stem cells and is frequently used as a marker for undifferentiated cells) has been associated with adult stem cells, as well as their immortalized and tumorigenic derivatives, but not with normally differentiated daughter cells (Trosko and Tai, 2006).

### 2.2.3 Mutation versus Epigenetic Theories of Carcinogenesis

Mutagenesis is the process that brings about a qualitative alteration of nuclear genetic information. An epigenetic process alters the expression of genetic information at the transcriptional, translational, or posttranslational levels. There can also be chromosomal mutations (i.e., a translation or a nondisjunction of a chromosome) that can induce an epigenetic event (i.e., the extra chromosome 21 in Down syndrome can alter gene expression without mutation) (Trosko and Ruch, 1998).

While the mutation theory of carcinogenesis has a long history, the idea that nonmutagenic events might play a role, during either the entire or a specific, phase of carcinogenesis, has not. That has caused some investigators to think that mutagenesis alone explains all of carcinogenesis. Both mutagenic and epigenetic mechanisms likely operate stochastically to generate a complex, multistage, and possibly multiphase process of carcinogenesis. Other theories of carcinogenesis, such as initiation/promotion/progression, stem cell, and "nature or nurture" theories could each integrate the mutation and epigenetic theory.

### 2.2.4 The Tissue Organization Field Theory of Carcinogenesis and Neoplasia

The tissue organization field theory of carcinogenesis and neoplasia was proposed by Sonnenschein and Soto (2000). Its components are altered cell-to-cell and tissue-to-tissue interactions (Alber et al., 2006). According to this theory, carcinogens disrupt the normal cell-cell interactions in the parenchyma and stroma of an organ thus initiating tumorigenesis (mechanism that reminds the phenomena of "morphogenetic fields" in a developing organism). Stroma appears to be the primary target of carcinogens. Carcinogenesis is assumed to operate at the tissue hierarchical level of organism complexity (Sonnenschein and Soto, 2000; Soto and Sonnenschein, 2004).

### 2.2.5 Telomere Dysfunction Theory

Telomere dysfunction theory was suggested primarily for radiation carcinogenesis. In this model, the radiation exposure contributes to the induction of telomeric instability, which may lead to the breakage-fusion-bridge cycle that potentially drives genome rearrangements. Thus, telomere dysfunction initiates and promotes chromosomal instability that is critical at an early step of radiation carcinogenesis (Kodama et al., 2006).

## 2.3 An Overview of Formal Quantitative Models of Carcinogenesis

The theoretical models may be divided into two very broad classes: deterministic models and stochastic models. Deterministic models are supposed to model/predict the average behavior of systems according to precise rules, while stochastic models are supposed to describe the probability of very specific behaviors of individuals rather than average behavior of the population, which is potentially more informative in that it considers rare events and not just average properties. Newtonian physics is deterministic, and quantum mechanics is stochastic (Beckman and Loeb, 2005). Parameters are the variables which are expected to influence the model outcome. In cases when their values are unknown or could be expected to vary over a known range, these parameters are adjustable, and they should be used in theoretical modeling with care: if the model can fit the data, that does not always mean its validity, since nearly any model could be fit to the data by adjusting the adjustable parameters, especially when there are a large number of these sort of parameters included in the model. It is important, in general, to verify that the number of fitted or predicted experimental data points exceeds the number of adjustable parameters in the model: the greater the excess of independent experimental data point over adjustable parameters, the more valid the experimental confirmation of the theory (Beckman and Loeb, 2005).

The multiple-stage, or multiple-"hit", nature of carcinogenesis has been demonstrated experimentally by Barrett with colleagues using the cell culture method on rat tracheal epithelial cells (Nettesheim and Barrett, 1984) and on Syrian hamster embryo fibroblasts (Barrett, 1979; Barrett and Fletcher, 1987). Under normal environmental conditions at least two phases – immortalization and transformation – are required for the process of carcinogenesis, but often these two stages are not enough for conversion of "normal" cells to cancer. It has been suggested that tumor genesis typically involves alteration of 5–10 genes (Hopkin, 1996). Studies on tumor tissue biopsies from colon cancer patients showed that 5–7 mutations were most often presented (Fearon and Jones, 1992; Wagener, 2001). Colorectal cancer is recognized by many researches as a good model for the study of stages in cancer progression. Mutation of the APC regulatory pathway (adenomatous polyposis coli gene – a tumor suppressor gene that is inactivated in most colorectal cancers) appears to be the first step, which regulates $\beta$-catenin, thus influencing an expression of *c-myc* and other proteins promoting cell division, and affecting the stickiness of the epithelial cells surface, thus leading to adenomatous growth (Kinzler and Vogelstein, 2002). One of the next genetic events of progression is supposed to be a *ras* gene (e.g., *K-ras*, *N-ras*, *H-ras*) mutation. With continued growth, adenomas tend to lose part of the long arm of chromosome 18q (Kinzler and Vogelstein, 2002; He et al., 1998), with the possible role genes *DCC*, *SMAD4*, and *SMAD2* playing here in development of late adenomas (Frank, 2007). Transition to colon cancer is driven by the loss of functional *p53* by damage to both alleles, accompanied by the acceleration of genetic changes due to chromosomal aberrations, such as loss of heterozygosity (Nowak et al., 2002) [however, it is still unclear, whether chromosomal instability appears early in carcinogenesis, thus playing a key role in driving genetic changes, or if it develops at the later stages when the genome is increasingly disrupted – supposed that it might be chromosomal instability-dependent and independent pathways of progression (Frank, 2007)]. Several alternative pathways are supposed existing in colorectal carcinogenesis, such as microsatellite instability (Rajagopalan et al., 2003; Jass et al., 2002a), hypermethylation of promoter regions of *p14*, *p16*, *hMLH1*, *hMSH2*, *TIMP3*, *MINT2*, *MGMT*, *HPP1/TPEF*, etc. (Jass et al., 2002a; Issa, 2004; Niederhuber et al., 2004; Kim et al., 2008; Jass et al., 2003). The hereditary nonpolyposis colorectal cancer (HNPCC) (a component of Lynch's syndrome) pathway differs from the "classical" colorectal cancer pathway by the number of losses of heterozygosity (LOH), by the ratio of *BRAF* to *K-ras* mutations, fewer *p53* mutations, and more mutations in various growth-related genes such as *TGFβIIr*, *IGF2*, *Bax* and others (Boland, 2002; Jass et al., 2002b; Rajagopalan et al., 2002; Storm and Rapp, 1993). It is unlikely that it will be possible to obtain all details of colon carcinogenesis from experimental studies or to perform the precise measurements of cancer's age-onset patterns. However, it seems possible to formulate and test the hypotheses, such as that pathways with fewer rate-limiting stages or faster transitions between stages differ in age-onset patterns when compared with pathways that have more stages or slower rates of transition (Frank, 2007).

The typical restrictions of some models to two stages is partly based on mathematical convenience with two-stage models, which have more easily identifiable model parameters and are partly based on biological arguments about mutation rates (Little, 1996; Tan et al., 2004). Many of those two-stage models, however, suffer from the assumption that the initiated cell "instantaneously" grows into a malignant tumor, thus ignoring tumor progression (Tan and Chen, 1998).

Animal experiments first showed that at least three phases of carcinogenesis existed: initiation, promotion, and progression. Each of these three phases may involve many steps. *Initiation* is the first stage of tumor induction, when cells are altered by the exposure to a carcinogenic agent, thus they are more likely to form a tumor when being exposed to a promoting agent ( see "promotion"). Cells' alterations in the genetic expression believed to be irreversible and produce the cell's phenotypes changes. Genetic changes may be as gene mutations, as well the other genetic changes (e.g., specific chromosomal aberrations).

*Promotion* is the stimulation of tumor induction. It follows initiation and promotes an agent which may itself be even noncarcinogenic. This phase may be reversible. In this phase the clonal expansion of initiated cells occurs. Promotion may be direct or indirect, potentially involving many epigenetic factors (e.g., chromatin participating in defining nuclear structure). Most promoting agents are not mutagenic. It is supposed that tumor promotion in vivo may be inhibited by antioxidants and other inhibitors of reactive oxygen species.

The *progression* phase involves the development of metastatic tumor cells, formation of groups of tumor cells of various sizes, their migration through the circulatory system, nesting at distant tissues/organs, and developing the distant metastases by penetration of capillary walls and tissue "seeding" and microvascularization (Nowell, 1986). Various proteins/enzymes are involved in this stage, such as extracellular proteases (matrix metalloproteinases), chemokines, growth factor signaling molecules, cell–cell adhesion molecules (cadherins, integrins), etc. It is increasingly apparent that the stromal microenvironment, in which cancer cells develop, influences cancer progression: the influence of the microenvironment in carcinomas are mediated by bi-directional interactions (including adhesion, survival, proteolysis, migration, immune escape mechanisms, lympho- and angiogenesis, homing on target organs) between epithelial tumor cells and neighboring stromal cells (e.g., fibroblasts, endothelial and immune cells) (Bogenrieder and Herlyn, 2003). The formation of metastases at distant sites may depend on local gene mutations, which may be turned on/off by DNA hypomethylation and by disruption of immunocompetency (Ling et al., 1985), and invasive potential of tumor might be acquired early in carcinogenesis, already presenting at the time of cell transformation (van't Veer et al., 2002; Ramaswamy et al., 2003).

The carcinogenesis process, its pathways and the number of its stages depend not only on cancer genes but also on environmental risk factors which affect the rates of mutation. Different individuals may experience different pathways of carcinogenesis. We can expect a mixture of different models for carcinogenesis for the same cancer type to be found in the members of a human population. The multilevel Gibbs sampling model was recently applied to the British physician data on lung cancer with smoking: the obtained results indicated that the tobacco nicotine was an initiator, but at ages 60 and older it was also a promoter (Tan et al., 2004).

There are three basic types of cancer genes which participate in carcinogenesis process: oncogenes, antioncogenes (suppressor genes), and accessory genes (modifier genes). *Oncogenes* are regulatory genes which regulate cell proliferation and differentiation. Mutation of oncogene functionally releases cells from regulated growth control. *Antioncogenes* suppress the expression of oncogenes or other genes, so that their inactivation or deletion would lead to carcinogenesis. Unlike oncogenes, which are dominant, antioncogenes are recessive, so that only homozygotes or hemizygotes for the gene can give rise to the cancer phenotype. *Accessory cancer genes* relate to cancers indirectly by increasing mutation rates of oncogenes and antioncogenes, and/or by facilitating cell proliferation of intermediate cells and/or cancer progression. Strachan and Read (1999) used the analogy of a bus to picture the oncogenes as the accelerator and the tumor suppressor genes as the brake: "Jamming the accelerator on (a dominant gain of function of an oncogene) or having all the brakes fail (a recessive loss of function of a tumor suppressor gene) will make the bus run out of control. Alternatively, a saboteur could simply loosen nuts and bolts at random (inactivate the tumor suppressor genes that safeguard the integrity of the genome) and wait for a disaster to happen".

To make theories of carcinogenesis useful for the modeling of cancer effects on human populations, it is necessary to develop quantitative models embedding those theories in parametric mathematical forms. Their values can be mathematically identified and statistically estimated by combining a variety of human population and clinical data, in vitro studies, and results of specific studies of model systems. We briefly review some of these models below.

One of the suggested solutions to the multistage model can be derived from Bateman's solution of successive radioactive decays (Bateman, 1910): the isotope radium $C$ decays into radium $C'$ and radium $C''$, which both decay into lead, so the diagram describing the process contains a loop. It is highly likely that these more complex forms are also relevant to cancer incidence (Ritter et al., 2003) (however, not all "initiated" cells proceed to cancer, and some may undergo clonal expansion and multiply).

### 2.3.1 Nordling, and Armitage and Doll

In 1953–1954, Nordling, and Armitage and Doll, working with national mortality data in the United Kingdom, proposed a multistage/"multihit" theory of cancer to describe cancer mortality age distributions. This model assumes that cancer develops from a single cell by going through a series of irreversible, heritable mutation events in nuclear DNA (see Fig. 2.1). This model laid out the foundational principles of cancer progression and epidemiology in mathematical form long before other studies discovered the molecular basis of somatic mutation and the key role of genes such as *p53* and APC. The main question they asked was "What can be said about the dynamical process of progression within individuals that would explain the aggregate patterns of epidemiology observed in population?" (Frank, 2004).

Nordling (1953) has proposed that seven genetic changes would be required to produce a cancer cell: one preexisting mutation and six subsequent mutations, with limiting latter rate, leading to dependence on the sixth power of age. Fisher and Hollomon interpreted this data in a different way, suggesting that at least six cells must each acquire one mutation to form a sufficient cluster of genetically altered cells to result in a tumor (Fisher and Hollomon, 1951). Armitage and Doll supposed that if the factors leading to various genetic changes varied over time, then the risk of acquiring the next genetic change was not constant and observed incidence could deviate from the sixth power law; and fewer than seven steps could still lead to a sixth power law if one or more of those steps increased in greater than linear proportion with age (Armitage and Doll, 1954). Later more cancer types were examined from the dataset from different countries, and it has been found out that the sixth power relationship worked precisely only in a minority of cases, especially if the whole lifespan was taken into analysis (Cook et al., 1969). It has been hypothesized that it might be because of the delay between exposure to a risk factor and cancer development, thus older people might die from another cause prior the development of cancer.

This process may be efficiently described by the power law formula $I(t) = at^{k-1}$, or $lnI = lna + (k-1)lnt$, where $k$ is the number of stages (identified as produced by specific genetic mutations), and $a$ includes the effects on specific gene mutation rates of various risks representing environmental risk factors, such as smoking, alcohol consumption, diet, as well as genetic susceptibility factor. It has been estimated that this model fits the age-specific mortality rates satisfactorily for many solid tumors in adults younger than 85, approximately representing age-specific power law for incidence rates with values of $k$ between 4 and 8 (Cook et al., 1969; Manton and Stallard, 1988). Cook et al. (1969) found that while $k$ varied for different tumor site, it was constant between countries,

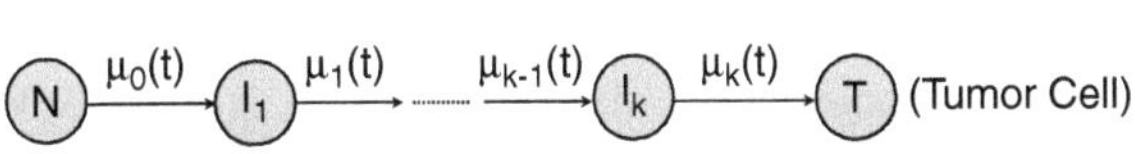


**Fig. 2.1** The Armitage–Doll model of carcinogenesis

while the constant $a$ varied between countries (that might be expected from the differences in environmental risk factors).

In the analyses of their initial mathematical models, Armitage and Doll observed a logarithmic increase in cancer mortality with age in cancers of stomach, esophagus, colon, and pancreas – sites for which there had not been recent historic changes in mortality in the mid-1950s. They specifically excluded cancers of the lung and bladder from their analysis because of the large "proportion of the cases of the lung is believed to be related to cigarette smoking, which has become more prevalent in the last 50 years, and a proportion of the cases of cancer of the bladder was due to occupational hazards, to which men have been exposed for various periods at various ages". They also excluded "hormonally related cancers" (breast, corpus uteri, ovary, cervix uteri, and prostate) because they believed hormone levels, and therefore cancer risks, would be heterogeneous over age (due to the age dependence of hormonal production) and differ by gender (Armitage and Doll, 1954).

Some biological data indicate that proliferation of normal stem cells and immortalized cells are important aspects of carcinogenesis, which may suggest why the Armitage–Doll multistage model, while empirically useful, may not be biologically completed because of not representing the balance of cell replication and death at each stage (Tan, 1991). In some analyses, such as of lung cancer and smoking, the Armitage and Doll model gives quantitatively different results than use of an exact incidence function where direct adjustment of cell birth and death is considered (Hazelton et al., 2005).

The modification of the Armitage and Doll multistage/hit model with Weibull hazard function has been empirically successful in explaining incidences for many types of solid tumors at ages from 30 to 85 (Cook et al., 1969; Heidenreich and Paretzke, 2001; Manton and Stallard, 1988). Studies of tumor tissues from patients with solid cancers (e.g., colon cancer) showed the need for, on average, from five to seven mutations (Fearon and Vogelstein, 1990; Hopkin, 1996) to initiate a tumor. It has been suggested to reduce the tension of applicability of Armitage–Doll model by taking into account a clonal expansion, allowing that some mutations may considerably increase the rate of subsequent mutations by impairing repair mechanisms, and also other types of genomic instability. When the clonal expansion of some intermediate cells plays an important role, the observed age dependence of cancer incidence can be fitted with fewer steps, and when one of the early steps increases mutation rates, the subsequent steps may happen faster and may not be rate limiting (Heidenreich and Paretzke, 2001). Because two mutational events may be not sufficient for most of solid tumors, models based on two rate-limiting steps may not reflect what is known about tumor initiation and growth from molecular biology and from the direct evaluation of human tumor tissue in clinical histological studies (Fearon and Jones, 1992; Kinzler and Vogelstein 1997, 1998; Wagener, 2001; Hopkin, 1996).

Steps to bridge the gap between multihit and two-stage cancer models require resolving the effects of a clonal expansion, of the fact that some mutations may impair gene repair, and of other sources of genomic instability. What is generally being argued between the multistage and the two-stage models is that the number of rate-limiting stages of cancer initiation may be smaller than the number of gene mutations directly observed in human tumor tissue. Cell birth and death dynamics in two-stage models are then used to mimic the effects of some gene mutations in the model. Attribution of empirical effects to one or the other mechanism, however, may not be directly identifiable from human population data on cancer mortality or incidence.

### 2.3.2 *The Moolgavkar–Venzon–Knudson Two-Stage Model*

As a background for developing a model, Knudson studied the incidence function differences between familial and nonfamilial forms of retinoblastoma (a rare type of cancer of the eye typically affecting children) and proposed that two successive mutations/"hits" were required to turn a normal cell into a tumor cell and that in familial forms of this tumor one of the "hits" was inherited (Knudson, 1971). The model assumes that a malignant tumor develops from a single normal stem cell by clonal expansion and views carcinogenesis as the result of two discrete, heritable, and irreversible events in normal cells (Moolgavkar and Knudson, 1981). Each "mutational" event occurs during a single cell division. A distinct feature of this model is that the first event may occur either in germ line cells or in somatic cells, but the second event always occurs in somatic cells.

According to this model, there are three types of cells: normal cells, intermediate cells (initiated cells), and cancer cells. A schematic of this model is shown in Fig. 2.2. Moolgavkar and Luebeck (1992) assumed two or three gene mutations were necessary to describe the incidence of colon cancer in a general population and in patients with familial adenomatous polyposis, with the role of mutation at the FAP gene locus not one rate-limiting in colon carcinogenesis. They found that both models gave good fits to select datasets, but that the

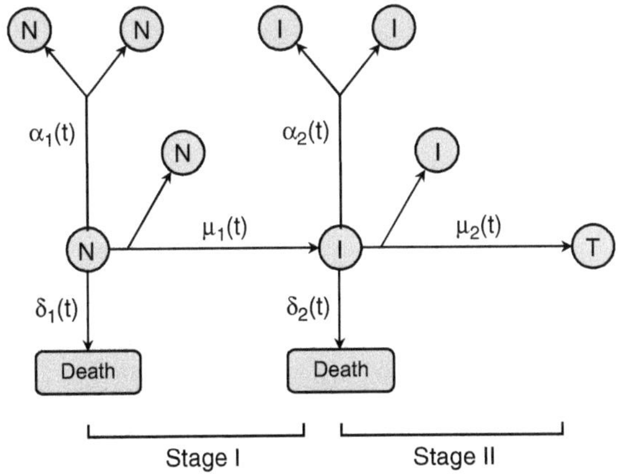


**Fig. 2.2** The two-stage model of carcinogenesis

model with two mutations implied implausibly low mutation rates, so that the three-mutation model was preferred on theoretical grounds.

This model is not absolutely free from limitations: e.g., it provides a mathematical description of the biological mechanism of carcinogenesis for a type of tumor, which involves only a single antioncogene (as the Rb gene for retinoblastoma). This is appropriate in those tissues where tumorigenic conversion of normal stem cells involves only immortalization and transformation. The most serious limitation is that this model is inconsistent with the direct experimental and clinical observation of multiple (i.e., more than two) stages that were found in a various solid tumors, from 5 to 10 genes may be involved (Hopkin, 1996).

According to the United Nations Scientific Committee on the Effects of Atomic Radiation (UNSCEAR), a major concern, identified with the two-stage MVK model, was the instantaneous increase in risk after ionizing radiation exposure. One resolution of the problem was to assume a fixed latency (e.g., 3.5 years). To achieve the observed gradual increase in excess relative risk shortly after exposure a stochastic process must then be used to model the transition from the first malignant cell to clinically detected cancer (UNSCEAR, 2000, p. 151). This is why Little (1996) generalized the MVK model in certain analyses to include 3 or 4 mutations. Clearly a resolution of this problem probably involves successfully modeling the tumor latency period as a separate tumor growth process involving additional biological factors (Yakovlev and Tsodikov, 1996).

The most popular version of the two-stage model is the two-stage clonal expansion (TSCE) model which additionally assumes that (1) the number of susceptible normal cells is either constant or described by a deterministic function, and (2) all rates are time independent (see Fig. 2.3). An attractive property of the model is that the spontaneous hazard rate can be expressed analytically in terms of only three parameters (Heidenreich and Paretzke 2001):

$$h(t) = \frac{X(e^{(\gamma+2q)t} - 1)}{q(e^{(\gamma+2q)t} + 1) + \gamma}; \quad X = N\mu\mu_1, \quad \gamma = \alpha - \beta - \mu, \quad q = \frac{1}{2}(-\gamma + \sqrt{\gamma^2 + 4\alpha\mu}).$$

The main disadvantage is that not all biological parameters (i.e., the number of stem cells $N$, first $\mu_I$ and second $\mu$ mutation rates, and proliferation $\alpha$ and

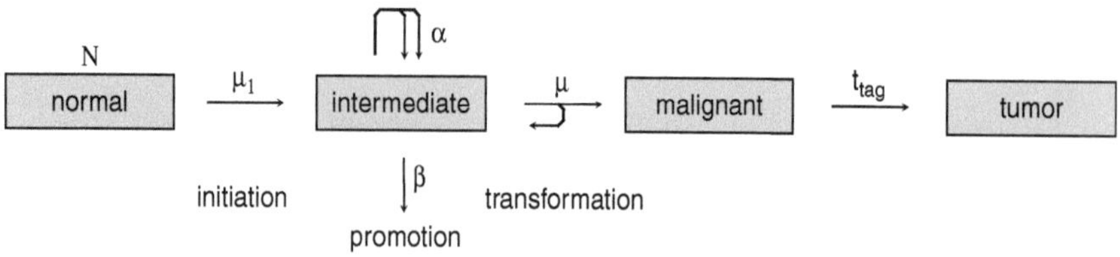


**Fig. 2.3** The two-stage clonal expansion model of carcinogenesis

death/differentiation $\beta$ rates) can be identified using the data on age-specific incidence rates.

Recently the TSCE model was applied for description of IR induced carcinogenesis, which is a specific and important subtask of carcinogenesis modeling. IR can induce specific mutations or epigenetic events in stem cells, therefore increasing the number of intermediate cells susceptible to further stages of carcinogenesis. IR can also have a promoting effect to carcinogenesis. The basic argument is that stem cells inactivated by IR may be replaced by the division of stem cells in which intermediate cells have a growth advantage (Heidenreich et al., 2001). A typical way to incorporate these effects into mechanistic models is to assume that rates of initiation, promotion, and progression become dose dependent. Recently, such effects were analyzed and discussed for radon-induced lung cancer in Colorado Plateau uranium miners (Little et al., 2002), and French and Czech miner cohorts (Brugmans et al., 2004; Heidenreich et al., 2004). Further discussions (Bijwaard et al., 2005; Heidenreich, 2005a, b, Laurier et al., 2005) covered several aspects: biological viability of the models, testing hypotheses about the processes of radiation carcinogenesis, selection of the best fitting model, comparison to the empirical approach which uses statistical modeling in describing the data, etc. One conclusion of the discussion was that "even if biologically motivated mathematical models of carcinogenesis are necessarily a crude simplification of the biological reality, such models constitute a complementary approach to empirical statistical models" (Laurier et al., 2005). Further analyses of TSCE properties for description of spontaneous and radiation carcinogenesis were performed by Heidenreich (2005a, b, 2006).

Even though TSCE is definitely one of the most popular models of IR induced carcinogenesis, it still has limitations. The first is the problem with parameter identifiability: only three combinations of biological parameters are identifiable from the age-specific hazard function. The second is that the biological mechanisms represented by TSCE are oversimplified. The next limitation is that the parameters used in this model (as well as in the more general ones) cannot be directly measured, thus restricting the capability of predicting the individualized risks. A possible solution is to combine data on the age-specific hazard function with additional measurements which are indirectly related to the model parameters, e.g., to measure the apoptosis rate. One promising approach to carcinogenesis modeling that is capable of overcoming this difficulty was recently suggested by Akushevich et al. (2007). In this approach, carcinogenesis is represented as a dynamic trade-off between two antagonistic forces or processes, promoting or hindering carcinogenesis at its different stages. Processes promoting the cell malignization are represented by mutations or adverse epigenetic events, while antagonistic processes preventing the neoplastic transformation of the cell and forthcoming its fixation in next cell generations are represented by barrier mechanisms, such as apoptosis, reparation, and antioxidant defense. One advantage of the modeling approach is in the natural combining of two types of measures expressed in terms of model parameters: age-specific hazard rate and states of barrier mechanisms, e.g., reparation efficiency. Another advantage is in the application to the case of protracted low-dose irradiation, when barrier mechanisms for repair of genetic damages or elimination of cells carrying unrepaired damages play a special role.

### 2.3.3 The Generalized MVK and Armitage–Doll Models

The first model of this class was proposed by Chu (1985) as an extension of the Armitage–Doll multistage model of carcinogenesis. The multievent model assumes that malignant tumors develop from cells by going through a fixed number $k$ ($k>1$) of heritable genetic mutations. It differs from the Armitage– Doll model in that the intermediate cells are assumed to be subjected to stochastic birth–death processes for cell proliferation and cell differentiation. A typical scheme of the class of multistage models generalizing approaches of MVK and Armitage–Doll is presented in Fig. 2.4.

If $k=2$, the multievent model reduces to the two-stage model considered by Moolgavkar and Venzon (1979), and Moolgavkar and Knudson (1981). However, since most solid tumors appear to involve at least from 5 to 7 mutations (Fearon and Vogelstein, 1990; Shen et al., 2001, 2004), the multi"hit"/stage models in Armitage and Doll and other multistage models with an appropriate tumor growth function might be better applied (see Section 2.4 of this chapter). It is important to realize that the multiple hit and multiple event model may lead to the same incidence/hazard rate function assuming a Markovity[1] condition for the mutational process (Little, 1996).

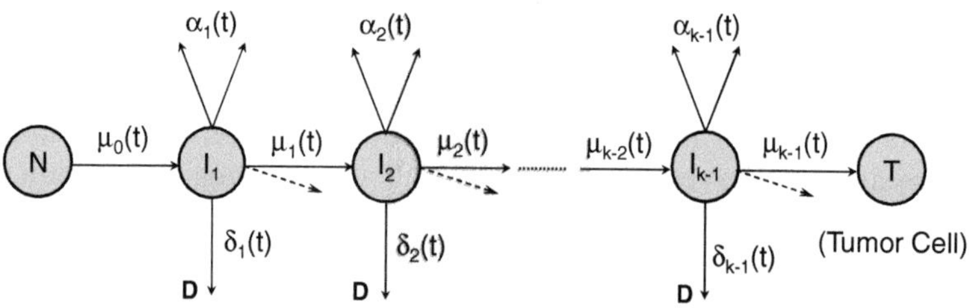


**Fig. 2.4** The generalized Armitage–Doll and MVK multistage models involving $k$ stages. *Dashed arrows* stand for the MVK multistage models only

[1] A Markov process, named after the Russian mathematician Andrey Markov (1856–1922), is a mathematical model for the random evolution characterized by having a "lack-of-memory" property, i.e., the conditional distribution of what happens in the future given everything up to now and depends only on the present state, so, the future and past are independent. This work founded a completely new branch of probability theory and launched the theory of stochastic processes. Markov also made some studies of poetry and poetic styles, applying the ideas of his theory to analysis of vowels and consonants in literary texts.

Great progress in further generalizations of the two-mutation carcinogenesis model of Moolgavkar–Venzon–Knudson (to allow for an arbitrary number of mutational stages), and of the model of Armitage and Doll, was achieved in a series of papers by Mark Little (1995, 1996), suggesting that the analytical solution for the hazard function in the generalized models is no longer possible. Instead, the hazard function is expressed in terms of a probability generating function which satisfies the Kolmogorov's backward equation. This equation is solved numerically. Although the computation becomes much more complicated, this approach allowed for many generalizations and specific applications, e.g., Little with co-authors (2002) applied these generalized models to data on humans exposed to IR. They compared predictive powers of the generalized models, identified optimal models for predictions of population risks of solid cancers and leukemia, and examined the behavior of the excess risk attributable to IR, when certain parameters were the subjects to small instantaneous perturbations. The most recent efforts on further generalizations of these models were directed to modeling the genomic instability and incorporating it into existing carcinogenesis models (Little and Wright, 2003; Little and Li, 2007; Little et al., 2008).

## 2.3.4 The Multiple Pathway Models of Carcinogenesis

The same cancer type may arise through different pathways. This leads to multiple pathway models of carcinogenesis. It was showed by Medina (1988), and Tan with co-authors (Tan, 1991; Tan and Chen 1991) that multiple pathway models of carcinogenesis provide a logical explanation of many biologically inconsistent findings from epidemiological data on human cancers. Figure 2.5

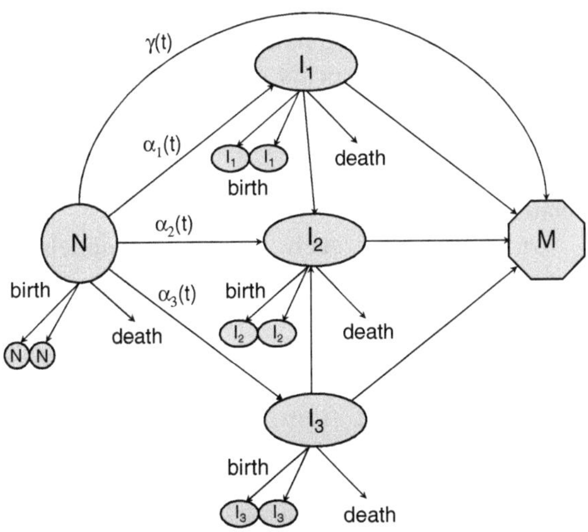


**Fig. 2.5** The multiple pathway model involving a one-stage model, three two-stage models, and two three-stage models of carcinogenesis

shows a model involving a one-stage model, three two-stage models, and two three-stage models. Roughly, the hazard rate is the sum of all possible pathways. For the simplest paths only (e.g., for one-stage path), the hazard rate can be calculated analytically.

A multiple pathway model may involve one-stage, two-stage, and multiple-stage models of carcinogenesis. Because most of these models are quite complicated and thus are far beyond the scope of the MVK two-stage model, the traditional Markov theory approach becomes too complicated to be of much use. It has been proposed as an alternative approach by using stochastic differential equations (Tan and Chen, 1998). These stochastic differential equations were used to develop state space models (Kalman filter models) for carcinogenesis. Tan and Chen (1998) also demonstrated how their formalism was related to classical formalism based on the probability generating function. Tan et al. (2004) have developed the advanced statistical procedures to estimate the unknown parameters of the state space model via the multilevel Gibbs sampling method (i.e., using the Markov Chain Monte Carlo method – MCMC) and applied these procedures to the British physician data on lung cancer due to smoking.

### *2.3.5 Mixed Models of Carcinogenesis*

The process of carcinogenesis, and the number of stages it involves, may depend not only on mutations in cancer genes but also on environmental factors. Mixed models of carcinogenesis also arise in cancers which involve both hereditary and nonhereditary factors. Consider a large population of individuals and suppose that the population is divided into a number of nonoverlapping subpopulations. For certain cancers, it often happens that these different subpopulations may involve different models of carcinogenesis. Various components of the mixed model may present in these cases, such as (a) a mixture of one-stage and two-stage models related to antioncogenes, (b) a mixture of two-stage models related to oncogenes, and (c) a mixture of multiple pathway models involving one and two-stage models related to oncogenes.

An example of such population models where specific components are characterized by different disease processes can be found in the two-disease model of female breast cancer, where early, aggressive disease occurs in a genetically distinct subgroup at relatively early ages (Manton and Stallard, 1979). The growth characteristics of early familial breast cancer are distinct from late-onset breast cancer: an early-onset breast cancer is more aggressive, more sensitive to ionizing radiation, and recently has been found to be characterized by specific genetic features leading to overexpression of specific tissue growth factors (Land, 1995).

### 2.3.6 *Cancer at Old Age and Approaches to Modeling: If the Cancer Incidence Rates Are Declining?*

Cancer has always been suggested as disease that is more prevalent in older people, with its risk increasing with aging. Many researchers focused their studies on increase of cancer mortality rates with age (Peto et al., 1975; Rainsford et al., 1985; Krtolica and Campisi, 2002; Dix, 1989; Volpe and Dix, 1986). These studies predominantly analyzed the data on age-specific cancer mortality (which is traditionally limited to age 75, thus not allowing the study of the rates at older ages) rather than incidence data (Arbeev et al., 2005). Several decades ago it had been suggested that the replicative ability of many types of cells could markedly decrease as they aged (Hart and Setlow, 1976). The loss of cell replication capacity at old ages is currently a controversial issue in aging studies: one of the questions is whether the correlation of residual cells' replicative capacity and age exists, and whether it depends on cell/tissue type and the health status of the person whose cells were donated (Cristofalo et al., 1998).

Based on epidemiological studies, there is considerable evidence that cancer incidence naturally slows, independent of prevention measures, at advanced ages: in recent decade data on cancer incidence in the United States, the Netherlands, and Hong Kong indicated a flattening and perhaps a turnover at advanced age (Pompei and Wilson, 2001). Several studies demonstrated that both the cancer incidence and the role of cancer as a cause of death might decline after age 95 (Stanta et al., 1997; Kuramoto et al., 1993). Experiments on mice found out that the cancer incidence rose as a function of age with the subsequent flattening, and even turnover, at an age of about 800 days – the old age in mice (Pompei et al., 2001).

The successfully applied model which will fit this data might provide an insight into the underlying biological mechanism. The three-parameter beta-function model fits both the mice and the human data well. This old ages turnover may occur either because of declining incidence, due to the pool of highly genetically susceptible individuals being depleted by mortality, or because of a mechanism that slows or arrests cancer development at older ages – processes that might be independent of an individual's life span (Manton and Stallard, 1988). Other conceptual models suggested that the slowing of tumor growth and expression at ages 85+ could be due to senescent-related changes in basal metabolism and mitotic index with age, that the growth rate of all tumors, enhanced by declines in the nutritional and vascular support of rapidly growing tissue, is reduced or suppressed. One of the tumors which is strongly characterized by slow growth at late ages is prostate cancer, which in many cases often has a lengthy (10–15 years) "indolent" period leading to the clinical strategy of "watchful waiting" [conservative or expectant management of prostate cancer, conducted with curative intent, that may be a reasonable approach for selected men older than 65 with a high likelihood of harboring small-volume prostate cancer based on serum prostate-specific antigen (PSA) and prostate biopsy criteria (Nelson et al., 2004)]. This decline in tumor progression rates with advanced age led to interest in studying how the processes of cancer and senescence might interact and be related (Cutler and Semsei, 1989; Ershler and Keller, 2000). Pompei and Wilson (2001) showed that for some sites, such as lung, larynx, breast, thyroid, and brain, declines in incidence were observed at the oldest ages. For other sites, such as pancreas, esophagus, melanoma, multiple myeloma, urinary bladder, a flattening in the age-incidence rate, but not a decline, was observed at ages 95 and older (Manton and Stallard, 1988). Interestingly, there were some cancer sites, such as stomach and colorectal cancers, which showed decline in some human populations, and not in others. Also, it has been shown that male and female cancer incidence rates at older ages differ, being higher in males (it might be due to differences in strategies of "fighting external stress" and "fighting physiological aging" in males and females) (Arbeev et al., 2005).

Early models described cancers as being clinically identified when the number of tumor cells reached a certain critical volume or mass. The assumption that cancers may be initiated throughout the life span leads to an empirical age-specific cancer incidence rate $I(t)$ initially increasing exponentially with age $t$ as $I(t) = Ae^{bt}$ (i.e., implying Gompertzian cell growth dynamics). Olkin et al. (1978) found that a good fit of many adult cancers can be made with a beta-distribution assumed for age-specific incidence: $I(t) = (\alpha t)^{k-1}\ (1-\beta t)$. The beta-distribution model fit to the SEER data for all cancer types produces a very different fit than the curves calculated either from the Armitage–Doll or the MVK clonal expansion model. The SEER data (assumed to be reliable and free from bias) do not extend to a high enough age to fully test this model prediction, but cancers of the lung, larynx, brain, and corpus uteri did show a marked downturn of the age-specific incidence rate within the observed age range, with the evidence of certain uniformity of adult cancers peaking in incidence at about the same age, including cancers in the United States and in other cancers (Pompei and Wilson, 2001).

The beta-distribution model can be viewed as the superposition, at each age, of two types of cell dynamics: (1) cancer creation, which is most simply modeled with the usual power law (Weibull) multistage assumption; and (2) cancer extinction, which is modeled as a cumulative probability linearly increasing to age 100 (Pompei and Wilson, 2002). The first factor may be interpreted as caused by somatic mutation and promotion steps from genetic and environmental risks/exposures. As to the second factor, apoptosis is a candidate for the mechanism of extinction, and a second possibility is the cell senescence (e.g., loss of proliferative ability due to senescence). If the rate of telomere shortening was uniform over tissue type and time, this mechanism could be modeled as causing cell senescence with the age-dependent probability $\beta t$, and thus could become the $(1-\beta t)$ cancer extinction age factor in the beta distribution model.

However, the biological explanation of cancer rates decline in older ages should take into account other important aspects due to which the overtime dynamics of age-specific cancer rates (e.g., incidence rate) reflect the combined effects of social, behavioral, environmental, medical factors, etc. (Liang et al., 2002). Traditional analyses of mortality time trend by year of death, even when stratified by age, fail to describe differences in mortality between generations, while age–period–cohort analysis measures the component explained by birth cohort and risk of dying vis-à-vis circumstances prevailing at the time of death (i.e., cohort effects relates to lifelong exposure to risk factors shared by whole generation, and period effect relates to factors that simultaneously affect the whole population, such as introduction of widely available medical care and changes in diagnostic or cause-of-death coding practices) (Medrano et al., 1997). Age–period–cohort models are widely used in epidemiological data for trend analyses in disease incidence and mortality over age, time, and birth cohort, while the other models, such as describing a differential selection in a heterogeneous population, analyze the mixture of two populations, one of which is prone to cancer and the other is not, resulting in a decline in cancer incidence rate in the entire population due to the dying-off of the susceptible subpopulation (Vaupel and Yashin, 1988). However, neither age– period–cohort nor heterogeneity models could describe the underlying biological processes (Arbeev et al., 2005). Among the possible causes of this decline are (1) the effects of cross-sectional data that transform cohort dynamics into age patterns (e.g., age patterns of lung, colorectal, breast, stomach, and ovarian cancers differ over time and place, probably reflecting differences in time and place of exposure to carcinogens, thus masking the organ/tissue-specific dependence of cancer risk changes with age), (2) population heterogeneity that selects susceptible to cancer individuals (Vaupel and Yashin, 1988), (3) a decline with aging in some carcinogenic exposure [e.g., smoking (Peto et al., 1985)], and (4) underdiagnosed cancers in older patients (e.g., some diagnostic procedures may be restricted in the oldest old ages due to severe chronic diseases) (Ukraintseva and Yashin, 2003; Arbeev et al., 2005). More elaborate models may include all possible causes of cancer rate declines in the elderly to analyze their contributions to changes of observed trends.

Recent studies hypothesized several possible consequences of “cancers fall” at old ages. If the action of a drug or an environmental agent is to increase senescence to reduce cancer, then it might be accompanied by the serious side effect of reduction in longevity (e.g., alterations in the *p53* gene can do this) (Ritter et al., 2003). For example, melatonin, which is known to reduce DNA damage, has been shown be able to increase cancers and increase longevity in mice (that might suggest that antioxidants might require more careful consideration). Some possibilities for resolution of this dilemma might exist: when a drug or an environmental agent could be targeted to a specific cancer stage, then, probably, a reduction in cancer might be achieved without life span reduction (Ritter et al., 2003).

### 2.3.7 Complexity and Chaos Theory

Complexity and chaos theory, such as implied in Lotka–Volterra equations (Lotka, 1925; Volterra, 1926), is potentially useful mathematical modeling approaches to the analysis of carcinogenic biological processes related to cell population selection and competition. The Lotka–Volterra difference equations were originally developed to model competition between predator and prey populations in a specific ecological setting. In certain situations, these equations give chaotic, nonlinear, and nonpredictable results for the relative sizes of the predator and prey populations. Given what is already known about the enormous complexity of the human carcinogenic process, use of models such as these may be too simplistic to provide the theoretical framework for understanding human carcinogenesis (Garte, 2006). Specifically, there is no direct analog to predatory and prey species in cell model systems within an individual, i.e., it is not clear how cancer and normal cells would fulfill predator and prey roles in cell populations in a highly structured multitissue environment. One analog might be the relation of immunological response to tumor burden with macrophage population being the predator and tumor cells the prey. A dynamic which does show oscillatory behavior over time as suggested by predator/prey models is the angiogenesis process: the nonlinear dynamic is driven by the creation of metastasis greater than 1–2 mm, which then drive the generation of vascular endothelium growth factor, which then allows vascular remodeling to support tumors of larger volumes (Agur et al., 2004).

For a model to be biologically useful, the distribution of events (e.g., incidence of tumors in a population) should have a consistent relation to intraindividual physiological processes (Tan et al., 2004). Instead of the Lokka–Volterra equations, below we expand on these nonlinear population dynamics using nonlinear forms of the Fokker–Planck equations (Risken, 1996, 1999), where biological "field" effects are built into the deterministic drift term in the differential equations (Shiino, 2003).

### 2.3.8 Statistical/Empirical Cancer Models

Cancer incidence and mortality has been often modeled by epidemiologists using a sort of biologically naive statistical model (e.g., Cox regression without a latency parameter) to test for risk factors and disease risk associations in different populations. The results of this kind of statistical analyses are based on the assumption of linear or linear-quadratic dose–response functions (Little, 2004; Krestinina et al., 2005). Use of biologically uninformed dose–response functions (e.g., Cox regression assuming hazard rates are proportional over time) limits the degree to which parameter estimates can be used for analyzing disease mechanisms, to determine the absolute burden of disease in a population, and to make accurate forecasts, as well as for selecting informed and effective health interventions. The rationale for using an empirical model was often to be able to compare results across different population studies by using the same statistical model relying on the LNT assumption (linear no-threshold dose–response function). This was argued to be conservative, and thus to best protect the public health.

### *2.3.9 The Other Common Modeling Approaches for Carcinogenesis*

In 1966, Pike developed the basic statistical methodology to apply to analysis of different types of experimental animal data on carcinogenesis (e.g., accelerated life testing models, sacrificing animals to determine tumor burden for fixed times from the carcinogen exposure) (Pike, 1966). Klebanov with co-authors (Klebanov et al., 1993) proposed a stochastic model of radiation carcinogenesis that had much in common with the ideas originally suggested by Pike and was based on several biological assumptions (1) that the immediate biological consequence of irradiation is the formation of precancerous lesions in the nuclear genome of irradiated cells [according to the Armitage–Doll "hit and target" principle, the number of such lesions is a Poisson random variable with expectation proportional to the dose rate (Whittemore, 1977)], (2) that the primary lesions are subject to repair processes [to specify the probability for the DNA lesions to be misrepaired it has been considered later that the repair system might be described as "a queue with loss", i.e., using the applied mathematical apparatus of queueing theory (Kalashnikov, 1994)], and (3) that each of promoted lesions can ultimately give a rise to tumor after a certain period of time, which is also considered as a random variable. Under these assumptions, it is possible to relate the promotion time distribution with a survival function which is experimentally observable.

A generalization of Pikes' model was also suggested by Yakovlev and Polig (Yakovlev and Polig, 1996). The key feature of their model was that it allowed for radiation-induced killing of cells to compete with the process of tumor promotion at the cellular level. This new model described and explained a wide range of experimental findings. Assuming that the process of cell initiation can be described as a Poisson's by nature and that primary lesions are generated independently, the multihit model with constant parameters can be also represented in terms of the promotion time of cumulative distribution function. Using a formal "goodness of fit" test, designed to accommodate censored observations, Gregori with co-authors (2002) demonstrated that such models of carcinogenesis with competing causes of cell death provide a good fit to the data for many types of solid tumors.

## 2.4 Modeling for Populations with Heterogeneous Mutational Events and Tumor Growth Rates

### *2.4.1 Model Innovations, Fuzzy State Processes, Heterogeneous Tumor Risks, and Tumor Growth Rates*

One of the problems with the models of carcinogenesis described above is that they often assume that the population of interest, except for explicit dose measurements, is genetically and risk exposure homogenous. This is unlikely, due to either various genetic or latent risk exposure differences. Failure to deal with heterogeneity (including both latent risk and tumor growth potential heterogeneity) in the population means that parameter estimates made from the above models will be confounded with the effects of population heterogeneity (Aalen, 1988; Hougaard, 1984).

One of the problems raised by traditional models of carcinogenesis based on analyses of Markov processes is that they are difficult to apply, unless done for biologically very simple formulations (e.g., a two-event model) which may produce an aberrant behavior, i.e., that each tumor cell grows immediately into a malignant tumor (Tan and Chen, 1998). As a consequence it has been suggested that the use of stochastic differential equations defining changes in state may be a better approach because growth of the size of tumor can be explicitly modeled with Kalman filters (state space models). It is also suggested that no two tumors will be exactly alike when a very heterogeneous mixture of processes is involved (Klein and Klein, 1984). In addition, the above models are subject to the assumption that each gene mutation is a discrete, homogenous event. It is unclear whether mutation of every gene is an identical discrete event in each individual, with precisely the same physiological effect. The effects of a mutation may also be altered by differences in gene expression processes and epigenetic factors as described in the first section of this chapter.

One of the advantages of fuzzy state models is its flexibility in describing the state distribution that means that forecasts will not be strictly dependent upon having the "correct" model for the initial state distribution. This information can be used in the extended Kalman filter to make robust forecasts of the outcome of multivariate fuzzy state processes (Manton et al., 1994).

### *2.4.2 Intracellular Processes: Interactions Complicate Modeling*

The set of mutations that can trigger tumor initiation is a subset of all possible gene mutations that could lead to initiation of a tumor in a specific tissue. For example, disablement of the *p53* gene may disable intracellular protein messengers that may cause the cell deconstruction into its components (Mihara et al., 2003). A failure to follow the specific sequences of mutations may result in a disorganized type of cell death (necrosis), where no cell constituents are preserved after "dismantling" for their possible internal reuse.

The core model, assuming initially that mutations are independent discrete events, can be empirically approximated, within specific age limits (Cook et al., 1969), by the Weibull hazard function,

$$\mu(t) = \alpha t^{m-1} \tag{2.1}$$

where $\mu(t)$ is the risk of tumor growth initiating at age $t$ after $m$ nuclear gene mutations. The status of mutations should not be "frozen", i.e., the $m$ nuclear gene mutations can be "repaired" and a Weibull hazard function may still be applicable (Watson, 1977; Jewell, 1982). Equation 2.1 reflects the independent genetic events triggering the initiation of tumor growth within the individual in a specific organ or tissue type. The biological events underlying the initiation of a tumor can be made more transparent by expanding the Weibull scale term $\alpha$. This term reflects the product of the probability of each of the $m$ mutational events within a tissue type composed of $m$ cells, where either mutations (a) have to occur in a fixed order (the multistage model) or (b) do not have to occur in a fixed order (the multihit model). The structure of $\alpha$, where $N$ is the number of cells at risk of a specific mutation, and $p$ is the average probability of the $m$th mutation can be written as,

$$Np^m/(m-1)! \tag{2.2}$$

for the multistage form of the model, and by,

$$Np^m \tag{2.3}$$

for the multi-hit form of the model (Armitage and Doll, 1954, 1961). Clearly the difference between equations (2.2) and (2.3) is in the internal structure of the combinatorial term, $\alpha$, which requires that mutations occur in specific, complete or partial, sequences in equation (2.2), i.e., the genes that determine the functional loss have to be mutated in certain sequences to allow the cell to survive and move through the initiation step in the multiple stage model. In equation (2.3) such an order is not assumed. An important implication of the difference between equations (2.2) and (2.3) is that mutation rates will be higher in equation (2.2) for a given level of risk and a common tumor growth process.

What is not well reflected in this formulation is that (a) the cell has an internal, highly organized organellas and molecular structure affecting gene expression, and (b) the cell exists in a complex stochastic environment of other cells, with their communication controlled by various biochemical messengers. For example, equation (2.1) does not describe the rate at which cells in a given tissue divide, which is regulated by growth control factors (e.g., cellular maturation, and differentiation, in part controlled by retinoid compounds, operating in the nuclear and mitochondrial genome) and by metabolic rate (e.g., in part it may be regulated by thyroid hormones) (Wrutniak-Cabello et al., 2001). To represent the internal (intracellular) and external (extracellular) environments of the nuclear gene mutation process, additional linked stochastic equations are needed. In physics, this would be characterized as the problem of determining the "mean field" effect of multiple interacting subsystems, with the target system of the most interest.

First, we will show how the distribution of tumor incidences, or deaths, in a risk heterogeneous population is determined. This involves mixing the individual level processes in a population, according to the distribution of traits that affect the rates at which specific nuclear mutations occur. If we assume that such distributions are purely determined by genetic endowment, we can use a fixed, state distribution to mix the individual processes. This can be described as Weibull tumor hazard function mixed by a gamma distribution with parameters $\mu$ and $\gamma$ (parameters are defined by the way that mean equals $\mu$ and variance equals $\mu^2/\gamma$),

$$h(x; \mu, \gamma) = \frac{\mu \cdot x^{m-1}}{1 + \mu \cdot x^m/(m \cdot \gamma)} \tag{2.4}$$

where $X$ is age at tumor onset. The scale parameter of the Weibull, $\alpha$, differs over individuals, with the $\alpha_i$ having a standard distribution; e.g., the gamma as in equation (2.4) or inverse Gaussian distribution. This implies that either (a) the set of mutations differ over individuals or (b) the probability of a given mutation differs over individuals.

A limitation of this model is that the mixing distribution is static (Hougaard, 1984; Vaupel et al., 1979; Aalen, 1992). A more realistic model is one that allows the susceptibility distribution function to evolve with time and age (i.e., dynamic heterogeneity). This leads to more complex distributions of individual mutational risks, known as Levy distributions, which have higher order moments than Gaussian processes (i.e., greater than second order) (Gjessing et al., 2003).

The development of the dynamic heterogeneity model of carcinogenesis can start with a general function for disease, assuming no specific form of the hazard function or mixing distribution (Manton et al., 1993),

$$\mu(x) = \frac{\mu_0(x)}{\left[1 + n\gamma \int_0^x du \cdot \mu_0(u)\right]^{1/n}}. \tag{2.5}$$

The denominator reflects a slowing of the age-associated increase in mortality as vulnerable persons (high $\alpha_i$ values) die first, changing the mean of the distribution of the $\alpha_i$. This decreases the age rate of increase of $\mu(t)$, from what it would have been if there were no heterogeneity and a pure hazard (e.g., the Weibull) described the age increase in mortality. The parameter $\gamma$ is the squared coefficient of variation of individual frailty, where $n$ controls the shape of the $\alpha_i$ distribution, and $n = 1$ – for a gamma, and $n = 2$ – for an inverse Gaussian.

The integral in equation (2.5) can be evaluated, if a functional specification is selected for the hazard rate's dependence on age, $\mu_0(x)$. An expression which is independent of the functional form of $\mu_0(x)$ [by evaluating the integral in the denominator in equation (2.5)] is,

$$h(x) = \frac{\mu_0(x)}{[1 + n\gamma(\Lambda(x) - \Lambda(0))]^{1/n}}, \qquad (2.6)$$

where model parameters are subsumed in $\mu_0(x)$ and $\Lambda(x)$. For the Weibull, $\Lambda(x) = x\mu_0(x) / m$. A disease-specific latency parameter, $\ell$, can be included as $x\mu_0(x-\ell) / m$. The parameter $\ell$ is usually the mean (or median) of the distribution of times from tumor initiation to its clinical detection.

The distribution of individual risks may be applicable over the entire age range of tumor initiation, so they must be generalized to be dynamic in the sense that the shape parameter of the hazard rate is distributed over individuals (i.e., individual differences in the rate of aging), as well as the scale parameter (i.e., the fixed heterogeneity factor). In most demographic models of frailty the heterogeneity distribution of individual risk differences is assumed static (Hougaard, 1984). This is an obvious approximation that fails for many types of cancers at late ages. To eliminate this constraint, the shape parameter $n$ in equation (2.6) was allowed to go to 0. Setting $n = 0$ generated a mixing distribution, which allows human mortality at extreme ages to be fit by a declining hazard rate. As discussed above, a declining hazard rate is often found in cancer mortality data at late ages.

Dynamic heterogeneity could mean, for example, that the number of mutations needed to trigger a cancer might be distributed over individuals, causing the hazard to increase more slowly at late ages. Alternately, loss of immune function with age might allow tumors to initiate growth with fewer mutations and less damage. This suggests that the dynamic mixing distribution (i.e., the mixture of hazard rate parameters) might generate a distribution with more persons, at both lower and higher levels of risk, than the gamma or inverse Gaussian mixing distribution, usually used in mortality models for heterogeneous population, i.e., the new (Levy's type) distribution of risk levels is "flatter" and has thicker "tails".

In such models, it is important to identify how a given external stress, such as IR, biologically alters the risk of certain cancer type (or other disease). Since $\alpha$ reflects the product of the probabilities of each of $m$ independent mutations that parameter should reflect the alteration of mutation rates either by chemical carcinogens or by IR. So, for female breast cancer, IR strongly elevated early tumor risk in both Russian studies of exposed by the Chernobyl accident population and Japanese studies of victims of Hiroshima and Nagasaki A-bombing. Thus, it appears that it is primarily early aggressive breast tumors, whose risk is elevated by IR exposure, accelerating the rate of accumulation of mutations in a disease that is already partly genetically predefined (Land, 1995) (an additional discussion of breast cancer risks is found in Chapter 7). To specify how particular types of stressors could produce tumor (or other disease), the initiation of elements of the models have to be individually considered, as well as the positive and negative feedbacks relative to other cell structures and functions.

Though the Weibull and its generalizations deal with either fixed or dynamic heterogeneity in susceptibility to tumor onset perform well empirically over broad age ranges (Manton et al., 1989, 1993), they are approximations of the underlying biological mechanisms, since they substitute independent mutation probabilities for mutation hazard rates by assuming the $p$ are small (i.e., that the probabilities well approximate the hazard rates) for a risk homogenous population. This may be true for relatively rare site-specific tumors, but not for more prevalent cancers (i.e., lung cancer), or for the descriptions of noncancer diseases (i.e., stroke, ischemic heart disease, and congestive heart failure). If certain $p$ is large, the Weibull no longer would describe the data with the mortality trajectory tending to rise less rapidly at late ages. The assumption that $p$ is independently generated is at variance with the existence of autocrine and paracrine effects: i.e., when the biological field is not homogeneous, and there is cell-to-cell communication, and the tissue has biologically meaningful structure (Prasad et al., 2004; Tubiana et al., 2005).

To deal with these potential violations of the standard assumptions for the Weibull failure process, one can use the so-called "exact form" of the multiple-stage model, where the probability of malignant transformation (i.e., an event when each of the $n$ necessary mutations has occurred) of a particular cell by time $x$, is

$$p(x;\alpha) = \prod_{j=1}^{m} (1 - e^{-\alpha_j x}). \tag{2.7}$$

If $X_1,\ldots,X_N$ and $X$ are random variables representing times to malignant transformations of $N$ susceptible cells and time to appearance of malignant tumor, then $X = \min\{X_1,\ldots,X_N\}$ is a minimum of order statistics. A computation shows that the respective hazard function for the tissue can be expressed in terms of probabilities (2.7) as $h(x) = Np'(x,\alpha)/(1 - p(x,\alpha))$ (Moolgavkar et al., 1999). This "exact" model still assumes that the population is homogenous for the rate of specific mutations. Both Weibull and exact models require the evaluation of the heterogeneity of individual susceptibility to cancer risk, whether it is static or dynamic (i.e., the probability that mutation rates vary over individuals). Heterogeneity slows the increase of cancer hazard with age if $\gamma$ is small. A similar effect is found in equation (2.7), assuming homogeneity if the number of mutations is large. The exact form can be extended to a risk heterogeneous population, by assuming that the hazard for each of the $m$ hits is independently statistically distributed as a gamma function over individuals,

$$p(x;\mu,\gamma)=\prod_{j=1}^{m}\{1-(1+\mu_j x/\gamma_j)^{-\gamma_j}\}. \tag{2.8}$$

When $\gamma_j$ becomes large (variability of the $j$th hit declines), the exact model approaches the Weibull.

The primary reason for using the exact equation (2.8), rather than the Weibull, is that it is applicable to a wider range of diseases, as long as they are affected by genetic factors. Consequently, this model could be used for diseases with a higher incidence rates than many cancers. It also is a potentially applicable when the different genetic mutations can interact with one another.

Biological complexity (i.e., interactions of biological subsystems) can be modeled in several ways. The disease may lead to correlated changes in the host's internal environment across the biological scales of organization. Burch (1976a, b) proposed to use a compound Weibull hazard function to reflect the growth of a tumor being due to $\rho$ mutations in the immune system, as well as a neoplastic transition in a target tissue. This can be represented for the exact model by addition of a parameter, $\theta$,

$$p(x;\alpha,\theta)=\prod_{j=1}^{m}\left(1-e^{-\alpha_j x}\left(\prod_{i=1}^{\rho}\left(1-e^{-\theta_i x}\right)\right)\right). \tag{2.9}$$

where $m$ represents errors in internal cell regulatory mechanisms, and $\rho$ represents errors in stem cells in the immune system. As discussed above, this formulation could also be used to describe the interaction of mitochondrial and nuclear DNA mutations, and apoptosis by the equation (2.9), which assumes that independent parallel processes have to be generalized to allow their interactions.

Another form of system dependence could result from complex extracellular interactions, affecting the probability of forming a viable multicellular clone. This could affect the ability of the clone to (a) vascularization (e.g., angiogenesis), (b) penetrate arterial walls to metastasize, or (c) produce cytokine-stimulating inflammatory mechanisms. Thus, equation (2.9) is a candidate mathematical form for the currently unspecified forms of the disease component processes described in the first section of this chapter.

### 2.4.3 Tumor Growth and Growth Heterogeneity

There are two necessary steps in describing the carcinogenesis and tumor expansion. The first, evaluated in the prior section, is to describe interactions between the nuclear genetic mutation process and the other intracellular processes. The second is to describe the growth (kinetics) from a single cell to a tumor with sufficient number of cells to pass various thresholds, such as a clinical detection, a functional dysregulation of other organs, and a disruption of overall intra- and interorgan homeostasis potentially leading to organism death. The different tumor sizes are associated with the tumor "aging" that defines specific types of tumor latency.

The linkage of tumor initiation and growth led Manton and Stallard (1988) to model carcinogenesis in a cohort as the convolution of three waiting time distributions: (1) a tumor initiation component described by a multi-hit/stage Weibull process in a risk-heterogeneous population; (2) growth, progression, and metastization described by a logistic growth function in a growth potential heterogeneous population, where the logistic parameters are determined by exponential growth of tumors restricted by linear constraints on tumor nutrition and vascularization; and (3) the modification of tumor growth and progression due to clinical intervention. This formulation has the advantage of being generally parameterized with the tumor growth and progression process parameters, being interpretable in terms of cell birth and death kinetics, and being estimable by combined use of several types of data (e.g., data on tumor growth rates obtained from the animal studies and human clinical trials). This model has been useful in describing the complex age dependence of several types of solid tumors [e.g., female lung cancer (Manton and Stallard, 1979)]. It does not, however, explicitly deal with other cellular mechanisms (e.g., mitochondrial function) affecting these various tumor stages. This stochastic, nonlinear compartment model with three waiting time distributions is the core of the more complete stochastic state variable process model of carcinogenesis using differential equations that we present below.

Once a tumor is initiated, a separate mathematical function (or stochastic process model) is assumed to describe its growth. The tumor growth function starting from a single cell may be described by a simple logistic cell kinetic function (Manton and Stallard, 1988), predicting the volume of cells viable in the tumor as a logistic function, e.g., a gamma-distributed mixture of Gompertz (exponential) tumor growth functions. The gamma mixing distribution represents differences in the survival time of individual cells in the tumor described by the Gompertz. The time of survival of individual cells is controlled by $\theta_i$ which controls how long a dedifferentiated tumor cell of type $i$, can expect to survive, and by the distribution of $\theta_i$ in the cells in the target organ of an individual. In fact, $\theta_i$ can be viewed as defining a process controlled by a cell kinetic model, governed by two-thirds of surface area–volume relation (Manton and Stallard, 1982, 1988). In this case, nutrients are passed through cell membranes with critical/growth limiting cell membrane surface area, being the capillary membrane's interface between the circulating blood compartment and the intracellular cytoplasmic compartment. The volume is determined by the number of cells in the tumor. Tumor growth is governed by the (a) rate of growth of the capillary surface area with respect to time, since tumor initiation (determined by cytokines influencing inflammatory responses and angiogenesis factors regulating microvascularization); (b) rate of growth of the number of cells in the tumor mass since tumor initiation; and (c) rate of change in the proportion of the tumor's cells that is viable since tumor initiation. The "bending over" of the Gompertz trajectory in the logistic tumor growth function reflects the decreasing proportion of variable tumor cells as tumor volume increases which decreases the rate of efficiency of transfer of nutrients over the existing capillary surface area.

### 2.4.4 Stochastic Multivariate Models of Carcinogenesis

In both cancer and circulatory diseases, there would be the interactions of intracellular components and/or of multiple organ systems that could produce "mean field" effects (Shiino, 2003). It means that in an equation, describing the change with time/age of an organ/tissue type, there is a new variable defined in the "drift" term (describing deterministic changes) that represents the average effects of all lower level biological functions, affecting the cell's specific function of interest (e.g., cancer onset due to the manifestation of specific mutations). For example, a cell may be affected by growth factors communicating with the surrounding cells. If the cell is affected by the close proximity of the first layer of surrounding cells, then the effect of the local "field" of cells can be represented by a simple average of the effects of each of those "neighboring" cells. The nature of the field, or interaction, effect may be organized by other ways than simple distance measures, e.g., the effects of cytokines on a specific cell. Depending on the biological scale of complexity of the effect, the equation describing change may have a fractional power.

If the factors affecting the likelihood of a mutation are stochastic and measurable, then the gamma distribution of the probability of a given mutation occurring in the exact incidence function may be replaced by a function of measures of individual differences affecting the likelihood of a mutation, i.e., the fixed gamma distribution is replaced by a solution of the Fokker–Planck equation, describing the change in the risk of a mutation as a function of the state of the cell and interacting tissues, stochastically and by using fuzzy set descriptions of uncertainty about the specification of the functional relations of the internal state of the cell to each of the candidate nuclear mutations. Such changes may be described by specialized nonlinear Fokker–Planck equations, i.e., equations describing the deterministic and stochastic changes in the probability distribution of certain traits (Risken, 1996). The Fokker–Planck equation for $J$ variables where the "field" influences the rate of progression of the process, may be written in univariate form (Shiino, 2003) as,

$$\frac{\partial p}{\partial t} = -\frac{\partial}{\partial x}\left[\left(-\frac{\partial \phi}{\partial x} + \varepsilon \int xpdx\right)p\right] + D\frac{\partial^2}{\partial x^2}p^q - \mu p \tag{2.10}$$

where the first term represents "drift", the second – diffusion, and the third – $\mu p$, represents a cellular mutation incidence process in either tumor initiation or cell death. That affects the probability distribution function $p$ by "removing" cells from the tissue. Such a cell loss is related to the loss of viability of the individual. The term in the square brackets reflects the average effect of other subsystems (integral $\int xpdx$) on the process of carcinogenesis (with effect parameter $\epsilon$).

For specific cases (i.e., Gaussian diffusion), the mortality function could reflect quadratic dependence on the state variable process, making $\mu$ a hazard rate process as suggested by Gjessing et al. (2003). The description of the cellular process of carcinogenesis, to be most useful, should translate into a related Fokker–Planck equation, describing the incidence of tumor onset or death in a population of individuals. In this case, the question about 2.10 is whether the entropy, or difference, of cellular function is extensive (additive) or nonextensive (requiring evaluation of interactions of processes). For equations that produce extensive (additive) statistics, the Fokker–Planck equation is relatively straightforward.

The stochastic state variable Weibull process can be written as (Manton and Stallard, 1988; Manton et al., 1992; Manton and Yashin, 2000),

$$\mu = (x_t Q x_t) t^m \tag{2.11}$$

where $x_t$ is the value of state variables at time $t$. We could replace $t^m$ in 2.11 by the exact model with heterogeneity for each mutation represented by a gamma distribution for each person. There could be interactions between the observed state variable process $x_t$ the gamma distributions describing genetic heterogeneity. For the cell components which do not yet have well-accepted formal models of structure and function, the use of fuzzy states offers a flexible modeling tool, which can be combined with stochastic differential equations modified to be appropriate to describe the evolution of cancer cell populations (Zhang and Wunsch, 2003). Thus, a comprehensive model of carcinogenesis necessarily requires a system of multiple-linked equations (i.e., the multivariate stochastic processes describing the function of cell structures that influence one another), describing the interaction of different cell components and processes. It seems like more general models would replace the tumor incidence function with a fixed susceptibility distribution by a stochastic state variable process that includes the changes of susceptibility to cancer biomarkers/risk factors.

## 2.5 Summary

After more than half of a century has passed, many papers were published discussing the Armitage–Doll model and next steps that should be done in the mathematical modeling of carcinogenesis. In future studies, mathematical modeling will be important in connecting genetic changes to the associated biochemical pathways and to the consequences for cellular birth and death rates. The contemporary mathematical models can be complex, with many details, to suggest certain new hypotheses about cancer initiation and progression. Most of the recent studies of carcinogenesis have focused on detailed analyses of molecular mechanisms. However, it seems like now interest has started to shift to studies of how various mechanisms of carcinogenesis may combine to determine the complex systems behavior in a quantitative perspective (Frank, 2004).

The variations in cure rates for cancer in the past 50 years suggest that it is now more important to discuss incidence rather than mortality and that incidence may now be much better determined than 50 years ago (Ritter et al., 2003) (the intensive screening strategy for some cancers, e.g., cervical cancer, should be also taken into account).

Recently it has been supposed that many more mutations are present in cancer than the 4–8 slow stages predicted from the epidemiological data: a test for mutated DNA for colon cancer includes 21 specific mutations (Tagore et al., 2003), and considerably more mutations are known to be consequences of genetic instabilities caused by early stage alterations, e.g., 11,000 are reported by Stoler et al. (1999) for colon cancer (Lengauer et al., 1998; Duensing and Munger, 2002). It has been hypothesized that the most of these alterations must occur very rapidly and thus do not affect the age distribution of cancer, which is determined by the much slower and more rare rate-limiting stages. As a strategy for reducing cancer incidence, it appears that it is much more productive to develop environmental (e.g., diet or behavioral) strategies which would further slow (thus making them less probable) the slow stages to reduce cancers, rather than strategies which make fast stages less probable, which would not reduce cancers appreciably (Ritter et al., 2003). The efforts of trying to identify the stages as slow and fast might help to develop the effective preventive strategies, however, this work is challenging.

The other important question that needs the detailed further studies is whether it is possible to describe cancer incidence at ages older than 80 by using the exact multistage model but no other assumptions. Some recent studies prove that it is unlikely to happen, however, with additional biological assumptions it is possible to fit the data (Ritter et al., 2003). One of these assumptions is senescence, and the other is many people are not susceptible to cancer.

In this chapter, we presented an overview of formal quantitative models of carcinogenesis and illustrated the further directions in the development of the substantive and mathematical basis for a generalized model of carcinogenesis, that recognizes the intrinsic structural and functional complexity of the human cell, and cell-to-cell communication (including the effects of cellular structures beyond that of the nuclear DNA that are considered influencing on the carcinogenesis). Furthermore, the possibilities of translational studies, that make it possible to model disease process across the biological spectrum from human population to inter- and intracellular levels, are discussed. Additionally, the arguments are discussed by which this model might be generalized to describe not only cancer but also noncancer aging-related diseases (e.g., CVD, cerebrovascular disease). This chapter sets the theoretical and analytic context for subsequent chapters, where we examine specific cancers models applied to specific data sets.

## References


Aalen O.O., 1988. Heterogeneity in survival analysis. Stat Med 7:1121–1137.

Aalen O.O., 1992. Modeling heterogeneity in survival analysis by the compound Poisson distribution. Ann Appl Probab 2:951–972.

Agur Z., Arakelyan L. et al., 2004. HOPF point analysis for angiogenesis models. Discrete Cont Dyn Syst B 4(1):29–38.

Akushevich I., Veremeeva G., Kulminski A., Ukraitseva S., Arbeev K., Akleev A.V., Yashin A.I., 2007. New perspectives in modeling of carcinogenesis induced by ionizing radiation. "The 13th International Congress of Radiation Research, San Francisco, California, July 08–12, 2007. Abstract PS4175. In abstract book, p. 246.

Alber M., Chen N. et al., 2006. Multiscale dynamics of biological cells with chemotactic interactions: from a discrete stochastic model to a continuous description. Phys Rev E 73:051901-1–051901-11.

Andersen M.H., Becker J.C., Straten P., 2005. Regulators of apoptosis: suitable targets for immune therapy of cancer. Nat Rev Drug Discov 4:399–409.

Arbeev K.G., Ukraintseva S.V., Arbeeva L.S. et al., 2005. Mathematical models for human cancer incidence rates. Demogr Res 12:237–272.

Armitage P., Doll R., 1954. The age distribution of cancer and a multi-stage theory of carcinogenesis. Br J Cancer 8(1):1–12.

Armitage P., Doll R., 1961. Stochastic models for carcinogenesis. In Proceedings of the Fourth Berkeley Symposium on Mathematical Statistics and Probability, University of California Press.

Artandi S.E., 2003. Complex roles for telomeres and telomerase in breast carcinogenesis. Breast Cancer Res 5:37–41.

Bateman H., 1910. Solutions of a system of differential equations occurring in the theory of radioactive transformation. Proc Camb Philol Soc 15:423–427.

Barrett J.C., 1979. The progressive nature of neoplastic transformation of Syrian hamster embryo cells in culture. Prog Exp Tumor Res 24:17–27.

Barrett J.C., Fletcher W.F., 1987. Cellular and molecular mechanisms of multistep carcinogenesis in cell culture models. In: Barrett J.C. (ed.). Mechanisms of Environmental Carcinogenesis, Vol. 2. Boca Raton, FL: CRC Press, pp. 73–116.

Beckman R.A., Loeb L.A., 2005. Genetic instability in cancer: theory and experiment. Semin Cancer Biol 15:423–435.

Bergers G., Javaherian K. et al., 1999. Effects of angiogenesis inhibitors on multistage carcinogenesis in mice. Science 284(5415):808–812.

Bijwaard H., Brugmans M.J., Rispens S.M., 2005. Comment on "Studies of radon-exposed miner cohorts using a biologically based model: comparison of current Czech and French data with historic data from China and Colorado" by Heidenreich W.F., Tomasek L., Rogel A., Laurier D., Tirmarche M., 2004, Radiat Environ Biophys 43:247–256. Radiat Environ Biophys 149–151; author reply 153–154.

Bisacchi D., Benelli R., Vanzetto C. et al., 2003. Antiangiogenesis and angioprevention: mechanisms, problems and perspectives. Cancer Detect Prev 27:229–238.

Bogenhagen D.F., 1999. Repair of mtDNA in vertebrates. Am J Hum Genet 64:1276–1281.

Bogenrieder T., Herlyn M., 2003. Axis of evil: molecular mechanisms of cancer metastasis. Oncigene 22:6524–6536.

Boland C.R., 2002. Heredity nonpolyposis colorectal cancer (HNPCC). In: Vogelstein B., Kinzler K.W. (eds.). The Genetic Basis of Human Cancer. 2nd edition. New York: McGraw-Hill, pp. 307–321.

Boveri T., 1929. The Origin of Malignant Tumors. Baltimore, MD: Williams & Wilkins.

Breckenridge D.G., Germain M., Mathai J.P. et al., 2003. Regulation of apoptosis by endoplasmic reticulum pathways. Oncogene 22:8608–8618.

Brissette J.L., Kumar N.M., Gilula N.B., Dotto G.P., 1991. The tumor promoter 12-*o*-tetradecanoylphorbol-13-acetate and the ras oncogene modulate expression and phosphorylation of gap junction proteins. Mol Vellul Biol 11:5364–5371.

Brouland J-P., Gelebart P., Kovasc T. et al., 2005. The loss of sarco/endoplasmium reticulum calcium transport APTase 3 expression is an early event during the multistep process of colon carcinogenesis. Am J Pathol 167:233–242.

Brugmans M.J., Rispens S.M., Bijwaard H., Laurier D., Rogel A., Tomasek L., Tirmarche M., 2004. Radon-induced lung cancer in French and Czech miner cohorts described with a two-mutation cancer model. Radiat Environ Biophys, 43(3):153–163.

Burch P.R., 1976a. Letter: application of the Weibull distribution. Br J Radiol 49(582):564.

Burch P.R., 1976b. Lung cancer and smoking: is there proof. Br Med J 2(6036):640.

Chambers A.F., Matrisian L.M., 1997. Changing views of the role of matrix metalloproteinases in metastasis. J Natl Cancer Inst 89:1260–1270.

Chu K.C., 1985. Multievent model of carcinogenesis: a mathematical model for cancer causation and prevention. In Carcinogenesis: A Comprehensive Survey. Vol. 8, New York: Raven Press, pp. 411–421.

Cifone M.A., Fidler I.J., 1980. Correlation of patterns of anchorage-independent growth with in vivo behavior of cells from a murine fibrosarcoma. Proc Natl Acad Sci USA 77:1039–1043.

Cook P.J., Doll R., Fellingham S.A., 1969. A mathematical model for the age distribution of cancer in man. Int J Cancer 4:93–112.

Cristofalo V., Allen R. et al., 1998. Relationship between donor age and the replicative lifespan of human cells in culture: a reevaluation. Proc Natl Acad Sci USA 95:10614–10619.

Cutler R., Semsei I., 1989. Development, cancer and aging: possible common mechanisms of action and regulation. J Gerontol 44(6):25–34.

Delfino A.B.M., Barreto E.C., da Silva E.T. et al., 1997. The involvement of genes and proteins in apoptosis – Carcinogenesis regulation. Rev Bras Cancerol 43(3). At: http://www.inca.gov.br/rbc/n_43/v03/english/article.html

Denmeade S.R., Jakobsen C.M., Janssen S. et al., 2003. Prostate-specific antigen-activated thapsigargin prodrug as targeted therapy for prostate cancer. J Natl Cancer Inst 95:990–1000.

Denning C., Pitts J.D., 1997. Bystander effects of different enzyme-prodrug systems for cancer gene therapy depend on different pathways for intercellular transfer of toxic metabolites, a factor that will govern clinical choice of appropriate regimens. Hum Gene Ther 8:1825–1835.

DiMauro S., Schon E.A., 2003. Mitochondrial respiratory-chain diseases. N Engl J Med 348:2656–2668.

Dix D., 1989. The role of aging in cancer incidence: an epidemiological study. J Gerontol 44(6):10–18.

Doege K., Heine S., Jensen I. et al., 2005. Inhibition of mitochondrial respiration elevates oxygen concentration but leaves regulation of hypoxia-inducible factor (HIF) intact. Blood 106:2311–2317.

Duensing S., Munger K., 2002. Human papillomaviruses and centrosome duplication errors: modeling the origins of genomic instability. Oncogene 21:6241–6248.

Duffy M.J., Maguire T.M., Hill A. et al., 2000. Metalloproteinases: role in breast carcinogenesis, invasion and metastasis. Breast Cancer Res 2:252–257.

Economos A., 1982. Rate of aging, rate of dying, and the mechanisms of mortality. Arch Gerontol Geriatr 1(1):3–27.

El-Omar E.M., Rabkin C.S., Gammon M.D. et al., 2003. Increased risk of non-cardia gastric cancer associated with proinflammatory cytokine gene polymorphisms. Gastroenterology 124(5):1193–1201.
Eng C., Kiuru M., Fernandez M.J., Aaltonen L.A., 2003. A role for mitochondrial enzymes in inherited neoplasia and beyond. Nat Rev Cancer 3:193–202.
Ershler W.B., Keller E.T., 2000. Age-associated increased interleukin-6 gene expression, late-life diseases, and frailty. Annu Rev Med 51:245–70.
Fearon E.R., Jones P.A., 1992. Progressing toward a molecular description of colorectal cancer development. FASEB J 6:2783–2790.
Fearon E.R., Vogelstein B., 1990. A genetic model for colorectal tumorogenesis. Cell 61(5):759–767.
Fesik S.W., 2005. Promoting apoptosis as a strategy for cancer drug discovery. Nat Rev Cancer 5:876–885.
Fisher J.C., Hollomon J.H., 1951. A hypothesis for the origin of cancer foci. Cancer 4:916–918.
Fliss M.S., Usadel H., Caballero O.L. et al., 2000. Facile detection of mitochondrial DNA mutations in tumors and bodily fluids. Science 287:2017–2019.
Folkman J., 1971. Tumor angiogenesis: therapeutic implications. N Engl J Med 285:1182–1186.
Folkman J., 1995. Angiogenesis in cancer, vascular, rheumatoid and other disease. Nat Med 1:27–30.
Frank S.A., 2004. Commentary: Mathematical models of cancer progression and epidemiology in the age of high throughput genomics. Int J Epidemiol 33:1179–1181.
Frank S.A., 2007. Dynamics of cancer. incidence, inheritance and evolution. In: Orr H.A. (ed.). Princeton Series in Evolutionary Biology. Princeton and Oxford: Princeton University Press, p. 378.
Garte S., 2006. Theory in carcinogenesis and epidemiology. J Epidemiol Community Health 57:85.
Gerl R., Vaux D.L., 2005. Apoptosis in the development and treatment of cancer. Carcinogenesis 26:263–270.
Girald-Rosa W., Vleugels R.A., Musiek A.C. et al., 2005. High-throughput mitochondrial genome screening method for nonmelanoma skin cancer using multiplexed temperature gradient capillary electrophoresis. Clin Chem 51:305–311.
Gjessing H.K, Aalen O.O., Hjort N.L., 2003. Frailty models based on Lévy processes. Adv Appl Probab 35:532–550.
Gregori G., Hanin L., Luebeck G., Moolgavkar S., Yakovlev A., 2002. Testing goodness of fit for stochastic models of carcinogenesis. Math Biosci 175:13–29.
Hanahan D., Weinberg R.A., 2000. The hallmarks of cancer. Cell 100:57–70.
Harding H.P., Calfon M., Urano F. et al., 2002. Transcriptional and translational control in the Mammalian unfolded protein response. Annu Rev Cell Dev Biol 18(575–599).
Harris C.C., 1987. Human tissues and cells in carcinogenesis research. Cancer Res 47:1–10.
Hart R.W., Setlow R.B., 1976. DNA repair in late-passage human cells. Mech Ageing Dev 5(1):67–77.
Hasty P., 2005. The impact of DNA damage, genetic mutation and cellular responses on cancer prevention, longevity and aging: observations in humans and mice. Mech Aging Dev 126(1):71–77.
Hazelton W., Clements M. et al., 2005. Multistage carcinogenesis and lung cancer mortality in three cohorts. Cancer Epidemiol Biomarkers Prev 14(5):1171–1181.
He Q., Huang Y., Sheikh M.S., 2004. Proteasome inhibitor MG132 upregulates death receptor 5 and cooperates with Apo2L/TRAIL to induce apoptosis in Bax-proficient and -deficient cells. Oncogene 23:2554–2558.
He Q., Lee D.I., Rong R. et al., 2002a. Endoplasmic reticulum calcium pool depletion-induced apoptosis is coupled with activation of the death receptor 5 pathway. Oncogene 21:2623–2633.

He Q., Luo X., Huang Y. et al., 2002b. Apo2L/TRAIL differentially modulates the apoptotic effects of sulindac and a COX-2 selective non-steroidal anti-inflammatory agent inBax-deficient cells. Oncogene 21:6032–6040.

He T.C., Sparks A.B., Rago C. et al., 1998. Identification of c-MYC as a target of the APC pathway. Science 281(5382):1509–1512.

Heffelfinger S.C., Gear R.B., Schneider J. et al., 2003. TNP-470 inhibits 7,12-dimethylbenz[$\alpha$]anthracene-induced mammary tumor formation when administered before the formation of carcinoma in situ but is not additive with tamozifen. Lab Invest 83:1001–1011.

Heidenreich W.F., 2005a. Response to the comment on "Studies of radon-exposed miner cohorts using a biologically based model: comparison of current Czech and French data with historic data from China and Colorado" by Heidenreich W.F., Tomasek L., Rogel A., Laurier D., Tirmarche M., 2004. Radiat Environ Biophys 43:247–256. Radiat Environ Biophys, 44(2):153–154.

Heidenreich W., 2005b. Heterogeneity of cancer risk due to stochastic effects. Risk Anal 25:1589–1594.

Heidenreich W.F., 2006. Heterogeneity of cancer risk due to stochastic effects. Risk Anal 25(6):1589–1594.

Heidenreich W.F., Atkinson M., Paretzke H.G., 2001 Radiation-induced cell inactivation can increase the cancer risk. Radiat Res 155:870–872.

Heidenreich W., Paretzke H., 2001. The two-stage clonal expansion model as an example of a biologically based model of radiation-induced cancer. Radiat Res 156:678–81.

Heidenreich W.F., Tomasek L., Rogel A., Laurier D., Tirmarche M., 2004. Studies of radon-exposed miner cohorts using a biologically based model: comparison of current Czech and French data with historic data from China and Colorado. Radiat Environ Biophys 43(4):247–256.

Heselmeyer K., Hellstrom A.C., Blegen H., Schrock E., Silversward C., Shah K., Auer G., Ried T., 1998. Primary carcinoma of the fallopian tube: comparative genomic hybridization reveals high genetic instability and a specific, recurring pattern of chromosomal aberrations. Int J Gynecol Pathol 17:245–254.

Heslin M.J., Jieming Y., Jonson M.R. et al., 2001. Role of matrix metalloproteinases in colorectal carcinogenesis. Ann Surg 233:786–792.

Hibi K., Nakayama H., Yamazaki T et al., 2001a. Detection of mitochondrial DNA alterations in primary tumors and corresponding serum of colorectal cancer patients. Int J Cancer 94:429–431.

Hibi K., Nakayama H., Yamazaki T. et al., 2001b. Mitochondrial DNA alteration in esophageal cancer. Int J Cancer 92:319–321.

Hiyama K., Hiyama E., Ishioka S. et al., 1995a. Telomerase activity in small-cell and non-small-cell lung cancers. J Natl Cancer Inst 87:895–902.

Hiyama E., Yokoyama T., Tatsumoto N. et al., 1995b. Telomerase activity in gastric cancer. Cancer Res 55:3258–3262.

Hopkin K., 1996. Tumor evolution: survival of the fittest cells. J NIH Res 8:37–41.

Hougaard P., 1984. Life table methods for heterogeneous populations: distributions describing the heterogeneity. Biometrika 71:75–83.

Huang H., Patel D.D., Manton K.G., 2005. The immune system aging: roles of cytokines, T cells and NKcells. Front Biosci 10:192–215.

Hunter T., 1986. Cell growth control mechanisms. Nature 322:14–15.

Issa J.P., 2004. Opinion: CpG island methylator phenotype in cancer. Nat Rev Cancer 4:988–993.

Jagat N., Kolodgie F., Virmani R., 2000. Apoptosis and cardiomyopathy. Molecular genetics. Curr Opin Cardiol 15(3):183–188.

Jakupciak J.P., Wang W., Markowitz M.E. et al., 2005. Mitochondrial DNA as a cancer biomarker. J Mol Diagn 7(2):258–267.

Jass J.R., Barker M., Fraser L. et al., 2003. APC mutation and tumor budding in colorectal cancer. J Clin Pathol 56:69–73.

Jass J.R., Whitehall V.L., Young J. et al., 2002a. Emerging concepts in colorectal neoplasia. Gastroenterology 123:862–876.

Jass J.R., Young J., Leggett B.A., 2002b. Evolution of colorectal cancer: change of pace and change of direction. J Gastroenterol Hepatol 17:17–26.

Jeronimo C., Nomoto S., Caballero O.L. et al., 2001. Mitochondrial mitations in early stage prostate cancer and bodily fluids. Oncogene 20:5195–5198.

Jewell N., 1982. Mixtures of exponential distributions. Ann Stat 10:479–484.

Jillella A.P., Day D.S., Severson K. et al., 2000. Non-Hodgkin's lymphoma presenting as anasarca: probably mediated by tumor necrosis factor alpha (TNF-alpha). Leuk Lymphoma 38(3–4):419–422.

Jones J.B., Song J.J., Hempen P.M. et al., 2001. Detection of mitochondrial DNA mutations in pancreatic cancer offers a 'mass'-ive advantage over detection of nuclear DNA mutations. Cancer Res 61:1299–1304.

Kagan J., Srivastava S., 2005. Mitochondria as a target for early detection and diagnosis of cancer. Crit Rev Clin Lab Sci 42(5–6):453–472.

Kalashnikov V.V., 1994. Mathematical Methods in Queuing Theory. Dordrecht, The Netherlands: Kluwer.

Kamat A.A., Fletcher M., Gruman L.M. et al., 2006. The clinical relevance of stromal matrix metalloproteinase expression in ovarian cancer. Clin Cancer Res 12:1707–1714.

Kerr J.F., Wyllie A.H., Currie A.R., 1972. Apoptosis: a basic biological phenomenon with wide-ranging implications in tissue kinetics. Br J Cancer 26:239–257.

Khuder S., Herial N. et al., 2005. Non-sterodial anti-inflammatory drug use and lung cancer: a meta-analysis. Chest 127:748–754.

Kim S., Kaminker P., Campisi J., 1999. TIN2, a new regulator of telomere length in human cells. Nat Genet 23:405–412.

Kim H.C., Lee H.J., Roh S.A. et al., 2008. CpG island methylation in familial colorectal cancer patients not fulfilling the Amsterdam criteria. J Korean Med Sci 23:270–277.

Kinzler K.W., Vogelstein B., 1997. Gatekeepers and caretakers. Nature 386:761–763.

Kinzler K.W., Vogelstein B., 1998. Landscaping the cancer terrain. Science 280(5366):1036–1037.

Kinzler K.W., Vogelstein B., 2002. Colorectal tumors. In: Vogelstein B., Kinzler K.W. (eds.). The Genetic Basis of Human Cancer. 2nd edition. New-York: McGraw-Hill, pp. 583–612.

Klebanov L.V., Yakovlev A.Yu., Rachev S.T., 1993. A stochastic model of radiation carcinogenesis: latent time distributions and their properties. Math Biosci 113(1):51–75.

Klein G., Klein E., 1984. Oncogene activiation and tumor progression. Carcinogenesis 5:429–435.

Klein S., McCormick F., Levitzki A., 2005. Killing time for cancer cells. Nat Rev Cancer 5:573–580.

Knudson A.G., 1971. Mutation and cancer: statistical study of retinoblastoma. Proc Natl Acad Sci 68:820–823.

Kodama S., Ariyoshi K., Watanabe S. et al., 2006. Telomere biology: implications for radiation carcinogenesis. Radiation Risk Perspectives: Proceedings of the Second Nagasaki Symposium of International Consortium for Medical Care of Hibakusha and Radiation Life Science, Nagasaki, Japan, 2006. Int Congr Ser 1299:242–247.

Kondo S., 1983. Carcinogenesis in relation to the stem cell mutation hypothesis. Differentiation 24:1–8.

Krestinina L.Y., Preston D.L., Ostroumova E.V., Degteva M.O., Ron E., Vyushkova O.V. et al., 2005. Protracted radiation exposure and cancer mortality in the Techa River Cohort. Radiat Res 164(5):602–611.

Krtolica A., Campisi J., 2002. Cancer and aging: a model for the cancer promoting effects of the aging stroma. Int J Biochem Cell Biol 34(11):1401.

Kumimoto H., Yamane Y., Nishimoto Y. et al., 2004. Frequent somatic mutations of mitochondrial DNA in esophageal squamous cell carcinoma. Int J Cancer 108:228–231.

Kuramoto K., Matsushita S., Esaki Y. et al., 1993. Prevalence, rate of correct clinical diagnosis and mortality of cancer in 4,894 elderly autopsy cases. Nippon Ronen Igakkai Zasshi 30(1):35–40.

Land C., 1995. Studies of cancer and radiation dose among atomic bomb survivors. The example of breast cancer. J Am Med Assoc 274(5):402–407.

Land H., Parada L.F., Weinberg R.A., 1983. Tumorogenic conversion of primary embryo fibroblasts requires at least two cooperating oncogenes. Nature 304:596–602.

Laurier D., Rogel A., Tomasek L., Tirmarche M., 2005. Comment on "Studies of radon-exposed miner cohorts using a biologically based model: comparison of current Czech and French data with historic data from China and Colorado" by Heidenreich W.F., Tomasek L., Rogel A., Laurier D., Tirmarche M., 2004. Radiat Environ Biophys 43:247–256, and "Radon-induced lung cancer in French and Czech miner cohorts described with a two-mutation cancer model" by Brugmans M.J.P., Rispens S.M., Bijwaard H., Laurier D., Rogel A., Tomasek L., Tirmarche M., 2004. Radiat Environ Biophys 43:153–163. Radiat Environ Biophys, 44(2):155–156.

Lengauer C., Kinzler K.W., Vogelstein B., 1998. Genetic instabilities in human cancers. Nature 396:643–649.

Liang J., Bennett J., Krause N. et al., 2002. Old age mortality in Japan. J Gerontol Ser B: Psychol Sci Social Sci 57:S294–S307.

Lievre A., Chapusot C., Bouvier A.M. et al., 2005. Clinical value of mitochondrial mutations in colorectal cancer. J Clin Oncol 23:3517–3525.

Lightowlers R.N., Chinnery P.F., Turnbull D.M., Howell N., 1997. Mammalian mitochondrial genetics: heredity, heteroplasmy and disease. TIG 13:450–455.

Ling V., Chambers A.F., Harris J.F., Hill R.P., 1985. Quantitative genetic analysis of tumor progression. Cancer Metast Rev 4:173–194.

Little M.P., 1995. Are two mutations sufficient to cause cancer? Some generalizations of the two-mutation model of carcinogenesis of Moolgavkar, Venzon, and Knudson, and of the multistage model of Armitage and Doll. Biometrics 51:1278–1291.

Little M., 1996. Generalizations of the two-mutation and classical multi-stage models of carcinogenesis fitted to the Japanese atomic bomb survivor data. J Radiol Prot 16(7):7–24.

Little M., 2004. Risks associated with ionizing radiation. Br Med Bull 2003;68:259–275.

Little M.P., Haylock R.G., Muirhead C.R., 2002. Modelling lung tumour risk in radon-exposed uranium miners using generalizations of the two-mutation model of Moolgavkar, Venzon and Knudson. Int J Radiat Biol, 78(1):49–68.

Little M.P., Heidenreich W.F., Moolgavkar S.H., Schöllnberger H., Thomas D.C., 2008. Systems biological and mechanistic modelling of radiation-induced cancer. Radiat Environ Biophys 47(1):39–47.

Little M.P., Li G., 2007. Stochastic modelling of colon cancer: is there a role for genomic instability? Carcinogenesis 28(2):479–87.

Little M.P., Wright E.G., 2003. A stochastic carcinogenesis model incorporating genomic instability fitted to colon cancer data. Math Biosci 183(2):111–34.

Loewenstein W.R., 1966. Permeability of membrane junctions. Ann NY Acad Sci 137:441–472.

Lotka A.J., 1925. Elements of Physical Biology. Baltimore, MD: Williams & Wilkins Co.

Manton K.G., Lowrimore G., Yashin A.I., 1993. Methods for combining ancillary data in stochastic compartment models of cancer mortality: generalization of heterogeneity models. Math Popul Stud 4(2):133–147.

Manton K.G., Stallard E., 1979. Maximum likelihood estimation of a stochastic compartment model of cancer latency: lung cancer mortality among white females in the U.S. Comput Biomed Res 12:313–325.

Manton K.G., Stallard E., 1982. A population-based model of respiratory cancer incidence, progression, diagnosis, treatment and mortality. Comput Biomed Res 15:342–360.
Manton K.G., Stallard E., 1988. Chronic disease risk modeling: measurement and evaluation of the risks of chronic disease processes. In the Griffin Series of the Biomathematics of Diseases. London, England: Charles Griffin Limited.
Manton K.G., Stallard E., Singer B.H., 1992. Projecting the future size and health status of the U.S. elderly population. Int J Forecast 8:433–458.
Manton K.G., Volovyk S., Kulminski A., 2004. ROS effects on neurodegeneration in Alzheimer's disease and related disorders: on environmental stresses of ionizing radiation. Curr Alzhei Res (Lahiri D.K., ed.) 1(4): 277–293.
Manton K.G., Woodbury M.A., Stallard E., Riggan W.B., Creason J.P., Pellom A.C., 1989. Empirical Bayes procedures for stabilizing maps of U.S. cancer mortality rates. J Am Stat Assoc 84(407):637–650.
Manton K.G., Woodbury M.A., Tolley H.D., 1994. Statistical Applications Using Fuzzy Sets. New York: Wiley-Interscience Publication, John Wiley & Sons, p. 312.
Manton K.G., Yashin A.I., 2000. Mechanisms of aging and mortality: searches for new paradigms. Monographs on Population Aging, 7. Odense, Denmark: Odense University Press.
Markert C., 1968. Neoplasia: a disease of cell differentiation. Cancer Res 28:1908–1914.
Maruyama Y., Hanai H., Fujita M. et al., 1997. Telomere length and telomerase activity in carcinogenesis of the stomach. Jpn J Clin Oncol 27(4):216–220.
Maximo V., Soares P., Seruca R. et al., 2001. Microsatellite instability, mitochondrial DNA large deletions, and mitochondrial DNA mutations in gastric carcinoma. Genes Chromosomes Cancer 32:136–143.
Medina D., 1988. The preneoplastic state in mouse mammary tumorogenesis. Carcinogenesis 9:1113–1119.
Medrano M.J., Lopez-Abente G., Barrado M.J. et al., 1997. Effect of age, birth cohort, and period of death on cerebrovascular mortality in Spain, 1952 through 1991. Stroke 28:40–44.
Meeker A.K., Hicks J.L., Iacobuzio-Donahue C.A. et al., 2004. Telomere length abnormalities occur early in the initiation of epithelial carcinogenesis. Clin Cancer Res 10:3317–3326.
Meierhofer D., Mayr J.A., Fink K. et al., 2006. Mitochondrial DNA mutations in renal cell carcinomas revealed no general impact on energy metabolism. Br J Cancer 94:268–274.
Michor F., Nowak M. et al., 2006. Stochastic dynamics of metastasis formation. J Theor Biol 240:521–530.
Mihara M., Erster S., Zaika A., Petrenko O., Chittenden T., Pancoska P., Moll U.M., 2003. p53 has a direct apoptogenic role at the mitochondria. Mol Cell 11:577–590.
Minna J.D., Gazdar A.E., 1996. Translational research comes of age. Nat Med 2(9):974–975.
Miura N., Horikawa I., Nishimoto A. et al., 1997. Progressive telomere shortening and telomerase reactivation during hepatocellular carcinogenesis. Cancer Genet Cytogenet 93(1):56–62.
Moolgavkar S.H., Knudson A.G., 1981. Mutation and cancer: a model for human carcinogenesis. J Natl Cancer Inst 66(6):1037–52.
Moolgavkar S., Krewski D., Schwarz M., 1999. Mechanisms of carcinogenesis and biologically based models for estimation and prediction of risk. In: Moolgavkar S., Krewski D., Zeise L., Cardis E., Møller H. (eds.). Quantitative Estimation and Prediction of Human Cancer Risks. Scientific publications No. 131, Lyon: International Agency for Research on Cancer, pp. 179–237.
Moolgavkar S.H., Luebeck E.G., 1992. Multistage carcinogenesis: population-based model for colon cancer. J Natl Cancer Inst 84:610–618.
Moolgavkar S.H., Venzon D.J., 1979. Two event model for carcinogenesis: incidence curves for childhood and adult cancer. Math Biosci 47:55–77.
Nagy A., Wilhelm M., Sukosd F. et al., 2002. Somatic mitochondrial DNA mutations in human chromophobe renal cell carcinomas. Genes Chromosomes Cancer 35:256–260.

Neckelmann N., Li K., Wade R.P. et al., 1987. cDNA sequence of a human skeletal muscle ADP/ATP translocator: lack of a leader peptide, divergence from a fibroblast translocator cDNA, and coevolution with mitochondrial DNA genes. Proc Natl Acad Sci USA 84:7580–7584.

Nelson W.G., Carter H.B., DeWeese T.L. et al., 2004. Prostate cancer. In: Abeloff M.D., Armitage J.O., Niederhuber J.E., Kastan M.B., McKenna W.G. (eds.). Clinical Oncology. 3rd edition. London: Elsevier, Churchill Livingstone, pp. 1877–1942, 3205.

Nettesheim P., Barrett J.C., 1984. Tracheal epithelial cell transformation: a model system for studies on neoplastic progression. Crit Rev Toxicol 12(3):215–239.

Niederhuber J.E., Cole C.E., Grochow L. et al., 2004. Colon cancer. In: Abeloff M.D., Armitage J.O., Niederhuber J.E., Kastan M.B., McKenna W.G. (eds.). Clinical Oncology. 3rd edition. London: Elsevier, Churchill Livingstone, pp. 1877–1942, 3205.

Nishikawa M., Nishiguchi S., Shiomi S. et al., 2001. Somatic mutations of mitochondrial DNA in cancerous and noncancerous liver tissue in individuals with hepatocellular carcinoma. Cancer Res 61:1843–1845.

Nishitoh H., Matsuzawa A., Tobiume K. et al., 2002. ASK1 is essential for endoplasmic reticulum stress-induced neuronal cell death triggered by expanded polyglutamine repeats. Genes Dev 16:1345–1355.

Nordling C.O., 1953. A new theory on the cancer-inducing mechanism. Br J Cancer 7:68–72.

Nowak M.A., Komarova N.L., Sengupta A. et al., 2002. The role of chromosomal instability in tumor initiation. Proc Natl Acad Sci USA 99:16226–16231.

Nowell P.C., 1976. The clonal evolution of tumor cell populations. Science 194:23–28.

Nowell P.C., 1986. Mechanism of tumor progression. Cancer Res 46:2203–2207.

Olkin I., Gleser L.J., Derman C., 1978. Probability Models and Applications. New York: MacMillan Publishing, pp. 289–298.

Oyadomari S., Mori M., 2004. Roles of CHOP/GADD153 in endoplasmic reticulum stress. Cell Death Differ 11:381–389.

Paillard F., 1997. Bystander effects in enzymes/prodrug gene therapy – Commentary. Hum Gene Ther 8:1733–1735.

Pardoll D.W., 2004. Immunology and cancer. In: Abeloff M.D., Armitage J.O., Niederhuber J.E., Kastan M.B., McKenna W.G. (eds.). Clinical Oncology. 3rd edition. London: Elsevier, Churchill Livingstone, 2004.

Parr R.L., Dakubo G.D., Thayer R.E. et al., 2006. Mitochondrial DNA as a potential tool for early cancer detection. Hum Genomics 2(4):252–257.

Pejovic T., Ladner D., Intengan M. et al., 2004. Somatic D-loop mitochondrial DNA mutations are frequent in uterine serous carcinoma. Eur J Cancer 40:2519–2524.

Peto R., Parish S.E., Gray R.G., 1985. There is no such thing as adeing, and cancer is not related to it. IARC Sci Publ 58:43–53.

Peto R., Roe F.J., Lee P.N. et al., 1975. Cancer and ageing in mice and men. Br J Cancer 32(4):411–426.

Petros J.A., Baumann A.K., Ruiz-Pesini E. et al., 2005. myDNA mutations increase tumorogenicity in prostate cancer. Proc Natl Acad Sci USA 102:719–724.

Pettepher C.C., LeDoux S.P., Bohr V.A. et al., 1991. Repair of alkalilabile sites within the mitochondrial DNA of RINr 38 cells after exposure to the nitrosourea streptozotocin. J Biol Chem 266:3113–3117.

Pfeffer U., Ferrari N., Morini M. et al., 2003. Antiangiogenic activity of chemopreventive drugs. Int J Biol Markers 18:70–74.

Pike M.C., 1966. A method of analysis of a certain class of experiments in carcinogenesis. Biometrics 22:142–161.

Pitot H.C., Goldsworthy T., Moran S., 1981. The natural history of carcinogenesis: implications of experimental carcinogenesis in the genesis of human cancer. J Supramol Struct Cellul Biochem 17:133–146.

Polyak K., Li Y., Zhu H. et al., 1998. Somatic mutations of the mitochondrial genome in human colorectal tumours. Nat Genet 20:291–293.

Pompei F., Plkanov M., Wilson R., 2001. Age distribution of cancer in mice: the incidence turnover at old age. Toxicol Ind Health 17(1):7–16.
Pompei F., Wilson R., 2001. Age distribution of cancer: the incidence turnover at old age. Hum Ecol Risk Assess 7(6):1619–1650.
Pompei F., Wilson R., 2002. A quantitative model of cellular senescence influence on cancer and longevity. Toxicol Ind Health 18:365–376.
Potter V.R., 1973. Biochemistry of cancer. In: Holland J., Frei E. (eds.). Cancer Medicine. Philadelphia: Lea and Febiger Publishers, pp. 178–192.
Potter V.R., 1978. Phenotypic diversity in experimental hepatomas: the concept of partially blocked ontogeny. Br J Cancer 38:1–23.
Prasad K.N., Cole W.C., Hasse G.M., 2004. Health risks of low-dose ionizing radiation in humans: a review. Exp Biol Med 229:378–382.
Ramaswamy S., Ross K.N., Lander E.S. et al., 2003. A molecular signature of metastasis in primary solid tumors. Nat Genet 33:49–54.
Rainsford J., Cohen P., Dix D., 1985. On the role of aging in cancer incidence: analysis of the lung cancer data. Anticancer Res 5(4):427–430.
Rajagopalan H., Bardelli A., Lengauer C. et al., 2002. Tumorigenesis: RAF/RAS oncogenes and mismatch-repair status. Nature 418:934.
Rajagopalan H., Nowak M.A., Vogelstein B. et al., 2003. The significance of unstable chromosomes in colorectal cancer. Nat Rev Cancer 3:695–701.
Rao R.V., Ellerby H.M., Bredesen D.E., 2004. Coupling endoplasmic reticulum stress to the cell death program. Cell Death Differ 11:372–380.
Reddel R.R., 2000. The role of senescence and immortalization in carcinogenesis. Carcinogenesis 21:477–484.
Reed J.C., Pellecchia M., 2005. Apoptosis-based therapies for hematologic malignancies. Blood 106:408–418.
Rhim J.S., 2001. Molecular and genetic mechanism of prostate cancer. Radiat Res 155:128–132.
Richard S.M., Bailliet G., Paez G.L. et al., 2000. Nuclear and mitochondrial genome instability in human breast cancer. Cancer Res 60:4231–4237.
Risken H., 1996. The Fokker-Planck Equation: Methods of Solution and Application. 2nd edition. Amsterdam: Elsevier.
Risken H., 1999. The Fokker-Planck Equation. 2nd edition. New York: Springer.
Ritter G., Wilson R., Pompei F. et al., 2003. The multistage model of cancer development: some implications. Toxicol Ind Health 19:125–145.
Rudolph K.L., Chang S., Lee H.W. et al., 1999. Longevity, stress response, and cancer in aging telomerase-deficient mice. Cell 96:701–712.
Rutkowski D.T., Kaufman R.J., 2004. A trip to the ER: coping with stress. Trends Cell Biol 14:20–28.
Schmitt H., Blin N., Zankl H. et al., 1994. Telomere length variation in normal and malignant human tissues. Gene Chromosomes Cancer 11:171–177.
Schwartz H.S., Juliao S.F., Sciadini M.F. et al., 1995. Telomerase activity and oncogenesis in giant cell tumor of bone. Cancer 75:1094–1099.
Sell S., 1993. Cellular origin of cancer: de-differentiation or stem cell maturation arrest. Environ Health Perspect 101:15–26.
Sell S. (ed.), 2003. Stem Cells. Handbook. Totowa, NJ: Humana Press. p. 509.
Shen Z., Xu L. et al., 2001. A comparative study of telomerase activity and malignant phenotype in multistage carcinogenesis of esophageal epithelial cells induced by human papillomavirus. Int J Mol Med 8:633–639.
Shen Z., Xu L. et al., 2004. The multistage process of carcinogenesis in human esophageal epithelial cells induced by human papillomavirus. Oncol Rep 11:647–654.
Shiino M., 2003. Stability analysis of mean-field-type nonlinear Fokker-Planck equations associated with a generalized entropy and its application to the self-gravitating system. Phys Rev E 67:056118-1–056118-16.

Sommerfield H.J., Meeker A.K., Piatyszek M.A. et al., 1996. Telomerase activity: a prevalent marker of malignant human prostate tissue. Cancer Res 56:218–222.

Sonnenschein C., Soto A.M., 2000. Somatic mutation theory of carcinogenesis: why it should be dropped and replaced. Mol Carcinog 29:205–211.

Soto A.M., Sonnenschein C., 2004. The somatic mutation theory of cancer: growing problems with the paradigm? Bioessays 26:1097–1107.

Spencer S.L., Gerety R.A., Pienta K.J. et al., 2006. Modeling somatic evolution in tumorogenesis. PLoS Computational Biology 2:e108.

Stanta G., Campagner L., Cavallieri F. et al., 1997. Cancer of the oldest old. What we have learned from autopsy studies. Clin Geriatr Med 13(1):55–68.

Stoler D.L., Chen N., Basik M. et al., 1999. The onset and extent of genomic instability in sporadic colorectal tumor progression. Proc Natl Acad Sci 96:15121–15126.

Storm S.M., Rapp U.R., 1993. Oncogene activation: c-raf-1 gene mutations in experimental and naturally occurring tumors. Toxicol Lett 67:210–210.

Strachan T., Read A.P., 1999. Human Molecular Genetics. 2nd edition. New York: John Willey & Sons, Publishers, p. 576.

Sui G., Zhou S., Wang J. et al., 2006. Mitochondrial DNA mutations in preneoplastic lesions of the gastrointestinal tract: a biomarker for the early detection of cancer. Mol Cancer 5:73, doi:10.1186/1476-4598-5-73. At: www.molecular-cancer.com/content/5/73.

Tagore K.S., Lawson M.J., Yucaitis J.A. et al., 2003. Sensitivity and specificity of a stool DNA multitarget assay panel for the detection of advanced colorectal neoplasia. Clin Colorectal Cancer 3:47–53.

Tan W.-Y., 1991. Stochastic Models of Carcinogenesis. New York: Markel Dekker.

Tan D.J., Bai R.K., Wong L.J., 2002. Comprehensive scanning of somatic mitochondrial DNA mutations in breast cancer. Cancer Res 62:972–989.

Tan W.-Y., Chen C.W., 1991. A multiple pathway model of carcinogenesis involving one stage models and two-stage models. In: Arino O., Axelrod D.E., Kimmel M. (eds.). Mathematical Population Dynamics, Chapter 31. New York: Marcel Dekker, Inc., pp. 469–482.

Tan W., Chen C., 1998. Stochastic modeling of carcinogenesis: some new insights. Math Comput Model 28(11):49–71.

Tan W., Zhang L., Chen C., 2004. Stochastic modeling of carcinogenesis: state space models and estimation of parameters. Discrete Cont Dyn Syst Ser B 4(1):297–322.

Thompson H.J., McGinley J.N., Wolfe P. et al., 2004. Targeting angiogenesis for mammary cancer prevention: factors to consider in experimental design and analysis. Cancer Epidemiol, Biomarkers Prev 13(7):1173–1184.

Till J.E., 1982. Stem cells in differentiation and neoplasia. J Cell Physiol 1:3–11.

Trosko J.E., Chang C.C., 1979. Genes, pollutants and human diseases. Quart Rev Biophys 11:603–627.

Trosko J.E., Chang C.C., Madhukar B.V., Dupont E., 1993. Oncogenes, tumor suppressor genes and intercellular communication in the "oncogeny as partially blocked ontogeny" hypothesis. In: Iversen O.H. (ed.). New Frontiers in Cancer Causation. Washington, DC: Taylor and Francis Publishers, pp. 181–197.

Trosko J.E., Chang C.C., Medcalf A., 1983. Mechanisms of tumor potential role of intercellular communication. Cancer Invest 1:511–526.

Trosko J.E., Ruch R.J., 1998. Cell-cell communication in carcinogenesis. Front Biosci, 3, d208–d236.

Trosko J.E., Tai M.-H., 2006. Adult stem cell theory of the multi-stage, multi-mechanism theory of carcinogenesis: role of inflammation on the promotion of initiated stem cells. In: Zaenker D.T., Schmidt A. (eds.). Infection and Inflammation: Impacts on Oncogenesis. Contrib Microbio; Basel: Karger, 13:45–65.

Tsutsumi S., Gotoh T., Tomisato W. et al., 2004. Endoplasmic reticulum stress response is involved in nonsteroidal anti-inflammatory drug-induced apoptosis. Cell Death Differ 11:1009–1016.

Tubiana M., Aurengo A., Averbeck D., Bonnin A., Le Guen B., Masse R., Monier R., Valleron A.J., de Vathaire F., 2005. Dose-effect relationship and estimation of the carcinogenic effects of low doses of ionizing radiation: the joint report of the Academie Des Sciences (Paris) and of the Academie Nationale De Medicine. Int J Radiat Oncol Biol Phys 63(2):317–319.

Ukraintseva S.V., Yashin A.I., 2003. Individual aging and cancer risk: how are they related? Demogr Res 9(8):1163–196.

UNSCEAR, 2000. United Nations Scientific Committee on the Effects of Atomic Radiation Report to the General Assembly, with scientific annexes. At: www.unscear.org/unscear/en/publications/2000_1.html.

van't Veer L.J., Dai H., van de Vijver M.J. et al., 2002. Gene expression profiling predicts clinical outcome of breast cancer. Nature 415:530–536.

Vaupel J., Manton K. et al., 1979. The impact of heterogeneity in individual frailty on the dynamics of mortality. Demography 16:439–454.

Vaupel J.W., Yashin A.I., 1988. Cancer rates over age, time and place: insights from stochastic models of heterogeneous populations. WP#88-01-1 of the Center for Population Analysis and Policy, University of Minnesota, and MPIDR Working Paper WP1999-006.

Volpe E.W., Dix D., 1986. On the role of aging in cancer incidence: cohort analyses of the lung cancer data. Anticancer Res 6(6):1417–1420.

Volterra V., 1926. Variazioni e fluttuazionin del numero d'individui in specie animali conviventi. Mem R Accad Naz dei Lincei Ser VI, 2:31–113.

von Zglinicki T., 1996. Are the ends of chromosomes the beginning of tumor genesis? On the role of telomeres in cancer development. Fortschr Med 114:12–14.

Wagener C., 2001. Molecular oncology: prospects for cancer diagnosis and therapy. Oncology. www.roche.com/pages/downloads/company/pdf/rtpenzberg01e.pdf.

Wallace D.C., 1995. Mitochondrial DNA variation in human evolution, degenerative disease and aging. Am J Hum Genet 57(2):201–223.

Wallace D.C., Brown M.D., Lott M.T., 1997. Mitochondrial genetics. In: Rimoin D.L., Connor J.M., Pyeritz R.E., Emery A.E.H. (eds.). Emory and Rimoin's Principles and Practice of Medical Genetics. London: Churchill Livingstone, pp. 277–332.

Wallace D.C., Shoffner J.M., Watts R.L. et al., 1992. Mitochondrial oxidative phosphorylation defects in Parkinson's disease. Ann Neurol 32:113–114.

Warburg O., 1956. On the origin of cancer cells. Science 123:309–314.

Watson G., 1977. Age incidence curves for cancer. Proc Natl Acad Sci 74:1341–1342.

Whittemore A., 1977. The age distribution of human cancer for carcinogenic exposures of varying intensity. Am J Epidemiol 106:418–432.

Wrutniak-Cabello C., Casas F., Cabello G., 2001. Thyroid hormone action in mitochondria. J Mol Endocrinol 26:67–77.

Wu C.W., Yin P.H., Hung W.Y. et al., 2005. Mitochondrial DNA mutations and mitochondrial DNA depletion in gastric cancer. Genes Chromosomes Cancer 44:19–28.

Yakovlev A.Y., Polig E., 1996. A diversity of responses displayed by a stochastic model of radiation carcinogenesis allowing for cell death. Math Biosci 132:1–33.

Yakovlev A., Tsodikov A., 1996. Stochastic Models of Tumor Latency and Their Biostatistical Applications. New Jersey: World Scientific.

Yamada O., Oshimi K., Mizoguchi H., 1993. Telomere reduction in hematologic cells. Int J Hematol 57:181–186.

Yamaguchi H., Wang H-G., 2004. CHOP is involved in endoplasmic reticulum stress-induced apoptosis by enhancing DR5 expression in human carcinoma cells. J Biol Chem 279(44):45495–45502.

Yang X.Y., Kimura M., Jeanclos E., Aviv A., 2000. Cellular proliferation and telomerase activity in CHRF-288-11 cells. Life Sci 66:1545–1555.

Yeh J.J., Lunetta K.L., van Orsouw N.J. et al., 2000. Somatic mitochondrial DNA (mtDNA) mutations in papillary thyroid carcinomas and differential mtDNA sequence variants in cases with thyroid tumours. Oncogene 19:2060–2066.

Yu C., Pan K., Xing D. et al., 2002. Correlation between a single nucleotide polymorphism in the matrix metalloproteinase-2 promoter and risk of lung cancer. Cancer Res 62:6430–6433.

Zanssen S., Gunawan B., Fuzesi L. et al., 2004. Renal oncocytomas with rearrangements involving 11q13 contain breakpoints near CCND1. Cancer Genet Cytogenet 149:120–124.

Zanssen S., Schon E.A., 2005. Mitochondrial DNA mutations in cancer. PLoS Med 2(11):e401. DOI: 10.1371/journal.pmed.0020401.

Zeviani M., Tiranti V., Piantadosi C., 1998. Reviews in molecular medicine. Mitochondrial disorders. Medicine 77:59–72.

Zhang N., Wunsch D. II., 2003. An extended Kalman filter (EKF) approach on fuzzy system optimization problem. IEEE Int Conf Fuzzy Syst 2:1465–1470.

Zhivotovsky B., Orrenius S., 2006. Carcinogenesis and apoptosis: paradigms and paradoxes. Carcinogenesis Advance Access. New York: Oxford University Press.

Zhu W., Qin W., Bradley P. et al., 2005. Mitochondrial DNA mutations in breast cancer tissue and in matched nipple aspirate fluid. Carcinogenesis 26:145–152.

# Chapter 3
# Cancer Risk Factors

## 3.1 Overview of Cancer Risk Factors

Cancer has a complex etiology with multiple risk factors: the most accurate assessment of an individual's risk for developing cancer would be an estimate of behavioral and environmental exposures, together with an information on interindividual differences in genetic/epigenetic host susceptibility (including inheritable variations in carcinogen-metabolizing enzymes, germline mutations in tumor-associated genes, and inherited differences in DNA adduct formation and DNA repair mechanisms). The data on lifestyle factors (diet, exercise, smoking, etc.) can be collected and interpreted with some degree of confidence, but genetic assessments for most individuals are not readily available at present.

### *3.1.1 Biomarkers*

Biomarkers are indicators of events for physiological, cellular, subcellular, and molecular alterations in the multistage development of specific diseases. The development of disease resulting from exposure to an environmental agent or other toxic factors is multistage, starting with exposure, getting an internal dose (deposited body dose), biologically effective dose (dose at the site of toxic action), progressing to early biological effect (at the subcellular level), altered structure or function (subclinical changes), and manifestating as disease (National Research Council Committee on Biological Markers, 1987). Any step in this process may be modified by host susceptibility factors, including genetic traits and effect modifiers, such as diet or other environmental exposures.

*Biomarkers of exposure* are used to estimate current or past exposure to a specific environmental agent, measuring xenobiotics, their metabolites, or their interactive products found in blood, urine, sweat, feces, breast milk, etc. (Barrett et al., 1997). For example, specific metabolites of one of the tobacco-specific nitrosamines, 4-(methyl-nitrosamino)-1-(3-pyridyl)-1-butanone (NNK), which is

K.G. Manton et al., *Cancer Mortality and Morbidity Patterns in the U.S. Population*, Statistics for Biology and Health, DOI 10.1007/978-0-387-78193-8_3,
© Springer Science+Business Media, LLC 2009

a potent chemical carcinogen, have been detected and quantified in the urine of smokers, but not in nonsmokers (Carmella et al., 1995). In addition to the use of biomarkers for quantifying the impact of exogenous exposures, it is desirable to have a biomarker of endogenous oxidative damage, as it has been found that endogenous oxidative DNA damage correlates with the formation of chronic degenerative diseases and cancer (Ames and Gold, 1991). For example, among the many oxidatively damaged DNA bases formed, 8-oxo-2-deoxyguanosine can be sensitively measured (Wang et al., 2001).

*A biomarker of effect* is a measurable biochemical, physiological, behavioral, or other alteration within an organism that can be recognized as associated with established/possible health impairment or disease (International Programme on Chemical Safety, 1993). In practice, biomarkers of effect represent changes at the subcellular level, particularly at the chromosomal and molecular levels, such as cytogenetic alterations and gene mutations. For example, an exposure to ionizing radiation, alkylating cytostatics, tobacco smoking, benzene, and styrene has been found to induce chromosome aberrations in humans (Barrett et al., 1997). Tobacco smoking, alkylating cytostatics, and ethylene oxide can also induce sister chromatid exchanges in human lymphocytes (Tucker et al., 1993). Increased micronuclei frequencies in human lymphocytes have been found after exposure to ionizing radiation and formaldehyde (Ballarin et al., 1992; Norppa et al., 1993).

*Biomarkers of susceptibility* indicate an inherent or acquired organism's ability to respond to the challenge of exposure to specific xenobiotic substances or other toxicants. Genetic differences in the expression of the *enzymes involved in activation and detoxification of xenobiotics* could be a major source of inter-individual variation in susceptibility to disease: phase I enzymes are mainly the cytochrome P450 family with mixed functions of oxidase enzymes, and phase II enzymes mostly act on oxidized substrates to conjugate them with glucuronic acid, glutathione, and sulfate (some of them are presented in Table 3.1).

Another type of biomarker of susceptibility is *DNA repair capacity*: genetically determined individual DNA repair capacity may influence the rate of removal of DNA damage and of the repair of mutations. An age-related decline in DNA-repair capacity was detected. It has been shown that a reduced repair capacity is an important risk factor for young individuals with basal cell carcinoma and for those with a family history of skin cancer (Wei et al., 1994).

### *3.1.2 Genotoxic and Nongenotoxic Mechanisms*

For cancer risk assessment, it is important to distinguish between genotoxic mechanisms of carcinogen/risk factor effect (involving direct DNA changes, such as alkylation or chromosome breakage) and nongenotoxic mechanisms (acting on the sites in the cell that do not involve genetic material). The rationale for this distinction is the assumption that genotoxic actions may have no

**Table 3.1** Genetic polymorphism of the enzymes involved in activation and detoxification of xenobiotics and associated cancers

| Polymorphism of enzyme | Associated cancer |
|---|---|
| *N*-acetyltransferase (NAT) | Bladder cancer (slow acetylator phenotype) |
| | Colon cancer (rapid acetylator phenotype) |
| | Gastric adenocarcinoma |
| Cytochrome P450 family: | |
| CYP1A2 | Colorectal cancer |
| | Bladder cancer |
| CYP1A1 | Lung cancer |
| | Laryngeal cancer |
| | Colon cancer |
| | Pancreatic adenocarcinoma |
| CYP2D6 | Lung cancer |
| | Larynx cancer |
| | Breast cancer |
| CYP2C19 | Hepatocellular carcinoma (in patients with HCV cirrhosis) |
| | Breast cancer |
| CYP1B1 | Lung cancer |
| | Breast cancer |
| CYP2E1 | Esophageal dysplasia |
| | Lung cancer |
| | Nasopharyngeal cancer |
| | Laryngeal cancer |
| | Colorectal cancer |
| Human cytosolic glutothion-S-transferases (GST): | |
| GSTM1 | Lung cancer (tobacco related) |
| | Gastric adenocarcinoma |
| | Distal colorectal carcinoma |
| | Pancreatic adenocarcinoma |
| GSTT1 | Colorectal cancer |
| | Pancreatic adenocarcinoma |
| Alcohol dehydrogenase (ADH) | |
| ADH3 | Oropharyngeal cancer |
| | Laryngeal cancer |
| | Esophageal dysplasia |
| Aldehyde dehydrogenase (ALDH) | |
| ALDH2 | Colorectal cancer |
| Methylenetetrahydrofolate reductase (MTHFR) | Colorectal cancer |

effective threshold (a linear low-dose response), whereas nongenotoxic effects require a threshold dose to disturb a homeostatic system sufficient to elicit a toxic effect. The question of potential genotoxity is important for risk assessment of factor/substance shown to be carcinogenic in experimental animals or humans. If the substance is assumed to be genotoxic, the minimal risk level might be estimated at a level much lower than the estimated safe level based on a nonlinear dose response. This difference may, in turn, make an enormous difference in regulation of the substances in air, water, food, or toxic waste (Fan and Howd, 2001).

### 3.1.2.1 Mutations and DNA Damage: Exogenous and Endogenous Agents

Mutations and DNA damage that resulting of exposure to both exogenous and endogenous carcinogenes can cause damages at background level (see Table 3.2). The endogenous carcinogens can be subdivided into chemical, biochemical, physiological, and pathophysiological agents. Exogenous sources may be subdivided according to their avoidability/possibility to control.

For carcinogens to induce a mutation, a number of conditions have to be met. For cancer to arise, several steps have to be passed (see Fig. 3.1), including (1) whether the phase I of reaction of biotransformation results in the formation of a DNA-reactive intermediate, (2) does the reactive intermediate escape the various enzymatic and nonenzymatic detoxication processes, (3) does it react with DNA or with another molecule, (4) whether the carcinogen–DNA–adducts that are formed are repaired before DNA is replicated, (5) does the

**Table 3.2** Sources of background DNA damage (from Lutz, 2001)

| Source | Type | Example |
|---|---|---|
| *Endogenous* | | |
| Chemical | DNA instability | Depurination |
| Biochemical | Errors during DNA replication | Essential metal ions |
| | Errors during DNA repair | S-adenosylmethionine |
| | DNA-reactive chemicals | Aldehyde forms of carbohydrates |
| Physiological | Oxygen stress derived (ROS) | HO•, NO•, peroxides |
| | | Lipid peroxidation products |
| Pathophysiological | Formation of carcinogens in vivo | Nitroso-compounds (NOC) |
| *Exogenous* | | |
| Hardly avoidable | Radiation | UV, ionizing radiation |
| | Natural radioactive isotopes | $^{222}Rn$, $^{40}K$ |
| | Carcinogens in ambient air | PAH*, benzene |
| | Some therapies | Tumor therapy |
| Avoidable in part | Natural dietary carcinogens | Estragole** |
| | Carcinogens from food processing | Urethane |
| | | Arylamines, PAH, NOC |
| | Food pyrolysis products | Vinyl chloride |
| | Exposure at the workplace | Passive smoking |
| | Carcinogens in ambient air | |
| Avoidable in principle | Active smoking | |
| | Dietary, environmental and work-related exposures | |

*Polynuclear aromatic hydrocarbons; ***p*-allylanisole or methyl chavicol, an organic compound is used in perfumes and as a food additive for flavor.

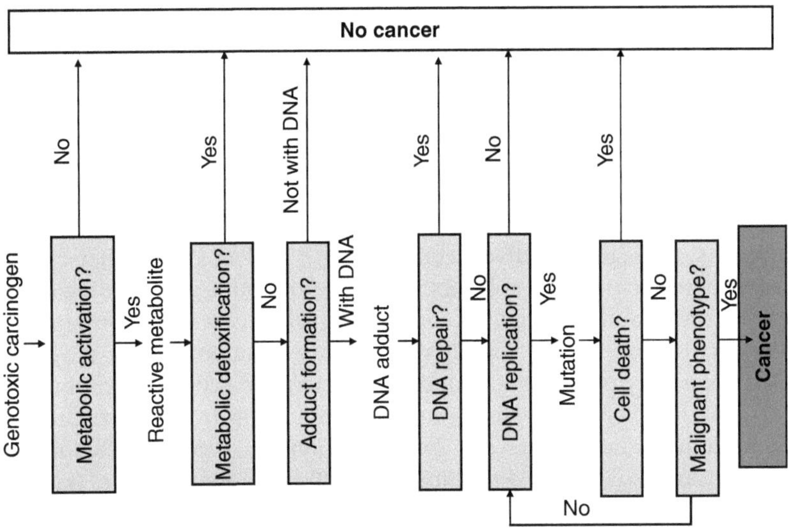


**Fig. 3.1** Sequence of events that modulate the probability of cancer induction from exposure to endogenous genotoxic carcinogens (from Lutz, 2001)

mutation lead to the death of the cell – if not, the cycle can be repeated until the necessary number of permanent changes would be reached, and the cell gains a fully malignant phenotype (Lutz, 2001).

#### 3.1.2.2 What Modulates the Rate of Carcinogenesis in Individuals?

The multi-stage/multi-"hit" conception of carcinogenesis can be better understood when most of germline mutations, environmental factors and their cellular targets will be clearly identified. Germline mutations play a key role in the relative risk of cancer predisposition, but they are limited to familial cases, and that indicates the importance of the environmental factors influencing carcinogenesis (Minamoto et al., 1999). Polymorphisms in metabolic activation and detoxification enzymes play an important role in individual susceptibility to environmental factors: an individual's genetic background influences the ability to "neutralize" certain carcinogens, determining individual cancer susceptibility and the mutation rate in multistep carcinogenesis. Genetic polymorphisms in tumor suppressor genes and proto-oncogenes can produce large differences in the susceptibility of individuals to develop cancer. This could lead to a reduction in the number of steps necessary for malignant transformation. Susceptibility factors can modulate the rate of the process, such as polymorphisms of DNA repair enzymes or polymorphisms of enzymes catalyzing metabolic activation and detoxication (D'Errico et al., 1996). Modulation by lifestyle-dependent factors can also be important. Exposure to one carcinogen can result in increased potency of another carcinogen. Examples are the supraadditive (synergistic) cancer risks from smoking plus alcohol for cancer of the oral cavity, larynx, and esophagus, or from smoking plus radon or asbestos exposure for lung cancer (Lutz, 2001).

#### 3.1.2.3 How Nongenotoxic Carcinogens Work

Nongenotoxic carcinogens differ from the genotoxic by not reacting, directly, or indirectly, through a reactive metabolite, with nuclear DNA. They can affect the cell's reproductive process by acting as cytotoxicants, inducing regenerative cell proliferation and producing secondary critical effects such as inflammation (Butterworth et al., 1995). Cytotoxicity releases nucleases that may induce DNA damage. Cytolethality can induce regenerative cell proliferation, which can also influence spontaneous tumor rates that result from preexisting mutations. Inflammation can increase the generation of oxygen radicals that may induce mutations. Chemicals that interact with hormone receptors (estrogens, androgens, growth hormone), or increase the synthesis of these hormones, can increase cell division in responsive tissues, and thus lead to increased growth of preexisting tumor cells.

### *3.1.3 Controllable and Noncontrollable Risk Factors*

While modeling the effects of one or a combination of risk factors on human health (e.g., cancer morbidity and mortality), the purpose is to find the model flexible enough and realistic enough to make it possible to reflect the main processes that real human populations live with (birth, reproduction, diseases, death, and various risk factors/combination of risk factors influence with their stable or changeable rates during the life). One of the ways to achieve this aim is to make interventions, changing risk factors' parameters in various age/sex groups of a population. For some risk factors, it is very difficult, or even impossible, to change risk factor parameters. For other risk factors, it is possible to make interventions. From this perspective, controllable risk factors may be included in model as changeable (partly avoidable in reality, or completely avoidable in an ideal scenario), and noncontrollable (hardly or non-avoidable). Among noncontrollable risk factors, the most common are age, sex, race/ethnicity, geographical location, heredity, poverty. Among the most common potentially controllable factors are smoking, alcohol consumption, obesity, occupation, diet, infection, stress, sun exposure, and precancerous lesion removal. We can make an intervention in a model by changing the intensity of a single controllable risk factor, or of a complex of factors. Below we will pay the most attention to controllable cancer risk factors as the objects of our interventions in models described in subsequent chapters. Table 3.3 presents risk factors associated with the most common cancers: part A includes noncontrollable and part B includes controllable risk factors.

**Table 3.3** Noncontrollable (A) and controllable (B) risk factors of various cancer localizations

*Part A*

| Cancer localization | Sex | Age, years | Race/ ethnic group | Family/personal disease history | Sex hormones/childbirth | Other |
|---|---|---|---|---|---|---|
| Lung | F[1] | > 40 | Black | Family history of lung cancer | NE[2] | • Radiation therapy to the chest area in anamnesis (Hodgkin disease or breast cancer treatment)<br>• Living in city for > 10 years (?)[3] |
| Breast | F | F > 40<br>M[4] > 60 | • Jewish Ashkenazi*<br>• In black – more aggressive tumor | • F, M: Family history of breast cancer<br>• F, M: Family and/or personal history of Li-Fraumeni syndrome, Cowden's syndrome.<br>• F: Mother or sister with ovarian cancer<br>• F: Personal history of breast, ovarian, or endometrial cancer<br>• F: Atypical hyperplasia<br>• M: Gynecomastia<br>• M: Klinefelter syndrome | • Early menarche (< 12 years)<br>• Late menopause (> 52 years)<br>• Never bearing children<br>• First child born after age 30<br>• Use of DES[5] in anamnesis<br>• M: Use of estrogen-related drugs for prostate treatment in anamnesis<br>• M: Use of high doses of estrogen as part of a sex change procedure | • Ionizing radiation (high dose)<br>• Height (tall women)<br>• In anamnesis, the treatment of chest area for another cancer (Hodgkin disease, non-Hodgkin lymphoma), especially in childhood<br>• M: liver cirrhosis |
| Prostate | M | > 50 | Black | • Father/brother with prostate cancer<br>• Mother/sister with breast or ovarian cancer<br>• Personal history of certain types of prostatic hyperplasia | – | • Height (tall men)<br>• Vasectomy performed at age < 35 |
| Esophagus | M | > 55 | • White – for adenocarcinoma<br>• Black – for squamous cell cancer | • Personal history of tylosis<br>• Plummer–Vinson syndrome<br>• Paterson–Kelly syndrome | NE | • Achalasia<br>• Lye** ingestion in children |
| Stomach | M | > 50 | • Hispanic<br>• Black | • Family history of stomach cancer<br>• Hereditary nonpolyposis colorectal cancer (HNPCC, or Lynch syndrome)<br>• Familial adenomatous polyposis (FAP)<br>• Personal history of adenomatous polyps<br>• Personal history of hypertrophic gastropathy (Menetrier disease) | NE | Having in anamnesis the Billroth II operation (ulcer treatment) (?) |

**Table 3.3** (continued)

*Part A*

| Cancer localization | Sex | Age, years | Race/ ethnic group | Family/personal disease history | Sex hormones/childbirth | Other |
|---|---|---|---|---|---|---|
| Colorectal | M(?) | > 50 | • Jews Ashkenazi<br>• Black<br>• Alaska natives | • Family history of colorectal cancer<br>• Family and/or personal history of adenomatous polyps<br>• Familial adenomatous polyposis (FAP)<br>• Hereditary nonpolyposis colon cancer (HNPCC)<br>• Personal history of Crohn's disease or ulcerative colitis for > 10 years<br>• F: Personal history of endometrial, ovarian or breast cancer | F: posmenopause status | • Cholecystectomy<br>• Height (tall individuals)<br>• Diabetes mellitus (?) |
| Ovarian | F | > 50 | • Jewish<br>• White | • Family history of ovarian, breast, or colorectal cancer<br>• Personal history of breast cancer | • Early menarche (< 12 years)<br>• Late menopause (> 52 years)<br>• Never bearing children<br>• First child born after 30 | NE |
| Corpus uteri | F | > 50 | White | • Mother/sister with uterine cancer<br>• Family history of inherited form of colon cancer | • Never bearing children<br>• Early menarche (< 12 years)<br>• Late menopause (> 52 years) | • Pelvic irradiation during treatment in anamnesis<br>• Diabetes mellitus (?)<br>• Endometrial hyperplasia<br>• Polycystic ovary syndrome |
| Cervical (cervix uteri) | F | 30–40*** | • Hispanic<br>• Black (?) | Family history of cervical cancer | Use of DES by person's mother (exposure in utero) | NE |
| Pancreatic | M | > 60 | • Black<br>• Jewish | • Family history of pancreas, breast, colorectal cancer, or melanoma<br>• Inherited chronic pancreatitis | NE | Diabetes mellitus (?) |
| Liver | M | > 65 | • Black (?)<br>• Asian (?) | • Hemachromatosis<br>• Tyrosinemia<br>• Alpha-1-antitripsin deficiency<br>• Porphyria cutanea tarda<br>• Wilson's disease | NE | Diabetes mellitus (?) |

**Table 3.3** (continued)

*Part A*

| Cancer localization | Sex | Age, years | Race/ ethnic group | Family/personal disease history | Sex hormones/childbirth | Other |
|---|---|---|---|---|---|---|
| Oral/ oropharyngeal | M | > 65 | Black | Plummer–Vinson syndrome | NE | Use of immunosuppressive drugs |
| Gallbladder | F | > 70 | • Native Americans in Southwest<br>• Mexican American | • Family history of gallbladder cancer<br>• Anomalous pancreatobiliary duct junction and other abnormalities of the bile ducts | Association trough the female predominance in incidence | NE |
| Brain | M | • < 15<br>• > 60 | White | Family and/or personal history of neurofibromatosis type 2, tuberous sclerosis, Li–Fraumeni syndrome, von Hippel–Lindau disease, neurofibromatosis type I (von Recklinghausen disease) | NE | • Severe head trauma<br>• Ionizing radiation exposure (during the treatment of other cancers, as children leukemia, scalp ringworm)<br>• Living in metropolitan areas (?) |
| Kidney | M | > 45 | Black | • Family history of kidney cancer<br>• Family or personal history of von Hippel–Lindau disease<br>• Hereditary papillary renal cell carcinoma<br>• Birt–Hogg–Dube syndrome<br>• Hereditary leiomyomatosis renal cell carcinoma syndrome<br>• Hereditary renal oncocytoma | NE | NE |
| Urinary bladder | M | > 65 | White | • Family history of bladder cancer<br>• Personal history of extrophy (birth defect) | Early menopause (< 45 years) | NE |
| Melanoma | • F- higher incidence<br>• M – higher mortality | Any age | White | • Family history of melanoma<br>• Personal history of melanoma in anamnesis<br>• Multiple moles<br>• Atipical moles<br>• Giant congenital moles<br>• Xeroderma pigmentosum | Pregnancy (?) | • Fair skin<br>• Blue eyes<br>• Red hair<br>• High density of freckles<br>• Taking immunosuppressive drugs |

**Table 3.3** (continued)

*Part B*

| Cancer localization | Smoking | Alcohol | Obesity | Diet | Occupational exposure | Infection | Sex hormones | Other |
|---|---|---|---|---|---|---|---|---|
| Lung | + | +**** | NE | • High fat<br>• Low fiber | • Diesel exhaust<br>• Gasoline<br>• Beryllium<br>• Radon<br>• Uranium<br>• Polycyclic hydrocarbons<br>• Nickel chromates<br>• Mustard gas<br>• Chlormethyl ethers<br>• Inorganic arsenic<br>• Chromium<br>• Vinyl chloride<br>• Tars<br>• Silica<br>• Coke oven fumes<br>• Coal gasification<br>• Cadmium<br>• Asbestos | NE | NE | • Residential radon exposure<br>• Residential asbestos<br>• Arsenic in drinking water<br>• Marijuana (?)[3]<br>• Silicosis<br>• Berylliosis |
| Breast | + | + | • M: +<br>• F: + in menopause | • High saturated fat<br>• Flame-broiled food | • Polycyclic aromatic hydrocarbons<br>• Magnetic fields<br>• Metalworking fluids (?) | NE | • Use of postmenopausal hormones > 5 years<br>• Oral contraceptive use after age 45 | Magnetic fields |
| Prostate | + (?) | + (?) | + | • High fat<br>• Low fiber<br>• A lot of red meat<br>• A lot of fatty dairy products | • Cadmium<br>• Herbicides (atrazine) | NE | – | NE |
| Esophagus | + | + | + for adenocarcinoma | • Low fiber<br>• Low carotene<br>• Low ascorbic acid<br>• Low niacin<br>• Low riboflavin | Solvents used for dry cleaning | NE | NE | • Barrett's esophagus<br>• Long-standing GERD[6] |

**Table 3.3** (continued)

*Part B*

| Cancer localization | Smoking | Alcohol | Obesity | Diet | Occupational exposure | Infection | Sex hormones | Other |
|---|---|---|---|---|---|---|---|---|
| | | | | • Low zinc<br>• Low selenium<br>• Low thiamin<br>• Frequent drinking of very hot liquids | | | | |
| Stomach | + | + (?) | + for very obese | • Low fiber<br>• Pickled vegetables<br>• Smoked food<br>• Salted fish<br>• Cured meat | Exposure to certain dusts and fumes (?) | • *H. pylori*[7]<br>• Epstein–Barr virus (?) | NE | NE |
| Colorectal | + | +* | + (?) | • High saturated fat<br>• Low fiber<br>• A lot of red meat<br>• Low calcium<br>• Pickled vegetables<br>• A lot of eggs | • Asbestos (?)<br>• Chromium<br>• Oil mist | HPV[8](?) | NE | • Laxatives<br>• Constipation<br>• Physical inactivity<br>• Night-shift work (?)<br>• Asbestos (?) |
| Ovarian | + (?) | NE | + | A lot of milk (?) | • Diesel exhaust (?)<br>• Herbicides (atrazine) | *Chlamydia trachomatis* | Use of estrogen replacement therapy (estrogen alone) in anamnesis (?) | Prolonged use of clomiphene citrate |
| Corpus uteri | NE | NE | + | High fat | Magnetic fields (?) | NE | Use of postmenopausal hormonal therapy for > 10 years | • Tamoxifen<br>• Arterial hypertention (?) |
| Cervix (cervix uteri) | + | NE | + | Low fiber | • Aliphatic and alicyclic solvents<br>• Aromatic solvents<br>• Chlorinated hydrocarbons<br>• Silica dust<br>• Wood dust | • HPV<br>• Chlamidia trachomatis (?) | • Multiple birth<br>• Long-term oral contraceptive use | • Beginning sexual life at earlier age<br>• Many sexual partners<br>• Tamoxifen |
| Pancreatic | + | + (?) | + | • High saturated fat<br>• A lot of red meat<br>• Processed meat | • Certain pesticides<br>• Dyes (textile manufacture) | *H. pylori* (?) | NE | NE |

**Table 3.3** (continued)

*Part B*

| Cancer localization | Smoking | Alcohol | Obesity | Diet | Occupational exposure | Infection | Sex hormones | Other |
|---|---|---|---|---|---|---|---|---|
| | | | | • A lot of fatty dairy products<br>• Heavy coffee consumption (?) | • Gasoline<br>• Chromium<br>• Petroleum products<br>• DDT | | | |
| Liver | + | + | + (?) | Food contaminated with aflatoxin (peanuts, wheat, soybeans, ground nuts, corn, rice) | • Thorium dioxide (thorotrast)<br>• Vinyl chloride<br>• Arsenic | • Hepatitis B virus<br>• Hepatitis C virus | Long-term use of oral contraceptives | • Liver cirrhosis<br>• Long-term use of anabolic steroids (athletes)<br>• Arsenic in drinking water |
| Oral/oropharyngeal | + (including smokeless tobacco) | + | NE | • Low fiber<br>• Low zinc<br>• Low vitamin A<br>• A lot of processed food | • Formaldehyde<br>• Mustard gas | HPV | NE | • Poor oral hygiene<br>• Trauma due to ill-fitting dentures or jaded teeth |
| Gallbladder | NE | NE | + | • High fat<br>• Low fiber<br>• High carbohydrates | • Azotoluene<br>• Nitrosamines | Typhoid (*Salmonella typhi*) (?) | • Use of estrogen-containing drugs<br>• Pregnancy | • Gallstones<br>• "Porcelain" gallbladder<br>• Choledochal cysts<br>• Gallbladder polyps |
| Brain | + | + | NE | • Cured meat<br>• Aspartam (?) | Electromagnetic fields (?) | NE | NE | Electromagnetic fields (?) |
| Kidney | + | NE | + | NE | • Asbestos<br>• Cadmium<br>• Certain herbicides<br>• Benzene<br>• Organic solvents (trichloroethylene) | NE | NE | • Physical inactivity<br>• High blood pressure (?)<br>• Phenacetin<br>• Certain diuretics (?) |

**Table 3.3** (continued)

| Part B | | | | | | | | |
|---|---|---|---|---|---|---|---|---|
| Cancer localization | Smoking | Alcohol | Obesity | Diet | Occupational exposure | Infection | Sex hormones | Other |
| Urinary bladder | + | + | NE | • Heavy coffee consumption (?)<br>• Cyclamate (?) | • Aromatic amines<br>• Pain manufacture<br>• Rubber manufacture<br>• Drivers<br>• Leather manufacture<br>• Barbers<br>• Printing manufacture<br>• Textile manufacture<br>• Aluminium<br>• Benzidine<br>• 2-Naphthylamine | Schistosoma haematobium (?) | NE | • Phenacetin (long-term use)<br>•Chlornaphazine<br>•Cyclophosamide<br>• Arsenic in drinking water |
| Melanoma | NE | + (?) | + (?) | Low zinc (?) | Sunlight exposure | NE | Oral contraceptives (?) | • Sun exposure<br>• History of severe repeated sunburns in childhood |

[1]Female.
[2]No evidence that this cancer is associated with this risk factor.
[3]Factor needs further study about association with certain cancer.
[4]Male.
[5]Diethylstilbestrol.
[6]Gastroesophageal reflux disease.
[7]*Helicobacter pylori.*
[8]Human papillomavirus.
*Ashkenazi – the branch of European Jews who settled in central and northern Europe; **Component of strong household cleaners, as drain cleaners); ***Age at risk depends on how widespread screening with Pap smear in country; ****Evidence is limited to a high alcohol consumption.

#### 3.1.3.1 Results of Meta-analyses for Controllable Risk Factors

We summarized the results of meta-analyses and large population-based case-control studies of controllable risk factors of cancer. Table 3.4 presents controllable risk factors described as being associated with certain cancers. Table 3.5 presents more detailed analyses of relative risks and odds ratios. Smoking and diet patterns are the leading risk factors associated with the widest spectrum of cancers, compared to other controllable risk factors (see Table 3.4).

#### 3.1.3.2 Limitations of Meta-analyses

The procedure of selection of studies that are included in meta-analyses, interpretation of results of meta-analysis, and a subsequent discussion of results, and making conclusions about the association between risk factor and certain cancer, still has several unresolved problems. Meta-analysis (as defined by Last in 1995) combines similar trials in order to obtain a larger number of patients to improve the evaluation of whether the statistically reliable differences exist between groups being compared. However, meta-analyses during the past decade have received some strongly critical reviews (Lau et al., 1998; Villar et al., 1995; Cappelleri et al., 1996; LeLorier et al., 1997; Egger et al., 1995). It has been shown that some problems can't be completely resolved using meta-analyses' results:

*Heterogeneity bias.* Critics have opined that meta-analyses combined studies which are very heterogeneous and therefore unreliable (Eysenck, 1994). A meta-analysis should attempt to evaluate heterogeneity, rather than just ignore differences by pooling data (Lau et al., 1998).

*Publication bias.* This is a well-known fact that studies described statistically significant outcomes have a higher chance to get published than non-significant studies (Naylor, 1997). Additionally, small trials are less likely to be published as compared to large trials. Negative studies (i.e., studies with negative results) also take longer to appear in print. Papers published in languages other than English are more likely to be excluded from meta-analyses (Juni et al., 2002). These facts can make it difficult to pretend that absolutely all necessary information concerning the subject of a specific meta-analysis has been published and, even if published, has been found and included in the meta-analysis.

*Discrepancies with megatrials.* One of the reasons for the recent skepticism about the value of meta-analyses is the discrepancy between the results of meta-analyses and subsequent large randomized controlled trials. Lorier et al. (1997) found that meta-analyses would have led to the adoption of an ineffective treatment in 32% of the cases and rejection of a useful treatment in 33%. Several efforts to resolve these discrepancies have been attempted, using various models (Lau et al., 1998; Woods, 1995; Ioannidis et al., 1998; Egger et al., 1995).

Presently, potential areas for bias in a meta-analysis include (1) inclusion/exclusion criteria used to select studies for the meta-analysis; (2) methods used

**Table 3.4** Controllable risk factors of selected cancers, based on results of meta-analyses (grey blocks mean the presence of association)

| Cancer site | Lung | Leukemia | Colorectal | Stomach | Brain | Breast | Esophagus | Prostate | Oropharyngeal | Cervical | Pancreas | Liver |
|---|---|---|---|---|---|---|---|---|---|---|---|---|
| **Smoking** | + | + | | + | | + | + | | + | + | + | + |
| Passive smoking | + | | | | + | | | | | | | |
| **Alcohol** | −/+[2] | | +[2] | | | + | +[1] | | + | | | + |
| **Obesity** | | | | | | + | + | | | | + | |
| **Diet:** | | | + | | + | | | | | | | |
| Meat | − | | + | | | + | | | | | | |
| Dietary saturated fatty | | | | | | | | + | | | | |
| acids | | | | | | | | + | | | | |
| Dietary linolenic acid | | | + | | | | | | | | | |
| Dairy products and | | | | | | + | | | | | | |
| calcium intake | | | + | | | | | | | | | |
| Eggs | | | | | | | | | | | | |
| Dietary oleic acid | | | | | | | | | | | | |
| Pickled vegetables | | | | | | | | | | | | |
| **Hormones:** | | | | | | + | | | | + | | + |
| Oral contraceptives | | | | | | + | | | | | | |
| Menopausal hormone use | | | | | | | | | | | | |
| Magnetic fields | | + | | | | + | | | | | | |
| Poor oral hygiene/poor dental care | | | | | | | | | + | | | |
| Constipation | | | + | | | | | | | | | |

[1]In genetically susceptible individuals.
[2]Evidence is limited to a high alcohol consumption.

**Table 3.5** Controllable risk factors associated with cancer: results of meta-analyses and large population-based studies

| Risk factor | | Associated cancer | Factor-related RR[1] or OR[2] (with 95% CI[3]) | Studies included in meta-analysis | Reference |
|---|---|---|---|---|---|
| **Smoking** | Smoking | Adult leukemia | RR = 1.3 (1.3–1.4) | 7 prospective studies | Brownson et al. (1993) |
| | | | RR = 1.1 (1.0–1.2) | 8 case–control studies | |
| | | Hepatocellular carcinoma | RR = 1.9 (N/A) | 3 studies | Villa et al. (1991) |
| | | Esophageal carcinoma with p53 alteration | OR = 1.64 (1.13–2.37) | 14 studies | Wang et al. (2004) |
| | Cigarette smoking | Colorectal cancer | OR = 1.40 (1.10–1.77) | 14 case–control studies | Chen et al. (2003) |
| | Smoking: 11–20 cigarettes per day | Oral and oropharyngeal squamous cell carcinoma | OR = 2.4 (1.3–4.1) | Population-based case–control study | Rosenquist (2005) |
| | Ever-smokers:<br>• pooled<br>• in premenopausal<br>• early age of starting smoking | Breast cancer | RR = 1.10 (1.02–1.18)<br>RR = 1.21 (1.08–1.36)<br>RR = 1.14 (1.06–1.23) | Peer-reviewed studies from MEDLINE and Cancer Abstract (1980–2001) databases, published in 1984–2001 | Khuder et al. (2001) |
| | Current smokers | Oral and oropharyngeal squamous cell carcinoma | OR = 1.47 (1.15–1.88) | 6 case–control studies | Berrington de Gonzalez et al. (2004) |
| | | | OR =2.30 (1.31–4.04) | 10 case–control studies | Plummer et al. (2003) |
| | | Stomach cancer | RR =1.84 (1.39–2.43) | 2 population-based prospective cohort studies | Koizumi et al. (2004) |
| | | Lung cancer | RR =8.96 (6.73–12.11) | 177 case–control, 75 cohort and 2 nested case–control studies | Gandini et al. (2008) |
| | | Laryngeal cancer<br>Pharyngeal cancer<br>Oral cancer | RR =6.98 (3.14–15.52) | | |

**Table 3.5** (continued)

| Risk factor | | Associated cancer | Factor-related RR[1] or OR[2] (with 95% CI[3]) | Studies included in meta-analysis | Reference |
|---|---|---|---|---|---|
| | | Cancer of upper digestive tract | RR =6.76 (2.86–15.98)<br>RR =3.43 (2.37–4.94)<br>RR =3.57 (2.63–4.84) | | |
| | Ex-smokers | Oral and oropharyngeal squamous cell carcinoma | OR =1.80 (0.95–3.44) | 10 case–control studies | Plummer et al. (2003) |
| | | Stomach cancer | RR =1.77 (1.29–2.43) | 2 population-based prospective cohort studies | Koizumi et al. (2004) |
| | Passive smoking (paternal) | Childhood neoplasms | RR =1.10 (1.03–1.19) | 12 studies | Boffetta et al. (2000) |
| | | Children brain tumors | RR =1.22 (1.05–1.40) | 10 studies | |
| | | Children lymphomas | RR =2.08 (1.08–3.98) | 4 studies | |
| | Passive smoking (paternal smoking during pregnancy) | Childhood brain tumor | RR =1.29 (1.07–1.53) | 7 observational studies | Huncharek et al. (2001) |
| | Passive smoking (women) | Lung cancer | RR =1.29 (1.17–1.43) | 43 studies | Taylor et al. (2001) |
| **Alcohol** | Alcohol:<br>• dose ≥30 g/day in men<br>• dose ≥30 g/day in women<br>• dose ≥15 g/day in never smokers men<br>• smoking adjusted, dose 1000–1999 g/month<br>• smoking adjusted dose ≥2000 g/month | Lung cancer | RR =1.21 (0.91–1.61);<br>RR =1.16 (0.94–1.43)<br>RR =6.38 (2.74–14.9)<br>RR =1.04 (0.88–1.22)<br>RR =1.53 (1.04–2.25) | 7 prospective studies<br><br><br>4 studies | Freudenheim et al. (2005)<br><br><br>Korte et al. (2002) |
| | Alcohol:<br>• in ALDH2*1*1 homozygotes[4] | Esophageal cancer | OR =3.19 (1.86–5.47) | 7 studies | Lewis and Smith |

**Table 3.5** (continued)

| Risk factor | | Associated cancer | Factor-related RR[1] or OR[2] (with 95% CI[3]) | Studies included in meta-analysis | Reference |
|---|---|---|---|---|---|
| | • in ALDH2*2*2 homozygotes | | OR =0.36 (0.16–0.80) | | (2005) |
| | Alcohol:<br>• 30–45 g/day<br>• >45 g/day<br>• beer<br>• wine<br>• liquor | Colorectal cancer | RR =1.16 (0.99–1.36)<br>RR =1.41 (1.16–1.72)<br><br><br>RR =1.26 (1.13–1.41)<br>RR =1.11 (0.91–1.36)<br>RR =1.13 (0.99–1.29) | 8 cohort studies<br><br><br><br>27 studies | Cho et al. (2004)<br><br><br><br>Longnecker et al. (1990) |
| | Alcohol: >80 g/day | Hepatocellular carcinoma | RR =1.65 (N/A) | 3 studies | Villa et al. (1991) |
| | Alcohol: women, 12 g/day in average | Breast cancer | RR =1.10 (1.06–1.14) | 42 studies | Ellison et al. (2001) |
| **Hormones** | Oral contraceptives use<br>• ≤8 years of use<br>• >8 years of use | Hepatocellular carcinoma | RR =2.8 (N/A)<br>RR =9.9 (N/A) | 3 studies | Villa et al. (1991) |
| | Oral contraceptive use<br>• <45 years old<br>• nulliparous women<br>• >8 years of use | Breast cancer | RR =1.16 (1.07–1.25)<br>RR =1.21 (0.99–1.47)<br>RR =1.27 (1.12–1.44) | 27 epidemiological studies | Rushton and Jones (1992) |
| | Menopausal hormone therapy:<br>• estrogen therapy<br>• estrogen/progestin therapy | Ovarian cancer | RR =1.28 (1.18–1.40)<br>RR =1.11 (1.02–1.21) | 42 studies | Greiser et al. (2007) |
| **Obesity:** | • weight at birth > 4000 g | Childhood leukemia: | RR =1.26 (1.08–1.20) | 18 studies | Hjalgrim et al. (2003) |

**Table 3.5** (continued)

| Risk factor | Associated cancer | Factor-related RR[1] or OR[2] (with 95% CI[3]) | Studies included in meta-analysis | Reference |
|---|---|---|---|---|
| | • acute lymphoblastic leukemia<br>• acute myeloid leukemia | OR =1.27 (0.73–2.20) | | |
| • weight at birth > 4000 g<br>• weight at birth <2500 g<br>• weight at birth <2500 g | Testicular cancer<br>Seminoma | OR =1.12 (1.02–1.22)<br>OR =1.18 (1.01–1.38)<br>OR =1.44 (1.11–1.88) | 13 studies | Michos et al. (2007) |
| • BMI>30 kg/m² vs. BMI[5] = 25–30 kg/m² | Esophageal adenocarcinoma | OR =1.52 (1.15–2.01)<br>OR =2.78 (1.85–4.16) | 7 studies | Hampel et al. (2005) |
| • Risk increases with increasing waist circumference (per 10-cm increase)<br>• Risk increases with increasing waist–hip ratio (per 0.1-unit increase) | Colon cancer | Male: RR =1.33 (1.19–1.49)<br>Female: RR =1.16 (1.09–1.23)<br>Male: RR =1.43 (1.19–1.71)<br>Female: RR =1.20 (1.08–1.33) | 30 prospective studies | Larsson and Wolk (2007a) |
| • Overweight vs. normal body weight<br>• Obese vs. normal body weight | Liver cancer | RR =1.17 (1.02–1.34)<br>RR =1.89 (1.51–2.36) | 11 cohort studies | Larsson and Wolk (2007b) |
| • BMI≥30 kg/m² vs. BMI[5]<25 kg/m² | Colorectal cancer | Male: RR =1.41 (1.30–1.54)<br>Female: 1.RR =08 (0.98–1.18) | 23 cohort and 8 case–control studies | Moghaddam et al. (2007) |

**Table 3.5** (continued)

| Risk factor | | Associated cancer | Factor-related RR[1] or OR[2] (with 95% CI[3]) | Studies included in meta-analysis | Reference |
|---|---|---|---|---|---|
| | • Highest vs. lowest quantiles of BMI<br>• Highest vs. lowest quantiles of waist circumference<br>Highest vs. lowest quantiles of waist–hip ratio | Colon cancer | Male: RR =1.59 (1.35–1.86)<br>Female: RR =1.22 (1.08–1.39) | PubMed, EMBASE, and Cochrane Library search | Dai et al. (2007) |
| | | Rectal cancer | Male: RR =1.16 (0.93–1.46)<br>Female: RR =1.23 (0.98–1.54) | | |
| | | Colon cancer | Male: RR =1.68 (1.36–2.08)<br>Female: RR =1.48 (1.19–1.84) | | |
| | | Rectal cancer | Male: RR =1.26 (0.90–1.77)<br>Female: RR =1.23 (0.81–1.86) | | |
| | | Colon cancer | Male: RR =1.91 (1.46–2.49)<br>Female: RR =1.49 (1.23–1.81) | | |
| | | Rectal cancer | Male: RR =1.93 (1.19–3.13)<br>Female: RR =1.20 (0.81–1.78) | | |
| **Magnetic fields** | Magnetic fields:<br>• magnetic level 0.2–0.5 μT<br>• magnetic level >0.5 μT | Childhood leukemia | RR =2.0 (1.0–4.1)<br>RR =5.1 (2.1–12.6) | 2 case–control studies | Feychting et al. (1995) |
| | Magnetic fields | Children leukemia<br>Children lymphoma | RR =1.49 (1.11–2.0)<br>RR =1.58 (0.91–2.76) | 13 epidemiological studies | Washburn et al. (1994) |

**Table 3.5** (continued)

| Risk factor | | Associated cancer | Factor-related RR[1] or OR[2] (with 95% CI[3]) | Studies included in meta-analysis | Reference |
|---|---|---|---|---|---|
| | | Children nervous system tumors | RR =1.89 (1.34–2.67) | | |
| | Magnetic field:<br>• women<br>• men | Breast cancer | RR =1.12 (1.09–1.15)<br>RR =1.37 (1.11–1.71) | 43 studies | Erren (2001) |
| **Flights*** | Flights:<br>• female flight attendants<br>• male cabin attendants<br>• civil pilots<br>• military pilots | Melanoma<br>Breast cancer<br>Melanoma<br>Other skin cancer<br>Melanoma<br>Other skin cancers<br>Prostate cancer<br>Melanoma<br>Other skin cancers | SIR[6] = 2.15 (1.56–2.88)<br>SIR =1.40 (1.19–1.65)<br>SIR =3.42 (1.94–6.06);<br>SIR =7.46 (3.52–15.89)<br>SIR =2.18 (1.69–2.80);<br>SIR =1.88 (1.23–2.88)<br>SIR =1.47 (1.06–2.05)<br>SIR =1.43 (1.09–1.87)<br>SIR =1.80 (1.25–2.58) | 7 studies | Buja et al. (2005; 2006) |
| | Flights: in female flight attendant | Melanoma of skin<br>Breast cancer | RR =2.13 (1.58–2.88)<br>RR =1.41 (1.22–1.62) | 5 cohort studies | Tokumaru et al. (2006) |

**Table 3.5** (continued)

| Risk factor | | Associated cancer | Factor-related RR[1] or OR[2] (with 95% CI[3]) | Studies included in meta-analysis | Reference |
|---|---|---|---|---|---|
| **Diet** | Fat intake: in women in the highest quintile of energy-adjusted total fat intake compared with women in their lowest quintile | Breast cancer | RR =1.05 (0.94–1.16) | 7 prospective studies | Hunter et al. (1996) |
| | Fat intake | Colorectal cancer | OR =3.16 (2.22–4.51) | 14 case–control studies | Chen et al. (2003) |
| | Fat and cholesterol intake<br>• total fat<br>• saturated fat<br>• monounsaturated fat<br>• polyunsaturated fat<br>• dietary cholesterol 100 mg/day | Lung cancer | RR =1.01 (0.98–1.05)<br>RR =1.03 (0.96–1.11)<br>RR =1.01 (0.93–1.10)<br>RR =0.99 (0.90–1.10)<br>RR =1.01 (0.97–1.05) | 8 prospective studies | Smith-Warner et al. (2002) |
| | Dairy products and calcium:<br>• highest vs. lowest dairy products intake<br>• highest vs. lowest calcium intake | Prostate cancer | RR =1.33 (1.00–1.78)<br>RR =1.46 (0.65–3.25) | 12 studies | Gao et al. (2005) |
| | • milk consumption | OR =1.68 (1.34–2.12) | 11 case–control studies | Qin et al. (2004) | |
| | • calcium intake >2000 mg/day | Increased risk | 2 cohort studies | Dagnelie et al. (2004) | |
| | • highest vs. lowest quantile of consumption | RR =1.13 (1.02–1.24) | 13 studies | Qin et al. (2007) | |
| | Dairy products | Breast cancer | No association | 8 prospective cohort studies | Missmer et al. (2002) |
| | Pickled vegetables | Colorectal cancer | OR =1.86 (1.67–2.07) | 14 case–control studies | Chen et al. (2003) |
| | Dietary ALA[7]: high intake | Prostate cancer | RR =1.70 (1.12–2.58) | 9 cohort and case–control studies | Brouwer et al. (2004) |

**Table 3.5** (continued)

| Risk factor | | Associated cancer | Factor-related RR[1] or OR[2] (with 95% CI[3]) | Studies included in meta-analysis | Reference |
|---|---|---|---|---|---|
| | Meat:<br>• red meat – highest quantile vs. lowest<br>• processed meat<br>• red and white meat | Colorectal cancer | RR =1.35 (1.21–1.51)<br>RR =1.31 (1.13–1.51) | 34 case–control and 14 cohort studies | Norat et al. (2002) |
| | Breast cancer | No association | 8 prospective cohort studies | Missmer et al. (2002) | |
| | Cured meat**:<br>• in children of women consumed cured meat during pregnancy<br>• bacon<br>• ham | Childhood brain tumor | RR =1.68 (1.30–2.17) | 6 observational studies | Huncharek and Kupelnik (2004) |
| | | Adult glioma | RR =1.31 (1.0–1.71)<br>RR =1.64 (1.27–2.14) | 9 observational studies | Huncharek et al. (2003) |
| **Occupational** | Pesticides exposure for >10 years | Non-Hodgkin lymphoma | OR =1.65 (1.08–2.51) | 13 studies | Merhi et al. (2007) |
| | | Leukemia and multiple myeloma | OR =2.18 (1.43–3.35) | | |
| | Night-shift work*** | Breast cancer | RR =1.48 (1.36–1.61) | 13 studies | Megdal et al. (2005) |
| **Stress:**<br>• maternal death in childhood<br>• chronic depression with severe episodes<br>• severe emotional trauma | | Breast cancer | OR =2.56 ($p<0.001$)<br>OR =14.0 ($p<0.001$) | Baltimore Epidemiologic Catchment Area Study | Jacobs and Bovasso (2000) |
| | | Colorectal cancer | OR =2.95 (2.81–3.09) | 14 case–control studies | Chen et al. (2003) |

**Table 3.5** (continued)

| Risk factor | | Associated cancer | Factor-related RR[1] or OR[2] (with 95% CI[3]) | Studies included in meta-analysis | Reference |
|---|---|---|---|---|---|
| **GI* anamnesis** | Cholecystectomy:<br>• pooled<br>• in women<br>• right-side cancer localization | Colorectal cancer | OR =1.11 (1.02–1.21)<br>OR =1.14 (1.01–1.28)<br>OR =1.86 (1.31–2.65) | 35 studies | Reid et al. (1996) |
| Constipation | Colorectal cancer | OR =1.48 (1.31–1.66) | 14 case–control studies | Sonnenberg and Muller (1993) | |
| **Cosmetic/ hygiene** | Hair dyes | Hematopoietic cancer | RR =1.15 (1.05–1.27) | 40 studies | Takkouche et al. (2005) |
| | Sunscreens use:<br>• latitude: at >40 degrees from equator | Melanoma of skin | OR =1.6 (1.3–1.9) | 17 studies | Gorham et al. (2007) |

[1]Relative risk.
[2]Odds ratio.
[3]Confidence interval.
[4]Aldehyde dehydrogenase.
[5]Body-mass index.
[6]Meta-standardized incidence ratio.
[7]Alpha-linolenic acid – the (*n*–3)-fatty acid in vegetable oil.
*Complex of factors as exposure to cosmic ionizing radiation, chemicals, including cabin air pollutants, electromagnetic fields from cockpit instruments, disrupted sleep patterns; **Cured meat is an important source of dietary *N*-nitroso-compound; ***Night-shift work is a surrogate for exposure to light at night with subsequent melatonin suppression; GI – gastroenterological.

to perform the meta-analysis; (3) conclusions which are reached; (4) statements by the authors, regarding the reliability of the results of their meta-analysis; (5) declarations of broad applicability for the conclusions of a particular meta-analysis.

Several bias indicators have recently been suggested: (1) Egger et al. (1997) proposed a test for asymmetry of the funnel plot, (2) Begg and Mazumdar (1994) proposed testing the interdependence of variance and effect size using Kendall's method (Kendall, 1990)[however, with many studies in the meta-analysis, the Begg method has very low power to detect biases (Sterne et al., 2000)], (3) Harbord with coauthors (2005) have developed a test that maintains the power of the Egger test by reducing the false positive rate. Some other statistical methods can be used to investigate the effects of study characteristics other than sample size upon effects (Sterne et al., 2002). Meta-analysis is focused on mean effects and differences between studies, but what really matters are the effects on individuals. It will be expedient to quantify individual responses as a standard deviation (i.e., a second-order moment), which itself can be meta-analyzed. To predict the individual responses, such characteristics as age, gender, genotype, and other need to be known. The current approach of using mean subject characteristics as covariates in the analysis does not seem to work well.

Alternative approaches to meta-analysis involve empirical Bayesian regression procedures, where the effective parameters are assumed to be distributed. In this case, the parameters of the model include not only the mean of the effect but also the effect variability over individuals. For example, Manton et al. (1989) used a nested negative binomial model. In this negative model, the individual event rates were assumed to be generated from a heterogeneous population, where the individual Poisson event rates were assumed to have a gamma distribution, conditional on the measured distribution of the risk factors. Adjustment of such models for bias in the selection of studies requires the use of methods for dealing with informative missing data. Adjustments can be made by noting that the likelihood of measurement of the effects in an analysis is systematically correlated with factors of interest in the analysis. The terms, representing the interaction due to the data being missing, and the value of data, that is not missing, should be included in the model.

#### 3.1.3.3 Surrounded by Cancer Risk Factors: Rumors or Facts?

During our everyday activity at home, in office, and even on vacations, we often can be exposed to various cancer risk factors. Nearly all of civilization's innovations were blamed at least once for being associated with one or more cancers. Below we present some information about human "casual life" risk factors, which have been discussed recently in both, scientific journals and nonscientific mass media.

*Oral health and alcohol-containing mouthwashes.* Eight studies have been analyzed; the lowest OR for oral cancer was 0.87, and the highest OR was

1.26; at present, no relationship between the use of alcohol-containing mouthwashes and the development of oral cancer has been established. Future studies should take into consideration the evaluation of the metabolic susceptibility genes (e.g., alcohol dehydrogenase genotype), as well as genetic and molecular markers of tobacco metabolism as concurrent risk factor of oral cancer (Carretero-Pelaez et al., 2004).

*Oral health and the allergy to metal dental restorations.* Contact allergy to metal dental restorations (gold, mercury, silver, copper) may be a risk factor for development of intraoral squamous cell carcinoma (Hougeir et al., 2006).

*Antiperspirants/deodorants.* Certain preservatives (such as parabens) that are used in deodorants and antiperspirants, as well as in many cosmetic and pharmaceutical products and food, can mimic the activity of estrogen in human cells (Harvey and Everett, 2004). Because estrogen promotes the growth of breast cancer cells, it has been suggested that the use of antiperspirants or deodorants could cause the accumulation of parabens in breast tissue, and that may contribute to the development of breast cancer. A study by McGrath (2003) found that the age of breast cancer diagnosis was significantly lower in women who used antiperspirants/deodorants and shaved their underarms more frequently, but this study does not demonstrate a conclusive link between these underarm hygiene habits and breast cancer. Darbre and coauthors (2004) found parabens in tissues from human breast tumors. However, the National Cancer Institute is not aware of any conclusive evidence linking the use of underarm antiperspirants or deodorants and the subsequent development of breast cancer (see: National Cancer Institute, Antiperspirants/Deodorants and Breast Cancer). Additional research is needed to investigate this relationship and other factors that may be involved.

*Smoking intensity and habit duration.* A large case–control study of lung cancer showed that smoking at a lower intensity for longer period is more deleterious than smoking at a higher intensity for shorter duration: the excess OR per pack-year increases with intensity for subjects who smoke $\leq 20$ cigarettes per day, and it decreases with intensity for subjects who smoke more than 20 cigarettes per day (Lubin and Caporaso, 2006).

*Radon exposure in buildings.* Radon, a radioactive gas released from the normal decay of uranium in rocks and soil, can enter homes through cracks in floors, walls, or foundations, and can be collected indoors, as well as being released from building materials, or from water obtained from wells that contain radon. In the United States, radon is the second-ranked (after tobacco) cause of lung cancer: approximately 15,000–22,000 lung cancer deaths per year are related to radon exposure (see: National Cancer Institute, Radon and Cancer).

*Cell phones.* There are no long-term studies of the effects of radiofrequency (RF) energy from cellular phones on humans. RF energy produces heat, which can increase local body temperature and damage organs/body parts exposed to it. A cellular phone's main source of RF energy is its antenna, so hands-free kits may help reduce the amount of RF energy exposure to the head. Plus, the distance of cellular phone from the base station antenna determines the amount of RF energy exposure to the user (depending on the higher power level). Studies funded by Wireless Technology Research LLC and the National Cancer Institute performed in 1994–1998 showed that the use of hand-held cellular phones was unrelated to the risk of brain cancer, but additional studies covering longer periods of observation were recommended (Muscat et al., 2000). Studies performed for the INTERPHONE project showed no increased risk for acoustic neuroma, meningioma, and glioma in long-term (10 years) cellular phone users compared to short-term users (Lonn et al., 2004; Christensen et al., 2005). But brain tumors develop over many years, so scientists have been unable to follow cellular phone users consistently for the amount of time it might take for a brain tumor to develop (Ahlbom et al., 2004). It is possible that children may be at the greatest risk of RF health effects because their nervous systems are still developing; if RF energy from cellular phones is proved to cause cancer, researchers would expect children to be more susceptible than adults. Further research is needed to determine what effects, if any, RF energy has on the human health, and whether it is dangerous (see: National Cancer Institute, Cellular Telephone Use and Cancer).

*Menopausal hormone therapy*. The Women's Health Initiative (WHI) estrogen-plus-progestin study concluded that hormonal therapy increases the risk of breast cancer: after 5 years of follow-up, it has been shown that risk increased up to 26% compared with women who took placebo (Rossouw et al., 2002). In the Million Women Study, it has been shown that current use of estrogen, or estrogen-plus-progestin, significantly increased the risk of breast cancer in women aged 50–64 (Beral, 2003). It has been also shown that women who used estrogen for 10 years or more were twice as likely to develop ovarian cancer compared with women who did not use menopausal hormones (Lacey et al., 2002).

### 3.1.4 Advanced Age as Cancer Risk Factor

Advanced age is one of the most important cancer risk factors: about 76% of all cancers were diagnosed in persons aged 55 and older. In the United States, the lifetime risk to develop cancer is 1 in 2 for men and 1 in 3 for women (Cancer Facts and Figures, 2006), and this dramatic age-dependent escalation in cancer risk is fueled largely by a marked increase in epithelial carcinomas from ages 40 to 80, as opposed to cancers of mesenchymal or hematopoietic origin (DePinho, 2000). It has been suggested that older people may have cancer-prone phenotype due to mutations accumulated throughout their lives, increased probability of epigenetic silencing, higher rate of telomere dysfunction, and alterations of stromal structures caused by existing chronic disorders and diseases.

The observation that rates of spontaneous mutations [the spontaneous mutation rate in cultured human cells is approximately $2 \times 10^{-7}$ per gene per cell division, so, each cell would accumulate only a few mutations over a lifetime (Jackson and Loeb, 1998)] seem to be insufficient for the extensive tumor-associated genomic changes led to the concept of the "mutator phenotype" which is the result of mutations in genes which themselves govern a stability of human genome (Loeb, 1991; 2001). It has been suggested that the probability of this phenotype may increase with age when inactivation of "governing" genes due to mutations leads to an accelerated rate of mutations overall and increased cancer risk. Although mutator and clonal expansion mechanisms are likely to contribute to the increase in cancer as a function of age, these mechanisms have not yet been documented in the context of aging tissues and fail to provide a unifying principle that accounts for this tissue distribution and cytogenetic profiles of most adult cancers.

An alternative estimation of mutation rates has been proposed. When mutation rates are calculated not based on the number of cell divisions but on the person's biological age, it has been suggested that this approach could closer estimate the number of mutations presumed necessary for initiation of tumorigenesis (Turker, 1998). Tissue-specific differences in mutational rates were revealed (e.g., intestine and liver > heart > brain), as well as radically different mutational spectra in these aged tissues, the latter pointing to probably organ-specific differences in genome maintenance mechanisms (Vijg and Dolle, 2001).

The epigenetic mechanisms that predominantly operate on the levels of DNA methylation and chromatin structure can modulate (i.e., change activity) various genes, and these mechanisms are likely age-dependent and tissue-specific (DePinho, 2000). A recent study using the telomerase-knockout mouse has indicated that differences in the telomere length and regulation might impact dramatically on both the spectrum and cytogenetics of tumors during aging (Artandi et al., 2000). The cytogenetic and telomerase data from human epithelial cancers, particularly breast and colon cancers, make the telomere–carcinoma connection an intriguing concept.

So, cancer risk the oldest group of population (aged 85 and older) seems to differ from those at younger ages. More detailed discussion of this phenomena could be found in Section 2.3.6 and Section 3.1.5.

### *3.1.5 Factors to Consider in Cancer Risk Analysis and Cancer Risk Prediction*

There are a number of sources of errors and biases in low-dose risk analysis and prediction, some of which can be partially addressed through adjustment factors or through statistical modeling:

*Age at exposure and observation.* Age can impact cancer risk in different ways. (1) There can be inherent differences in susceptibility at different ages resulting, e.g., from the changes that tissues undergo through the course of development, including differences in cell proliferation rates, hormone responsiveness, immunological activity, and development and maturation of enzyme systems that activate or detoxify chemicals. (2) There is also the issue of timing or latency: an individual exposed early in life simply has a longer period for the damage to be expressed. (3) Differences in exposure are associated with food consumption patterns, behavioral factors, and physiological differences. The significantly greater breast cancer susceptibility of teenage and prepubescent girls compared with adults exposed to ionizing radiation has been noted for atomic bomb survivors (Tokunaga et al., 1994) and for patients treated for cancer (Bhatia et al., 1998) and ankylosing spondilitis (Nekolla et al., 1999). The substantially greater risk associated with exposure during childhood compared to adulthood has been described for other tumors (e.g., for thyroid carcinoma). This effect explains two major factors: the greater number of remaining years of life and increasing spontaneous cancer initiation rate with age (Zeise, 2001) (more detailed discussion of that subject could be find in Chapter 7). The multistage model provides a basic framework for mathematically modeling age-dependent carcinogenesis, taking into account that early exposure is correlated with greater remaining years of life. When the first stage of two-stage/multistage model of carcinogenesis was dose-dependent and initiated cells proliferated rapidly, cancer risk could be considerably underestimated by using average dose as the dose metric (Murdoch and Krewski, 1988). The results may differ substantially with the model chosen for predicting age-dependent risk from single exposure: therefore, predictions of lung cancer risk at ages 70 and 100 from ionizing radiation exposure in survivors of the atomic bomb detonations differed significantly, depending on the model used (Kai et al., 1997; National Research Council BEIR V Committee, 1990).

*Pharmacokinetics*. The effects of metabolic pathways, that factors/substances are undergoing, can be taken into account through physiologically based pharmacokinetic (PBPK) modeling. The aim of this model in cancer risk prediction is to obtain a better estimate of effective dose than the administrated dose, using model parameters measured in, or estimated from experiments (Zeise, 2001). Use of PBPK modeling in cancer risk prediction requires the identification of the activation pathway associated with carcinogenesis, and of a kinetic model (e.g., Michaelis–Menten).

*Human heterogeneity*. Individuals differ in their susceptibility to cancer because of genetic, environmental, occupational, and lifestyle factors. A variety of genetic disorders have been identified as conferring substantially greater risk for certain types of cancer (e.g., familial retinoblastoma, ataxia telangiectasia, xeroderma pigmentosum), but these syndromes explain only a small fraction of human cancers (Zeise, 2001). Large interindividual differences in activities of a number of cytochrome P450 enzymes, glutathione-*S*-transferase, and other enzymes involved in carcinogen activation and detoxification have been described. The difference also may be due to race, gender, and exposure to inducers. A PBPK model was used in one study to quantify the formation of proximate carcinogen (*N*-hydroxy-4-aminobiphenyl) and DNA binding in the bladder: metabolic parameters for N-oxidation and N-acetylation and for physiological factors, including urine pH and frequency of urination, were varied via Monte Carlo simulation (Bois et al., 1995). Another approach to describe human heterogeneity, a hierarchical Bayesian modeling approach, was applied in a PBPK modeling framework to data in healthy human volunteers exposed to perchloroethylene (Bois, 1999).

*Influence of background exposures.* The degree that an exposure adds to and modulates background processes is important in assessing risk at low doses. In cases where a low-dose nonlinear mode of action is suspected, background exposures may be sufficiently large to move certain subgroups into a level at which risk is roughly proportional to dose (Zeise, 2001). Estimates of background exposure are difficult to achieve, but a screening-level analysis may be helpful in providing a qualitative finding as to whether one should proceed with a nonlinear dose–response assumption.

## 3.2 Environmental Cancer Risk Factors

The term "environment" refers not only to air, water, and soil but also to substances and conditions at home and at the workplace, including diet, smoking, alcohol, drugs, exposure to chemicals, sunlight, ionizing radiation, electromagnetic fields, infectious agents, etc. Lifestyle, economic and behavioral factors are all aspects of our environment. Human cancer risk is influenced not only by factors which are present in our environment but also by factors that are absent from our environment. People are exposed to a wide variety of environmental factors for varying times. These factors may interact in ways that are still not fully understood. Also individuals differ by their susceptibility to these factors.

The roles of genetic constitution and environmental exposure in the causation of cancer have been debated for long time. Nearly 30 years ago, the widely accepted estimate came that 80–90% of human cancers are due to environmental risk factors (Higginson and Muir, 1977, 1979). The tricky question was always how to distinguish the contributions of genetic predisposition and of environmental factors in cancer risk. The gold standard for making that distribution was the study of monozygotic (MZ) and dizygotic (DZ) twins: comparing the incidence of disease in unrelated people, fraternal (DZ) twins and identical twins (MZ) allow estimates to be made in the heritable and environmental components of risk. Findings in twin studies in the United States and Europe suggest that inherited predisposition does not explain a large proportion of either the risk of specific cancers or cancer mortality (Braun et al., 1995; Carmelli et al., 1996). A report of breast and ovarian cancer in only one of the two identical twins who both had an inherited BRCA1 mutation clearly illustrates the possibility of differences in phenotypic expression of such mutations (Diez et al., 1997). In a study by Buckley et al. (1996), the concordance for all cancers combined (excluding leukemia and retinoblastoma) was 1.2% overall and 2.2% in monozygotic twins, indicating greater environmental, than genetic, influences (concordance for leukemia in monozygotic twins was 5%).

Cancer prevalence may be analyzed in immigrants from areas with specific cancer rates. For example, the rates of breast cancer among recent female immigrants to the United States from rural Asia are similar to those in their homeland, and about 80% lower than the rates among third-generation Asian-American women, who have rates similar to or even higher than those among white women in the United States (Ziegler et al., 1993). This pattern is consistent with a study where 73% of breast cancer causation is environmental and 27% heritable, particularly if a portion of the effect of heritable factors relates to genetic modification of environmental risk factors (Lichtenstein et al., 2000). Several aspects are important while analyzing genetic and environmental factors in the causation and control of cancer: (1) information about types of environmental exposure that affects the risk of cancer should point to genes that might modify this risk, and the identification of genes associated with risk could help to indicate previously unrecognized environmental risk factors; (2) when genes and environment interact to produce a risk greater than the sum of their independent effects, this interactive component can be eliminated by removing either the genetic or the environmental factor (Hoover, 2000).

Most of the known and probable/potential environmental human carcinogens are presented in Table 3.6.

Some of the environmental cancer risk factors are discussed here and in Chapter 8 (where the microsimulation approach and risk factors interactions are discussed).

### 3.2.1 Radiation Exposure

Radiation exposure is obviously not one of the most widespread cancer risks (i.e., compared to smoking) (see Table 3.7). Why do we pay our attention to this risk factor, especially ionizing radiation (IR) exposure, in this chapter? At present, radiation remains one of the most mysterious risk factors for both cancer and noncancer diseases. Its role is likely to increase as nuclear energy production increases. Many difficulties exist in directly studying the effects of IR exposure, especially low doses, on human health in "real-time scenario".

Humans may be exposed to several types of radiations, and different types of exposure may be associated with different types of cancer. The IR (it is subdivided into two types: electromagnetic – X-rays and gamma-rays, and particulate – alpha-particles, beta-particles, neutrons, protons) is mostly associated with solid cancers and leukemia. UV radiation is a known skin carcinogen. Long-wave length radiation showed no clear evidence of carcinogenicity at present. Microwave radiation needs further research to confirm its carcinogenicity for the central nervous system. Humans may be exposed to various sources of radiation: radon gas from the ground – 50% (of all radiation sources human exposed), gamma rays from the ground and buildings – 14%, medical – 14%, internal – 11.5%, cosmic rays – 10%, occupational – 0.3%, fallout – 0.2%,

**Table 3.6** Carcinogenic and potentially carcinogenic for human agents, mixtures, and exposure circumstances (based on data from the IARC Monographs, A; B)

| | Carcinogenic to humans (Group 1) | Probably carcinogenic to humans (Group 2A) |
|---|---|---|
| Agents and groups of agents | • Aflatoxins (naturally occurring mixtures of)<br>• 4-Aminobiphenyl<br>• Arsenic and arsenic compounds (Note: This evaluation applies to the group of compounds as a whole and not necessarily to all individual compounds within the group)<br>• Asbestos<br>• Azathioprine<br>• Benzene<br>• Benzidine<br>• Beryllium and beryllium compounds<br>• *N,N*-Bis(2-chloroethyl)-2-naphthylamine (chlomaphazine)<br>• Bis(chloromethyl)ether and chloromethyl methyl ether (technical-grade)<br>• 1,4-Butanediol dimethanesulfonate (busulphan; Myleran)<br>• Cadmium and cadmium compounds<br>• Chlorambucil<br>• 1-(2-Chloroethyl)-3-(4-methylcyclohexyl)-1-nitrosourea (methyl-CCNU; semustine)<br>• Chromium [VI] compounds<br>• Cyclophosphamide<br>• Cyclosporin (ciclosporin)<br>• Diethylstilbestrol (DES)<br>• Epstein–Barr virus<br>• Erionite<br>• Estrogen therapy, postmenopausal<br>• Estrogens, nonsteroidal (Note: This evaluation applies to the group of compounds as a whole and not necessarily to all individual compounds within the group)<br>• Estrogens, steroidal (Note: This evaluation applies to the group of compounds as a whole and not necessarily to all individual compounds within the group)<br>• Ethylene oxide | • Acrylamide<br>• Adriamycin<br>• Androgenic (anabolic) steroids<br>• Aristolochic acids (naturally occurring mixtures of)<br>• Azacitidine<br>• Benx[a]anthracene<br>• Benzidine-based dyes<br>• Benzo[a]pyrene<br>• Bischloroethyl nitrosoutea (BCNU)<br>• 1,3-Butadiene<br>• Captafol<br>• Chloramphenicol<br>• A-Chlorinated toluenes (benzal chloride, benzotrichloride, benzyl chloride) and benzoyl chloride (combined exposure)<br>• 1-(2-Chloroethyl)-3-cyclohexyl-1-nitrosourea (CCNU)<br>• 4-Chloro-ortho-toluidine<br>• Chlorozotocin<br>• Cisplatin<br>• *Clonorchis sinensis* (infection with)<br>• Dibenz[a,h]anthracene<br>• Diethyl sulfate<br>• Dimethylcarbamoyl chloride<br>• 1,2-Dimethylhydrazine<br>• Dimethyl sulfate<br>• Epichlorohydrin<br>• Ethylene dibromide<br>• *N*-Ethyl-*N*-nitrosourea<br>• Etoposide<br>• Glycidol<br>• Human papillomavirus type 31<br>• Human papillomavirus type 33<br>• Indium phosphide<br>• IQ (2-amino-3methylimidazo[4,5-f]quinoline)<br>• Kaposi's sarcoma herpervirus/human herpesvirus B (KSHV/HHV-8)<br>• Lead compounds, inorganic<br>• 5-Methoxypsoralen<br>• 4,4-Methylene bis(2-chloroaniline) (MOCA)<br>• Methyl methanesulfonate |

**Table 3.6** (continued)

| | Carcinogenic to humans (Group 1) | Probably carcinogenic to humans (Group 2A) |
|---|---|---|
| | • Etoposide in combination with cisplatin and bleomycin | • *N*-Methyl-N'-nitro-N-nitrosoguanidine (MNNG) |
| | • Formaldehyde | • *N*-Methyl-*N*-nitrosourea |
| | • Gallium arsenate | • Nitrogen mustard |
| | • Gamma radiation | • *N*-Nitrosodiethylamine |
| | • *Helicobacter pylori* (infection with) | • *N*-Nitrosodimethylamine |
| | • Hepatitis B virus (chronic infection with) | • Phenacetin |
| | • Hepatitis C virus (chronic infection with) | • Procarbazine hydrochloride |
| | • Herbal remedies containing plant species of the genus *Aristolochia* | • Styrene-7,8-oxide |
| | • Human immunodeficiency virus type 1 (infection with) | • Teniposide |
| | • Human papillomavirus type 16 | • Tetrachloroethylene |
| | • Human papillomavirus type 18 | • *ortho*-Toluidine |
| | • Human T-cell lymphotropic virus type I | • Trichloroethylene |
| | • Melphalan | • 1,2,3-Trichloropropane |
| | • 8-Methoxypsoralen (methoxsalen) plus ultraviolet A radiation | • Tris(2,3-dibromopropyl)phosphate |
| | • MOPP and other combined chemotherapy including alkylating agents | • Ultraviolet radiation A |
| | • Mustard gas (sulfur mustard) | • Ultraviolet radiation B |
| | • 2-Naphthylamine | • Ultraviolet radiation C |
| | • Neutrons | • Vinyl bromide |
| | • Nickel compounds | • Vinyl fluoride |
| | • *Opisthorchis viverrini* (infection with) | |
| | • Oral contraceptives, combined (Note: There is also conclusive evidence that these agents have a protective effect against cancers of the ovary and endometrium) | |
| | • Oral contraceptives, sequential | |
| | • Phosphorus-32, as phosphate | |
| | • Plutonium-239 and its decay products (may contain plutonium-240 and other isotopes), as aerosols | |
| | • Radioiodines, short-lived isotopes, including iodine-131, from atomic reactor accidents and nuclear weapon detonation (exposure during childhood) | |
| | • Radionuclides, alpha-particle-emitting, internally deposited | |

Table 3.6 (continued)

| | Carcinogenic to humans (Group 1) | Probably carcinogenic to humans (Group 2A) |
|---|---|---|
| | (Note: Specific radionuclides for which there is sufficient evidence for carcinogeneity to humans are also listed individually as Group 1 agents)<br>• Radionuclides, beta-particle emitting, internally deposited (Note: Specific radionuclides for which there is sufficient evidence for carcinogenity to humans are also listed individually as Group 1 agents)<br>• Radium-224 and its decay products<br>• Radium-226 and its decay products<br>• Radium-228 and its decay products<br>• Radon-222 and its decay products<br>• *Schistosoma haematobium* (infection with)<br>• Silica, crystalline (inhaled in the form of quartz or cristobalite from occupational sources)<br>• Solar radiation<br>• Talc-containing asbestiform fibers<br>• Tamoxifen (Note: There is also conclusive evidence that this agent (tamoxifen) reduces the risk of contralateral breast cancer)<br>• 2,3,7,8-Tetrachlorodibenzo-paradioxin<br>• Thiotepa<br>• Thorium-232 and its decay products, administered intravenously as a colloidal dispersion of thorium-232 dioxide<br>• Treosulfan<br>• Vinyl chloride<br>• X- and Gamma radiation | |
| Mixtures | • Alcoholic beverages<br>• Analgesic mixtures containing phenacetin<br>• Areca nut<br>• Betelquid with tobacco | • Creosotes (from coal-tars)<br>• Diesel engine exhaust<br>• Hot mate (drink like tea which is made using leaves of South American holly) |

**Table 3.6** (continued)

| | Carcinogenic to humans (Group 1) | Probably carcinogenic to humans (Group 2A) |
|---|---|---|
| | • Betel quid without tobacco<br>• Coal-tar pitches<br>• Coal-tars<br>• Mineral oils, untreated and mildly treated<br>• Salted fish (Chinese style)<br>• Shale oils<br>• Soots<br>• Tobacco products, smokeless<br>• Wood dust | • Non-arsenical insecticides (occupational exposure in spraying and application of )<br>• Polychlorinated biphenyls |
| Exposure circumstances | • Aluminum production<br>• Arsenic in drinking water<br>• Auramine, manufacture of<br>• Boot and shoe manufacture and repair<br>• Coal gasification<br>• Coke production<br>• Furniture and cabinet making<br>• Hematite mining (underground) with exposure to radon<br>• Involuntary smoking<br>• Iron and steel founding<br>• Isopropanol manufacture (strong-acid process)<br>• Magenta, manufacture of<br>• Painter (occupational exposure as a)<br>• Rubber industry<br>• Strong inorganic acid mists containing sulfuric acid (occupational exposure to)<br>• Tobacco smoking | • Art glass, glass containers and pressed ware (manufacture of)<br>• Cobalt metal with tungsten carbide<br>• Hairdresser or barber (occupational exposure as a)<br>• Petroleum refining (occupational exposures in)<br>• Sunlamps and sunbeds (use of) |

discharges – <0.1%, other products – <0.1% ("Living with radiation", 1998). These percentages reflect an average situation, but for certain groups/populations at certain time, these percentages may change dramatically (e.g., in case of a civil or military accident, or a terrorist attack), and large human populations may be exposed in short time, externally and internally. In Table 3.8, the graduation of the average incorporated doses is presented for various exposed groups, including occupational exposure and exposure resulting from nuclear accidents.

#### 3.2.1.1 Ionizing Radiation and Cancer Risk

Ionizing radiation is an environmental risk factor for cancer that several decades ago was meant to be controllable, but consequences of nuclear power plant/ reactor accidents, such as the Chernobyl nuclear power plant accident, Windscale

**Table 3.7** Causes of cancer in the United States: estimated percentage of total cancer deaths attributable to established causes of cancer. (From the Harvard Reports on Cancer Prevention, 1996)

| Risk factor | Percentage |
|---|---|
| Tobacco | 30 |
| Adult obesity | 30 |
| Sedentary lifestyle | 5 |
| Occupational factors | 5 |
| Family history of cancer | 5 |
| Viruses/other biologic agents | 5 |
| Perinatal factors/growth | 5 |
| Reproductive factors | 3 |
| Alcohol | 3 |
| Socioeconomic status | 3 |
| Environmental pollution | 2 |
| Ionizing/ultraviolet radiation | 2 |
| Prescription drugs/medicine procedures | 1 |
| Salt/other food additives/contaminants | $< 1$ |

(Sellefield), Three-Mile Island, nuclear weapon tests (Semipalatinsk, Marshalls Islands, Nevada), military action (Hiroshima and Nagasaki atomic bombing), large environmental releases ("Mayak", Hanford), orphan IR sources (Goiania, Taiwan), and transportation accidents (Palomares, Thule), as well as the recent threatening of terrorists' attacks with involvement of nuclear facilities, showed that IR in some cases is a risk factor that is hard to control. It is difficult to control IR from the position of preventive measures. This is why it is so important to have a strategy of how to manage the consequences, short- and long-term, of IR accidents, including cancer and noncancer diseases (e.g., cardiovascular and cerebrovascular disease, CNS disorders, cataract).

The current estimates of cancer risk from exposure to IR in humans are most often based on epidemiological studies of the exposed atomic bomb survivors of Hiroshima and Nagasaki. This approach has provided relatively reliable estimates of risk for high doses and high-dose rate exposes, yet it is the effect of low doses and low-dose rates that is of major importance for the general population. Risk estimates for low doses are usually calculated by extrapolations from existing high-dose data (Rothkamm and Lobrich, 2003). This model assumes that cellular responses, including DNA repair, operate equally efficiently at low and high IR doses. The most biologically significant IR-induced damages that generally believed cause cancer and noncancer hereditary diseases are DNA double-strand breaks (DSBs). DSBs are considered to be the most relevant lesion for the deleterious effects of IR, and a single radiation track can produce this kind of damage. Surprisingly, DSBs induced in cultures of nondividing primary human fibroblasts by low radiation doses ($\approx$1 mGy) remain unrepaired for many days, in strong contrast to the efficient DSB repair that is observed at higher doses. If the cells are allowed to proliferate after irradiation, the level of

**Table 3.8** Some comparative radiation doses and their effects (from The World Nuclear Association, 2004)

| | |
|---|---|
| 2 mSv[1]/year | Typical background radiation experienced by everyone (1.5 mSv in Australia, 3 mSv in North America). |
| 1.5–2.0 mSv/year | Average dose to Australian uranium miners, above background and medical. |
| 2.4 mSv/year | Average dose to US nuclear industry employees. |
| up to 5 mSv/year | Typical incremental dose for aircrew in middle latitudes. |
| 9 mSv/year | Exposure by airline crew flying New York–Tokyo polar route. |
| 10 mSv/year | Maximum actual dose to Australian uranium miners. |
| 20 mSv/year | Current limit (averaged) for nuclear industry employees and uranium miners. |
| 50 mSv/year | Former routine limit for nuclear industry employees. It is also the dose rate which arises from natural background levels in several places in Iran, India, and Europe. |
| 100 mSv/year | Lowest level at which any increase in cancer is clearly evident(*). Above this, the probability of cancer occurrence (rather than the severity) increases with dose. |
| 350 mSv/lifetime[2] | Criterion for relocating people after Chernobyl accident. |
| 1000 mSv/ cumulative | Would probably cause a fatal cancer many years later in 5 of every 100 persons exposed to it (i.e., if the normal incidence of fatal cancer was 25%, this dose would increase it to 30%). |
| 1000 mSv/single dose | Causes (temporary) radiation sickness such as nausea and decreased white blood cell count, but not death. Above this, severity of illness increases with dose. |
| 5000 mSv/single dose | Would kill about half those receiving it within a month. |
| 10,000 mSv/single dose | Fatal within a few weeks. |

*The lowest limit of "cancer-safe" IR dose is still under the study in large human populations exposed to IR, with long-term follow-up.

[1]Sv (Sievert) – the equivalent dose; that is equal to "absorbed dose" multiplied by a "radiation weighting factor" (this factor depends on the type of IR and energy range).

[2]The Soviet National Committee on Radiation Protection (NCRP) proposed this lifetime dose for the relocation of population groups from the areas polluted by Chernobyl nuclear plant accident. This value was lower by a factor of 2–3 than recommended by the International Commission on Radiological Protection (ICRP) for the same countermeasure. The NCRP proposal was not adopted by the Supreme Soviet. A special Commission was established later, and it developed new recommendations, based on the levels of ground contamination by Cs-137, Sr-90, and Pu-239. As a result, the level of 1480 kBq/m$^2$ (40 Ci/km$^2$), based on Cs-137 level, was used as the criterion for permanent resettlement of population, and of 555–1480 kBq/m$^2$ (15–40 Ci/km$^2$) for temporary relocation (NEA Report, 2002).

DSBs in them decreases to the level of unirradiated cells, and cells with unrepaired DSBs are eliminated (Rothkamm and Lobrich, 2003).

Not only nuclear DNA but also mtDNA has been proposed to be involved in IR induced carcinogenesis, with its high susceptibility to mutations and limited repair mechanisms in comparison to nuclear DNA. Because of lack of introns in the mtDNA, it is likely that most mutations will occur in coding regions, and accumulation of these mutations may lead to cancer formation

(Penta et al., 2001). Gene mutations may not be necessary at all, if IR alters protein production (e.g., by structural disruption of the Golgi apparatus or endoplasmic reticulum) (Bennett et al., 2001).

The two-mutation model has been shown to be a more effective predictor of cancer risk than linear interpolation. However, both the mechanisms of radiation action and the cellular changes that lead to malignancy need a clear understanding to be used in the development of reliable radiobiological models. Biophysical studies strongly favor "no" threshold models, since they suggest that DSBs in DNA are potentially inducible by a single electron in a cell (Dendy and Brugmans, 2003). While talking about IR exposure effects, it is necessary to mention the phenomenon that may increase the complexity of cancers caused by IR exposure – bystander effect, a biological response in cells that do not themselves receive any energy deposition from IR, but which responds to signals produced by cells that do (Mothersill and Seymur, 2001). This effect to date has only been demonstrated *in vitro*, and most convincingly with alpha particles [it is not known if this is a general effect with all IR exposure, or, at low doses, it is restricted to high linear-energy-transfer (LET) radiation, or whether the effect is significant for *in vivo* systems] (Dendy and Brugmans, 2003). The bystander effect may have implications for extrapolation models of low-dose radiation risk, but till now it is not clear if that may reduce the low-dose radiation risk due to extra cell killing, or that may increase the risk due to, e.g., enhanced gene mutations and cell transformations in cells that would be killed at higher doses (Ballarino and Ottolenghi, 2002).

Presently, the Life Span Study (LSS) of atomic bomb survivors in Hiroshima and Nagasaki has the longest follow-up of more than 60 years. In this cohort, leukemia showed the earliest increase in incidence. Analysis of site-specific excess relative risks (ERRs) for solid cancers in 1950–1997 showed the highest ERR/Sv for bladder, breast, and esophagus cancers (1.25, 1.0, and 0.95, respectively), and lower ERR/Sv for such solid cancers, as stomach, colon, rectum, liver, gall bladder, pancreas, lung, uterus, ovary, and prostate (Preston et al., 2003). But results of LSS cohort cannot answer all questions about an association of cancer (and noncancer) health risks in human because they represent only one type of IR exposure. Other studies, e.g., the "Mayak" plant accident, Chernobyl nuclear power plant accident, provide information about low-dose IR exposure effects, combination of internal and external IR exposure, with a spectrum of radionuclides, and without other factors (e.g., thermal, chemical), compared with LSS.

Figures 3.2, 3.3, and 3.5 present several models of the virtual scenarios of the consequences of terrorist attacks using "dirty bomb" scenarios in Washington, DC and New York (Kelly, 2002): the rings on these pictures are showing the forecasted number of cancer deaths in certain exposed population groups, and Fig. 3.4 presents the modeled contaminated areas which are compared with the "real-life" scenario from contaminated Chernobyl accident areas in the Ukraine, Belarus, and Russia. While radiological attack would result in some deaths, it would not result in the hundreds of thousands of fatalities that could be caused by a crude nuclear weapon, such as an A-bomb, but it could

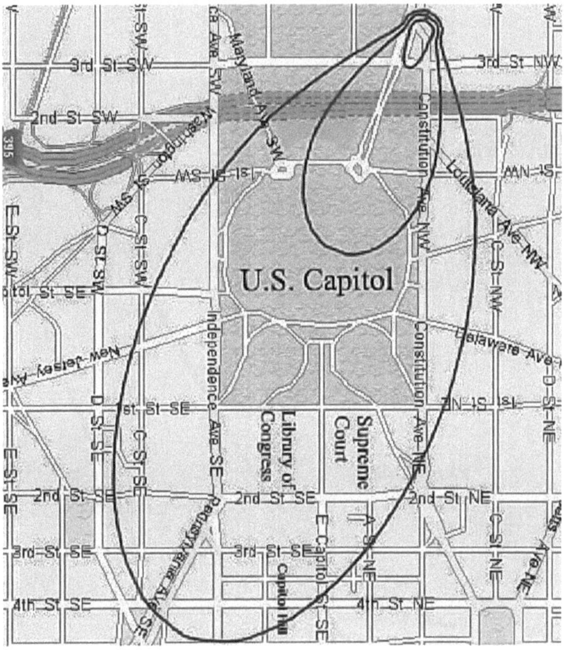


Fig. 3.2 Long-term contamination due to cesium bomb in Washington, DC (From Kelly, 2002)

contaminate large urban areas with populations exposed to IR levels, which cause predominantly low-dose health consequences (uncertainties inherent in the complex models used in predicting the effects of a radiological weapon mean that it is only possible to make crude estimates of impact).

#### 3.2.1.2 Ionizing Radiation and Thyroid Cancer

Here, we will describe IR as risk factor for thyroid cancer in more detail because

(1) IR is the only currently well-established risk factor for thyroid carcinoma. Several studies showed the increase of thyroid cancer incidence in populations exposed to IR (Ron et al., 1995; Heidenreich et al., 1999; Gilbert et al., 1998; Kodama et al., 1996; Ivanov et al., 1997).
(2) The thyroid gland exhibits a high degree of sensitivity to both external and internal ionizing irradiations.
(3) Thyroid cancer was the first solid tumor found to have a significantly increased incidence among Japanese A-bomb survivors (Wood et al., 1969).
(4) According to the WHO 2006 report (EGH, 2006) of the health effects of the Chernobyl nuclear power plant accident, thyroid cancer is the only cancer proved to have occurred as a consequence of this accident (but that does not eliminate the necessity of future studies of other cancers exposed in this accident population with long-term follow-up).

**Fig. 3.3** Long-term contamination due to cobalt bomb in NYC – EPA Standards (From Kelly, 2002)

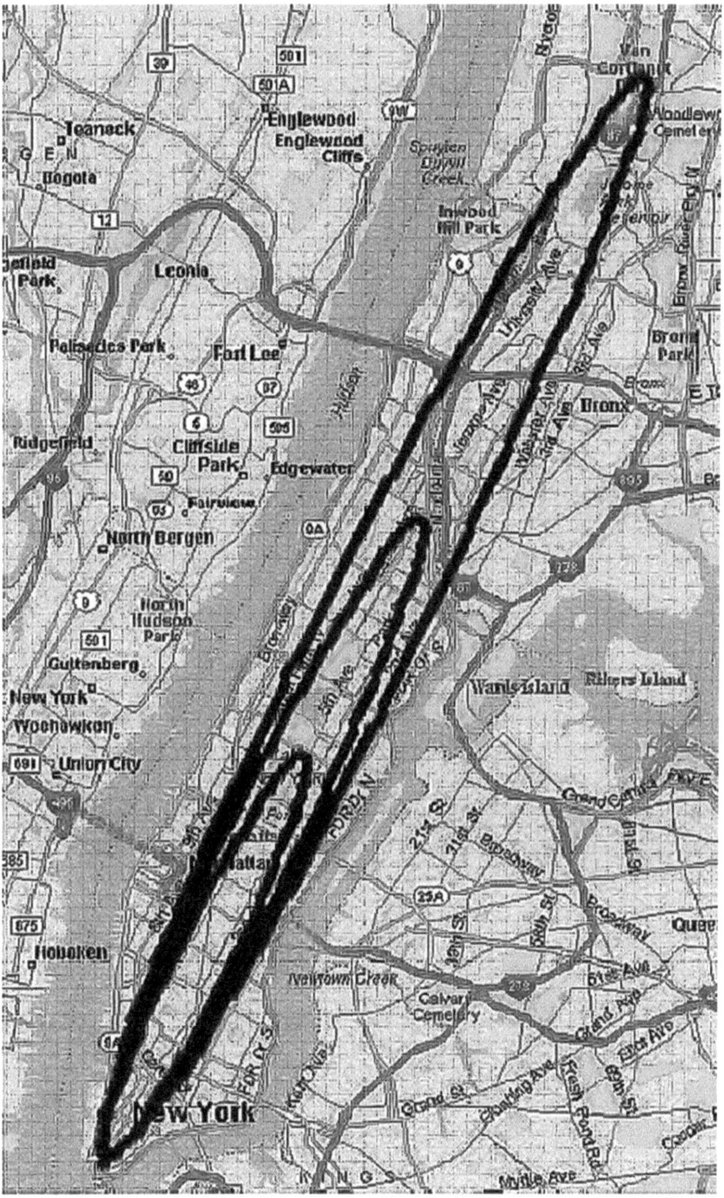


**Fig. 3.4** Contamination due to cobalt bomb in NYC – compared to Chernobyl accident (From Kelly, 2002)

Thyroid cancer has recently manifested increases in its global incidences. In the United States, incidence rates increased approximately from 4.8 per 100,000 in 1973 to 9.1 per 100,000 in 2005 (in white females, the incidence rate in 2005 was 14.1) (SEER, 1999; SEER, 2005). A preclinical form of thyroid carcinoma might be not so rare in humans: it has been described as carcinoma *in situ* in approximately 10% of all autopsy specimens (Bisi et al., 1989). After years of latency, there may be a sudden spontaneous transformation of thyroid carcinoma into an aggressive tumor with the development of widespread metastases. The mechanisms of thyroid carcinoma, responsible for its onset and growth, are still unknown, as well as which additional to IR risk factors may be important at different ages in males and females. Thyroid carcinoma is not a very easy cancer to analyze and model. Since the rate of spontaneous thyroid cancer increases with age, although not as steeply as some other types of cancer, it is important for the purpose of projecting lifetime risk to determine whether a relative risk model (i.e., one which multiplies the background age-specific rates), or an absolute risk model (which adds a constant increment at all ages) is more appropriate. A longer period of observation in the population helps reduce these difficulties. The following is important to take into account while modeling thyroid carcinogenesis: different age and sex patterns of thyroid carcinoma risks, variations in latent period, different pathomorphologic forms of thyroid carcinoma (two most prevalent are papillar and follicular thyroid carcinomas), and areas of iodine endemicity/deficiency. Below we discuss some of these parameters in our study of thyroid carcinoma incidence in a large population exposed to low-dose internal IR as consequence of Chernobyl nuclear power plant accident in 1986.

The IR risk of thyroid carcinoma in studies with large numbers of cases has been quantitatively analyzed for external irradiation, but not for biologically incorporated radionuclide exposure. The risk associated with exposure to radioactive iodine ($^{131}$I) is important in public health because of the medical uses of radioactive iodine and also because the radioactive iodine is potentially one of the quantitatively more important releases of radionuclides from nuclear facility accidents. Most studies of internal IR exposure have generally been based on studies of small numbers of persons medically exposed to iodine radionuclides. The Chernobyl nuclear power plant accident caused the exposure of a large population to iodine radionuclides resulting in contamination. That produced mostly internal IR exposure due to biological incorporation of $^{131}$I in populations which now have been followed for more than 20 years. We studied the population of 5.17 million residents of four oblasts of Russia, contaminated with iodine radionuclides resulting from the Chernobyl accident, with a total of 4650 thyroid-confirmed cancer cases diagnosed in 1982–2000 (Ivanov et al., 2005). The procedure for dose estimation due to internal radionuclides exposure was developed in 2000 (Ramzaev et al., 2000). Three types of radiological data were used to reconstruct the IR dose to the thyroid gland: (1) the direct measurements of the level of $^{131}$I in residents' thyroid glands; (2) the measurements of the $^{131}$I concentration in milk consumed by the local population (Balonov et al., 2002); and (3) if neither measurements were available, thyroid doses were estimated using statistical models relating $^{131}$I dose to $^{137}$Cs contamination levels, measured in small population areas – rayons (small administrative areas like US counties).

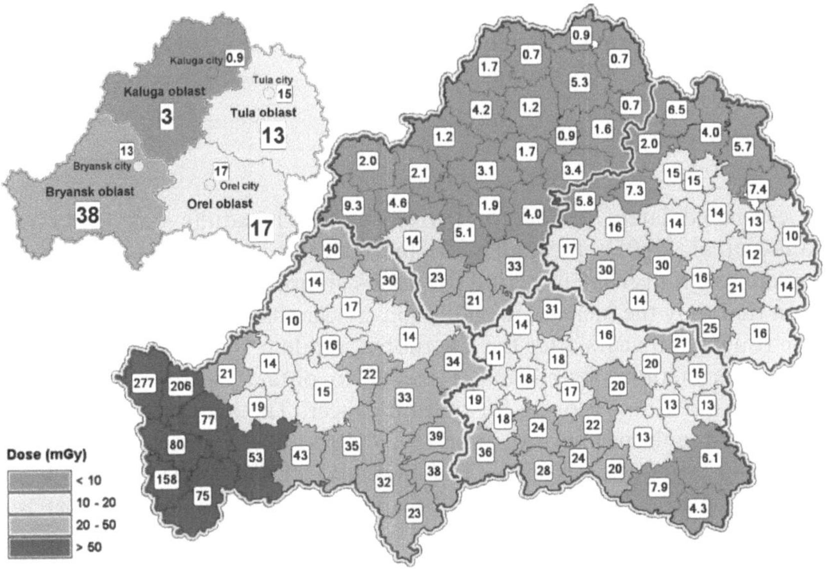


**Fig. 3.5** Map of distribution of mean-rayon thyroid doses in the Bryansk, Kaluga, Orel, and Tula oblasts (Ivanov et al., 2005)

Figure 3.5 presents the estimates of mean-rayon thyroid doses for four oblasts (the administrative unit which is consisted of rayons).

## Age-Dependent Prevalence

A statistically significant risk of thyroid cancer has been found in all age groups. The number of persons with diagnosed thyroid carcinoma in 1991–2000 increased 4-fold, as compared to 1982–1990 (pre-exposure and first 4 years after exposure), with the most significant increase in 10–14-year-old children (age at the time of exposure; we didn't analyze this ratio for younger children because of the small number of diagnosed thyroid cancer cases at those ages) (Ivanov et al., 2005). The question about the difference of the effect of IR exposure on children and adult thyroids is still under discussion. In Ukraine, the most highly affected group was children aged $\leq 5$ years old at exposure (Tronko et al., 1999). A trend for decreasing risk with increasing age at exposure has frequently (Shore, 1992; Thompson et al., 1994; Ron et al., 1989), but not always (Fjalling et al., 1986), been reported. One of the possible explanations for the critical importance of age at exposure to IR for thyroid cancer may be related to the differences in radiation dose between children and adults (other possible underlying reasons of increased risk of thyroid cancer will be discussed below in Section Age–Sex Differences in Thyroid Cancer Risk). Children's thyroid glands have high sensitivity to IR due to high iodine uptake by the thyroid, high mitotic rate, and high remaining growth potential. Several estimates and direct measurements concluded that absorbed IR doses after

Chernobyl accident were 3–10 times higher in children than in adults, and were greater in younger, rather than older children (Ilyin et al., 1990; Gavrilin et al., 1992; Castronovo, 1987; Malone et al., 1991).

### Latent Period

The estimated latent period for thyroid cancer in adults was 13–14 years (Ivanov et al., 2005). In studies of thyroid cancer risk in case of internal exposure, the longer latent period for adults compared to children may be the reason that only the increase in childhood cancers have shown up significantly in the roughly 10 years after the Chernobyl accident (Leenhouts et al., 2000). Continuous lifetime observation is needed to obtain information about the latency period after exposure for thyroid cancer. The highest risk for thyroid cancer in A-bomb survivors was seen 15–29 years after exposure, and it was still elevated 40 years after exposure (UNSCEAR, 2000). The length of the risk persistence period is still unknown because no population has not yet been followed throughout its lifetime (it has been shown in several studies that an excess risk in IR exposed groups was observed after 50 years or even more) (Shore, 1989).

### Age–Sex Differences in Thyroid Cancer Risk

Our statistical analysis was based on a general excess-relative risk model, for which linear dependence on dose is assumed. The region of relatively small doses observed in this study is very important because the results can shed light on the complicated problem of estimation of the fraction of tumors induced by small radiation doses. This is why we preferred to use nonparametric methods and to calculate dose trend, rather than excess of relative risk per 1 Sv. Improved analytic sensitivity is possible, using Bayesian principles, on the basis of biologically motivated and mathematically consistent models, such as empirical Bayes, quadratic hazard, and/or stochastic process models (Manton et al., 1989, 1992; Manton and Yashin, 2000; Manton and Stallard, 1988) (detailed description of each of these models is found in Chapter 5).

According to a majority of studies, thyroid cancer shows different patterns in young children, adolescents, and adults: while in adults it has a significantly higher prevalence in females, in young children disease prevalence does not differ significantly in girls and boys (Harach and Williams, 1995; Ron et al., 1987; Manole et al., 2001). Our results agree with that pattern: 6-times higher thyroid cancer prevalence in females was found compared to males in pre-exposed and latent periods, as well as more than a 5-times higher thyroid cancer prevalence for females compared to males in the post-latent period. There were no sex differences in thyroid carcinoma prevalence in children. However, not all researchers share the opinion that females are more susceptible to thyroid cancer than males (Jaklic et al., 1995; Thorvaldsson, 1992).

Our data showed that not only thyroid cancer prevalence in population exposed to internal $^{131}$I irradiation is sex-dependent but the ERR for thyroid cancer also differs by gender: the ERR was significantly higher in females during their reproductive period (12–50 years old), but ERR for the postmenopausal females did not differ from the ERR for males (Ivanov et al., 2005). More detailed analysis of age dependence of ERR allows us to reveal two peaks in females: at ages 13–17 and 44–48. These ages are the average ages of menarche and peri-menopause periods in Central Russia (Iampol'skaia, 1997; Balan, 1995). Immune system functioning is known to be strongly dependent on sex hormones (Pacifici et al., 1989; Deswal et al., 2001; Nguyen et al., 2003). During the periods of age-associated sex hormone fluctuation, a so-called cytokine storm may occur, causing an increased risk of immune-relevant disorders. Thyroid carcinoma may be contained by the immune-surveillance system involving host or tumor factors, and that either a secondary neoplastic event, or the breakdown of normal immune function is necessary for its progression to a clinically significant state (Baker and Fosso, 1993). Kingsmore and Patel (2003) showed that many cytokines are strongly expressed immediately before adolescence: the levels of many of 78 cytokines increased in serum between ages 9 and 13. Estradiol is involved in the regulation of cytokine production, including IL-1 (Pacifici et al., 1989) and TNF (Deswal et al., 2001). Cytokines can modulate the expression of the ecto-nucleotide pyrophosphate/phosphodiesterase family member autotoxin, which plays a regulator's role in thyroid carcinoma cell motility (Kehlen et al., 2004), also the cytokines play an important role in host defense against malignant progression (Baker and Fosso, 1993) and in the regulation of the growth of thyroid cells and their immunological functions (Lahat Sheinfeld et al., 1992). Thus, we can suppose that the two ERR age peaks in females may be at least partly explained by the way the thyroid gland may express a secondary neoplastic event after the incorporated $^{131}$I irradiation, coincident with a hormonal shift (puberty and menopause) associated with "cytokines storm" – as a result, these events resulting in thyroid carcinoma (we suggested that this mechanism might be involved in carcinogenesis of other cancer sites – some details are discussed in Chapter 7).

The male also has age-dependent hormonal changes during puberty and climacterium, and these changes, as in females, may influence the male's thyroid susceptibility to IR-induced carcinogenesis, but the lower incidence of diagnosed thyroid cancers in males and often "subclinical" preclimacteric/climacteric conditions create an additional difficulty in statistical analysis of age–sex associated changes in thyroid susceptibility to IR in males (we have also observed certain trends at puberty and climacteric periods in males – that finding required further detailed analysis).

These results are preliminary due to the limited follow-up period relative to the latencies usually found for solid tumors. The follow-up should be continued to check the assumption about the peak incidence of children's thyroid cancer being over. Forthcoming efforts among others may be directed toward (a) the detailed sensitivity analysis with special attention to dose reconstruction methods and screening effect; (b) the mathematical modeling intended to use indirect information (e.g., oblasts endemic for iodine deficiency, papillary/follicular thyroid cancer ratio, etc.); and (c) analysis of large male population to confirm or reject associated with the certain age periods increases in thyroid gland's susceptibility to IR exposure, etc.

### *3.2.2 Nutrition as a Cancer Risk Factor*

Although at the cellular level, cancer is often recognized as a disease of genes, there is good epidemiological evidence that it is strongly modulated by environmental factors such as diet. For example, after Japan adopted a westernized diet, the Japanese population, having possibly increased susceptibility to it, had rates of colon cancer increase rapidly, being low in 1960, but now exceeding European rates (Bingham, 2005). Dietary factors are estimated to account for approximately 30% of cancers in industrialized countries (making diet a second only to tobacco as a theoretically preventable cause of cancer), and in developing countries, dietary factors are associated with almost 20% of cancers (Junien and Gallou, 2004). Results of meta-analyses demonstrated that diet is associated with increased risk of colorectal, breast, brain, and prostate cancers (see Tables 3.3, 3.4, and 3.5) (the role diet plays in cancer prevention is discussed in Chapter 9). Several large studies were dedicated to studies of the nutritional effects on human health, including the role of diet as a cancer risk factor. The European Prospective Investigation Into Cancer (EPIC) study is the largest prospective study ever undertaken to specifically investigate the link between diet and cancer. This project began in 1992, and almost 4000 cases of breast cancers and over 1000 colorectal cancers have been registered. In this cohort (including more than 521,000 persons from 10 European countries), it has been estimated that red and processed meat intakes were associated with an increased risk of gastric noncardia cancer, especially in *Helicobacter pylori* antibody-positive subjects (Gonzalez et al., 2006). The other study in EPIC cohort showed that high consumption of red meat increases the risk of colorectal cancer. Several hypotheses have been proposed to explain this association, among which the most strongly supported is that red meat increases the endogenous formation of carcinogenic *N*-nitroso-compounds in the gut (Bingham, 2005). The other large study takes place in the United States – the National Health and Nutrition Examination Survey (NHANES). The Epidemiologic Follow-up Study (NHEFS) was designed to investigate the relationship between clinical, nutritional, and behavioral factors assessed in the first NHANES, and subsequent morbidity, mortality, and hospital utilization, as well as changes in risk factors, functional limitation, and institutionalization.

Dietary factors can modify cancer risk in several different ways at multiple stages of the carcinogenic process with both genetic and epigenetic mechanisms involved. (1) Direct carcinogens, or initiators, cause structural damage or malfunction of the genes that regulate cell proliferation, DNA repair, the survival of damaged cells, and the ability of cells to invade and migrate into close proximity and distant tissue/organs (they may occur naturally or be produced during cooking, digestion, or the metabolism of certain foods: e.g., aflatoxin is a carcinogenic fungus that occurs naturally in moldy grains; heterocyclic amines are produced by frying meat and fat at high temperatures; acrylamide is generated from carbohydrates when cooking foods such as French fries and pancakes; polycyclic aromatic hydrocarbons are produced by grilling and charring foods; nitrosamines are generated in the stomach during digestion). (2) Dietary factors promote tumor development (tumor promoters) by accelerating cell turnover, so that genetically damaged cells multiply more rapidly and have greater likelihood of acquiring additional mutations needed for malignant transformation. They include diverse chemical classes, such as phorbol ester derivates, non-TPA-type tumor promoters, chlorinated hydrocarbons from industrial or agricultural sources, alcohol, salt, and others: e.g., an increased ingestion of fats enhancing tumor promotion in experiments in skin, breast, colon, and liver. (3) Dietary components which can improve cellular defense mechanisms (e.g., the bioactive compounds found in plants are able to increase expression/induction of crucial detoxification enzymes, as glutathione synthetase, glutathione transferase, and glucuronyl transferase, resulting in decreased bioavailability of potentially DNA-damaging carcinogens) (Minamoto et al., 1999) (the third group of dietary component is discussed in Chapter 9). Many hypothetical and actual ways of how dietary factors may increase cancer risk have been described, and most are still relevant: (1) ingestion of powerful, directly acting carcinogens or their precursors, including carcinogens in natural foods (e.g., pyrrolizidine alkaloids in *Senecio* plant, safrole in sassafras, bracken fern), carcinogens produced in cooking (e.g., benzo[a]pyrene and other polycyclic hydrocarbons while cooking meat or fish by broiling, smoking, or frying reusing fat), carcinogens produced in stored food by microorganisms (e.g., aflatoxin produced by *Aspergillus flavus* in peanuts); (2) affected carcinogen formation in the body, including providing substrates for the carcinogen formation (e.g., nitrites, nitrates, secondary amines), altering intake or excretion of cholesterol and bile acids, and altering the bacterial flora of the bowel; (3) affected transport, activation and/or deactivation of carcinogens, including altering concentration in feces or duration of contact with feces, induction or inhibition of enzymes which affect carcinogen metabolism, and deactivation or prevention of formation of short-lived intracellular species; (4) effects on "promotion" of cells, including vitamin A deficiency, retinol-binding protein, and other factors affecting stem cell differentiation; (5) overnutrition, including effects on age at menarche, adipose-tissue-derived estrogens and others (Doll and Peto, 1981).

Hypotheses and supporting evidence relating dietary factors to cancer can be obtained from the following: (1) *in vitro* studies, and animal experiments, are helpful in directing human research and mechanisms of action, but they cannot by themselves provide information that is directly relevant to humans (Ames et al.,

1987); (2) metabolic or biochemical studies in humans (e.g., studying association of diet with estrogen profiles or markers of DNA damage). These studies do not address the relations between dietary intake and the occurrence of cancer directly, but they can be invaluable in the interpretation of other forms of evidence (Willet, 2006). There are several aspects of epidemiological studies of diet effects on cancer: (1) the etiology of cancers at different sites varies, and so many different nutrients or food constituents need to be assessed; (2) bias in dietary recall is a serious problem for case–control studies, and so prospective studies are needed; (3) age-specific rates of cancers are low, and for rare cancer sites are extremely low, and so prospective studies will need to be very large to accumulate enough cases; (4) study design problems are compounded by measurement error, both in assessment of dietary exposures and in genotyping (Bingham, 2005).

During the last 30 years, approaches for assessing dietary intake (e.g., standardized questionnaires to assess intakes of foods from which nutrient intakes can be calculated, biochemical determinations of body tissues, anthropometric measurements) have been developed and have been shown to be informative. Early epidemiological studies of diet and cancer were mostly correlational, comparing the disease rates in populations with the population per capita consumption of specific dietary factors (while strong in some aspects, this approach did not take into account the potential determinants of cancer other than the dietary factor, thus has been considered the weakest form of evidence) (Willet, 2006). Another type of studies is based on (1) the studies of subgroups within a population that consume unusual diets (often defined by religious or ethnic characteristics) or (2) on migrants studies (which can be useful when correlations observed in the ecological studies are due to genetic factors, or for examining the latency or relevant times of exposure). Many of the weaknesses of correlational studies are potentially avoidable in case–control or cohort studies (however, in case–control studies the biases due to selection or recall could often occur, as well as selection of an appropriate control group might be a problem: diet may influence the incidence of many diseases, and it is often difficult to identify disease groups that are unrelated to the aspect of diet under investigation). Prospective cohort studies avoid most of the potential sources of methodological bias associated with case–control studies, providing the opportunity to obtain repeated assessments of diet over time and to examine the effects of diet on a wide variety of diseases (but these studies need an enrollment of tens of thousands of individuals, even for common cancers). The randomized trial (optimally, double blind) is a useful approach in evaluation of dietary hypotheses, however, preliminary data should ensure that benefit is reasonably probable and that an adverse outcome is unlikely. Experimental studies are particularly useful for evaluating hypotheses about minor dietary components, such as if specific micronutrient can reduce cancer risk (but some limitations, such as uncertainty of time between changes in dietary factor levels and expected changes in cancer incidence, a decreased compliance of diet during an extended trial, the enrolled participants being highly selected on the basis of health consciousness and motivation, and others, should be taken into account). Due to current uncertainty about measuring diets in early life, whether either study design will be able to address the influence of childhood diet on disease occurring decades later is currently unclear (Willet, 2006).

Food components have a fundamental influence on health that can be explained by the investigation of changes in epigenetic marking of the genome in gene expression, in the translation of messages into proteins and then into metabolites. Numerous processes, including cell division (e.g., AP-1, p53), bioactivation (e.g., Nrf2), inflammation (e.g., NF-kB), apoptosis (e.g., caspases, BAX, Bcl-2), angiogenesis, and metastasis (e.g., matrix metalloproteases, ICAM1, and VEGF) can be modified by food components (Davis and Milner, 2004; Milner, 2004; Aggarwal and Shishodia, 2006).

Nutrigenomics recently has emerged as new technologies, such as transcriptomics (using microarrays, it studies how nutritional exposure influences gene expression on a genomic scale), proteomics (using protein separation followed by quantification and identification, it investigates different protein expression under different conditions, or in different pathological processes), metabolomics (examines the global patterns of metabolites present in the cell or in body fluids in response to specific dietary exposure), and epigenetics (studies the genome modification, which do not involve changes to the primary sequence, but mediated through modification of chromatin proteins, such as histones, and through the methylation of DNA, regulating gene expression, in response to dietary and other exposures, and leading to altered cellular phenotypes associated with chronic disease or aging) (Mathers, 2005).

Certain *genetic polymorphisms,* influencing food metabolites, may be important factors influencing cancer risk of certain macro- and micronutrients. Polymorphisms of gene coding of methylenetetrahydrofolate reductase (this enzyme is critical to the regulation of factors in DNA methylation and synthesis) can influence colorectal cancer risk in affected individuals by altering cellular response to dietary folate and methionine (this polymorphism is important at the stage of transformation of adenoma to carcinoma, Chen et al., 1998). Risk of colon cancer is increased in persons who are rapid–rapid phenotype (Nat-2 and CYP1A2) and prefer well-done meat – they have RR = 6.45, compared to persons who are rapid–slow metabolizers and who prefer well-done meat (RR = 1.87); for those preferring rare/medium-cooked meat, the OR for the rapid–rapid phenotype was 3.13, and for rapid–slow phenotype – 0.91 (Lang et al., 1994).

Dietary factors may also have a *direct impact on the gastrointestinal mucous,* causing chronic irritation or inflammation, resulting in increased cellular proliferation, local production of growth factors, and oxidative stress, and this way enhancing carcinogenesis. So, besides alcohol consumption, higher risks of oral and esophageal cancers are associated with increased intake of salt-preserved meat and fish, smoked food, and charcoal-grilled meat, as well as with the consumption of very hot beverages.

#### 3.2.2.1 Cooking Process

Recent studies evaluated the association of specific cancers with methods of cooking the meat: e.g., frying or broiling meat may produce potential carcinogens, including polynuclear aromatic hydrocarbons (PAHs) and heterocyclic amines (HCAs), such as 2-amino-1-methyl-6-phenylimidazol (4,5-β) pyridine (PhIP). HCAs are formed from the cooking of muscle meats, such as beef, pork, fowl, and fish, by amino acids and creatine reacting at high cooking temperatures. Researches have identified that 17 different HCAs may be associated with human cancer risk (see: National Cancer Institute, Heterocyclic Amines in Cooked Meats). It has been shown that an increased risk of developing stomach, colorectal, pancreatic, and breast cancer is associated with high intakes of well-done, fried, or barbequed meat. Individuals who eat their beef medium well or well done had more than three times the risk of stomach cancer than those who ate their beef rare or medium rare. A recent study by Gallicchio et al. (2006) showed that flame-broiled food may be a risk factor for breast cancer among women with benign breast disease, who have genotypes consistent with rapid acetylation: the OR of breast cancer among rapid acetylators was 2.62 (95% CI = 1.06–6.46).

Nonmuscle-meat sources of protein (e.g., milk, eggs, tofu, liver) have very little or no HCAs (either, naturally or when cooked). Lower temperatures are used for oven roasting and baking and lower levels of HCAs could form (but gravy prepared from meat drippings contains a lot of HCAs). Meats that were microwaved for 2 min prior to cooking had a 90% decrease in HCA content (see: National Cancer Institute, Heterocyclic Amines in Cooked Meats). At present, no Federal Agency monitors HCAs' content in cooked meats, and there is no precise measure of how much HCAs would have to be eaten to increase cancer risk. No guidelines exist about consumption of foods with HCAs.

#### 3.2.2.2 Acrylamide

It has been determined that heating some foods to a temperature of 120°C (248°F) can produce acrylamide, a chemical compound whose primary use is to make polyacrylamide and acrylamide copolymers. High levels of acrylamide has been found in potato chips and French fries (see: National Cancer Institute, Acrylamide in Foods). Amino acid asparagines (it is present in many vegetables, with higher amounts in some varieties of potatoes) can form acrylamide when heated to high temperatures in the presence of certain sugars. Acrylamide is formed when high-heat cooking methods are used, such as broiling, frying, and baking. A longer cooking time may increase the amount of acrylamide produced when the temperature is high enough. The safety of acrylamide in food was evaluated in the 64th meeting of the Joint FAO/WHO Expert Committee on Food Additives (JECFA). Acrylamide was classified as "probably carcinogenic to humans (Group 2A)" by the International Agency for Research on Cancer

(IARC) from evidence of carcinogenicity in experimental animals and from evidence that acrylamide is metabolized to a genotoxic compound, glycidamide, in both rodents and humans. JECFA calculated margin of exposure (MOE) – the new approach to risk assessment for compound that are both genotoxic and carcinogenic – values of 300 for the general population and 75 for consumers of large quantities of food containing high levels of acrylamide (Toda et al., 2005).

#### 3.2.2.3 Artificial Sweeteners

Some food additives, e.g., artificial sweeteners, have been studied from the aspect of their carcinogenicity to humans. Artificial sweeteners are often used instead of sucrose (table sugar) to sweeten foods and beverages. Artificial sweeteners are regulated by the U.S. Food and Drug Administration (FDA).

*Saccharin*. It was shown in animal experiments that saccharin increased the incidence of urinary bladder cancer in rats. However, it has been delisted from the U.S. National Toxicology Program's Report on Carcinogens in 2000 (saccharin has been listed there since 1981) based on the studies that showed bladder tumors in rats had an irrelevant mechanismto humans, thus not providing enough evidence of saccharin's role as a human carcinogen (see: National Cancer Institute, Artificial Sweeteners and Cancer).

*Aspartame*. The safety of aspartame for humans was discussed in the 1996 report, when it was suggested that increased incidence of brain tumors in 1975-1992 might be associated with a wide aspartame use. But these data did not establish a clear link between the consumption of aspartame and the development of brain tumors (see: National Cancer Institute, Artificial Sweeteners and Cancer).

*Acesulfame potassium, Sucralose, and Neotame*. These three artificial sweeteners are currently permitted for use in food in the United States. The results of more than 100 safety studies performed for these sweeteners showed no evidence that these sweeteners cause cancer in humans (see: National Cancer Institute, Artificial Sweeteners and Cancer).

*Cyclamate*. Cyclamate was believed to increase the risk of bladder cancer in humans, and the FDA banned the use of cyclamate in 1969. The most recent studies concluded that cyclamate was not a carcinogen, and a food additive petition is currently filed with FDA for the reapproval of cyclamate (see: National Cancer Institute, Artificial Sweeteners and Cancer).

Although many associations of dietary compounds with various cancers have been found, the association does not always mean causation, which is very difficult to establish. Many studies were published on specific foods, nutrients, and lifestyle factors and specific cancer risks, but until recently no one study provides clear results and recommendations on this subject, and the single new report may sometimes overemphasize contradictory or conflicting results. Nevertheless, recent studies of dietary risk factors in cancer provided an opportunity for deeper insight into carcinogenesis mechanisms: e.g., there are likely to be genes that confer susceptibility to cancer through their effect on intermediate mechanisms, such as those involved in pathways of nutrient metabolism (Bingham, 2005), or the effect of diet on somatic DNA damage or by influencing other mechanisms involved in inherited cancer mutations. It is supposed that these effects may be modulated by diet to either increase or decrease cancer risk.

## 3.3 Summary

The efforts of many researchers on cancer for decades have been directed at identifying cancer risk factors, exposure to which could be considerably reduced or completely eliminated, thus preventing tumors. The more recent studies developed the concept about the combination effects of environmental factors and a genetic susceptibility on cancer risk. Over the last two decades, more evidence was found that a substantial proportion of cancer comes from environmental factors (the term "environmental" has been extended and now includes such lifestyle factors as smoking, alcohol, diet) and thus they may be potentially avoidable. In the United States and some other countries, there are many laws and regulations protecting the public from exposure to environmental carcinogenes, balancing of risk, feasibility of control, costs, and the force of political and other societal pressures. Large efforts are made in encouraging people to have a "healthier" lifestyle. Interdisciplinary approaches may be useful instruments for "multidimensional" analysis of various risk factors and their roles in carcinogenesis in humans (at different levels, from cell to large human populations), as well as for making forecasts of possible outcomes of certain risk factors' intervention on human health at different levels. The model used should be flexible and realistic enough to make it possible to reflect the main processes that real human populations live with (i.e., birth, reproduction, disease, death), and various risk factors or their complex influencing these parameters during an individual's life. Controllable cancer risk factors may be included in models as changeable (partly avoidable in reality, or completely avoidable – in an ideal scenario), or fixed, or noncontrollable (i.e., hardly or nonavoidable) (the examples of such interventions in controllable risk factors that allow to forecast the potential future changes in population characteristics are discussed in Chapter 8).

## References

Aggarwal B.B., Shishodia S., 2006. Molecular targets of dietary agents for prevention and therapy of cancer. Biochem Pharmacol 71(10):1397–1421.

Ahlbom A., Green A., Kheifets L. et al., 2004. Epidemiology of health effects on radiofrequency exposure. Environ Health Perspect 112(14):1741–1754.

Ames B.N., Gold L.S., 1991. Endogenous mutagens and the causes of aging and cancer. Mutat Res 250:3–16.

Ames B.N., Magaw R., Gold L.S., 1987. Ranking possible carcinogenic hazards. Science 236:271–280.

Artandi S.E., Chang S., Lee S.L. et al., 2000. Telomere dysfunction promotes non-reciprocal translocations and epithelial cancers in mice. Nature 406:641–645.

Baker J.R., Fosso C.K., 1993. Immunological aspects of cancers arising from thyroid follicular cells. Endocr Rev 14(6):729–746.

Balan V.E., 1995. Epidemiology of the climacteric period in a large city. *Akush Ginekol* (Mosk). 3:25–28.

Ballarin C., Sarto F., Giacomelli L. et al., 1992. Micronucleated cells in nasal mucosa of formaldehyde-exposed workers. Mutation Res 280:1–7.

Ballarino F., Ottolenghi A., 2002. Low-dose radiation action: possible implications of bystander effects and adaptive response. J Radiol Prot 22:A39–A42.

Balonov M.I., Zvonova I.F., Bratilova A.A. et al., 2002. Mean thyroid doses for inhabitants of different age, living in 1986 in settlements of Bryansk, Tula, Orel and Kaluga oblasts, contaminated by radionuclides as a result of the Chernobyl accident. Radiation and Risk. Special issue (in Russian)

Barrett J.C., Vainio H., Peakall D., Goldstein B.D., 1997. Susceptibility to environmental hazards. 12th meeting of the scientific group on methodologies for the safety evaluation of chemicals. Environ Health Perspect 105 (Suppl 4):699–737.

Begg C.B., Mazumdar M., 1994. Operating characteristics of a rank correlation test for publication bias. Biometrics 50:1088–1101.

Bennett C.B., Lewis L.K., Karthikeyan G., Lobachev K.S., Jin Y.H., Sterling J.F., Snipe J.R., Resnick M.A., 2001. Genes required for ionizing radiation resistance in yeast. Nat Genet 29:426–434.

Beral V., 2003. Breast cancer and hormone-replacement therapy in the Million Women Study. Lancet 362(9382):419–427.

Berrington de Gonzalez A., Sweetland S., Green J., 2004. Comparison of risk factors for squamous cell and adenocarcinomas of the cervix: a meta-analysis. Br J Cancer 90(90):1787–1791.

Bhatia S., Meadows A.T., Robison L.L., 1998. Second cancers after pediatric Hodgkin's disease. J Clin Oncol 16(7):2570–2572.

Bingham S., 2005. Diet and cancer. In: Nutrigenomics. Report of a Workshop Organized by the Public Health Genetics Unit. The Nuffield Trust.

Bisi H., Fernandes V.S., deCamargo R.Y.A., Koch L., Abdo A.H., deBrito T., 1989. The prevalence of unsuspected thyroid pathology in 300 sequential autopsies with special references to the incidental carcinoma. Cancer 64:1888–1893.

Boffetta P., Trendaniel J., Greco A., 2000. Risk of childhood cancer and adult lung cancer after childhood exposure to passive smoke: a meta-analysis. Environ Health Perspect 108(1):73–82.

Bois F.Y., 1999. Analysis of PBPK models for risk characterization. Ann NY Acad Sci 895:317–337.

Bois F.Y., Krowech G., Zeise L., 1995. Modeling interindividual variability in metabolism and risk: the example of 4-aminobiphenyl. Risk Anal 15:205–213.

Braun M.M., Caporaso N.E., Page W.F., Hoover R.N., 1995. A cohort study of twins and cancer. Cancer Epidemiol Biomark Prev 4:469–473.

Brouwer I.A., Katan M.B., Zock P.L., 2004. Dietary alpha-linolenic acid is associated with reduced risk of fatal coronary heart disease, but increased prostate cancer risk: a meta-analysis. J Nutr 134(4):919–922.

Brownson R.C., Novotny T.E., Perry M.C., 1993. Cigarette smoking and adult leukemia. A meta-analysis. Arch Intern Mad 153(4):425–427.

Buckley J.D., Buckley C.M., Breslow N.E., Draper G.J., Roberson P.K., Mack T.M., 1996. Concordance for childhood cancer in twins. Med Pediatr Oncol 26:223–229.

Buja A., Lange J.H., Perissinotto E., Rausa G., Grigoletto F., Canova C., Mastrangelo G., 2005. Cancer incidence among male military and civil pilots and flight attendants: an analysis on published data. Toxicol Ind Health 21(10):273–282.

Buja A., Mastrangelo G., Perissinotto E. et al., 2006. Cancer incidence among female flight attendants: a meta-analysis of published data. J Womens Health 15(1):98–105.
Butterworth B.E., Conolly R.B., Morgan K.T., 1995. A strategy for establishing mode of action of chemical carcinogens as a guide for approaches to risk assessment. Cancer Lett 93:129–146.
Cancer Facts and Figures, 2006. American Cancer Society. www.cancer.org/downloads/STT/CAFF2006PWSecured.pdf
Cappelleri J.C., Ioannidis J.P., Schmid C.H. et al., 1996. Large trial versus meta-analysis of smaller trials: how do their results compare? JAMA 276:1332–1338.
Carmella S.G., Akerkar S.A., Richie J.P., Hecht S.S., 1995. Intraindividual and interindividual differences in metabolites of the tobacco-specific lung carcinogen 4-(methylnitrosamino)-1-(3-pyridyl)-1-butanone (NKK) in smokers' urine. Cancer Epidemiol Biomarkers Prev 4:35–62.
Carmelli D., Page W.F., 1996. Twenty-four year mortality in World War II U.S. male veteran twins discordant for cigarette smoking. Int J Epidemiol 1996; 25:554–713.
Carretero-Pelaez M.A., Esparza-Gmez G.C., Figuero-Ruiz E., Cerero-Lapiedra R., 2004. Alcohol-containing mouthwashes and oral cancer. Critical analysis of literature. Med Oral 9:116–123.
Castronovo F.P., 1987. Iodine-131 thyroid uptake results in travelers returning from Europe after the Chernobyl accident. *J Nucl Med* 28:535–541.
Chen J., Giovannucci E., Hankinson S.E., Ma J., Willett W.C., Spiegelman D., Kelsey K.T., Hunter D.J., 1998. A prospective study of methylenetetrahydrofolate reductase and methionine synthase gene polymorphisms, and risk of colorectal adenoma. Carcinogenesis 19:2129–2132.
Chen K., Qiu J-L., Zhang Y., Zhao Y-W., 2003. Meta analysis of risk factors for colorectal cancer. World J Gastroenterol 9(7):1598–1600.
Cho E., Smith-Warner S.A., Ritz J., van den Brandt P.A., Colditz G.A. et al., 2004. Alcohol intake and colorectal cancer: a pooled analysis of 8 cohort studies. Ann Intern Med 140(8):603–613.
Christensen H.C., Schuz J., Koslejanetz M et al., 2005. Cellular telephones and risk for brain tumors: a population-based, incident case–control study. Neurology 64(7):1189–1195.
D'Errico A., Taioli E., Chen X., Vineis P, 1996. Genetic metabolic polymorphisms and the risk of cancer: a review of the literature. Biomarkers 1:149–173.
Dagnelie P.C., Schuurman A.G., Goldbohm R.A., Van den Brandt P.A., 2004. Diet, anthropometric measures and prostate cancer risk: a review of prospective cohort and intervention studies. BJU Int 93(8):1139–1150.
Dai Z., Xu Y.C., Niu L., 2007. Obesity and colorectal cancer risk: a meta-analysis of cohort studies. World J Gastroenterol 13(31):4199–4206.
Darbre P.D., Aljarrah A., Miller W.R. et al., 2004. Concentrations of parabens in human breast tumors. J Appl Toxicol 24:5–13.
Davis C., Milner J., 2004. Frontiers in nutrigenomics, proteomics, metabolomics and cancer prevention. Mutat Res 551:51–64.
Dendy P.P., Brugmans M.J.P., 2003. Low dose radiation risks. Br J Radiobiol 76:674–677.
DePinho R.A., 2000. The age of cancer. Nature 408(9):248–254.
Deswal A., Petersen N.J., Feldman A.M. et al., 2001. Cytokines and cytokine receptors in advanced heart failure. An analysis of the cytokine database from the Vesnarinone Trial (VEST). Circulation 24:2055–2059.
Diez O., Brunet J., Sanz J., del Rio E., Alonso M.C., Baiget M., 1997. Differences in phenotypic expression of a new BRCA1 mutation in identical twins. Lancet 350:713.
Doll R., Peto R., 1981. The causes of cancer: quantitative estimates of avoidable risks of cancer in the United States today. J Natl Cancer Inst 66(6):1191–1308.
Egger M., Smith G.D., 1995. Misleading meta-analysis. Lessons from "an effective, safe, simple" intervention that wasn't. BMJ 310:752–754.
Egger M., Smith G.D., Schneider M., Minder C., 1997. Bias in meta-analysis detected by a simple, graphical test. Br Med J 315:629–634.

EGH, 2006. Health effects of the Chernobyl accident and special health care programmes. Report of the UN Chernobyl Forum Expert Group "Health" (EGH), working draft, 2005. Edited – January 2006.

Ellison R.C., Zhang Y., McLennen C.E., Rothman K.J., 2001. Exploring the relation of alcohol consumption to risk of breast cancer. Am J Epidemiol 154(8):740–747.

Erren T.C., 2001. A meta-analysis of epidemiologic studies of electric and magnetic fields and breast cancer in women and men. Bioelectromagnetics, suppl 5:S105–S119.

Eysenck H.J., 1994. Meta-analysis and its problems. BMJ 309:789–792.

Fan A.M., Howd R.A., 2001. Quantative cancer risk assessment of non-genotoxic carcinogens. In: Genetic Toxicology and Cancer Risk Assessment. Choy W.H. (Ed). New-York, Basel: Marcel Dekker, Inc.

Feychting M., Schulgen G., Olsen J.H., Ahlbom A., 1995. Magnetic fields and childhood cancer – a pooled analysis of two Scandinavian studies. Eur J Cancer 31A(12):2035–2039.

Fjalling M., Tisell L.E., Carlsson S., Hansson G., Lundberg L.M., Oden A., 1986. Benign and malignant nodules after neck irradiation. Cancer 58, 1219–1224.

Freudenheim J.L., Ritz J., Smith-Warner S.A., Albanes D., Bandera E.V., van den Brabdt P. A. et al., 2005. Alcohol consumption and risk of lung cancer: a pooled analysis of cohort studies. Am J Clin Nutr 82(3):495–496.

Gallicchio L., McSorley M.A., Newschaffer C.J., Thuita L.W., Argani P., Hoffman S.C., Helzlsouer K.J., 2006. Flame-broiled food, NAT2 acetylator phenotype, and breast cancer risk among women with benign breast disease. Breast Cancer Res Treat Mar 16.

Gandini S., Botteri E., Iodice S. et al., 2008. Tobacco smoking and cancer: a meta-analysis. Int J Cancer 122(1):155–164.

Gao X., LaValley M.P., Tucker K.L., 2005. Prospective studies of dairy product and calcium intakes and prostate cancer risk: a meta-analysis. J Natl Cancer Inst 97(23):1768–1777.

Gavrilin I.I., Gordeev K.I., Ivanov V.K. et al., 1992. Characteristics and results of the determination of the doses of internal irradiation of the thyroid gland in the population of contaminated districts of the Byelorussian Republic. *Vestnik Akad Medicinsk Nauk* 2:35–43.

Gilbert E.S., Tarone R., Bouville A., Ron E., 1998. Thyroid cancer rates and I-131 doses from Nevada atmospheric nuclear bomb tests. J Natl Cancer Inst 1998 Nov 90:1654–1660.

Gonzalez C.A., Jakszyn P., Pera G., Agudo A., et al., 2006. Meat intake and risk of stomach and esophageal adenocarcinoma within the European Prospective Investigation Into Cancer and Nutrition (EPIC). J Natl Cancer Inst 98(5):345–54.

Gorham E.D., Mohr S.B., Garland C.F. et al., 2007. Do sunscreens increase risk of melanoma in populations residing at higher latitudes? Ann Epidemiol 17(12):956–963.

Greiser C.M., Greiser E.M., Doren M., 2007. Menopausal hormone therapy and risk of ovarian cancer: systematic review and meta-analysis. Hum Reprod Update 13(5):453–463.

Hampel H., Abraham N.S., El-Serag H.B., 2005. Meta-analysis: obesity and the risk for gastroesophageal reflux disease and its complications. Ann Intern Med 143(3):199–211.

Harach H.R., Williams E.D., 1995. Childhood thyroid cancer in England and Wales. Br J Cancer 72:777–783.

Harbord R.M., Egger M., Sterne J.A.C., 2005. A modified test for small-study effects in meta-analyses of controlled trials with binary endpoints. Stat Med 25:3443–3457.

Harvard Report on Cancer Prevention, 1996. Cancer Causes Control vol. 7 (suppl. November), ISSN 0957-5243. At: www.hsph.harvard.edu/cancer/resources_materials/reports/HCCPreport_1summary.htm

Harvey P.W., Everett D.J., 2004. Significance of the detection of esters of *p*-hydrobenzoic acid (parabens) in human breast tumors. J Appl Toxicol 24:1–4.

Heidenreich W.F., Kenigsberg J., Jacob P., Buglova E., Goulko G., Paretzke H.G., Demidchik E.P., Golovneva A., 1999. Time trends of thyroid cancer incidence in Belarus after the Chernobyl accident. Radiat Res 151:617–625.

Higginson J., Muir C.S., 1977. Determination del'importance des facteurs environnementaux dans le cancer human: role de l'epidemilogie. Bull Cancer 64:365–384.

Higginson J., Muir C.S., 1979. Environmental carcinogenesis: misconceptions and limitations to cancer control. J Natl Cancer Inst 63:1291–1298.

Hjalgrim L.L., Westergaard T., Rostgaard K., Schmiegelow K., Melbye M., Engels E.A., 2003. Birth weight as a risk factor for childhood leukemia: a meta-analysis of 18 epidemiologic studies. Am J Epidemiol 158(8):724–35.

Hoover R.N., 2000. Cancer: nature, nurture, or borth. Editorial N Engl J Med 343(2).

Hougeir F.G., Yiannias J.A., Hinni M.L., Hentz J.G., El-Azhary R.A., 2006. Oral metal contact allergy: a pilot study on the cause of oral squamous cell carcinoma. Int J Dermatol 45(3):265–71.

Huncharek M., Kupelnick B., Klassen H., 2001. Paternal smoking during pregnancy and the risk of childhood brain tumors: results of a meta-analysis. In Vivo 15(6):535–541.

Huncharek M., Kupelnick B., Wheeler L., 2003. Dietary cured meat and the risk of adult glioma: a meta-analysis of nine observational studies. J Environ Pathol Toxicol Oncol 22(2):129–37.

Huncharek M., Kupelnick B., 2004. A meta-analysis of maternal cured meat consumption during pregnancy and the risk of childhood brain tumors. Neuroepidemiology 23(1–2):78–84.

Hunter D.J., Spiegelman D., Adami H.O., Beeson L., van den Brandt P.A., Folsom A.R., Fraser G.E., Goldbohm R.A., Graham S., Howe G.R. et al., 1996. Cohort studies of fat intake and the risk of breast cancer – a pooled analysis. N Eng J Med 334(6):356–61.

Iampol'skaia I., 1997. Dynamics of puberty levels in girls of Moscow. Gin Sanit 3:29–30.

IARC Monographs, A. Volumes 1–99. Group 1: Carcinogenic to humans (105). Agents and groups of agents. At: monographs.iarc.fr/ENG/Classification/crthgr01.php

IARC Monographs, B. Volumes 1–99. Group 2A:Probably carcinogenic to humans (66). Agents and groups of agents. At: monographs.iarc.fr/ENG/Classification/crthgr02.php

Ilyin L.A., Balonov M.I., Buldakov L.A. et al., 1990. Radio-contamination patterns and possible health consequences of the accident at the Chernobyl nuclear power station. J Radiol Prot 10:3–29.

International Programme on Chemical Safety, 1993. Biomarkers and Risk Assessment: Concepts and Principles. In: World Health Organization, ed. Environmental Health Criteria 155:3–40.

Ioannidis J.P.A., Cappelleri J.C., Lau J., et al., 1998. Meta-analyses and large randomized controlled trials. NEJM 338:59.

Ivanov V.K., Manton K.G., Akushevich I. et al., 2005. Risk of thyroid cancer after irradiation in children and adults. Curr Oncol 12(2):55–64.

Ivanov V.K., Tsyb A.F., Gorsky A.I. et al., 1997. Leukemia and thyroid cancer in emergency workers of the Chernobyl accident: estimation of radiation risks (1986–1995). Radiat Environ Biophys 36:9–16.

Jackson A.L., Loeb L.A., 1998. The mutation rate and cancer. Genetics 148:1483–1490.

Jacobs J.R., Bovasso G.B., 2000. Early and chronic stress and their relation to breast cancer. Psychol Med 30(3):669–678.

Jaklic B.R., Rushin J., Ghosh B.C., 1995. Estrogen and progesterone receptors in thyroid lesions. Ann Surg Oncol 2(5):429–434.

Juni P., Holenstein F., Sterne J., et al., 2002. Direction and impact of language bias in meta-analyses of controlled trials: empirical study. Int J Epidemiol 31:115–123.

Junien C., Gallou C., 2004. Cancer nutrigenomics. In: Nutrigenetics and Nutrigenomics. Simopoulos A.P., Ordovas J.M. (Eds.). World Rev Diet. Basel: Karger 93:210–269.

Kai M., Luebeck E.G., Moolgavkar S.H., 1997. Analysis of the incidence of solid cancer among atomic bomb survivors using a two-stage model of carcinogenesis. Radiat Res 148:348–358.

Kehlen A., Englert N., Seifert A. et al., 2004. Expression, regulation and function of autotaxin in thyroid carcinomas. Int J Cancer 109:833–838.

Kelly H., 2002. Testimony of Dr. Henry Kelly, President Federation of American Scientists, before the Senate Committee on Foreign Relations. At: www.fas.org/ssp/docs/030602-kellytestimony.htm

Kendall M.G., Gobbons J.D., 1990. Rank Correlation Methods (5th edition). London: Arnold.
Khuder S.A., Mutgi A.B., Nugent S., 2001. Smoking and breast cancer: a meta-analysis. Rev Environ Health 16(4):253–261.
Kingsmore S.F., Patel D.D., 2003 Multiplexed protein profiling on antibody-based microarrays by rolling circle amplification. Curr Opin Biotechnol 14:74–81.
Kodama K., Mabuchi K., Shigematsu I., 1996. A long-term cohort study of the atomic bomb survivors. J Epidemiol 6:S95–S105.
Koizumi Y., Tsubono Y., Nakaya N., Kuriyama S., Shibuya D., Matsuoka H., Tsuji I., 2004. Cigarette smoking and the risk of gastric cancer: a pooled analysis of two prospective studies in Japan. Int J Cancer 112(6):1049–1055.
Korte J.E., Brennan P., Henley S.J., Boffetta P., 2002. Dose-specific meta-analysis and sensitivity analysis of the relation between alcohol consumption and lung cancer risk. Am J Epidemiol 155(6):496–506.
Lacey J.V. Jr., Mink P.J., Lubin J.H. et al., 2002. Menopausal hormone replacement therapy and risk of ovarian cancer. JAMA 288(3):334–341.
Lahat Sheinfeld M., Sobel E., Kinarty A., Kraiem Z., 1992 Divergent effects of cytokines on human leukocyte antigen-DR antigen expression of neoplastic and non-neoplastic human thyroid cells. Cancer 69:1799–1807.
Lang N.P., Butler M.A., Massengill J., Lawson M. et al., 1994. Rapid metabolic phenotypes for acetyltransferase and cytochrome P4501A2 and putative exposure to food-borne heterocyclic amines increase the risk for colorectal cancer or polyps. Cancer Epidemiol Biomarkers Prev 3:675–682.
Larsson S.C., Wolk A., 2007a. Obesity and colon and rectal cancer risk: a meta-analysis of prospective studies. Am J Clin Nutr 86(3):556–565.
Larsson S.C., Wolk A., 2007b. Overweight, obesity and risk of liver cancer: A meta-analysis of cohort studies. Br J Cancer 97(7):1005–1008.
Last J.M., 1995. A Dictionary of Epidemiology, 3rd edition. Oxford: Oxford University Press.
Lau J., Ioannidis J.P.A., Schmid H.C., Gregoire G., Benhaddad A. et al., 1998. Summing up evidence: one answer is not always enough. Lancet 351:123–127.
Leenhouts H.P., Brugmans M.J., Chadwick K.H., 2000. Analysis of thyroid cancer data from the Ukrainian after Chernobyl using a two-mutation carcinogenesis model. Radiat Environ Biophys 2000 Jun 39:89–98.
LeLorier J., Gregoire G., Benhaddad A., et al., 1997. Discrepancies between meta-analyses and subsequent large randomized controlled trials. NEJ 337:536–542.
Lewis S.J., Smith G.D., 2005. Alcohol, ALDH2, and esophageal cancer: a meta-analysis which illustrates the potentials and limitations of a Mendelian randomization approach. Cancer Epidemiol Biomarkers Prev 14(8):1967–1971.
Lichtenstein P., Holm H.V., Verkasalo P.K. et al., 2000. Environmental and heritable factors in the causation of cancer – analyses of cohorts of twins from Sweden, Denmark, and Finland. N Engl J Med 343:78–85.
Living with radiation, 1998. The National Radiological Protection Board. 5th edition. 70 pp.
Loeb L.A., 1991. Mutator phenotype may be required for multistage carcinogenesis. Cancer Res 51:3075–3079.
Loeb L.A., 2001. A mutator phenotype in cancer. Cancer Res 61:3230–3239.
Longnecker M.P., Orza M.J., Adams M.E., Vioque J., Chalmers T.C., 1990. A meta-analysis of alcoholic beverage consumption in relation to risk of colorectal cancer. Cancer Causes Control 1(1):59–68.
Lonn S., Ahlbom A., Hall P., Feychting M., 2004. Mobile phone use and the risk of acoustic neuroma. Epidemiology 15(6):653–659.
Lubin J.H., Caporaso N.E., 2006. Cigarette smoking and lung cancer: modeling total exposure and intensity. Cancer Epidemiol Biomarkers Prev 15(3):517–23.

Lutz W.K., 2001. Dose-response relationships in chemical carcinogenesis and cancer risk assessment. In: Genetic Toxicology and Cancer Risk Assessment. Choy W.H. (Ed.). New York, Basel: Marcel Dekker, Inc.

Malone J., Unger J., Delange F. et al., 1991. Thyroid consequences of Chernobyl accident in the countries of the European community. J Endocrinol Invest 14:701–717.

Manole D., Schildknecht B., Gosnell B. et al., 2001. Estrogen promotes growth of human thyroid tumor cells by different molecular mechanisms. J Clin Endocrinol Metab 86(3):1072–1077.

Manton K.G., Stallard E., 1988. Chronic Disease Risk Modeling: Measurement and Evaluation of the Risks of Chronic Disease Processes. In the Griffin Series of the Biomathematics of Diseases, London, England: Charles Griffin Limited.

Manton K.G., Stallard E., Singer B.H., 1992. Projecting the future size and health status of the U.S. elderly population. Int J Forecast 8:433–458.

Manton K.G., Woodbury M.A., Stallard E., Riggan W.B., Creason J.P., Pellom A.C., 1989. Empirical Bayes procedures for stabilizing maps of U.S. cancer mortality rates. J Am Stat Ass 84: 637–650.

Manton K.G., Yashin A.I., 2000. Mechanisms of Aging and Mortality: Searches for New Paradigms. *Monographs on Population Aging*. 7. Odense, Denmark: Odense University Press.

Mathers J., 2005. The science of nutrigenomics. In: Nutrigenomics. Report of a Workshop Organized by the Public Health Genetics Unit. The Nuffield Trust.

McGrath K.G., 2003. An earlier age of breast cancer diagnosis related to more frequent use of antiperspirants/deodorants and underarm shaving. Eur J Cancer Prev 12(6):479–485.

Megdal S.P., Kroenke C.H., Laden F., Pukkala E., Schernhammer E.S., 2005. Night work and breast cancer risk: a systematic review and meta-analysis. Eur J Cancer 41(13):2023–2032.

Merhi M., Raynal H., Cahuzac E et al., 2007. Occupational exposure to pesticides and risk of hematopoietic cancers: meta-analysis of case–control studies. Cancer Causes Control 18(10):1209–1226.

Michos A., Xue F., Michels K.B., 2007. Birth weight and the risk of testicular cancer: a meta-analysis. Int J Cancer 121(5):1123–1131.

Milner J., 2004. Molecular targets for bioactive food components. J Nutr 134:2492S–2498S.

Minamoto T., Mai M., Ronai Z., 1999. Environmental factors as regulators and effectors of multi-step carcinogenesis. Carcinogenesis 20(4):519–527.

Missmer S.A., Smith-Warner S.A., Spiegelman D., Yaun S.S., Adami H.O. et al., 2002. Meat and dairy food consumption and breast cancer: a pooled analysis of cohort studies. Int J Epidemiol 31(1):78–85.

Moghaddam A.A., Woodward M., Huxley R., 2007. Obesity and risk of colorectal cancer: a meta-analysis of 31 studies with 70,000 events. Cancer Epidemiol Biomarkers Prev 16(12):2533–2547.

Mothersill C., Seymour C., 2001. Radiation-induced bystander effects: past history and future directions. Rad Res 155:759–767.

Murdoch D.J., Krewski D., 1988. Carcinogenic risk assessment with time-dependent exposure patterns. Risk Anal 8:521–530.

Muscat J.E., Malkin M.G., Thompson S. et al., 2000. Handheld cellular telephone use and risk of brain cancer. JAMA 284(23):3001–3007.

National Cancer Institute, Acrylamide in Foods. Fact Sheet. www.nci.nih.gov/cancertopics/factsheet/acrylamideinfoods

National Cancer Institute, Antiperspirants/Deodorants and Breast Cancer: Questions and Answers. www.cancer.gov/cancertopics/factsheet/Risk/AP-Deo

National Cancer Institute, Artificial Sweeteners and Cancer. Questions and Answers. www.nci.nih.gov/cancertopics/factsheet/Risk/artificial-sweeteners

National Cancer Institute, Cellular Telephone Use and Cancer: Questions and Answers. www.cancer.gov/cancertopics/factsheet/Risk/cellphones

National Cancer Institute, Heterocyclic Amines in Cooked Meats www.nci.nih.gov/cancertopics/factsheet/Risk/heterocyclic-amines
National Cancer Institute, Radon and Cancer: Questions and Answers. www.cancer.gov/cancertopics/factsheet/Risk/radon
National Research Council BEIR V, 1990. Committee on the Biological Effects of Ionizing Radiation. Health Effects of Exposure to Low Levels of Ionizing Radiation. Washington D.C.: National Academy Press.
National Research Council Committee on Biological Markers, 1987. Biologic markers in environmental health research. Environ Health Perspect 74:3–9.
Naylor D.C., 1997. Meta-analysis and the meta-epidemiology of clinical research. BMJ 315:617–619.
NEA (Nuclear Energy Agency) Report, 2002. Chernobyl: Assessment of Radiological and Health Impact. 2002 Update of Chernobyl: Ten years on. www.nea.fr/html/rp/chernobyl/c03.html
Nekolla E.A., Kellerer A.M., Kuse-Isingschulte M., Eder E., Spiess H., 1999. Malignancies in patients treated with high doses of radium-224. Radiat Res 152 (suppl 6):S3–S7.
Nguyen L.T., Ramanathan M., Weinstock-Guttman B. et al., 2003. Sex differences in vitro pro-inflammatory cytokine production from peripheral blood of multiple sclerosis patients. J Neurol Sci 209(1–2):93–99.
Norat T., Lukanova A., Ferrari P., Riboli E., 2002. Meat consumption and colorectal cancer risk: dose-response meta-analysis of epidemiological studies. Int J Cancer 98(2):241–56.
Norppa H., Luomahaara S., Heikanen H., Roth S., Sorsa M., Renzi L., Lindholm C., 1993. Micronucleus assay in lymphocytes as a toll to biomonitor human exposure to aneuploidogens and clastogens. Environ Health Perspect 101 (suppl 3):139–143.
Pacifici R., Rifas L., McCraken R. et al., 1989. Ovarian steroid treatment blocks a postmenopausal increase in blood monocyte Il-1 release. Proc. Natl Acad Sci. USA 86(7):2398–2402.
Penta J.S., Johnson F.M., Wachsman J.T., Copeland W.C., 2001. Mitochondrial DNA in human malignancy. Mutat Res 488:119–133.
Plummer M., Herrero R., Franceschi S., Meijer C.J., Snijders P., Bosch F.X., de Sanjose S., Munoz N., IARC Multi-centre Cervical Cancer Study Group, 2003. Smoking and cervical cancer: pooled analysis of the IARC multi-centric case–control study. Cancer causes Control 14(9):805–14.
Preston D.L., Shimizu Y., Pierce D.A., Suyama A., Mabuchi K., 2003. Life Span Study Report 13. Studies of mortality of atomic bomb survivors. Report 13: Solid cancer and non-cancer disease mortality: 1950–1997. Radiat Res 160(4):381–407.
Qin L.Q., Xu J.Y., Wang P.Y. et al., 2007. Milk consumption is a risk factor for prostate cancer in Western countries: evidence from cohort studies. Asia Pac J Clin Nutr 16(3):467–476.
Qin L.Q., Xu J.Y., Wang P.Y., Kaneko T., Hochi K., Sato A., 2004. Milk consumption is a risk factor for prostate cancer: meta-analysis of case–control studies. Nutr Cancer 48(1):22–7.
Ramzaev P.B., Balonov M.I., Zvonova I.F. et al., 2000. Methodology for reconstruction of thyroid doses from iodine radioisotopes in residents of the Russian Federation exposed to radioactive contamination as a result of the Chernobyl accident in 1986 [Russian]. Moscow: Russian Ministry of Public Health. [Guideline MU-2.6.1000-00].
Reid F.D., Mercer P.M., Harrison M., Bates T., 1996. Cholecystectomy as a risk factor for colorectal cancer: a meta-analysis. Scand J Gastroenterol 31(2):160–169.
Ron E., Kleinerman R.A., Boice J.D. et al., 1987. A population-based case–control study of thyroid cancer. J Natl Cancer Inst 79:1–12.
Ron E., Lubin J.H., Shore R.E., Mabuchi K., Modan B., Pottern L.M., Schneider A.B., Tucker M.A., Boice J.D., 1995. Thyroid cancer after exposure to external radiation: a pooled analysis of seven studies. Radiat Res 141:259–277.
Ron E., Modan B., Preston D., Alfandary E., Stovall M., Boice J., 1989. Thyroid neoplasia following low-dose radiation in childhood. Radiat. Res. 120:516–531.

Rossouw J.E., Anderson G.L., Prentice R.L. et al., 2002. Risks and benefits of estrogen plus progestin in healthy postmenopausal women: principal results from the Women's Health Initiative randomized controlled trial. JAMA 288(3):321–333.
Rothkamm K., Lobrich M., 2003. Evidence for a lack of DNA double-strand break repair in human cells exposed to very low X-ray doses. PNAS 100(9):5057–5062.
Rushton L., Jones D.R., 1992. Oral contraceptive use and breast cancer risk: a meta-analysis of variations with age at diagnosis, parity and total duration of oral contraceptive use. Br J Obstet Gynaecol 99(3):239–46.
SEER (Surveillance, Epidemiology and End Results) Program of the USA 1973–1999 (1999). National Cancer Institute, Bethesda. seer.cancer.gov/csr/1973_1999/thyroid.pdf
SEER (Surveillance, Epidemiology and End Results) Program of the USA 2001–2005 (2005). National Cancer Institute, Bethesda. seer.cancer.gov/statfacts/html/thyro.html
Shore R.E., 1989. Radiation epidemiology: old and new challenges. Environ Health Perspect 81:153–156.
Shore R.E., 1992. Issues and epidemiological evidence regarding radiation-induced thyroid cancer. Radat Res 131:98–111.
Smith-Warner S.A., Ritz J., Hunter D.J., Albanes D., Beeson W.L., van den Brandt P.A., Colditz G et al., 2002. Dietary fat and risk of lung cancer in pooled analysis of prospective studies. Cancer Epidemiol Biomarkers Prev 11(10Pt1):987–992.
Sonnenberg A., Muller A.D., 1993. Constipation and cathartics as risk factors of colorectal cancer: a meta-analysis. Pharmacology 47, Suppl 1:224–233.
Sterne J.A.C., Gavaghan D., Egger M., 2000. Publication and related bias in meta-analysis: Power of statistical tests and prevalence in literature. J Clin Epidemiol 53:1119–1129.
Sterne J.A.C., Juni P., Schulz K.F., Altman D.G., Bartlett C., Egger M., 2002. Statistical method for assessing the influence of study characteristics on treatment effects in "meta-epidemiological" research. Stat Med 21:1513–1524.
Takkouche B., Etminan M., Montes-Martinez A., 2005. Personal use of hair dyes and risk of cancer: a meta-analysis. JAMA 293(20):2516–25.
Taylor R., Cumming R., Woodward A., Black M., 2001. Passive smoking and lung cancer: a cumulative meta-analysis. Aust N Z Public Health 25(3):203–211.
Thompson D.E., Mabuchi K., Ron E. et al., 1994. Cancer incidence in atomic bomb survivors. PII: Solid tumors, 1958–1987. Radiat Res 1994; 137 (suppl 2):S17–S67.
Thorvaldsson S.E., Tulinius H., Bjornsson J. et al., 1992. Latent thyroid carcinoma in Iceland at autopsy. Pathol Res Pract 188(6):747–750.
Toda M., Uneyama C., Yamamoto M., Morikawa K., 2005. Recent trends in evaluating risk associated with acrylamide in food. Focus on a new approach (MOE) to risk assessment by JECFA. Kokuritsu Ivakuhin Shokuhin Eisei Kenkyusho Hokoku 123:63–7.
Tokumaru O., Haruki K., Bacal K., Katagiri T., Yamamoto T., Sakurai Y., 2006. Incidence of cancer among female flight attendants: a meta-analysis. J Travel Med 13(3):127–132.
Tokunaga M., Land C.E., Tokuoka S., Nishimori I., Soda M., Akiba S., 1994. Incidence of female breast cancer among atomic bomb survivors. Hiroshima and Nagasaki, 1950–1985. Radiat Res 138:209–223.
Tronko M.D., Bogdanova T.I., Komissarenko I.V. et al., 1999. Thyroid carcinoma in children and adolescents in Ukraine after the Chernobyl nuclear accident. Statistical data and clinico-morphlogical characteristic. Cancer 86:149–156.
Tucker J.D., Auletta A., Cimino M.C., Dearfield K.L., Jacobson-Kram D., Tice R.T., Carrano A.V., 1993. Sister-chromatid exchange: Second report of the Gene-Tox program. Mutat Res 297:101–180.
Turker M.S., 1998. Estimation of mutation frequencies in normal mammalian cells and the development of cancer. Semin Cancer Biol 8:407–419.
UNSCEAR, 2000. United Nations Scientific Committee on the Effects of Atomic Radiation. 2000 Report to the General Assembly. Annex J. Exposures and effects of the Chernobyl accident. Int J Radiat Med 2000; 2:3–109.

Vijg J., Dolle M., 2001. Handbook of the Biology of Ageing. 5th edition (eds. Masoro E.J., Austad S.N.), San Diego, CA: Academic Press, 534 pp.

Villa E., Melegari M., Scaglioni P.P., Trande P., Cesaro P., Manenti F., 1991. Hepatocellular carcinoma: risk factors other than HBV. Ital J Gastroenterol 23(7):457–460.

Villar J. Carroli G., Belizan J.M., 1995. Predictive ability of meta-analyses of randomized controlled trials. Lancet 345:772–776.

Wang B., Zhang Y., Xu D.Z., Wang A.H., Zhang L., Sun C.S., Li L.S., 2004. Meta-analysis of the relationship between tobacco smoking, alcohol drinking and p53 alteration in cases with esophageal carcinoma. Zhonghua Liu Xing Bing Xue Za Zhi 25(9):775–8.

Wang J.-S., Links J.M., Groopman J.D., 2001. Molecular epidemiology and biomarkers. In: Genetic Toxicology and Cancer Risk Assessment. Choy W.H. (Ed.). New-York, Basel: Marcel Dekker, Inc.

Washburn E.P., Orza M.J., Berlin J.A., Nicholson W.J., Todd A.C., Frumkin H., Chalmers T. C., 1994. Residential proximity to electricity transmission and distribution equipment and risk of childhood leukemia, childhood lymphoma, and childhood nervous system tumors: systematic review, evaluation, and meta-analysis. Cancer Causes Control 5(5):487.

Wei Q., Matanoski G.M., Farmer E.R., Hedayti M.A., Grossman L., 1994. DNA repair related to multiple skin cancers and drug use. Cancer Res 54:437–440.

Willet W.C., 2006. Diet and nutrition. In: Schottenfeld, D., Fraumeni, J.F.Jr. (Eds). Cancer Epidemiology and Prevention. 3rd edition, pp. 405–421. Oxford: Oxford, University Press.

Wood J.W., Tamagaki H., Neriish S., Sato T., Sheldon W.F., Archer P.G, 1969. Thyroid carcinoma in atomic bomb survivors, Hiroshima and Nagasaki. Am J Epidemiol 89:4–14.

Woods K.L., 1995. Mega-trials and management of acute myocardial infarction. Lancet 346:611–614.

World Nuclear Association, 2004. Radiation and the nuclear fuel cycle. At: www.solarstorms.org/WorldNuclear.html

Zeise L., 2001. Risk assessment of genotoxic carcinogens. In: Genetic Toxicology and Cancer Risk Assessment. Choy W.H. (Ed.). New York, Basel: Marcel Dekker, Inc.

Ziegler R.G., Hoover R.N., Pike M.C. et al., 1993. Migration patterns and breast cancer risk in Asian-American women. J Natl Cancer Inst 85:1819–1827.

# Chapter 4
# Standard and Innovative Statistical Methods for Empirically Analyzing Cancer Morbidity and Mortality

In Chapter 2, we examined the substantive and mathematical basis of a number of theories and models of the mechanisms underlying human carcinogenesis and suggested some innovative modeling strategies to deal with various informational limitations of specific observational and measurement plans. In Chapter 3, we discussed the various types of controllable and uncontrollable risk factors that have been evaluated to determine if they increase cancer risk in humans. Both of these aspects are important to taken into account while modeling cancer risks and outcomes on population level, as well as for individualizing them. In this chapter, we discuss certain approaches to cancer analyses, such as the features of cancer survival and incidence and tumor growth models applied to national population cancer mortality and tumor registry data, and various techniques for assessing the quality and content of various types of data.

In the first section of this chapter, we describe the standard survival analysis procedures frequently employed in epidemiological and demographic studies of specific disease processes, and review their strengths and weaknesses in analyzing various types of longitudinal data. In the second section, we discuss the analysis of two presumable forms of female breast cancer using a two-disease model, assuming that "early" forms of female breast cancer may differ by several aspects of onset and progression from those at older ages (an additional discussion of that is presented in Chapter 7). In the third section, we apply the quadratic hazard stochastic process model for tasks related to analysis of cancer risk. We generalize standard approaches to incorporate nonlinear autoregression equations describing the covariate dynamics. In the fourth section, we discuss approaches to evaluate characteristics of individual and grouped data. Specifically, this section reviews procedures for evaluating the quality of information in certain data sets (e.g., on age reporting), and how various data sets may be linked and integrated in a statistical model. In the final section, dynamic heterogeneity is discussed for conditions when it is driven by latent influential variables, known as frailty. This approach allows us to develop a new class of model capable of describing the leveling-off in age patterns of cancer incidence and even its decline at advanced ages.

K.G. Manton et al., *Cancer Mortality and Morbidity Patterns in the U.S. Population*, Statistics for Biology and Health, DOI 10.1007/978-0-387-78193-8_4,
© Springer Science+Business Media, LLC 2009

In subsequent chapters (from 5 to 9), we present detailed applications of different types of models, estimated not only from population cancer mortality and tumor registry data, but also from various types of longitudinal data sets, where persons are followed for a significant period of time before death and, during that time, multiple state variables and their temporal changes are observed and measured.

## 4.1 Survival Analysis and Life Table Models

The most basic model for analyzing human survival and mortality is the life table, which describes the time to the occurrence of specific health changes – most often death due to a specific cause – in a human population. Life tables of different types have been constructed since Roman times. Ulpian (Dometius Ulpianus, died circa 228 AD), a Roman jurist and the emperor's chief adviser and Praetorian, has developed what now is known as Ulpian's life tables, which were preserved in the *Digest* (a massive collection of Roman legal thought). Those life tables were the series of calculations to determine the taxes on annuities based on what seemed to be predictions of life expectancy. It may represent a very early attempt of estimating mortality rates; however, it is obscure as to what these data were based on (McGeough, 2004; Frier, 1982; Kertzer and Laslet, 1995). The very early life tables were predominantly created to analyze mortality caused by infectious disease, especially the bubonic plague. John Graunt (1620–1674), "a citizen of London", merchant and haberdasher, who was an amateur scientist, in 1662 has published his work "Natural and Political Observations . . . Made upon the Bills of Mortality", where he analyzed data on London mortality beginning from 1592 (Jones, 1945) and included mostly deaths caused by epidemic diseases, such as plaque, etc. Daniel Bernoulli (1700–1782), a Swiss mathematician, and a member of a famous family of mathematicians, physicists, and philosophers, did pioneering work in probability statistics and developed sophisticated life table procedures for analyzing the smallpox morbidity and mortality, involving censored data to demonstrate the efficacy of vaccination, and by reanalyzing Halley's[1] life tables for Breslau. In those life tables, he attempted to estimate the heterogeneity of the exposed population with respect to its susceptibility in death from specific infections.

Classically the life table model (Chiang, 1968) and its parameters are constructed from the cohort mortality experience of a population born at a specific

[1] Edmond Halley (1656–1742), an English astronomer, geophysicist, mathematician, meteorologist and physicist, who published in 1693 an article on life annuities, in which he analyzed the age-at-death on the basis of the city of Breslau statistics provided by Caspar Nemann (a clergyman from Breslau who had a special interest in mortality rates). Thus Halley influenced the developing of actuarial science. His work followed a more primitive work by Graunt, and is one of the most important studies in the history of demography.

date followed till the last member of the cohort dies. The basic parameter of a life table is the age-specific mortality probability, $q_x$, which can be defined as,

$$q_x = \frac{n_x - n_{x+1}}{n_x} = \frac{d_x}{n_x} \tag{4.1}$$

where $n_x$ is the size of the population at age $x$, and $d_x$ is the number of deaths occurring at age $x$. All other life table parameters (e.g., $l_x$ proportion of the cohort surviving to age $x$; $J_x$ person-years lived after age $x$) may be calculated from the $q_x$. Life tables may be calculated in an abridged fashion by grouping a set (e.g., 5 or 10) of ages together. For example, the mortality probability for a 5-year age category, ${}_5q_x$, can be calculated as,

$${}_5q_x = \frac{n_x - n_{x+5}}{n_x} = \frac{{}_5d_x}{n_x} \tag{4.2}$$

The life table, as commonly defined and applied, makes no assumptions about the relation of the risk of death across the age categories, i.e., it treats time–age nonparametrically. Some assumptions, however, have to be made about the time dependence of the rate of deaths occurring within an age category, i.e., the distribution of the ages at death within each age interval. An individual hazard rate may also be viewed as directly related to a person's biological frailty (e.g., Simms, 1942).

For complete (single year of age) life tables within age category, distributional assumptions generally have little effect, except on life expectancy estimates at very young ages, where there is significant infant mortality during the first year of life. Often model-based smoothing assumptions over age categories are imposed where data are subject to significant error: e.g., the observed U.S. morality experience above the age 95 in SSA life tables (Life Tables for the United States Social Security Area) was used to fit parameters of a Gompertz-like hazard function at those very late ages (SSA Actuarial Study No.116, 2003).

From the age-specific probabilities of death, a number of other useful age-specific life table parameters can be calculated to examine changes in mortality and population survival over the life span. For example, life expectancy at age $x$ can be calculated as,

$$e_x = \frac{T_x}{n_x} = \frac{1}{n_x}\int_x^{x'} n(a)da, \tag{4.3}$$

where $T_x$ is the total number of years of life remaining for the considered cohort, and $x'$ is age when the last member of the cohort dies. The life expectancies are often compared cross-nationally at ages 65 and 85 to examine the rate of population aging and to compare the efficiency of the health-care system for different age groups and demographic groups in different countries (e.g., Japan and the United States).

Basic life tables can be generalized by dividing the set of observed deaths at an age group into $m$ cause of death-specific categories for that age. This is usually done using the health condition or disease reported on the death certificate as "the underlying cause of death" (Manton and Stallard, 1984). The total probability of death can be calculated by summing the $K$ cause-specific probabilities of death for a given age, or,

$$q_x = \sum_{k=1}^{k} q_x(k). \tag{4.4}$$

The life table calculated from the mortality experience of a population, all of whom have died from a specific cause (e.g., cancer), commonly is called a single decrement table.

In cause-specific analyses of mortality and life expectancy, it is often useful to make competing risk adjustments for specific causes of death, i.e., to determine the changes in the probability of death, assuming that the effect of a given disease (e.g., cancer) on mortality may be completely or partially eliminated. Also one can calculate changes in mortality from a cause (e.g., CVD), when another disease (e.g., cancer) is eliminated. Specifically assume that each person in the population is at risk of dying from each of $k$ possible independent causes of death. Each of the $k$ disease-specific risk processes generates a theoretical distribution of times to death. The set of $k$ cause-specific theoretical failure time distributions are assumed to be mutually independent in standard competing risk computations (Chiang, 1968).

Assuming the independence of the force of mortality of each of $k$ individual causes of death means that the persons, whose deaths are prevented at age $x$ by the elimination (or partial elimination) of a specific cause, are subjected to the risks of death from all remaining causes of death. The probability of death net of the effects of the $k$th cause is then simply,

$$q_{x.k} = 1 - (1 - q_x)^{(q_x - q_x(k))/q_x} \tag{4.5}$$

From this formula, one can see that if two risks are such that $q_x(k) > q_x(m)$, then $q_{x.k} < q_{x.m}$. The assumption of independence implies that a person observed to die of the eliminated condition will die from the cause with the second (theoretical) lowest age at death (Chiang, 1968).

In fact, cause of death processes may be dependent on each other or interact with each other within individuals, thus generating $k$-correlated time to death distributions and mortality probabilities. For example, diabetes mellitus type II is a risk factor for stroke – so, the elimination of diabetes as a cause of death (implicating the modification of the, possibly, multidimensional disease process associated with diabetes) may also reduce the risk of death from stroke.

Estimating the precise degree of correlation between the processes underlying specific causes is difficult using mortality data where only the first (earliest) underlying cause of death is observed for each person (i.e., the correlation of pairs of causes are unobserved because only the cause with the first time to death is observed for any given person, and the time to death from the $k$–1 other conditions are missing data). Tsiatis (1975) showed that one could calculate the potential range of competing risk effects assuming different degrees of correlation between each of the $k$ causes.

More information on that multidimensional cause of death distribution is available, if one has the multiple causes of death (usually recorded on the death certificate for each death), i.e., contributory diseases modulate the underlying cause effects on the observed time of death (Manton and Stallard, 1984) (e.g., when the age at death caused by cancer is accelerated because the person has developed pneumonia).

Such multiple cause of death recording can be used to produce more informed estimates of the degree of correlation of specific causes of death. One way such multiple cause data can be coded to conduct such multivariate life table analyses is by defining a $k$ element vector of 0's and 1's for which a subset of $k$ causes of death are found (code 1.0 – for absent causes, and code 0.0 – for presented causes) in the death certificate for sets of individuals. These vectors may be called "patterns of failure" and, since they explicitly encode all permutations of $k$ causes, they encode all dependence for causes observed at the time of death. Using the $2^k$ patterns of failure as independent individual causes, life tables can then be calculated. When health changes prior to death are observed for a significant period of time, the correlation of those diseases can be directly calculated using various explicit stochastic process models for state variable dynamics.

### 4.1.1 Useful Modification of Standard Life Table Computations

Four variations of population life table calculations are often used.

(1) The first is to calculate a period life table from all deaths occurring over a fixed time period, such as a year. For example, the period life table is frequently calculated, for year $y$, using,

$$L_{ay} = \int_0^1 n_{(a+t)(y+t)} dt \tag{4.6}$$

This assumes that the health experience of the persons in different cohorts, from the date of birth to death at age $x$ in year $y$, does not change markedly over time, i.e., mortality risks vary only over age and period. This is clearly a strong assumption and unlikely to be practically fulfilled. However, period-specific calculations of life expectancy at a given age still has considerable public health value as an index of overall population changes in health and mortality over time. It is still frequently used to compare the relative health state of different national populations (e.g., CIA World Factbook – an annual publication of the Central Intelligence Agency of the United States with almanac-style information about the countries of the world) because it can be calculated using vital statistics and Census data for most countries for recent time points.

(2) The second useful modification of standard life table computations is when instead of eliminating a cause of death, the age at death from a given condition is shifted upward a fixed number of years (say $l_k$), reflecting observed or expected improvements in survival for specific causes of death (e.g., a given number of additional years of survival for women with breast cancer due to improved treatment of early forms of breast cancer). This problem is conveniently considered in terms of the cumulative force of mortality. The probability of death $q_{i,delay}$ $(l_k)$ in this "cause delay" life table can then be calculated as (Manton et al., 1980)

$$q_{i,delay}(l_k) = 1 - (1 - q_{i-m})^{q_{i-m}(k)/q_{i-m}}(1 - q_i)^{1-q_i(k)/q_i} \tag{4.7}$$

Here, we assumed that that delay time $l_k$ is multiple of age interval size ($\Delta x$) in the life table, i.e., $m = l_k/\Delta x$; a general result is given in Manton et al. (1980). For $m = 0$ equation (4.7) is the identity, i.e., $q_{i,delay}$ $(l_k) = q_i$, and for $m > i$ it reproduces equation (4.5). The authors of this paper applied the approach to the U.S. population as follows: (i) age–race–cause–specific mortality counts were tabulated using multiple causes of death data, taking into account all data occurring in 1969 and (ii) age–race–sex–population data were derived by back-dating the adjusted census data to January 1, 1969. Table 4.1 presents the effects on the life expectancy at selected ages from delaying specific causes. Specifically, for the white male population, the net effect on life expectancy at birth of delaying the age at death from cancer by 5 years is 0.8 years increase in life expectancy. If cancer were wholly eliminated as a cause of death at birth, the increase would be 2.3 years.

(3) The third type of life table is useful when one wishes to calculate the probabilities of death at extreme ages, where population age reporting may be viewed as not totally reliable, but where ages reported on death certificates may be assumed reliable (Kestenbaum et al., 1992; Kestenbaum and Ferguson, 2002). This synthetic "extinct" cohort life table approach (Manton and Stallard, 1996) involves calculating the age-specific probabilities of death as

$$q_x^{ext} = \frac{D_x}{P_x^{ext}} = \frac{D_x}{\sum_{t=x}^{118} D_t} \tag{4.8}$$

**Table 4.1** Calculated effects on life expectancy at selected ages from delaying

| | White males' life expectancy with delay at | | | | | White females' life expectancy with delay at | | | | |
|---|---|---|---|---|---|---|---|---|---|---|
| At age | 0 years | 5 years | 10 years | 15 years | $\infty$ years | 0 years | 5 years | 10 years | 15 years | $\infty$ years |
| 0 | 68.0 | 68.8 | 69.3 | 69.7 | 70.3 | 75.6 | 76.3 | 76.8 | 77.3 | 78.2 |
| 30 | 41.2 | 41.9 | 42.5 | 42.8 | 43.5 | 47.8 | 48.5 | 49.0 | 49.4 | 50.3 |
| 65 | 13.1 | 13.4 | 13.7 | 14.0 | 14.6 | 17.1 | 17.3 | 17.6 | 17.7 | 18.4 |

where the estimate of the population count at age $x$ (here set at a maximum at age 118) is formed by summing back to age $x$ all deaths observed from the highest age of death (i.e., 118-year-old) observed to age $x$ (Manton and Stallard, 1996). The problem with these calculations is that the population estimated by the summation of deaths at age $x$ may be survivors from cohorts, each of which differed significantly in size at birth. Cohort sizes differences can be adjusted, if estimates of initial differences in birth cohort sizes are available. Demographic sources on birth statistics may be more reliable than the estimates of the population at specific ages above 95.

(4) The fourth useful modification of the life table is to calculate the amount of time expected to be spent in a specific health state prior to death. Such "active" life expectancy (ALE) calculations are thought to better reflect differences in health status in developed countries, where life expectancies have reached comparatively high levels. Calculations of ALE require a second set of data (usually from a nationally representative survey) to estimate the prevalence of a specific health state (such as persons with a specific set of functional impairments) among survivors to a specific age. If this estimate, as it is often done, is made from an independent health survey for the same time (Robine et al., 2003a, b), one can use that prevalence estimate normalized by the proportion surviving to that age (from vital statistics) to calculate ALE using the cross-sectional methods (originally was due to Sullivan (1971)).

The results may be plotted as shown in Fig. 4.1 (Manton et al., 2006).

As we can see, the proportion of disabled individuals survived to a given age tends to decrease in the United States over time, i.e., ALE is growing relatively faster than LE. We have provided the plots for years 1965, 1982, and 1999 for males aged 65 and above to illustrate how this measure may be a useful summary of the change in the proportion of the remaining lifetime, that is expected to be spent either disabled or not disabled (Manton et al., 2006).

Of substantive interest is that the relative increase in ALE (and decrease in chronic disability prevalence in the United States) was more rapid at ages 85 and above. If one actually has the disability and mortality experience of a cohort of individuals followed over time, then the health transitions for a cohort of individuals can be directly studied (Manton and Land, 2000a, b). This dynamic analysis will be discussed in more details below – in the discussion of stochastic process models. One useful relation to using ALE measures is to determine how the proportion of life expectancy remaining at age $x$ is related to

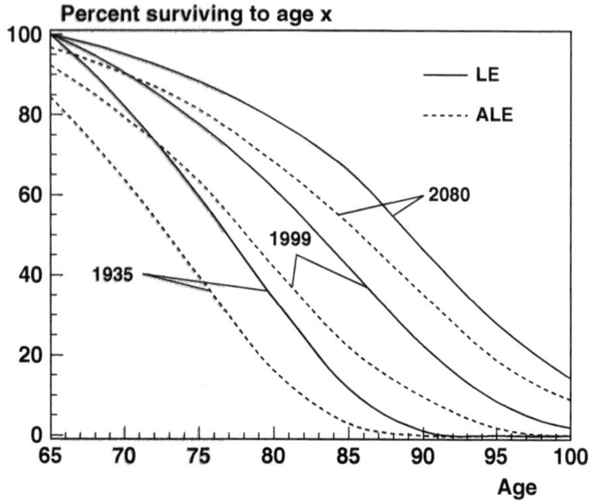


**Fig. 4.1** Survival curves (LE and ALE) at ages from 65 to 100 for years 1935, 1999, and projected to 2080, U.S. population (Manton et al., 2006)

the age-specific proportion of the total force of mortality that is due to cancer as a cause (i.e., to examine the correlation of cancer morbidity and mortality processes to the total risk of chronic disability in a population). This allows one to see how cancer mortality and its total impact on a population's health affect the overall, age-specific distribution of functional impairments.

The life table analyses described above are often done for the complete experience of an observed national population at a specific date (i.e., using decennial Census data and annual vital statistics enumeration), so that statistical inference may not be required (i.e., one is directly calculating realized population parameters). However, often such survival statistics are based on relatively small survey samples (e.g., ALE estimates), in which case the sampling properties of the statistical parameters estimates are relevant for making decisions. The total life table can be viewed as a chain binominal model with parameter variance estimates having the distributions as described in Chiang (1968). Cause of death specific life tables might be viewed as chain multinomial models. For some situations, categorical (grouped or contingency table) data modeling procedures are appropriate, such as the log linear and other grouped or aggregate data methods described by Bishop et al. (1975).

### *4.1.2 Standard Regression Procedures and Some of Their Limitations in Describing Longitudinal Data*

It is often the case that the available sample is either too small or the number of variables one wishes to statistically control is too large to support the standard categorical or grouped regression analyses of the type discussed above. This is because the number of model parameters rapidly increases as the number of higher order interactions that need to be evaluated increases. In these cases, the use of simplifying parametric assumptions in regression models are often necessary.

One approach is to use the parametric regression procedures without higher order interactions. However, since the (discrete) dependent variables are usually binomially, or multinominally, distributed, a standard linear regression model for individual data may not be appropriate, i.e., the dependent health outcome probabilities in a linear regression may fall outside the 0–1.0 range. One approach in this case which prevents out of range values is to use a logistic ($p/1-p$; an approximation of the odds ratio) transformed to make each of the survival probabilities a logistic function of covariates. The problem with using the logistic function as a model of chronic disease risk over time is that products of logistic variables for aggregations, or disaggregations, of survival experience over different time intervals will not be logistically distributed. Thus, the regression model is estimated assuming a logistic distribution is not closed under the aggregation of time period estimates of event rates (Woodbury et al., 1981), i.e., when logistic functions for each subinterval are multiplied they do not reproduce a logistic function for a total interval. Therefore, the unfavorable conclusion is that if logistic function is assumed to describe the relation of disease risk to risk variables for a study of a given length, then the coefficients of logistic multiple regression estimated from that study cannot be applied to a different length study (Woodbury et al., 1981). In this sense, the logistic is not a natural function to describe the results of dynamic processes for arbitrary time length.

A second model that is frequently used in survival (time to event) modeling is the Cox (1972) proportional hazard regression. In Cox regression, it is assumed that the effects of covariates are proportional to an effect on an unknown base hazard function. The base hazard is treated in a partial likelihood function as a nuisance parameter. One problem with the Cox proportional hazard model is that the assumption of proportional hazards is a substantive hypothesis that needs to be directly evaluated in each empirical application, i.e., statistical inferences about effect parameters are only valid if the survival curves for each risk state are proportional to the base hazard. For lengthy longitudinal studies, such as the Framingham Heart study, where risk factors may change over time, or at risk subgroups may be exhausted, such an assumption may be questionable. For extremely small clinical studies, where one may not have adequate power to evaluate whether the assumptions of proportionality holds, one might use nonparametric life table procedures, such as the Kaplan–Meier or the log rank test (Kaplan and Meier, 1958). A major recent study where this effect was evident was in the trial of hormonal replacement therapy in the Women's Health Initiative study (Writing Group for the Women's Health Initiative Investigators, 2002), in which the major health benefits and risks of the most commonly used combined hormone replacement therapy in healthy postmenopausal women (i.e., estrogen-plus-progestin therapy) has been assessed. That was the first randomized controlled trial to confirm that combined estrogen plus progestine increased the risk of breast cancer. The overall health risks obtained from this study exceeded the benefits from use of this type of therapy in postmenopausal women: the substantial risks for cardiovascular disease and breast cancer must be weighed against the benefit while choosing the medicine to prevent osteoporosis in postmenopause. An examination of the tables of outcomes presented in the paper suggests that the hazard rates are not proportional over the time of follow-up. Also assumed in the Cox model is that the population is heterogeneous only on the measured risk factors. Otherwise, over time bias due to mortality selection on latent risk factors could distort the results (e.g., Hougaard, 1988).

A third procedure that is sometimes employed in analyzing such longitudinal health data is an event history modeling (Allison, 1984). In event history modeling, the episode-specific transition rates are constructed for periods between measurements of covariates using logistic or Cox regression functions. In longitudinal studies with multiple risk factor assessments made over time, information on co-variable changes can be utilized. However, since the covariate vectors are fixed at the time of measurement, the parameters of the covariate dynamics are not described by the model.

A further complication of event history models is when measurements are triggered by changes in the state of interest, i.e., the measurement process is not fixed but is driven by health changes [e.g., as in the Social Health Maintenance Organization (S/HMO) (Manton et al., 1994)]. In this case, missing data (e.g., the failure to detect a health transition, so covariates are not measured relative to significant changes in health) is problematic because there is no parametric structure to infer covariate values at those times. In models based on stochastic differential equations, the problem of missing data does not occur because the process's parameters are estimated directly (Yashin and Manton, 1997).

In addition to Cox and logistic regressions, there are regression models useful for modeling cancer mortality which employs other types of error assumptions. Two most useful options are Poisson and negative binomial regressions. The Poisson regression for grouped or count data assumes homogeneity of the population at risk in categories so that the event rate is a Poisson rate parameter, $\lambda$, with variance $\lambda$. The negative binomial allows for individual risk heterogeneity within cells so that the event rate, $\lambda$, has super-Poisson variability $\lambda + \lambda^2/r$, where $r$ reflects the variability/dispersion of individual rate parameters within cells. This means that the rate parameter for a cell can be calculated using the shape and scale parameter estimates (Manton et al., 1981). Life tables could then be constructed by using maximum likelihood estimates of shape and scale parameters fixed for covariates in the model.

In Manton et al. (1989), the negative binomial regression function was generalized to allow for aerial or geographic variation in risk over all 3,000 U.S. counties. This is illustrated in the example discussed in Chapter 5 for kidney cancer. In this model, several components of variability are adjusted for dispersion in the negative binomial dispersion parameter due to difference in the size of U.S. county population. Rate estimates can be produced that are "shrunken" toward the overall population rate estimates to eliminate excess variability due to wide difference in the U.S. county population size.

The logistic and Cox regression models are procedures to analyze observed time to failure distributions and the effects of fixed covariates on those distributions. They do not directly represent assumptions about the nature of the risk factor processes being analyzed, or of how changes in risk factors are associated with health risks. Such evaluations must be substantively driven. For example, an exponential failure rate implies that the failure process has a particular nature. Gompertz functions are often used to smooth life tables where the data are sparse and the age-to-age variation of the survival parameter is irregular. The Gompertz hazard is frequently justified in terms of specific biological models of aging and mortality. Different theories of aging and senescence leading to different failure models (e.g., Gompertz and Weibull) are discussed by Strehler (1977). All of these models are estimated using univariate time to failure distribution. Yashin and Iachine (1997) described other cases where there are bivariate correlated outcomes, such as in twin studies where common genetic inheritance assumed to follow a specific distribution (e.g., a gamma) generates a correlation in the bivariate hazard models. Review of modern approaches to modeling mortality and aging based on the existing theoretical reliability models and approaches, which are helpful in understanding the mechanisms and age dynamics of systems failure, is presented by Gavrilov and Gavrilova (2006a, b).

## 4.2 Multiple Disease Stochastic Compartment Models for Complex Cancer Population Mortality Curves: A Two-Disease Analysis of U.S. Female Breast Cancer

In Chapter 2, we discussed specific hazard models based on the assumption that different failure processes led to tumor initiation at a given age. Most models are based on the Weibull hazard function applied to either cancer mortality (usually with a lag or latency parameter) or cancer incidence data. Interpretation of the parameters of such models depends upon the specific theory of carcinogenesis being applied. For any specific application, it is often necessary to modify the basic Weibull model using assumptions about the heterogeneity over persons of the parameters of genetic mutations and tumor growth processes, to reflect the real world conditions of applications to studying specific tumor types. In these applications, we proposed using the technique of nonlinear stochastic compartment models under semi-Markov conditions (convoluted waiting time distributions with different time frames) about time dependence to infer the effects of unobserved biological processes on (a) the initiation of and (b) growth and dissemination of tumors (e.g., Jacquez, 1972; Matis and Wehrly, 1979; Manton and Stallard, 1988).

This is the first type of cancer process heterogeneity modeled due to the effects of latent influential variables. In this situation, the observed time to death from cancer and other causes, is decomposed based on unobserved state residences in health state-specific life tables (as waiting time distributions for the time resident in each of those latent health states, or "compartments"). The total time to cancer death from birth, $f_c\ (a)$, is constituted from the evolution of the time from birth to tumor initiation $f_1(a)$, the time from tumor initiation to clinical manifestation of the tumor, $f_2(a)$, and the survival time with a diagnosed cancer without or with possible treatment $f_3(a)$. The results of the condition (as in Manton and Stallard, 1988) is

$$f_c(a) = \int_0^{\min(a,w)} f_3(u)\left\{\int_0^{a-u} f_1(a-u-t)f_2(t)dt\right\}du, \tag{4.9}$$

where $w$ is time at which a cure or recovery is assumed. The three waiting times are modeled as three distinct processes, each with their own waiting time vector, with two latent processes (tumor initiation and tumor progression) and one component vector being an observed waiting time distribution (e.g., a survival life table function estimated directly from SEER or other cancer survival statistics).

Breast cancer is sufficiently complex to illustrate a number of generalizations and features of the multistage model of human carcinogenesis (Armitage and Doll, 1954, 1961), when attempting to specialize the model to deal with tumor initiation processes in specific tissues. The base model assumes the risk of death from a tumor of each of two possible forms: "early" cancer and sporadic cancer (with onset usually at older age). This can be described by a mixture of two Weibull hazards modified to take into account (1) individual unobserved susceptibility (or frailty) modeled by the gamma distribution and (2) nonzero lag period, $l_d$. After these modifications, each of these two hazards takes the form

$$\mu_d(x;\bar{\alpha}_d,m_d,s_d,l_d) = \frac{\bar{\alpha}_d(x-l_d)^{m_d-1}}{1+\bar{\alpha}_d(x-l_d)^m/m_d s_d}, \tag{4.10}$$

where $d = 1$ for "early" cancer onset (predominantly these patients had a family history of breast cancer) and $d = 2$ for late cancer onset (predominantly without family history of breast cancer), $\bar{a}_d$ is the disease specific mean hazard at age $x=0$ (i.e., mean of gamma distribution representing individual frailty), $s_d$ is the variance of this distribution, and $m_d$ is the presumable number of mutational events that are necessary for tumor initiation. One reason why mixture of two hazards is necessary is that the simple or modified Weibull hazard does not apply to female breast cancer in Western female populations, where a slowing, or dip, of the hazard rate occurs about the age of menopause – the so-called Clemmensen's "hook" (De Waard et al., 1964; Anderson et al., 1974), with the age increase in cancer incidence increasing after the "hook." One of the possible explanations of the existence of Clemmensen's "hook" might be in different genes penetrance. The possibility of the higher prevalence of mutation carriers among women with early breast cancer onset is discussed by researchers during the recent two decades (more detailed discussion is presented in Chapter 7). That has similarity with the assumptions required in the two-stage model of familial retinoblastoma, where one mutation occurs after birth and one is fixed in the germ cell (Knudson, 1971), causing the hazard to rise linearly ($m = 1.0$). In nonfamilial retinoblastoma, the hazard increases quadratically ($m = 2.0$), suggesting both mutations are induced postnatally by environmental exposure.

Epidemiological data showed that breast cancer risk at later ages is negatively associated with age at first pregnancy (MacMahon et al., 1969). This suggests that the risk of this "late" form of breast cancer is a function of the length of estrogen exposure – especially estrogen exposure influenced by pregnancy and/or nursing. Some evidences indicate the slope of the log hazard rate for premenopausal breast cancer is higher (implying additional mutations) than the slope of breast cancer risk after menopause. This might suggests the probability of two forms of breast cancer.

We fit the complex U.S. female breast cancer mortality curve observed for 1969 (Fig. 4.2) by assuming there are two such functions, one for early and one for late cancer onset, with a proportionality factor, $\theta$ and $1-\theta$, weighting the two disease-specific functions (Manton and Stallard, 1980).

In our example, the total breast cancer mortality curve was well explained by the weighted combination of the total outcomes of two disease processes, each supposed to have different latencies ("early" form has a shorter latency – approximately 7 years, than sporadic form – around 20 years), but with the early disease having possibly a larger number of events resulting in cancer development, also suggesting that some genes might being predetermined to be more susceptible to environmental carcinogenes. The fit to 1969 U.S. breast cancer (total and for every suggested form) specific mortality rates generated using the two-disease breast cancer model are shown in Fig. 4.2. The total U.S. breast cancer mortality rate is, at each age, the sum of mortality from the early,

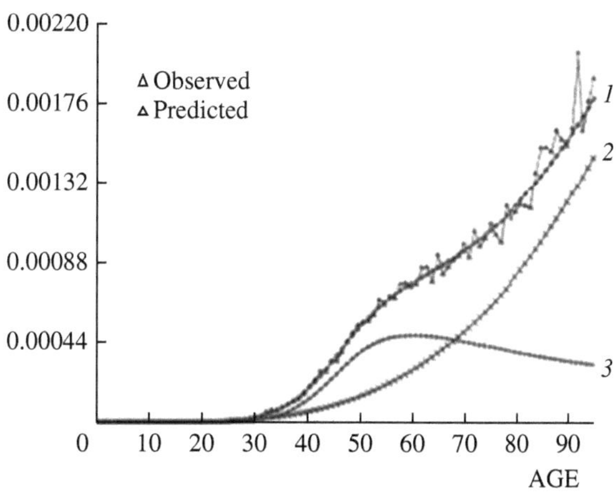


**Fig. 4.2** The age-specific trajectory of breast cancer for U.S. females in 1969 composed from Weibull processes

more aggressive form [with a short latency (rapid growth) and a high level of heterogeneity due to the relative rareness of its genetic determinants], and determined predominantly by noninherited factors sporadic form (with a longer latency, but with little evidence of individual heterogeneity of risk). The two curves in the figure under the total curve reflect the age trajectories of the two disease components. The curves could have been fit by different applications of some of the other models presented in Chapter 2, but the two disease formulation has a fairly direct biological interpretation and seems plausible in interpreting therapeutic results where early breast cancer tends to be so aggressive even when the primary tumor is still quite small, i.e., micro metastases begin early.

Cancer mortality is not an ideal characteristic for modeling by the multistage models of carcinogenesis. Strictly speaking, these models predict only cancer incidence rate, and extrapolation for mortality requires the consideration of the effects of therapy. Since cancer-specific death rates are measured much better than the respective incidence rates, application of these models to death rates is typical and broadly used. However, in this case obtained estimates of biological parameters in these models have to be carefully interpreted. In Chapter 7, we will further develop and apply the two-disease model for a set of sex–race–site–histology–specific incidence rates extracted from SEER data.

## 4.3 Stochastic Process Models of Cancer Risk: Latent and Observed State Variables Dependence

A more complete type of life table model can be constructed of the study of cancer mortality by extending the table so that it gives not only the distribution of times to death and disease incidence but also to describe cancer influencing individual's health status, as reflected by direct measurement of biomarkers or risk factors made over time. This life table model is based on analytic assumptions defining special forms of the Fokker–Planck equation (Risken, 1996) from the master equation and the quadratic hazard model.

In the life table model based on the Fokker–Planck equation (Woodbury and Manton, 1977, 1983), an individual's health state is described by the values of a set of risk factors or covariates observed at a specific time. Individual dynamics of covariates, and the rate of death specific to position in the health state space, have to be modeled to describe population or cohort state and mortality changes with age or time. The state dynamics are described by a system of autoregressive equations, in which the future health state is described in terms of current state, a drift function which can be dependent on age, and diffusion which represents latent stochastic, dynamic heterogeneity in the population. Mortality is assumed to be a quadratic function of covariates that provides biologically justified U- or J- shaped hazard functions of covariates (Witteman et al., 1994). The minimum of these quadratic hazard functions identifies a point (or, more generally, a multidimensional domain) in the covariate state space, corresponding to optimal health status (for total mortality, or – in a disease-specific model – a specific cause of death, like cancer) at a specific age. The probability of death increases in all directions going away from the optimal risk factor domain. This is illustrated at specific ages (e.g., age 65 and 95) in Fig. 4.4. Note that for certain risk factors only the right side of the parabola works. This is so-called J-shaped risks for which $\bar{x}$ can equal to 0 (e.g., for smoking as a risk factor for lung & bronchus cancer).

This model generalizes the concepts of individual frailty and dependent competing risks, and, in addition, allows one to directly consider the dynamics of observed age-dependent and unobserved covariates (risk factors).

With this model, it is possible to calculate life table functions and to forecast the characteristics of the health state distribution over age and time (Woodbury and Manton, 1977, 1983; Manton et al., 1992; Akushevich et al., 2005a). Thus, human aging and mortality are modeled by a random walk of an individual over a biological state space, with the distribution of "manholes" representing death risks at each point in the space assumed to be distributed quadratically over the state space. At the individual level, this model can be described by a system of stochastic differential equations for the individual's random walk over in the multivariable state space,

$$dx_w(t) = u(x_w, t)dt + d\xi(x_w, t), \tag{4.11}$$

with the survival probability $P(x_w)$ for individual $w$,

$$dP(x_w) = -\mu(x_w, t)P(x_w)dt. \tag{4.12}$$

The equation (4.11) represents changes for organism $w$ on each of $n$ health dimensions, $x = (x_j, j = 1,2,\ldots,n)$ at age $t$, while $u(x_w,t)$ is a deterministic function describing drift as a function of position in the state space. The $\xi(x_w,t)$ is a Gaussian process describing diffusion. Alternatively, this model can be represented for a population as a Kolmogorov–Fokker–Planck equation with a force of mortality term $\mu$, (Woodbury and Manton, 1977).

$$\frac{\partial f}{\partial t} = -\sum_j u_j \frac{\partial f}{\partial x_j} - f\sum_j \frac{\partial u_j}{\partial x_j} + \frac{1}{2}\sum_i \sum_j \sigma_{ij}^0 \frac{\partial^2 f}{\partial x_j x_j} - \mu f, \tag{4.13}$$

for a probability distribution $f = f(x,t)$ of individuals in the state space generalized to include the mortality term $\mu$. For linear drift, and a quadratic mortality function, $\mu$, this formulation provides direct analytic solutions. An advantage of these Fokker–Planck models for estimation is that the likelihood function, under certain general conditions, can be factored into three independent terms, i.e. the initial state distribution, state variable dynamics, and the parameters of the hazard function (Manton and Stallard, 1988; Manton et al., 1992; Akushevich et al., 2005a).

There are two types of random variables necessary to factor the overall likelihood: discrete age ($T$) at last survey or measurement before death and the covariate $J \times (T - t_0 + 1)$ matrix $X \stackrel{def}{=} x_i = x_{jt}$, where $j$ runs over covariates and discrete age $t$ changes from the age of forming the cohort ($t_0$) to $T$. The probability density is defined to reflect both survival and risk factor dynamics:

$$p(X,T) = p_1(\mathbf{x}_{t_0})\{\prod_{t=t_0}^{T-1} S(t|\mathbf{x}_t)\phi(\mathbf{x}_{t+1}|\mathbf{x}_t)\}\{1 - S(T|\mathbf{x}_T)\}, \tag{4.14}$$

where $p_1(\mathbf{x}_{t0})$ is the probability density of the risk factor distribution at $t_0$, and $\phi(\mathbf{x}_{t+1}|\mathbf{x}_t)$ is the transition probability matrix for state variable processes, e.g., $\mathbf{x}_{t+1} = u_0 + R\mathbf{x}_t^* + \varepsilon$, where $R$ are the transition coefficients. Different specifications of this function can represent different assumptions about the nature of state variable dynamics. For example, extending $\mathbf{x}_t$ to represent two prior times of measurement can describe more general types of state variable processes.

Because of the assumed normal distribution of residuals $\varepsilon$ in these formulas, $\phi(\mathbf{x}_{t+1}|\mathbf{x}_t)$ has the form of a multivariate Gaussian distribution. The normalization property for $p(X,T)$ now reads:

$$\begin{aligned}\lim_{T'\to\infty} \sum_{T=t_0}^{T'} \int dx_{t_0} \ldots dx_T\, p(X,T) = \lim_{T'\to\infty} \{\int d\mathbf{x}_{t_0} p(X,t_0) + \\ + \int d\mathbf{x}_{t_0} d\mathbf{x}_{t_0+1} p(X,t_0+1) + \ldots + \int d\mathbf{x}_{t_0} d\mathbf{x}_{t_0+1} \ldots d\mathbf{x}_{T'} p(X,T')\} = 1.\end{aligned} \tag{4.15}$$

The corresponding likelihood can be written as

$$L = \prod_i p(X^i, T^i). \tag{4.16}$$

Here, and below, the superindex $i$ marks the data ($\mathbf{x}_t^i$, $t^i$, $t_0^i$ and $T^i$) measured for person $i$. The likelihood (4.16) can then be written as a product of three independent terms, $L = L_1 L_2 L_3$ (Manton et al., 1992), where $L_1$ contains $p_1(\mathbf{x}_{t_0})$, $L_2$ includes transition probability densities $\phi(\mathbf{x}_{t+1}|\mathbf{x}_t)$, and $L_3$ includes only the survival probability $S(t|\mathbf{x}_t)$:

$$L_1 = \prod_i p_1(\mathbf{x}_{t_0^i}^i), \tag{4.17}$$

$$L_2 = \prod_i \prod_{t=t_0^i}^{T^i-1} \phi(\mathbf{x}_{t+1}^i|\mathbf{x}_t^i), \tag{4.18}$$

$$L_3 = \prod_i \{\prod_{t=t_0^i}^{T^i-1} S(t|\mathbf{x}_t^i)\}\{1 - S(T^i|\mathbf{x}_{T^i}^i)\}^{\delta_{T^i}}, \tag{4.19}$$

$\delta_{T^i} = 1$ if the death of individual $i$ is detected after the last time of measurement, and $\delta_{T^i} = 0$ when an individual is still alive or vital status is unknown. Since these three likelihoods contain nonoverlapping sets of parameters, each of them can be maximized separately. The first is maximized analytically assuming the normality of the initial distribution of risk factors. Maximization of $L_2$ can be done using linear least-squared methods with different specifications of the dynamic equation, used for linear and nonlinear dynamic models, respectively. Maximization of the survival likelihood term $L_3$ is more complicated due to the need for constraints [details of this optimization task are provided in Manton et al. (1992) and Akushevich et al. (2005a)].

Such an independent likelihood factorization requires data with small and fixed time intervals, which is not always the case. Furthermore, the application of this model to measurements, which are right censored or include informative missing data, requires additional assumptions. All these problems can be addressed without additional assumptions by using a modification of this approach, known as the stochastic process model (discussed in Chapter 5).

Assumptions about linear drift and constant diffusion restrict the predictive power of the model. Akushevich et al. (2005a) described how to generalize the model for dynamics of any complexity (e.g., violation of the Markovity condition, anomalous diffusion) by using a formalism based on the microsimulation of individual state trajectories. Applications of microsimulation models to estimation, intervention analyses, and forecasting are discussed below (and will be illustrated for different examples in later chapters).

After dynamic and mortality function parameter estimation, the calculation of life table parameters from those estimates has to be performed. The best life table solution is analytical, which is possible only under the specific assumptions addressed above. When these assumptions cannot be applied, another scheme, based on microsimulation estimation procedures, has to be applied.

An analytic solution of the stochastic Cauchy problem formulated in the form of the Fokker–Planck equation (4.13) can be found under specific conditions (Woodbury and Manton, 1977). The first assumption is that the population distribution can be described at time $t$ as a multivariate normal distribution $N(l_t,v_t,V_t)$, whose three parameters represent the population size ($l_t$), the vector of physiological variable means ($v_t = E(\mathbf{x}_t)$), and the variance–covariance matrix ($V_t = Var(\mathbf{x}_t)$). This is equivalent to assuming linear drift over $\mathbf{x}_t$ and constant diffusion. The last term in equation (4.13) ($\mu f$), corresponding to the force of mortality, changes the normalization ($l_t$) of the multivariate distribution function over time. The second assumption is that mortality is a quadratic function of risk factors or state variables at time $t$,

$$\mu(\mathbf{x}_t) = \mu_0 + b_t\mathbf{x}_t + \frac{1}{2}\mathbf{x}_t^T B_t\mathbf{x}_t. \tag{4.20}$$

This function describes the probability of dying, conditional on the multivariate health state described at time $t$ by the vector of risk factors $\mathbf{x}_t$. The

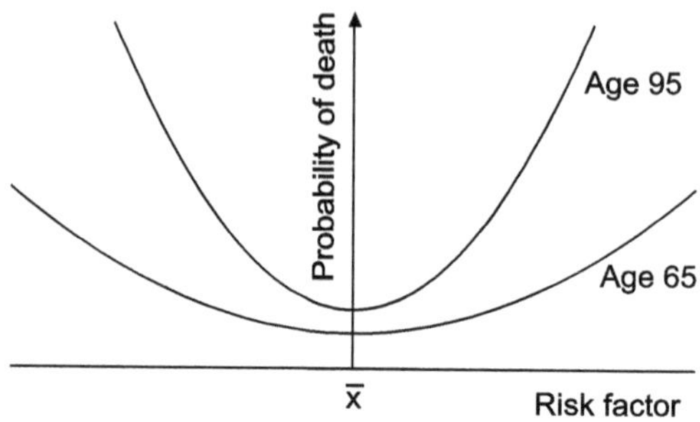


**Fig. 4.3** Probability of death

quadratic form of $\mu_t$ helps to model a situation when there is a point (or a domain) in the state space optimal for health (e.g., the vertex of the age-specific paraboloid; see Fig. 4.3), where mortality increases going away from this domain. Hence, matrix $B_t$ is assumed to be positive definite. For discrete time $\mu_0$, $b_t$ and $B_t$ parameters have the meaning of coefficients of the cumulative force of mortality for the age interval $(t,t + 1)$, i.e., risk in this function is expressed in absolute (e.g., number of events per some number cases), not relative, terms. The numerical estimation of these parameters in data with different sample and measurement characteristics are discussed below in this section.

An additional useful property of the quadratic hazard function is its ability to deal with the dependent competing risks. This is because the total force of mortality can be composed from the sum of the $K$ disease-specific hazards, or,

$$\mu_{TOT} = \sum_{k=1}^{K} \left(x_{it}^T Q_k x_{it}\right) \tag{4.21}$$

where, if the hazard function coefficients were proportional, there could be a decomposition of $\left(x_{it}^T Q x_{it}\right)$ into $k$ components. There may, however, be unexplained variation regarding the $Q_k$ function that is age dependent. This can be dealt with by writing the stochastic hazard function as age dependent in the expanded form,

$$\mu_{TOT} = \sum_{k=1}^{K} \left(x_{it}^T Q_k x_{it}\right) e^{\theta(k)-t} \tag{4.22}$$

If each of the $k$ cause-specific quadratic functions is estimated directly, then the component quadratic functions could be summed to form the total hazard. Each of the $k$ components represents how the force of mortality for the $k$th disease depends on the risk factors and age. Thus, by eliminating the $k$th hazard component from the sum in equation (4.22), one can calculate the net risk of death for the $k$–1 conditions in a way which reflects the correlations of disease $k$ with all other $k$–1 conditions, where the correlations of the two disease sets is modulated by the age and temporal dynamics of the state variables (Yashin et al., 1986). Because the $Q_K$ is, in practice, separately estimated, this allows for contradictory effects of a risk factor between two causes (e.g., the predominantly negative effect of smoking on human health, especially dramatically increases lung cancer risk, however, it probably could decrease the risk of certain immunologically related diseases, such as ulcerative colitis, which, in its turn, has an increased colon cancer risk). Some of these potentially antagonistic effects may operate through the nonlinearity of the hazard function.

If the first and second assumptions hold, the covariate distribution of survivors over the time period will remain normal $(N(l_t^*, v_t^*, V_t^*))$. This allows us to calculate (see equation (4.23)) the characteristics of the surviving population (vectors with asterisk) using the characteristics of the total population and the hazard coefficients b and B (Woodbury and Manton, 1983),

$$v_t^* = v_t - V_t^*(b_t + B_t v_t), \quad V_t^* = (V_t^{-1} + B_t)^{-1}. \tag{4.23}$$

Another consequence is that an explicit expression for the normalization parameter $l_t^* = l_{t+1}$ of the risk factor distribution, related to the survival function $l_t/l_0 = S_t \stackrel{\text{def}}{=} E(S(t|\mathbf{x}_{t_0}, ..., \mathbf{x}_t))$, can be written. The survival function starting as $S_{t_0} = 1$ is

$$S_{t+1} = S_t|I + V_t B_t|^{-1/2} \exp\{\frac{\mu(v_t) + \mu(v_t^*)}{2} - 2\mu(\frac{v_t + v_t^*}{2})\}. \tag{4.24}$$

Formula (4.24) is obtained without additional assumptions about the form of the dynamic equations. The time period appearing in the left-hand side of equation (4.24) exactly corresponds to the time period of the cumulative force of mortality in the right-hand side.

Next, we need to construct a dynamic model describing changes in risk factors over time. Below we discuss the different models, keeping in mind that the parameter estimation needed for covariates is measured at discrete times. The simplest model used by Woodbury and Manton (1977) is first order, linear auto regression with respect to variables $\mathbf{x}_t^*$,

$$\mathbf{x}_{t+1} = u_0 + R\mathbf{x}_t^* + \varepsilon. \tag{4.25}$$

In the development here, we use Gaussian assumptions about the initial covariate distribution with probability density $p_G(\mathbf{x}_{t_0}, v_{t_0}, V_{t_0})$, where mean $v_{t_0}$ and variance–covariance matrix $V_{t_0}$ are directly estimated from data. The vector $u_0$ and the regression matrix $R$ are calculated using ordinary least-square methods, and $\varepsilon = \sigma(\mathbf{x}, t)dW_t$ will be approximated by a time-independent normally distributed random variable. The asterisk for $\mathbf{x}$ in the right-hand side of equation (4.25) emphasizes the fact that $\mathbf{x}^*$ belongs to the distribution of individuals who are alive at time $(t + 1)$. A valuable feature of the linear model is that it preserves the normality of covariate distribution over time. The vector of means and the variance–covariance matrix can be calculated as:

$$v_{t+1} = u_0 + Rv_t^*, \quad V_{t+1} \sum +RV_t^* R^T, \tag{4.26}$$

where $\Sigma = Var(\varepsilon)$ is the empirical diffusion matrix. Thus, this model is based on the assumption that changes in each risk factor are related to the linear superposition on their prior values.

Assumptions about linear drift and constant diffusion may unduly restrict the predictive power of the model. The basic restriction is due to assumptions preserving the normal form of the covariate distribution, which is often only a first approximation of biological reality. Apart from the very strict analytical form required to use this distribution, it implies a finite probability of having unnatural (negative) values for covariates. Attempts to overcome this obstacle (e.g., to artificially keep covariates positive) require functional dependence of diffusion on covariates that may destroy the normality of the covariate distribution.

Another problem is related to assumptions about linear drift. Autoregressive models used to estimate covariate dynamics have $J$ parameters per a covariate (e.g., $J=11$ was used by Manton et al. (1992) and Akushevich et al. (2005a)), while longitudinal studies or surveys provide roughly $I \times N$ measurements per covariate. Thus, the linear model may not be rich enough to describe such data, i.e., many more parameters provided by the linear model can be estimated using such data.

Third, the analytical solution provides projections for population characteristics at the population (macro) level with limited possibilities for modeling at the individual level. Other factors affecting the assumptions of normality are nonlocal interaction effects and various possible biological scale effects.

Many of those factors may require dealing with more general, or anomalous, diffusion processes (Metzler and Klafter, 2000) and require generalization of stochasticity to include nonextensitivity, i.e., the interaction of stochastic perturbations, leading to Levy type diffusion processes and Tsallis entropy (see, also, discussion in Section 2.4). These concerns require modifying the model to represent more general stochastic processes, which evaluations may require microsimulation (nonanalytic) estimation procedures.

### 4.3.1 *Microsimulation Estimation of Stochastic Process Parameters*

The microsimulation estimation procedure is based on methods for simulating the trajectories for each person in the cohort. The mathematical basis of this technique is the theory of stochastic processes and the simulation methods of solving stochastic differential equations [Kloeden and Platen, 1992; Kloeden et al., 1994; see also Wolf (2001) on applications of microsimulation in the social sciences]. The individual trajectory is constructed as the solution of a deterministic system of differential equations, which is obtained after specific realizations of random variables: the Wiener process, the initial values of covariates, and random stopping (death) times.

To begin, a cohort of individuals at an initial time $t_0$ is constructed from the data. Cohort size is limited only by the required statistical accuracy and computational burden. The initial values of covariates for all individuals in the cohort are simulated assuming a multidimensional Gaussian distribution. Means $v_{t_0}$ and the coefficients of the variance–covariance matrix $V_{t_0}$ can be estimated from data at the initial age or can be based on model assumptions. Individual covariates $\mathbf{x}_{t_0}$ are simulated using a theorem about the decomposition of the multivariate normally distributed random vector:

$$\mathbf{x}_{t_0} = v_{t_0} + D^T Z, \tag{4.27}$$

where $Z$ is a vector of standardized normally distributed numbers $Z_i \sim N(0,1)$ and matrix $D$ is the square root of the matrix $V_{t_0}$; $(V_{t_0} = D^T D)$.

Generally, the assumption of a normal distribution is not necessary. Non-Gaussian corrections can be added to study the non-Gaussian effects of the initial population distribution on projections. Age at the initial time is fixed in calculations. Generalizations to the variable initial age are straightforward. For fixed age, the variance–covariance matrices ($V_t$ and $\Sigma$) have to be calculated conditional on age.

Having the initial distribution of individuals, we then model trajectories in the $J$ dimensional state space in two steps. First, the probability $S(t|\mathbf{x}_t) = E_{\mathbf{x}_{t_0},\ldots,\mathbf{x}_{t-1}}(S(t|\mathbf{x}_{t_0},\ldots,\mathbf{x}_t))$ of survival for each individual is calculated using the mortality rate $\mu_t$ for the interval $(t, t+1)$:

$$S(t|\mathbf{x}_t) = \exp(-\mu_t) = \exp(-\mu_0 - b_t\mathbf{x}_t - \frac{1}{2}\mathbf{x}_t^T B_t \mathbf{x}_t). \tag{4.28}$$

The integral over the time period $(t, t+1)$ does not appear explicitly because it is subsumed in the coefficients $\mu_0$, $b_t$, and $B_t$. Each individual in the cohort is simulated to survive or not, according the probability $S(t|\mathbf{x}_t)$. To do that, a uniformly distributed random number $r$ is generated. If $r > S(t \mid \mathbf{x}_t)$, the individual is assumed to have died, and thus is removed from the cohort. It is possible to demonstrate that the random removal of individuals from the cohort, with normally distributed covariates $\mathbf{x}_t$, gives a surviving population which is also normally distributed (relations between parameters describing total and surviving populations are given by equation (4.23)). In fact, the simulated number of survivors have risk factors in the neighborhood of $\mathbf{x}_t$, (i.e., in $(\mathbf{x}_t, \mathbf{x}_t \pm \frac{1}{2}\Delta\mathbf{x}_t)$), is $l_t S(t \mid \mathbf{x}_t)\, p_G(\mathbf{x}_t, v_t, V_t)\Delta\mathbf{x}_t$. In the limit of large cohort sizes, this number should exactly coincide with the number of individuals in the state space domain for the survival distribution $l_t^* p_G(\mathbf{x}_t, v_t^*, V_t^*)\Delta\mathbf{x}_t$. This means that the following equality has to be valid for any $\mathbf{x}_t$:

$$l_t^* p_G(\mathbf{x}_t, v_t^*, V_t^*) = \exp(-\mu_t) l_t p_G(\mathbf{x}_t, v_t, V_t), \tag{4.29}$$

which can be checked by direct calculation, using the explicit form of a multivariate Gaussian distribution $p_G$ and equations (4.23), (4.24), and (4.28).

New covariates $\mathbf{x}_{t+1}$ for surviving individuals are simulated using a linear equation (4.25) or nonlinear (Akushevich et al., 2005a) models. Diffusion is simulated assuming the Gaussian multivariate distribution, or $\varepsilon \square D'^T Z$, where vector components of vector $Z$ are distributed as $N(0,1)$ and the matrix $D'$ is the square root of the diffusion matrix $\Sigma$, which is estimated from data simultaneously with the regression parameters. This risk factor recalculation for individuals is the second stage of the numerical procedure.

The numerical procedure is repeated over time until all individuals in the cohort die. The results on the individual (micro) level, have to be averaged over the realized parameter, reduces for all surviving individuals in the cohort at each time period. This can be used for controlling and cross-checking the calculations. The mean values ($v_t$) of risk factors obtained during the microsimulation, constitute the dynamics reflecting mortality. These mean values, along with the calculated variance–covariance matrix ($V_t$), allow calculation of the characteristics of the survival distribution $N(l_{t+1}^*, v_{t+1}^*, V_{t+1}^*)$ for the next time period. The survival curve is the percentage of survivors in the cohort.

Figures 4.4 and 4.5 demonstrate the statistical quality of the quadratic hazard models, i.e., how the models being estimated reproduce the data used

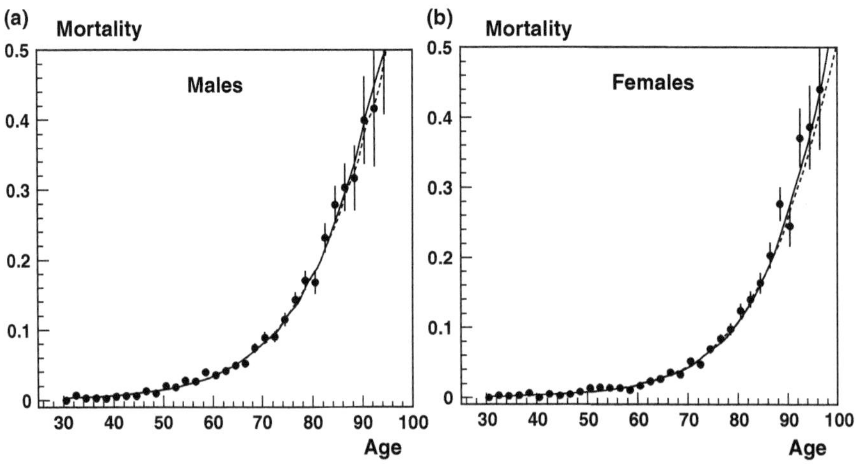


**Fig. 4.4** Probability of death within 2 years estimated from Framingham data (*filled dots* with *error bars*) for males (*left*) and females (*right*), with theoretical predictions given by linear (*dashed line*) and nonlinear (*solid line*) models

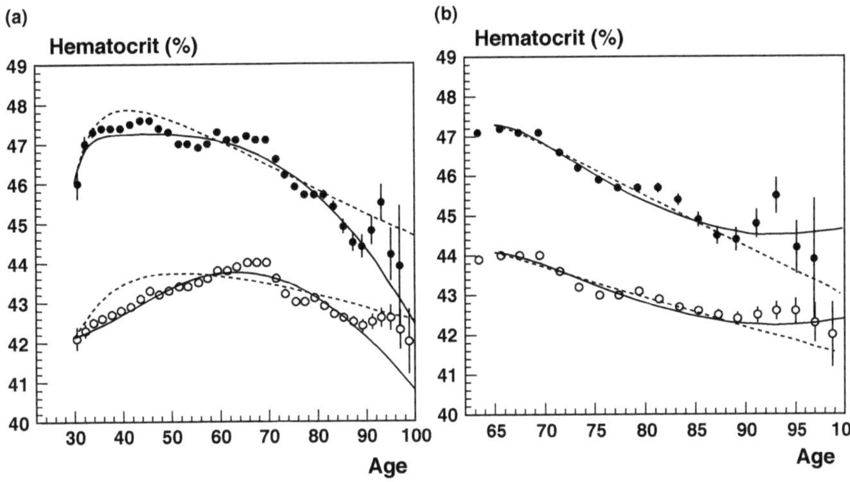


**Fig. 4.5** Projections from age 30 and from age 65 for hematocrit level for males (*filled dots, upper curves*) and females (*open dots, lower curves*)

for the estimation procedure. Versions of the dynamics with linear and non-linear equations were both considered in analyses of the Framingham data, with nonlinear dynamics performing better (Kulminski et al., 2004).

The effects of these calculations are also illustrated in Table 4.2 for males: (a) the total life table based on the 50 years follow-up of the Framingham Heart study, (b) the life table when cancer is eliminated as a cause of death (i.e., the life table for $Q_{t\text{-}d}$), and (c) the life table for persons expected to die from cancer (i.e., the single decrement table for cancer mortality).

Dynamics of nine covariates are also shown in the Table 4.2. These are pulse pressure (PP; in millimeters of mercury); diastolic blood pressure (DBP, in millimeters of mercury); body mass index (BMI) calculated as *weight/height*$^2$ [$kg/m^2$]; serum cholesterol (CHOL; in mg/100 ml); blood glucose (SUGAR, in mg/dl); hematocrit (HEMA, in %); vital capacity/height index (VCHI, calculated as $10 \times$ VC/height$^2$, where *VC* is vital capacity, in dl, and *height* in m); cigarette smoking – (CIG; cigarettes/day); and pulse rate (PR; beats/minute).

A comparison of life expectancy at age 30 shows that eliminating cancer increased $e_x$ by 2.95 years. Difference between this value and the value presented in Section 4.1.1 of this chapter is due to different underlying population (e.g., life expectancies at age 30 are 41.2 and 43.64 years) and statistical uncertainty in estimates based on the Framingham data. The $e_x$ for the cancer decrement table suggests that a significant proportion of all persons (~32.0%) survive to age 100 without experiencing cancer, thus suggesting some persons are not genetically susceptible to cancer. This analysis used a common age parameter θ for each disease, so that the cause-specific trajectory of the risk factors do not change with age.

**Table 4.2** The life tables for males: total, with cancer eliminated as a cause of death, and for persons expected to die of cancer

| $l_x$, % | $e_x$ yrs | Age | PP | DPB̀ | BMI | CHOL | SUGAR | HEMA | VCHI | CIG | PR |
|---|---|---|---|---|---|---|---|---|---|---|---|
| **Males. All causes** | | | | | | | | | | | |
| 100.0 | 43.64 | 30 | 49.6 | 81.8 | 25.9 | 209.0 | 81.1 | 46.0 | 138.7 | 13.1 | 70.8 |
| 98.15 | 34.35 | 40 | 42.4 | 86.2 | 27.3 | 243.3 | 79.3 | 47.9 | 136.8 | 14.2 | 77.2 |
| 93.72 | 25.71 | 50 | 48.0 | 84.8 | 27.6 | 240.1 | 86.5 | 47.6 | 129.4 | 12.9 | 76.4 |
| 84.01 | 18.04 | 60 | 55.0 | 82.2 | 27.4 | 228.8 | 94.5 | 47.1 | 121.1 | 10.5 | 74.6 |
| 65.10 | 11.69 | 70 | 62.1 | 79.2 | 26.8 | 216.2 | 102.1 | 46.5 | 112.9 | 7.7 | 72.5 |
| 36.05 | 6.92 | 80 | 69.1 | 76.2 | 26.1 | 203.7 | 109.4 | 45.8 | 105.2 | 4.7 | 70.3 |
| 9.10 | 3.77 | 90 | 76.1 | 73.2 | 25.6 | 192.4 | 116.3 | 45.2 | 98.3 | 1.9 | 67.9 |
| 0.38 | 2.02 | 100 | 83.3 | 70.6 | 25.4 | 184.1 | 122.6 | 44.6 | 93.3 | 0.0 | 65.2 |
| **Males. Cancer eliminated** | | | | | | | | | | | |
| 100.0 | 46.59 | 30 | 49.6 | 81.8 | 25.9 | 209.0 | 81.1 | 46.0 | 138.7 | 13.1 | 70.8 |
| 98.23 | 37.32 | 40 | 42.4 | 86.2 | 27.3 | 243.3 | 79.3 | 47.9 | 136.8 | 14.2 | 77.2 |
| 94.53 | 28.57 | 50 | 48.0 | 84.8 | 27.6 | 240.0 | 86.5 | 47.6 | 129.4 | 12.9 | 76.4 |
| 87.14 | 20.52 | 60 | 55.0 | 82.2 | 27.4 | 228.7 | 94.5 | 47.1 | 121 | 10.5 | 74.6 |
| 72.56 | 13.53 | 70 | 62.1 | 79.1 | 26.8 | 215.9 | 102.1 | 46.4 | 112.8 | 7.7 | 72.5 |
| 46.83 | 8.02 | 80 | 69.1 | 76.0 | 26.0 | 203.2 | 109.4 | 45.8 | 104.9 | 4.7 | 70.3 |
| 15.65 | 4.26 | 90 | 76.2 | 72.9 | 25.3 | 191.5 | 116.3 | 45.1 | 98.0 | 1.8 | 67.9 |
| 1.00 | 2.18 | 100 | 83.4 | 70.2 | 25.1 | 182.7 | 122.8 | 44.5 | 92.8 | 0.0 | 65.2 |
| **Males. Only cancer** | | | | | | | | | | | |
| 100.0 | 61.49 | 30 | 49.6 | 81.8 | 25.9 | 209.0 | 81.1 | 46.0 | 138.7 | 13.1 | 70.8 |
| 99.92 | 51.53 | 40 | 42.4 | 86.2 | 27.2 | 243.4 | 79.3 | 47.9 | 136.8 | 14.2 | 77.2 |
| 99.14 | 41.88 | 50 | 48.0 | 84.8 | 27.6 | 240.1 | 86.5 | 47.6 | 129.3 | 12.9 | 76.4 |
| 96.38 | 32.92 | 60 | 55.1 | 82.2 | 27.4 | 228.9 | 94.6 | 47.1 | 120.9 | 10.6 | 74.7 |
| 89.59 | 24.99 | 70 | 62.2 | 79.3 | 26.7 | 216.1 | 102.3 | 46.5 | 112.5 | 7.8 | 72.7 |
| 76.38 | 18.37 | 80 | 69.3 | 76.2 | 25.9 | 203.3 | 109.8 | 45.8 | 104.2 | 4.9 | 70.7 |
| 55.90 | 13.17 | 90 | 76.4 | 73.1 | 25.1 | 190.8 | 117.1 | 45.2 | 96.0 | 1.9 | 68.6 |
| 31.93 | 9.31 | 100 | 83.5 | 70.0 | 24.2 | 178.5 | 124.3 | 44.6 | 88.0 | 0.0 | 66.5 |

## 4.4 Evaluation of Characteristics of Individual and Grouped Data

In this section, we discuss how to screen and evaluate statistical and measurement uncertainties of the observed data patterns in specific studies, how to project results to the entire national population, and how to extract information about cancer incidence and mortality in the case of continuous medical history data. Sections will also be devoted to methods of sensitivity analyses and to two models, generalizing standard approaches to the modeling of carcinogenesis.

Modern demographic and epidemiological studies are designed to have both longitudinal and cross-sectional components. This can be illustrated by the six waves of the National Long Term Care Survey (NLTCS). The surveys were done roughly every 5 years (i.e., 1982–1984, 1989, 1994, 1999, and 2004) with a random sample of elderly Medicare enrollees drawn from Medicare enrollment files. Roughly 20,000 persons were screened in each NLTCS, with 15,000 survivors from the prior survey, and supplementary sample of roughly 5,000 persons who passed age 65–69 between NLTCS waves. Being a Medicare list sample it was possible to link all sampled enrollees to both, Medicare vital statistic files, and to Medicare Part A and Part B files recording their diagnoses, service use, and costs on a daily basis.

This allows a comprehensive range of empirical analyses, and to perform different types of mathematical modeling. Typically, NLTCS respondents enter the sample at different times and ages. Some of them stay healthy and alive during the local area study, or the national survey, so the effects of different censoring and missing data patterns have to be considered. In addition, there are frequently data quality problems at extreme ages (e.g., 85 + ) that need to be evaluated.

The incidence rate for a specific age group is calculated as the ratio of the number of new disease events observed in the group to the number of person-years accumulated by the members of the group. The calculation of incidence rates has to be performed taking into account right censoring. This means one has to calculate the individual duration of observation in survey and linked administrative records (e.g., in the NLTCS and Medicare), rather than in life tables. This allows one to perform calculations of event rates, which can be accurate to within one-day. For example, date of birth, onset, death, dis-/enrollment from/into the NLTCS and Medicare Parts A and B, and last record dates are known with one-day accuracy.

Projection of parameter estimates from the NLTCS data to the U.S. elderly population (65 + ) may be biased by sample design. To have estimates representative of the entire U.S. elderly population, sample design effects are represented using special weight functions, i.e., number of individuals in the population represented by an individual in the sample. The cross-sectional sample-weight function for each individual can "jump" when, e.g., a new wave starts. The date when the sample weight function jumps is known with one-day accuracy. Therefore, each individual can be associated with the precise time interval under observation with a corresponding weight function and medical service use history extracted from Medicare files. Age patterns of incidence rates can be assessed by stratifying the sample into relevant age categories (a year, or several years). The richness of the linked NLTCS-Medicare data defines precise time intervals to produce statistically significant estimates of incidence rates. Empirical age-specific risks ($\lambda_a$) are calculated as a ratio of weighted numbers of cases to weighted person-years at risk:

$$\lambda_a = \frac{n(a)}{N(a)}; \quad n(a) = \sum_n w_n(a); \quad N(a) = \sum_i w_i(a),$$

where $w_i$ $(a)$ is the individual weight at age $a$; $n$ runs over all disease onsets detected in the age group, and $i$ runs over all individuals at risk in $a^{th}$ age group.

Standard error (SE) and confidence interval (CI) calculations must also be adjusted for sample design effect (Manton et al., 1997; Manton and Gu, 2001). The approach suggested by the Census Bureau for the NLTCS uses generalized variance function methods (Wolter, 1985), in which the SE are adjusted for sample design using

$$SE = \sigma_E = \sqrt{\frac{b}{N(a)}\lambda_a(1-\lambda_a)}. \tag{4.30}$$

Parameter $b$ is an adjustment factor for studying sample design effects. In the case of NLTCS data, numerical values of $b$ produced by the U.S. Census Bureau are available for each NLTCS wave (U.S. Census Bureau, 2001). This factor is close to the mean individual weight for a specific wave. To calculate age-specific effects, equation (4.29) has to be generalized. First, we have to take into account that individuals contribute to person-years of the same age group from different waves, so the factor $b$ might be not a constant. Second, equation (4.30) as well as the standard Wald's CI, do not work when $N$ $(a)$ is small (Brown et al., 2001). A generalization for small samples based on Wilson's approach (Brown et al., 2001) uses

$$CI_w = \frac{N_b(a)\lambda_a + \frac{1}{2}z_{a/2}^2}{N_b(a) + z_{a/2}^2} \pm \frac{z_{a/2}\sqrt{N_b(a)}}{N_b(a) + z_{a/2}^2}\sqrt{\lambda_a(1-\lambda_a) + \frac{z_{a/2}^2}{4N_b(a)}}, \tag{4.31}$$

where $N_b(a) = \sum_i w_i(a)/b(w_i(a))$ and $b(w_i(a))$ can vary with time (i.e., from wave to wave). For large $N_b(a)$ and constant $b(w_i(a)) = b$ (one wave), equation (4.31) recovers the standard Wald's estimates of CIs, i.e., $CI_s = p_c \pm z_{\alpha/2}\sigma_E$; $z_{\alpha/2} = \Phi^{-1}(1\text{-}\alpha/2)$, where $\Phi(x)$ is the standard normal distribution function, and $\alpha$ denotes the confidence level.

### 4.4.1 Detection of Disease Onset

A data set, including continuous medical history information, which is linked on the individuals' level to the NLTCS, is the Continuous Medicare history files which contain information about costs, treatments, and diagnoses on service delivery dates, as well as the date of death. Medical information (disease diagnoses and service dates) can be used from the following administrative subfiles: clinical labs, durable medical equipment regional carrier (DMERC), home health agency, hospice, inpatient/Skilled Nursing Facility (SNF), outpatient, carrier (other than DMERC), and SNF claim records. Certain demographic and biomedical data sets, such as NLTCS, are linked with such Medicare information. It means that for each beneficiary from the surveys the continuous medical history information is available for certain time periods, usually before the starting date of the survey.

To calculate age-specific disease incidence rates we need to know the date of disease onset. Continuous Medicare data provide dates of claims for medical service, which are accompanied by the corresponding ICD-9 numeric codes. Therefore it is reasonable to assume that an individual might experience an onset of a disease during the period of observation, if there is at least one record with the ICD-9-CM code corresponding to this disease on a single institutional claim (inpatient, skilled nursing facility, home health care, hospice, and outpatient) or noninstitutional claim [carrier/physician supplier/Part B (1991–2001 only), durable medical equipment, clinical labs]. Actually, Medicare data do not contain information on whether the appearance of an ICD-9-CM code is a "true" onset (first report) or just a visit to treat a disease, with its onset possibly first registered outside (before) the observation period. Therefore, to determine a date of onset we will assume that beneficiaries with a chronic condition receive medical care at least once within the first 6 months since his/her enrollment into Medicare. If certain diagnoses appear in Medicare files within an initial 6-month period, such an individual will be considered as chronically impaired at the time of enrollment in Medicare. Otherwise, the date of the first appearance of the corresponding diagnosis in the Medicare records will be considered the date of onset.

Although this scheme for time of disease onset identification is straightforward, modifications of this scheme for specific diseases for clinical reasons are possible. For example, another scheme for diabetes onset identification (Hebert et al., 1999) is often used. In this scheme, it is required that a second record with the ICD-9 code diabetes mellitus type II must be observed, if the first diagnosis was registered as an ambulatory claim (i.e., a physician/supplier or hospital outpatient claim).

### 4.4.2 Cancer Incidence in U.S. Elderly

The analysis of the incidence rate of cancer in the NLTCS sample, including all malignant neoplasms (ICD-9 codes 140–208), is preceded by examining the

Medicare sources for ICD codes and their relative contribution to age-specific incidence rates. Our goal is to identify conditions for obtaining the most stable results that can be considered as basic. Then we perform detailed sensitivity analysis on them.

Medicare claim data have certain limitations from the way of how diagnoses are determined. One is associated with lack of ICD codes from the Physician/Supplier/Part B source before 1991. Lack of this information might result in underestimation of incidence rates and shift the estimated time of onset to later ages. We tested the significance of this effect by comparing incidence patterns for the period from 1992 to 2001 with, and without, information from this source. We found that diagnoses made by physicians account for about 30–40% of total onsets. Differences between incidence rates calculated for different diseases for 1984–1990 (where Physician/Supplier/Part B information is not available) and 1992–2005 are of this order. The Part-B-associated loss of diagnoses is mostly disease independent. It is also independent of age, with the exception of very advanced ages (100+), where differences essentially vanish. The effect is sex dependent, with more loss of diagnoses for females for most of diseases and ages. This conclusion about the essential role of diagnoses from the Physician/Supplier/Part B source is also confirmed by analysis of the time distribution of new diagnoses, in which an abnormally large peak appears in 1991 for participants of the NLTCS cohort from the first three (1982, 1984, and 1989) waves. Consequently, we limit our analysis to data from 1992 to 2001. Estimates of age-specific cancer incidence rates are presented in Table 4.3 (Akushevich et al., 2006).

The table shows that incidences peaks at about age 95 for males and age 80 for females.

**Table 4.3** Age-specific cancer incidence rates (Inc) means and CI's of NLTCS/Medicare data for 1992–2001

| | Males | | Females | |
|---|---|---|---|---|
| Age | Inc | CI | Inc | CI |
| 66–68 | 57 | (52,63) | 40 | (36,44) |
| 69–71 | 64 | (59,68) | 38 | (35,41) |
| 72–74 | 66 | (61,71) | 43 | (40,47) |
| 75–77 | 67 | (61,73) | 48 | (44,52) |
| 78–80 | 71 | (64,78) | 49 | (45,53) |
| 81–83 | 90 | (81,99) | 54 | (49,59) |
| 84–86 | 92 | (81,104) | 54 | (49,60) |
| 87–89 | 106 | (91,124) | 49 | (43,56) |
| 90–92 | 101 | (80,126) | 55 | (47,64) |
| 93–95 | 116 | (85,157) | 53 | (42,66) |
| 96–98 | 115 | (67,192) | 57 | (41,79) |
| 99–101 | 113 | (39,288) | 53 | (31,91) |
| 102–104 | 41 | (2,492) | 53 | (19,140) |
| 105–107 | | | 10 | (0,183) |

### 4.4.3 Sensitivity Analysis

A disadvantage of large administrative databases is the presence of factors producing systematic over-/underestimation of the number of diagnoses or the age at onset. One reason for such uncertainties concerns incorrectly reported dates of onset. Other sources involve incorrect reporting of date of birth and date of death. While the first affects age at onset, the second tends to reduce the number of person-years at risk. To evaluate the effect of these uncertainties on estimates, we describe several methods of sensitivity analysis. First, we perform calculations with different definitions of disease onset; second, we use censoring schemes employing alternative data to define individual observation periods; and, finally, we simulate unobservable effects of errors in reporting date of birth. Comparison of the recalculated incidence rates with our basic results provides the estimates of uncertainties due to these sources of error.

### 4.4.4 Uncertainty in Onset Calculation

Sources for uncertainties related to overestimation and shift of dates of onsets are the enrollment of new beneficiaries, and alteration of coverage by the Medicare program of certain beneficiaries, within the observation period for legal (eligibility) or administrative (enrollment under another health insurance) reasons. In the example of a calculation based on NLTCS and Medicare data, enrollment of new beneficiaries does not lead to overestimates in the analysis because (a) a 6-month cut to determine disease onset is used, and (b) Medicare data cover a longer time domain than that of individual observations from 1992 to 2004 (Part A data on diagnoses was collected for the NLTCS back to 1982; Part B diagnosis collection was started in 1991). To project estimates of incidence rates to the U.S. population, we use the sample design weights, which become nonzero after the first survey in which the individual participates. Thus, participants of the fourth wave (1994) contribute to incidence patterns only beginning from 1994 (when their weights become nonzero), but their diagnoses can be analyzed from 1991. This provides a sufficient time to avoid such bias.

### 4.4.5 Medicare Coverage and Censoring Uncertainties

Here we consider two types of uncertainties. The first is related to partial coverage (i.e., having only Medicare Part A). However, we do not expect that this is important because of the relatively small fraction of individuals who are not under both part A and part B coverage. In January 2001, 3.46% of sample persons had Part A only, 0.69% had Part B only, and 95.85% had both Part A and B coverage. Second bias is related to the effects of censoring, when an individual is "under observation", however, his/her records cannot appear in Medicare files. A well-known example of how information can be missed is the relocation of elderly abroad and, as a result, death or disease onset occurs outside the survey areas. Details of how such effects can influence mortality are discussed by Kestenbaum (1992), and Kestenbaum and Ferguson (2002).

Effect of such uncertainties can be investigated by applying different censoring schemes with different definitions of censoring dates in simulations. In the basic calculations, the final date of observation is selected to be the earliest date among the dates of disease onset, death, and the last date of cohort observation. In the first alternative censoring scheme, the last day of observation is assumed to be the last day of part B coverage. Since part B coverage requires monthly payments, this is a good indicator that an individual is alive and being observed.

Comparison of these results with our original calculations shows that the ratio of incidence rates deviates noticeably from 1.0 only for ages ~100, and does not normally exceed 1.03. A measure of these uncertainties estimated in units of CI is presented in Table 4.3.

To evaluate a maximal level of censoring uncertainties, we used a second censoring scheme, in which the last observation day is defined by the last record in either the NLTCS or the Medicare files. Uncertainties in this case are about 4–7% (i.e., the ratio is 1.04–1.07), increasing to 15% for younger ages. Such age dependence of the uncertainty allows us to speculate that this effect arises due to a decreasing contribution of person-years, that is important for younger ages, rather than due to effective loss of individuals, which is more likely at advanced ages.

### *4.4.6 Age Reporting Uncertainties*

An important source of uncertainty, especially for analyses at advanced ages, might be represented by errors in the reported date of birth. Estimation of the percentage of individuals in sex- and race-specific populations reporting the wrong date of birth, and probability distribution over years added/subtracted from original date of birth was performed by Rosenwaike and Stone (2003), and Preston et al. (1996). The distribution of errors from Rosenwaike and Stone on the accuracy of reporting the age at death was determined by comparison of SSA records of 700 individuals who died from 1980 to 1999, purportedly at ages 110 and older, to records of the U.S. censuses of 1880 and 1900, conducted when these individuals were children. This error distribution is in Fig. 4.6.

This distribution depends upon the degree (in years) of error in both directions, i.e., errors in rates will reflect the net errors in person years of both, over and under reports of age. One can use these distributions in stochastic simulation models for two scenarios (presented in Fig. 4.7): (a) that this error, found for the 1875 birth cohort, was maintained without improvement to the 1925

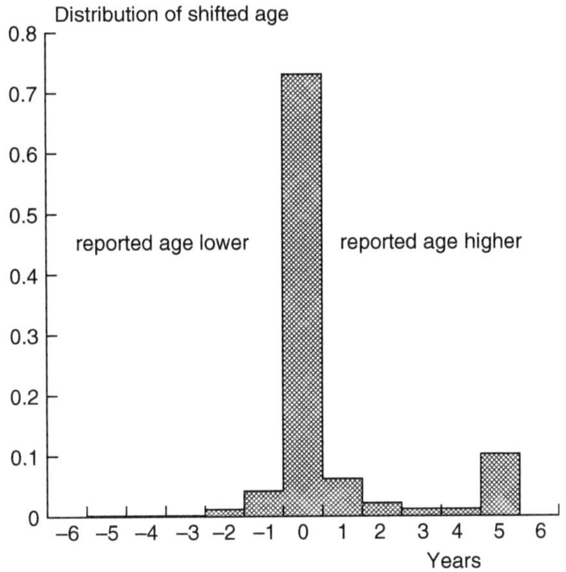


**Fig. 4.6** Distribution of age misreporting

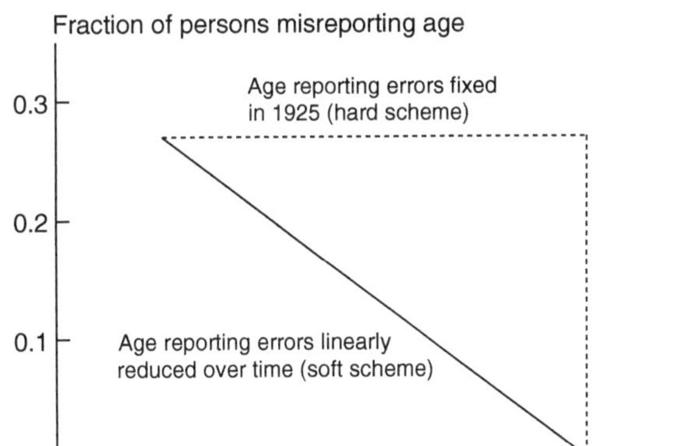


**Fig. 4.7** Fraction of individuals misreporting age

birth cohort (i.e., persons aged 65 in 1990, when Kestenbaum argued the Medicare data was of high quality), and then dropped to negligible levels (hard scheme), and (b) error declines linearly from 1875 to 1925 (soft scheme).

Both strategies should be evaluated. First, assuming a constant percentage of people misreporting age from 1875 to 1925, we assess the upper limit of uncertainty. A more realistic situation is that the percentage of people misreporting age is not constant and declines: being maximal for those born in 1875, and reaching a minimum for those born in 1925. Therefore a second strategy is to linearly interpolate this decline. The initial distribution of age-misreporting people, conditional on sex and race, is taken from Table 5 of Rosenwaike and Stone (2003). In both strategies, we keep this distribution fixed. For each individual, the offset in date of birth is simulated according to this distribution in calculating age-specific incidence rates. For the first strategy, we also keep fixed percentages for people who are misreporting age. In the second case, this percentage linearly decreases. We simulated 500 samples, which is sufficient to have statistically stable results. Averaging over these samples, we calculated final incidence rates with adjusted dates of birth. Comparison of the results obtained using these strategies with basic calculation will provide an estimation of the effect of age misreporting. Estimation of the upper level of the uncertainty using the model with a constant percentage of people misreporting age gives modest results: the incidence rate may change 2–5%; only for the last age interval (102–104 for males and 105–107 for females) does the correction factor reach 1.5–2.0.

## 4.5 Generalized Frailty Model

Above we presented stochastic process models, when risk factors on state variables are directly observed over time. When these variables are "latent," such as the case in national mortality data, we need models that can identify the trace of those latent state variables on mortality and health trajectories. This can be done with various latent frailty models.

"Frailty" models of mortality are described in terms of the joint probability distribution of two random variables: survival time ($T$) and frailty ($Z$) associated with population heterogeneity in the ability to resist forces of mortality (e.g., Vaupel et al., 1979; Beard, 1959, 1971). Population dynamics are governed by the conditional hazard rate, which is modeled as $\mu(t|Z = z) = z\mu_0(t) + c_0$ ($z \geq 0$, $\mu_0(t) \geq 0$, and $c_0 \geq 0$), where the age function $\mu_0(t)$ is independent of $z$, and $c_0$ is a constant.

Typical assumptions about the $T$- and $Z$-distributions in frailty models are (Manton et al., 1986) (1) that the shape of the frailty distribution can be modeled by the gamma or inverse Gaussian distribution; and (2) that the standard force of mortality $\mu_0(t)$ is modeled as either the Gompertz $\mu_G(t)$ or Weibull $\mu_W(t)$ hazard functions. A crucial problem is how to model age-related changes in the population frailty distribution. The observed age pattern of mortality rates is predicted by how the mean of the frailty distribution changes over time due to the selection of frail persons.

Manton et al. (1986, 1993) provided a general model, incorporating the effects of the heterogeneity of individual mortality risks,

$$\mu(t) = \frac{\mu_0(t)}{\left[1 + n\gamma \int_{t_0}^{t} du \cdot \mu_0(u)\right]^{\frac{1}{n}}}. \tag{4.32}$$

The denominator in equation (4.32) shows that the age-associated increase in mortality, as the most susceptible persons die first, changes the mean of the frailty distribution by the systematic selection of more robust individuals. The additional parameters in equation (4.32) are $\gamma$, which is the squared standard deviation of the distribution of individual frailty, and $n$, which controls the shape of the mixing distribution (e.g., $n = 1$ for a gamma, $n = 2$ for the inverse Gaussian). Since equation (4.32) is written in terms of $\mu_0(t)$ and $n$, no special assumptions have to be made about the form of either the age-dependent mortality risks for the individual or the mixing distribution. The initial mean of $Z$, being set to be equal to 1, does not lead to any further restrictions.

Recently Akushevich et al. (2005b) and Manton et al. (2005) generalized the model by developing mortality functions with (1) a location parameter ($t_0$); (2) a fixed (genetic) heterogeneity distribution ($\gamma$); and (3) parameters expressing empirical variability (e.g., age dependence) in the individuals' rate of aging ($\theta$ or $m$). New analytic formulae corresponding to the best fit were obtained for a generalized Weibull model,

$$\mu_W(t, t_0, m, \gamma) = \frac{m-1}{t_0 \gamma} \exp\left( -\left(\frac{t}{t_0}\right)^m \frac{m-1}{m} \right) \left(\frac{t}{t_0}\right)^{m-1}, \tag{4.33}$$

and a generalized Gompertz model:

$$\mu_G(t, t_0, \theta, \gamma) = \frac{b\theta}{\gamma} \exp(b(1 - e^{\theta t}) + \theta t), b = e^{-\theta t_0} \frac{\theta}{e^{\theta} - 1}. \tag{4.34}$$

Figure 4.8 gives the frailty distribution for the gamma ($n = 1$), inverse Gaussian ($n = 2$), and the new frailty distribution corresponding to ($n = 0$).

The new frailty distribution ($n = 0$) has a lower peak than the gamma and inverse Gaussian distributions, and a larger proportion of cases in the tail at the

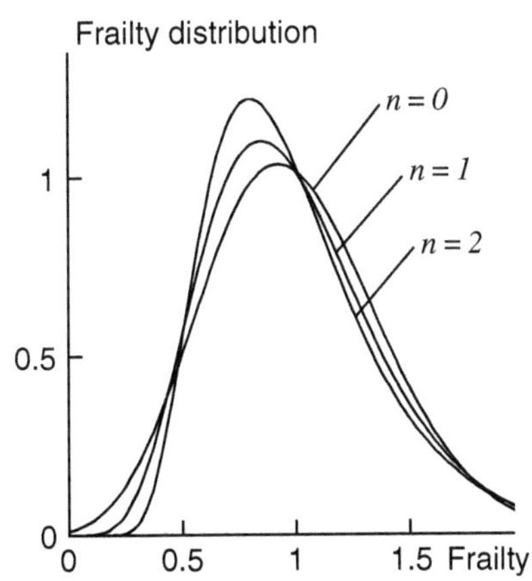


**Fig. 4.8** Frailty distribution for Gamma ($n = 1$), inverse Gaussian ($n = 2$), and newly proposed distribution (14) ($n = 0$)

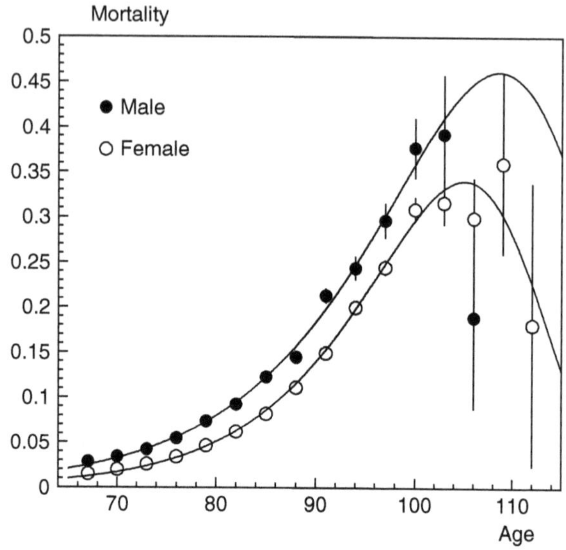


**Fig. 4.9** Age patterns of mortality for U.S. population using NLTCS data

low end of frailty. This new distribution can be used to describe both mortality plateaus and declines at extreme ages. This model could describe the complex mortality patterns observed in the NLTCS data: not only a plateau effect at age of about 95 but also the possible declines in the per annum hazard rate among survivors to ages 100+ (Fig. 4.9). The best fit corresponds to new frailty distribution ($n = 0$), properties of which were identified and analyzed.

In Table 4.4, we provide life table parameters ($\ell_x$, $e_x$) for 5-year age categories from age 65 to 115, using NLTCS data and compared them with recent predictions by Social Security Administration (SSA, 2003) and Society of Actuaries (SOA, 2000). The three sets of life tables show differences in the life expectancy at age 65 and at advanced ages. The use of the survival probabilities from NLTCS data with a mortality peak would lead to projections with considerably more centenarians ($\ell_{100}$ is larger) in the future than would the SSA male and female life tables. The SOA life tables produce more female centenarians. The SOA (2000) life tables provide the highest survival at age 65–80 for males and at age 65–110 for females. Thus, the NLTCS data with a mortality decline produced no more extreme life expectancy estimates than the SOA model with a plateau of 0.40 – especially for females. The primary advantage of our model is to help understand forces governing mortality at extreme ages, which would be especially beneficial for long-range projections, where the proportions surviving to late ages become large. In the case of cancer, this suggests that there is heterogeneity in both the Weibull shape and the scale parameters. This suggests that the rate of genetic mutations leading to cancer incidences may vary as a function of other background factors, such as metabolic changes and mitotic index.

**Table 4.4** Male and female life table parameters ($\ell_x$, $e_x$) from the SOA, SSA, and NLTCS data

| | SoA | | SSA | | NLTCS | |
|---|---|---|---|---|---|---|
| Age | $\ell_x$ | $e_x$ | $\ell_x$ | $e_x$ | $\ell_x$ | $e_x$ |
| **Males** | | | | | | |
| 65 | 100,000 | 18.11 | 100,000 | 16.24 | 100,000 | 16.70 |
| 70 | 92,167 | 14.38 | 87,745 | 13.07 | 88,808 | 13.41 |
| 75 | 80,099 | 11.07 | 72,281 | 10.20 | 73,547 | 10.55 |
| 80 | 62,825 | 8.25 | 53,544 | 7.70 | 54,591 | 8.15 |
| 85 | 41,101 | 5.99 | 32,654 | 5.70 | 34,215 | 6.22 |
| 90 | 19,612 | 4.36 | 14,540 | 4.23 | 16,624 | 4.74 |
| 95 | 5,765 | 3.34 | 4,083 | 3.22 | 5,609 | 3.69 |
| 100 | 971 | 2.75 | 630 | 2.56 | 1,182 | 3.04 |
| 105 | 96 | 2.50 | 48 | 2.06 | 155 | 2.81 |
| 110 | 8 | 2.49 | 1 | 1.64 | 17 | 3.18 |
| 115 | 1 | 2.38 | 0 | 1.31 | 3 | 3.62 |
| **Females** | | | | | | |
| 65 | 100,000 | 20.62 | 100,000 | 19.53 | 100,000 | 20.34 |
| 70 | 94,034 | 16.73 | 92,328 | 15.89 | 93,794 | 16.48 |
| 75 | 84,615 | 13.24 | 81,818 | 12.54 | 84,103 | 13.01 |
| 80 | 71,003 | 10.18 | 67,510 | 9.53 | 69,908 | 10.02 |
| 85 | 52,937 | 7.59 | 48,697 | 7.02 | 51,222 | 7.55 |
| 90 | 31,705 | 5.65 | 27,726 | 5.06 | 30,568 | 5.62 |
| 95 | 13,448 | 4.47 | 10,543 | 3.71 | 13,260 | 4.24 |
| 100 | 4,029 | 3.79 | 2,274 | 2.85 | 3,669 | 3.37 |
| 105 | 915 | 3.12 | 248 | 2.21 | 616 | 3.08 |
| 110 | 130 | 2.63 | 9 | 1.70 | 82 | 3.60 |
| 115 | 12 | 2.38 | 0 | 1.31 | 17 | 4.05 |

## 4.6 Summary

In this chapter, we described analytic strategies to be applied to analyses of the effect of cancer mortality on a population, and discussed how analytic strategies should be modified in specific data sets to answer specific analytic questions (e.g., cancer mortality trends at late ages). These methods include life table calculations and their extensions, approaches to modeling of age-specific cancer and noncancer mortality, and strategies that allow us to include in models both, observed and latent (i.e., unobserved) risk factors. Though the approaches described in this chapter are advanced, they are recognized and applied by both theoreticians and practitioners, and bring an important contribution to analysis and modeling of various demographic and biomedical aspects of cancer epidemiology.

In subsequent chapters, we will combine our theoretical modeling insights (Chapter 2) with data on certain cancer risk factors (Chapter 3), using the various statistical methods discussed above, to conduct the specific substantial analyses of cancer mortality and morbidity in the U.S. population (Chapters 6 and 7), and to perform simulation experiments to use the developed approaches for the tasks of projection and forecasting (Chapter 8).

## References

Akushevich I., Kulminski A., Manton K., 2005a. Life tables with covariates: life tables with covariates: dynamic model for nonlinear analysis of longitudinal data. Math Popul Stud 12(2):51–80.

Akushevich I., Kulminski A., Manton K.G., 2005b. Human Mortality and Chronic Disease Incidence at Extreme Ages: New Data and Analysis. Talk given at Session “New Direction on Mortality Research” in the Population Association of America Annual Meeting, Philadelphia, March 31–April 2, 2005.

Akushevich I., Kulminski A., Akushevich L., Manton K.G., 2006. Age Patterns of Disease Incidences in the U.S. Elderly: Population-Based Analysis. Trends Working Paper Series.

Allison P.D., 1984. Event History Analysis. Beverly Hills, CA: SAGE Publications.

Anderson D.E., 1974. Genetic study of breast cancer: identification of a high risk group. Cancer 34(4):1090–1097.

Armitage P., Doll R., 1954. The age distribution of cancer and a multi-stage theory of carcinogenesis. Br J Cancer 8(1):1–12.

Armitage P., Doll R., 1961. Stochastic models for carcinogenesis. In Proceedings of the Fourth Berkeley Symposium on Mathematical Statistics and Probability, Berkeley, CA: University of California Press.

Beard, R.E. (1959). Note on some mathematical mortality models. In: Wolstenholme, G.E.W. , O’Connor, M. (Eds.). The Lifespan of Animals (pp. 302–311). Boston: Little, Brown.

Beard, R.E. (1971). Some aspects of theories of mortality, cause of death analysis, forecasting and stochastic processes. In: Brass, W. (Ed.), Biological Aspects of Demography (pp. 57–68). London: Taylor and Francis.

Bishop Y.M.M., Fienberg S.E. Discrete Multivariate Analysis: Theory and Practice, Cambridge, MA: The MIT Press, 1975.

Brown L.D., Cai T.T., DasGupta A., 2001. Interval Estimation for a Binomial Proportion. Stat Sci 16:101–117.

Chiang C.L., 1968. Introduction to Stochastic Processes in Biostatistics. New York : John Wiley

Cox D., 1972. Regression models and life tables (with discussion). J R Stat Soc, Ser B 34:187–220.

De Waard F., Halewijn B., Huizinga J., 1964. The bimodal age distribution of patients with mammary carcinoma. Cancer 17:141–152.

Frier B., 1982. Roman life expectancy: Ulpian’s evidence. Harv Stud Classic Philol 86:213–251.

Gavrilov L.A., Gavrilova N.S., 2006a. Models of systems failure in aging. In: P. Michael Conn (Ed): Handbook of Models for Human Aging, Burlington, MA: Elsevier Academic Press, 45–68.

Gavrilov L.A., Gavrilova N.S., 2006b. Reliability theory of aging and longevity. In: Masoro E.J. & Austad S.N. (eds.): Handbook of the Biology of Aging, 6th edition. San Diego, CA, USA: Academic Press 3–42.

Hebert P.L., Geiss L.C., Tierney E.F. et al., 1999. Identifying person with diabetes using medicare claims data. Am J Med Qual 14(6):270–277.

Hougaard P., 1988. A boundary modification of kernel function smoothing, with application to insulin absorption kinetics. In Compstat Lectures 31–36. Physica, Vienna.

Jacquez J.A., 1972. Compartmental Analysis in Biology and Medicine, volume 50. New York: Elsevier.
Jones H.W., 1945. John Graunt and His Bills of Mortality. Bull Med Libr Assoc 33(1):3–4.
Kaplan E.L., Meier P., 1958. Nonparametric estimation from incomplete observations. J Am Stat Assoc 53:457–481.
Kertzer D.I., Laslet P. (eds.), 1995. Aging in the Past. Berkley-Los Angeles-Oxford: University of California Press. At: http://ark.cdlib.org/ark:/13030/ft096n99tf.
Kestenbaum B.A., 1992. Description of extreme aged population based on improved medicare enrollment data. Demography 29(4):565–580.
Kestenbaum B., Ferguson B.R., 2002. Mortality of the extreme aged in the United States in the 1990s, based on improved medicare data. North Am Actua J 6(3):35–44.
Kloeden P.E., Platen E., 1992. Numerical Solution of Stochastic Differential Equations. Application of Mathematics Series, volume 23. Heidelberg: Springer-Verlag.
Kloeden P.E., Platen E., Schurz H., 1994. Numerical Solution of SDE Through Computer Experiments. Berlin: Springer.
Knudson A.G. Jr., 1971. Mutation and cancer: statistical study of retinoblastoma. Proc Natl Acad Sci USA 68:820–823.
Kulminski A., Akushevich I., Manton K., 2004. Modeling nonlinear effects in longitudinal survival data: implications for the physiological dynamics of biological systems. Front Biosci 9:481–493.
MacMahon B., Cole P., Brown J., 1973. Etiology of human breast cancer: a review. J Natl Cancer Inst, 50:21–42.
Manton K.G., Stallard E., 1980. A two-disease model of female breast cancer: mortality in 1969 among white females in the United States. J Natl Cancer Inst. 64(1):9–16.
Manton K.G., Stallard E., 1984. Recent Trends in Mortality Analysis. Orlando, Florida, USA: Academic Press.
Manton K.G., Stallard E., 1988. Chronic Disease Modeling: Measurement and Evaluation of the Risks of Chronic Disease Processes. London: Charles Griffin.
Manton K.G., Stallard E., 1996. Longevity in the United States: age and sex-specific evidence on life span limits from mortality patterns: 1960–1990. J Gerontol Ser A-Biol Sci Med Sci 51(5):B362–B375.
Manton K.G., Land K.C., 2000a. Active life expectancy estimates for the U.S. elderly population: a multidimensional continuous-mixture model of functional change applied to completed cohorts, 1982 to 1996. Demography 37(3):253–265.
Manton K.G., Land K.C., 2000b. Multidimensional disability/mortality trajectories at ages 65 and over: the impact of state dependence. Soc Indic Res 51(2):193–221.
Manton K.G., Gu X., 2001. Changes in the prevalence of chronic disability in the United States black and non-black population above age 65 from 1982 to 1999. Proc Natl Acad Sci USA 98(11):6354–6359.
Manton K.G., Patrick C.H., Stallard E., 1980. Mortality model based on delays in progression of chronic diseases: alternative to cause elimination model. Public Health Rep. 95(6):580–588.
Manton, K., Woodbury, M., Stallard, E. 1981. A variance components approach to categorical data models with heterogenous mortality rates in North Carolina counties, Biometrics 37:259–269.
Manton K.G., Stallard E., Vaupel J.W., 1986. Alternative models for the heterogeneity of mortality risks among the aged. J Am Stat Assoc. 81(395):635–644.
Manton K.G., Woodbury M.A., Stallard E. et al., 1989. Empirical Bayes procedures for stabilizing maps of U.S. cancer mortality rates. J Am Stat Assoc. 84(407):637–650.
Manton K.G., Stallard E., Singer B.H., 1992. Projecting the future size and health status of the U.S. elderly population. Int. J. Forecast 8:433–458.
Manton K.G., Lowrimore G., Yashin A., 1993. Methods for combining ancillary data in stochastic compartment models of cancer mortality: generalization of heterogeneity models. Math Popul Stud 4(2):133–147.

Manton K.G., Woodbury M.A., Tolley H.D., 1994. Statistical Applications using Fuzzy Sets. New York: John Wiley and Sons.
Manton K.G., Corder L., Stallard E., 1997. Chronic disability trends in the U.S. elderly populations 1982 to 1994. Proc Natl Acad Sci USA 94:2593–2598.
Manton K.G., Akushevich I., Kulminski A., 2005. The Stochastic Linkage of Mortality Declines and Declines in Functional Disability. Invited paper for a meeting on "Projecting Mortality" at Brookings Institution, sponsored by the Office of Policy, Social Security Administration and the Center for Retirement Research at Boston College.
Manton K.G., Gu X., Lamb V.L., 2006. Long term trends in life expectancy and active life expectancy in the United States. Popul Develop Rev 32(1):81–105.
Matis J.H., Wehrly T.E., 1979. Stochastic models of compartmental systems. Biometrics 35:199–220.
McGeough K., 2004. The Romans: New Perspectives. Oxford: ABC-CLIO. 381 pp.
Metzler R., Klafter J., 2000. The random walk's guide to anomalous diffusion: a fractional dynamics approach. Phys Rep 339:1–77.
Preston S.H., Elo I.T., Rosenwaike I. et al., Hill M. 1996 African–American mortality at older ages: results of a matching study. Demography 33(2):193–209.
Risken H., 1996. The Fokker-Planck Equation: Methods of Solution and Applications, 2nd edition, Berlin: Springer-Verlag.
Robine J-M., Jagger C., Mathers C.D. et al. (eds.), 2003a. Determining Health Expectancies. West Sussex, UK: John Wiley and Sons.
Robine J-M., Romieu I., Michel J-P., 2003b. Trends in health expectancies. In: Robine J-M., Jagger C., Mathers C.D., Crimmins E.M., Suzman R.M. (eds.), Determining Health Expectancies. West Sussex, UK: John Wiley and Sons.
Rosenwaike I., Stone L.F., 2003. Verification of the ages of super-centenarians in the United States: results of a matching study. Demography 40(4):727–739.
Simms H., 1942. The use of measurable causes of death (hemorrhage) for the evaluation of aging. J Gen Physiol 26:169–178.
Social Security Association, 2003. Life Tables for the United States Social Security Area 1900–2100. Actuarial Study No. 116. Distributed by Social Security Association.
Society of Actuaries, 2000. RP-2000 Mortality Tables. In: http://www.soa.org/ccm/content/research-publications/experience-studies-tools/the-rp-2000-mortality-tables.
Strehler B., 1977. Time, Cells and Aging. New York: Academic Press.
Sullivan D.F., 1971. A single index of mortality and morbidity. HSMHA Health Rep 86:347–354.
Tsiatis A., 1975. A nonidentifiability aspect of the problem of competing risks. Proc Natl Acad Sci USA 72(1):20–22.
U.S. Bureau of the Census, 2001. 1999 LTC cross-sectional estimates: source and accuracy statement. Washington, DC: U.S. Census Bureau.
Vaupel J.W., Manton K.G., Stallard E., 1979. The impact of heterogeneity in individual frailty on the dynamics of mortality. Demography 16:439–454.
Witteman J.C.M., Grobbee D.E., Valkenburg H.A. et al., 1994. J-shaped relation between change in diastolic blood pressure and aortic atherosclerosis, Lancet 343:504–507.
Wolf D., 2001. The role of microsimulation in longitudinal data analysis. Can Stud Popul 28:165–179.
Wolter K.M., 1985. Introduction to variance estimation. New York: Springer-Verlag.
Woodbury M.A., Manton K.G., Stallard E., 1981. Longitudinal models for chronic disease risk: an evaluation of logistic multiple regression and alternatives. Int J Epidemiol 10: 187–197.
Woodbury M.A., Manton K.G., 1977. A random walk model of human mortality and aging. Theor Popul Biol 11:37–48.
Woodbury M.A., Manton K.G., 1983. A theoretical model of the physiological dynamics of circulatory disease in human populations. Hum Biol 55:417–441.

Writing Group for the Women's Health Initiative Investigators, 2002. Risks and benefits of estrogen plus progestin in healthy postmenopausal women: principal results From the Women's Health Initiative randomized controlled trial. *JAMA* 288:321–333.

Yashin A.I., Manton K.G., Stallard E. (1986) Dependent competing risks: a stochastic process model. J Math Biol 24(2):119–40.

Yashin A.I., Manton K.G., 1997. Effects of unobserved and partially observed covariate processes on system failure: a review of models and estimation strategies. Stat Sci 12(1):20–34.

Yashin A.I., Iachine I.A., 1997. How frailty models can be used for evaluating longevity limits: taking advantage of an interdisciplinary approach. Demography 34(1):31–48.

# Chapter 5
# Stochastic Methods of Analysis

## 5.1 Introduction

Many types of data are used to study carcinogenesis, e.g., data collected in case–control studies, tumor registries, follow-up data with covariate measurements made at regular, or irregular, time intervals, tracking of individual medical histories, and sample surveys. Data can also take the form of maps, where prevalence, incidence, or other quantities, characterizing the geographic distribution of cancer, is marked for administrative regions. These different forms of data require different statistical methods and models for their analysis.

Many population health models were developed by generalizing classical population and actuarial models. One of the first formal population models with an explicit biological rationale was the Bernoulli life table (see Chapter 4), used to describe the effects of the smallpox vaccination. The model was produced in 1825 by Benjamin Gompertz (1779–1865), a self-educated English mathematician, who became a fellow of the Royal Society, using a two-parameter expression to describe the age dependence of human mortality rates, i.e., $\mu_0(t) = \mu_G(t) = \alpha \exp \theta t$ (Gompertz, 1825). It arose by assuming a constant value for the slope of incidence versus age on log-linear scales, and it was logarithmic in incidence and linear in age. This model continues to be used since it works empirically to describe adult mortality in a wide range of applications. Strehler and Mildvan (1960) attempted to describe the effects of the combined action of aging and environmental stresses, also using an empirical Gompertz function.

One of the most popular statistical models for life data, Weibull distribution, was developed by Waloddi Weibull (1887–1979), an engineer and a prolific inventor from Denmark. It is a continuous probability distribution that describes the lifetime characteristics of parts and components. It is often used in life data analysis due to its flexibility: it can mimic the behavior of other statistical distributions, such as a normal and an exponential. A simple form of the Weibull model describes the failure rates versus age as a straight line on log-log scales, matching the simplest multistage model of carcinogenesis with constant log-log acceleration over all ages (Frank, 2007). The two-parameter

K.G. Manton et al., *Cancer Mortality and Morbidity Patterns in the U.S. Population*, Statistics for Biology and Health, DOI 10.1007/978-0-387-78193-8_5,

© Springer Science+Business Media, LLC 2009

Weibull function $\mu_0(t) = \mu_w(t) = \alpha t^{m-1}$ has often been used as a model of carcinogenesis (Armitage and Doll, 1954, 1961). About the same time, Sacher and Trucco (1962) postulated a stochastic aging mechanism with lethal fixed boundaries to describe mortality. Various aspects of historical and modern aspects in population modeling of aging and mortality were reviewed by Yashin et al. (2000), Akushevich et al. (2006), and Gavrilov and Gavrilova (2006a, b).

Modern population models must not only be applicable to a wide range of data type but also should be capable of performing the joint analyses of a combination of different data types, drawn from different data collection designs. In an ideal scenario, models will provide predictions not only at the population level but at the individual level. Population models currently used for data analyses and modeling include (1) frailty, generalized frailty, and extensions of frailty models; (2) logistic regression, proportional hazard, and event history models; (3) quadratic hazard models; (4) stochastic process models; (5) microsimulation models; (6) models of latent structure; and (7) Bayesian and empirical Bayesian models. The first three types of models were discussed in Chapter 4. We described their advantages and limitations, provided examples of their use, and discussed approaches for validation of these methods and the substantive interpretation of results. Briefly, frailty models can be used when the age patterns of a hazard rate has to be modeled, and the effects of population heterogeneity on the selected hazard function can be described by a "frailty" variable describing the distribution of individual risks (Manton et al., 2008). If, in addition to measurement of a hazard, covariates that represent a health state of an individual are measured, then one needs to model the dependence of the hazard on covariates. In the quadratic hazard model (Woodbury and Manton, 1977), individual dynamics are described by a system of autoregressive equations, and a hazard rate (e.g., death rate) is modeled as a quadratic function of covariates that can describe a biologically justified U- or J-shaped dynamic function of risk covariates (Witteman et al., 1994). In this chapter, we review the models of the last four types. These are advanced methods of population modeling, which were developed during recent decades. These methods, possessing numerous useful features, have a wide potential perspective in future cancer research applications. Currently, their application to cancer research is on different stages of development and will require additional efforts of researchers to result in discoveries that are almost impossible to make using more standard approaches.

The first model from this list, known as a stochastic process model, is a generalization of the binomial quadratic hazard models described in Chapter 4. An important advantage of the binomial quadratic hazard model is that the likelihood, under certain assumptions, can be factored into three independent terms: the initial distribution of risk covariates; covariate dynamics; and survival parameters (Manton and Stallard, 1988; Manton et al., 1992; Akushevich et al., 2005). This simplifies parameter estimation because each likelihood term can be separately maximized. However, such a likelihood factorization requires data generated for short and fixed time intervals, which is not always available. Furthermore, the application of this model to measurements, which are right censored, or which may have missing data, requires additional assumptions. These problems can be addressed without additional assumptions by using a modification of this approach, known as stochastic process model.

## 5.2 Stochastic Process Models

Most survival analyses ignore the dynamics of unobserved, or partly observed, stochastic covariates. The stochastic process model [i.e., the martingale version of the Manton–Woodbury–Yashin approach (Manton and Yashin, 2000)] has the necessary properties for estimating cancer incidence (as well as total and cancer-specific mortality) conditionally on the past trajectory of health characteristics (Yashin and Manton, 1997). When trajectories are unobserved, or incompletely observed, the conditional risk of the health events must be averaged over influential, unobserved variables. The computations can be difficult and burdensome. An advantage of the stochastic process model is that such averaging over process outcomes can be performed at the stage of model construction, without additional assumptions. This approach significantly extends the survival analyses, currently most often done in demography, epidemiology, and biostatistics, which assume that unobserved heterogeneity is fixed over time (e.g., models based on the concept of fixed frailty). The stochastic process model of aging and survival avoids many of the limitations of the hazard modeling approaches that currently are most often applied. Consequently, it can be used to analyze longitudinal data on aging, where stochastically changing covariates may be partly observed at time points sampled over an individual's life. The basic rationale of the generalization is to explicitly introduce the dynamics of unobserved processes into the model, for which there is a belief (coming from prior research, or from earlier empirical applications of the model) that the process satisfies certain stochastic differential equations. For example, such unobserved processes may describe risk factor dynamics measured not necessary regularly, i.e., with fixed time periods between measurements. It can, e.g., track these trajectories by considering the underlying physiological mechanisms of carcinogenesis.

The stochastic process model uses the same assumptions about dynamics of risk covariates and the form (quadratic) of the hazard (i.e., mortality or incidence functions) as the quadratic hazard model discussed in Chapter 4. Specifically, in this approach, the system of stochastic differential equations might be

$$\mathrm{d}x(t) = (a_0(t) + a_1(t)x(t))\mathrm{d}t + a_2(t)\mathrm{d}W_t, \tag{5.1}$$

and

$$\mu(x(t),t) = \mu_0(t) + 2b(t)x(t) + x^*(t)B(t)x(t). \qquad (5.2)$$

The hazard function in equation (5.2) describes the probability of dying, conditional on the health state described by the risk factor values at time $x(t)$. The quadratic form of $\mu(x(t),t)$ allows one to model a situation where there is a point (or a multidimensional domain) in the state space, which is optimal for health (e.g., the vertex of a paraboloid), with mortality risks increasing going away from this domain. The matrix $B(t)$ is thus assumed to be positive definite. For discrete times, $\mu_0(t)$, $b(t)$, and $B(t)$ are coefficients of the cumulative force of mortality for age interval $[t,t+1)$. Parameters of this model may be estimated using the likelihood,

$$L = \prod_{i=1}^{N} \hat{\mu}(\tau_i, \hat{x}(\tau_i))^{\delta_i} exp\left( -\int_0^{\tau_i} du\hat{\mu}(u, \hat{x}_i(u)) \right) \times \prod_{j=1}^{k_i} f(x_i(t_j)|\, \hat{x}_i(t_{j-1})), \qquad (5.3)$$

where $f(x_i(t_j)\backslash\hat{x}_i(t_{j-1}))$ is a pdf conditional on prior observations, $\tau_i$ are ages of death or disease onset, $\delta_i$ are indicators of censoring, $t_j$ are observation times, $\hat{x}_i(t_j)$ are discrete time observations. Indexes $i$ and $j$ run over individuals and exams of each individual, respectively. The equation,

$$\hat{\mu}(\hat{x}(t),t) = m^*(t)B(t)m(t) + 2b(t)m(t) + \mathrm{tr}(B(t)\gamma(t)) + \mu_0(t),$$

has the sense of a right-continuous mortality rate. The vector of covariate means $m(t)$ and the covariance matrix $\gamma(t)$ are defined by systems of ordinary differential equations at intervals $[t_j,t_{j+1})$:

$$\begin{aligned} dm(t)/dt &= a_0(t) + (a_1(t) - 2b(t))m(t) - 2\gamma(t)B(t)m(t), & m(t_j) &= \hat{x}(t_j), \\ d\gamma(t)/dt &= a_1(t)\gamma(t) + \gamma(t)a_1^*(t) + a_2(t)a_2^*(t) - 2\gamma(t)B(t)\gamma(t), & \gamma(t_j) &= 0. \end{aligned}$$

All parameters are defined in the joint likelihood in equation (5.3), so one numerical procedure can be used to simultaneously estimate all parameters. It is not necessary to use numerical procedures to fill in missing data values, because model-generated values $m(t)$ are used to replace them. Furthermore, the projection is obtained as a solution of differential equations, so the time interval between measurements does not have to be fixed or regular.

This model has a natural generalization if, in addition to a risk covariate, an environmental risk factor is measured. Consider the situation where a risk of cancer is studied for a population cohort chronically exposed to ionizing radiation (IR). Below we discuss how this model might be applied to data on IR-exposed populations. There are two questions which have to be addressed in order to choose a suitable population model for the analysis of IR-induced health effects. First, a researcher has to identify the structure of the available information because different models require different data structures for estimation. Second, a researcher has to generalize a model to represent the biological effects of radiation exposure. This may require additional assumptions.

To be specific we consider, as an example, a cohort of a longitudinally followed population of IR-contaminated regions of the Techa River in the Southern Urals region of Russia; i.e., the Extended Techa River Cohort (ETRC). As of 2005, the ETRC included 29,873 individuals. A detailed description of the ETRC composition, the procedure for data collection, the data structure, and IR dose estimation methods were recently published in two papers (Kossenko et al., 2005, Krestinina et al., 2005). Specific features of the data include (as documented in the UNCSEAR report, 2001): (1) the large size of the exposed population with a relatively long (~55 years) follow-up; (2) the wide range of accumulated IR doses (from 0 to 2 Sv); (3) the unselected nature of the population and the availability of local populations to construct comparison (nonexposed, control) groups; and (4) the possibility of examining ethnic differences in cancer risks.

The ETRC data are longitudinal, i.e., there are repeated measurements of health status and other indices obtained from medical examinations. The set of measured parameters included serum lipids spectrum, arterial blood pressure, glucose tolerance test, C-peptide level, bone densitometry (or X-ray-based densitometry) and complete blood count, as well as data on gynecological anamnesis, benign prostatic hyperplasia (fast growing), cholecystectomy, gout, age-related maculopathia, detached retina, and also a family history of CVD, stroke, diabetes, osteoporosis, cataract, and cancers of colorectum, endometrium, ovary, lung, prostate, and breast. Cancer Registry included information on cancer detection, diagnosis (including the way of its verification), histomorphology, and radiotherapy (if was used). Because time intervals between measurements in ETRC are not fixed, and there are missing data, the use of a stochastic process model and data on the exact date of birth and death, or other censoring events are important (Yashin and Manton, 1997). To understand the connection of mechanisms regulating the age dynamics of physiological covariates, mortality by cause, and longevity with IR dose, this model has to be further elaborated upon to analyze the health effects caused by chronic IR.

IR dose can be analyzed by several ways. First, it can be considered as an independent covariate. In this case, equation (5.1) is used, and assumptions for the dependence of coefficients on time (or age) are made. Assumptions for coefficients $a_0(t)$ and $a_1(t)$ in equation (5.1) have to reflect the averaged dynamics of IR dose accumulation in the human body, the reduction of the burden of biologically incorporated radionuclides during later life, and the decline of external exposures. The assumptions for $a_2(t)$ have to reflect physiological heterogeneity in the cohort (ETRC), with respect to their susceptibility to these factors. Parameters in the assumed functional forms are estimated from data using likelihood maximization procedures (see equation (5.3)). The simplest assumption for these coefficients to be estimated is that they are time constant.

If several covariates are used, then coefficients $a_0(t)$, $a_1(t)$, and $a_1(t)$ become matrices, the nondiagonal terms of which describe the mutual impact of covariates. The nondiagonal terms, corresponding to interactions of the accumulated IR dose with other covariates, can be set to zero or modeled employing additional assumptions, if such effects are evident from earlier studies. Other assumptions have to be made for $\mu_0$, $b(t)$, and $B(t)$. As discussed above, the risk function is a positive definite quadratic function of health covariates. Such a form is represented in expression (5.2) for the hazard. Therefore, the parameters (or vector $b(t)$ and matrix $B(t)$) can be chosen as numeric constants estimable from data. As follows from analysis of relative risks (e.g., Little, 2003), the dose-response function will have linear and quadratic terms, so nothing in the equation has to be changed if the IR dose is added as a covariate.

If data available in the ETRC are insufficient to statistically estimate all parameters, then another approach can be used, in which IR dose variables are not used as additional covariates, but the dose dependence of all model parameters $a_{0,1,2}(t)$, $\mu_0$, $b(t)$, and $B(t)$ is modeled instead. For such models, the same arguments as for the use of IR dose as a covariate can be made. For example, standard dose-response analysis assumes $\mu_0\,(d) = \mu_0 + \beta d$. If an effect of dose-response modification by age, or other covariate, has to be taken into account, then similar definitions have to be created for $b(t)$ or even $B(t)$.

These approaches essentially extend the standard epidemiologic analyses of data on cancer morbidity and mortality collected in cohorts (or populations) exposed to chronic or acute ionizing radiation. The main advantage of the extension is that such models allow the incorporation of auxiliary clinical information that is often irregularly measured, in the model, without additional assumptions.

## 5.3 Microsimulation and Interventions

There are many examples where the task of interest requires modeling approaches operational at several biological levels, where effects at different levels can interact. The simplest situation is when there are only two levels: the first – an aggregate level and the second – the level where the “behavior” of individuals (e.g., persons, patients, cells) is modeled. A researcher is typically interested in averaging effects over individuals at the microlevel, where specific physiological mechanisms are known or assumed, to obtain characteristics at an aggregate level, where data are typically collected. The averaging of characteristics at the microlevel required to complete this task can be performed analytically (by specifying particular functions) in some cases. However, even moderate complications in the laws governing the behavior of individual mechanisms are often required to approximate reality, e.g., if one wants to use an additional information on individuals, such analytical averaging immediately becomes an intricate procedure, often requiring the solution of a complex system of nonlinear equations. An alternative to the analytical approach is to use microsimulation (or Monte Carlo simulation). In this approach, a researcher simulates large representative populations at the micro-level, and numerically averages (assuming independence of outcome) micro-characteristics, in order to draw conclusions on trends at higher levels of aggregation. Simulated individuals typically have a number of attributes, such as age, sex, marital status, health state, physiological and behavior risk factors, and a number of transition probabilities, such as cancer incidence or cancer specific mortality.

Dynamic microsimulation is often used in epidemiology and demography. What is modeled at the microlevel are life histories or individual trajectories in the space representing individual health status. One step of the procedure usually corresponds to 1 year. The modeled population is changing due to aging and death of individuals according to life tables and, if necessary, birth of new individuals. Attributes of individuals can change at the each step of the procedure (Gilbert and Troitszsch, 1999).

Often the conditions required for the application of the stochastic process model may not be valid, e.g., when covariates are not measured, but transfer rates between discrete states are available, or when the drift function is not linear, or includes information about the history of the process. In such situations, microsimulation models can be used to generate projections. Technically, the construction of such projections includes (1) a simulation of the initial population, e.g., 100,000 individuals with normally distributed risk factors; (2) a calculation of the mortality and other transfer rates (e.g., increases risk factor levels) for each individual, conditional on its health state (e.g., set of risk factors); (3) a random definition of whether an individual survives (or another transfer event occurs) in a time interval, or not; and (4) a simulation of the risk factors (or sets of variables describing a person's health state) for individuals surviving to the next age. Such models are quite flexible, which makes them appropriate for the analysis of medical (i.e., screening, diagnosis, treatment of disease), economical/financial (including cost-effectiveness analysis), public health/behavioral (e.g., primary and secondary prevention strategy), and technological (new approaches to diagnosis and treatment) interventions. Transfer rates or parameters that describe the stochastic heterogeneity of a population (diffusion parameters) can be changed in accordance with the formulation of alternative strategies. Examples of applications of a microsimulation approach are presented in Chapters 4 and 8. In Chapter 4, we used mortality rates modeled as a quadratic function of risk factors estimated in binomial quadratic hazard models from Framingham data. Survival functions, with/without interventions are shown in Fig. 5.1 for males and females for nonlinear models developed by Akushevich et al. (2005). In Chapter 8, two microsimulation experiments for analyses of interventions are presented. The first is based on a microsimulation model, allowing one to forecast short- and long-term population changes conditional on the prevalence of a risk factor in population. In this model, population changes result from the aggregation of changes in individual

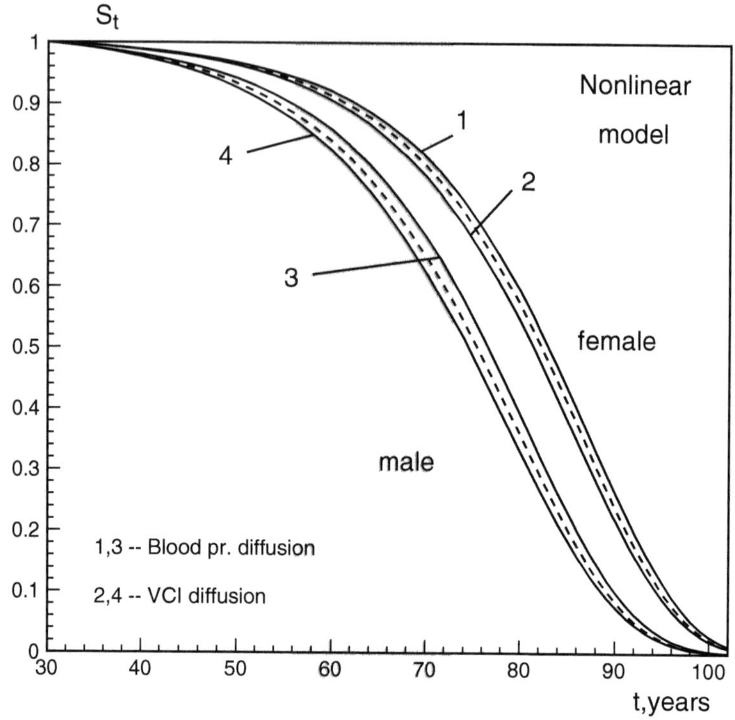


**Fig. 5.1** Survival functions for males (3, 4) and females (1, 2) for calculations without (*dashed line*) and with (*solid line*) interventions: two times increased diffusion for VCI, two times decreased diffusion for arterial blood pressure and pulse pressure

event histories, which, in turn, result from recalculated mortality and infertility rates, in accordance with the known relative risks in exposed population groups. Smoking, being the most widespread and influential public health risk factor, was chosen to demonstrate the abilities of the model to forecast the population mortality and fertility effects of hypothetical levels of smoking prevalence. Such projections were made using data of three types: (1) fertility and mortality rates; (2) relative risks of primary and secondary infertility, and incidence of smoking-associated diseases; and (3) mortality from smoking-associated diseases (Akushevich et al., 2007).

The second simulation experiment in Chapter 8 is to make a prognosis about the possible impact of progenitor cell therapy on CVD-associated mortality, life expectancy, and survival, as compared to the scenario of the "ideal" lifetime control of conventional risk factors (Kravchenko et al., 2005). Projections of CVD mortality (Fig. 8.4) were constructed analytically, using the same rates for risk factor dynamics and mortality. We considered three types of interventions: (1) keeping "classic" risk factors for CVD within select limits to model current clinical recommendations; (2) an age shift of 10 years to model the virtual effects of progenitor cell therapy of atherosclerosis; and (3) the elimination of the cancer-competing risk term (i.e., creating a hypothetical situation when cancer might be completely "beaten"). In these scenarios, dynamics of risk factors were modeled using autoregressive functions. Estimation of such a model requires detailed longitudinal epidemiological information, e.g., to model the intervention effects of stem cell therapy on adult atherosclerosis, we used the 46-year follow-up of the Framingham Heart Study with exams done biannually.

Generalization of these forecasting models to describe the national populations, and to include the effects of risk factors on fertility, is hampered by the lack of longitudinal data with detailed measurements of covariates. Therefore, simplifying assumptions are required for the practical implementation of forecasts for the U.S. population. The principal requirement for such assumptions is that all model parameters and rates used have to be defined by observation.

Apart from the microsimulation approach illustrated in these examples, this method has the potential for developing hierarchical models of carcinogenesis. In Chapter 2, we reviewed models based on different currently studied mechanisms. However, these models use only simplified versions of these mechanisms because of the computational difference between micro- and macrolevel models. Microsimulation allows us to numerically connect these two levels in a way that is both more biologically realistic and feasible for modern computers than in macrosimulation.

## 5.4 Grade of Membership and Other Latent-State Analysis Methods

Survey data typically represent sample-based collections of measurements made with discrete outcomes for individuals. Continuous risk factors, such as blood pressure and serum cholesterol level, could also be involved in these datasets after their categorization. Common property of such datasets is high dimensionality (i.e., large values of used variables – about several hundred), and the measured variables are highly correlated. In methods of latent analyses, it is assumed that the observed structure of multiple categorical variables are generated by the small number of latent (i.e., unobserved) variables. The task of latent analyses is to find these latent variables, estimate parameters of their distribution, and describe their properties using a sample of high-dimensional categorical variables. Generally, it is necessary to find the properties of a population, associated with latent variables, and properties of individuals, based on those multiple categorical measurements. It appears that both goals may be achieved simultaneously. To the increase precision of population and individual estimates, one has to increase both the sample size (i.e., the number of individuals) and the number of measurements (i.e., questions asked for each individual).

One example of the applicability of the methods of latent analyses is in making a medical diagnosis. A physician has to make a conclusion about a patient's state of health based on a number of measurements, which include both, objective (e.g., blood pressure) and subjective (specific questions about health) measurements. Decision making requires knowledge about the population (what it means "to be healthy" or "to have this disease"), derived from the results of similar measurements performed on other individuals. The ability to derive the properties of a population is provided by sampling a sufficient number of individuals, whereas the ability to derive properties of an individual is provided by a sufficiently large number of measurements on each individual (in practice, it may be several dozens). From a modeling point of view, this situation is complicated because the measurements, on one hand, should relate to the same underlying problem (individual state), while, on the other hand, they *must* be different (to avoid bias for a particular individual).

Often a mixture of distributions is used to describe the heterogeneous population, defined by a large number of discrete conditions, no combination of which occurs with high frequency. One fundamental weakness of mixture models is the presumption that the underlying population can be decomposed into distinct, well-defined categories at some level of precision. Methods dealing with such tasks, where the structures of such decompositions are also estimated from the data, are known as latent analysis. One of the best known of such methods is the latent class model (LCM), which can be characterized as a statistical method for finding discrete subtypes of related cases (latent classes) from multivariate categorical data (Lazarsfeld and Henry, 1968; Goodman, 1974; Clogg, 1995). In LCM, individuals are assigned to one of the several homogeneous classes. This requires the estimation of the individual latent variable (class number). Other models of this type, such as item-response theory and Rasch models, discrete latent class models (Heinen, 1996), and latent distribution analysis (Mislevy, 1984; Uebersax and Grove, 1993; Uebersax, 1997; Qu et al., 1996), differ by the assumptions made about the latent variable(s). One method for identifying the latent structure in large categorical datasets with a simultaneous evaluation of individual scores in a state space is Grade of Membership (GoM) analysis. Historically, GoM was introduced in a publication of Woodbury and Clive (1974). Manton et al. (1994) provided a detailed exposition of different version of this approach and reviewed its properties. Statistical properties of GoM models were rigorously analyzed by Tolley and Manton (1992), Singer (1989), Berkman et al. (1989), and Wachter (1999).

Recently, linear latent structure (LLS) analysis has been proposed to model high-dimensional categorical data (Kovtun et al., 2006, 2007). The LLS model was formulated using the mixing distribution theory. Similar to other latent structure analyses, the goal of LLS analysis is to derive simultaneously the properties of a population and individuals, using discrete measurements. The LLS, however, does not use maximization of a likelihood for parameter estimation. Instead, it uses an estimator, where the LLS parameter estimates are solutions of a quasilinear system of equations.

A cornerstone of latent structure analysis is called "the local independence" assumption. Roughly it says that individual responses are independent, conditional on individual latent variables. There are many excellent books and articles devoted to latent structure analysis; we refer to Bartholomew and Knott (1999), Clogg (1995), Goodman (1978), Heinen (1996), Langeheine and Rost (1988), Lazarsfeld and Henry (1968), and Marcoulides and Moustaki (2002) for discussion of the meaning and applicability of this assumption.

The results of $J$ measurements made on $N$ individuals can be presented as in the scheme (Fig. 5.2).

Mathematically, the task of GoM/LLS analysis is briefly described as follows. Individual outcomes are described by categorical random variables $X_1,\ldots,X_J$, where $X_j$ takes values in $\{1,\ldots,L_j\}$. The joint distribution of these variables is given by elementary probabilities: $p_\ell = \Pr(X_1 = \ell_1 \text{ and}\ldots\text{and } X_J = \ell_J)$, where $\ell = (\ell_1,\ldots,\ell_J)$ is the response patterns, including individual responses to all survey questions. These, and only these, values $p_\ell$ are directly estimable from observations. Frequencies $f_\ell = \frac{N_\ell}{N}$ are consistent estimators for $p_\ell$. In addition to probabilities $p_\ell$, the set of marginal probabilities can be considered. The most important cases are $\Pr(X_j = \ell_j)$ and $\Pr(X_j = \ell_j \text{ and } X_{j'} = \ell_{j'})$, i.e., the marginal probabilities of the first and second order.

The number of frequencies $p_\ell$ corresponds to the number of possible response patterns. This number equals to $\prod_j L_j$ and typically is very large (e.g., for a dataset restricted to 20 binary questions, this number is 1,048,576). As a result, in practical applications $p_\ell$ is estimated to be very inaccurate. Usually, there is no way to estimate all the probabilities, $p_\ell$, even to put all respective numbers into computer memory. However, since these random variables $X_1,\ldots,X_J$ are correlated (and highly correlated, e.g., for close survey questions), we can approximate probabilities $p_\ell$ by a model that

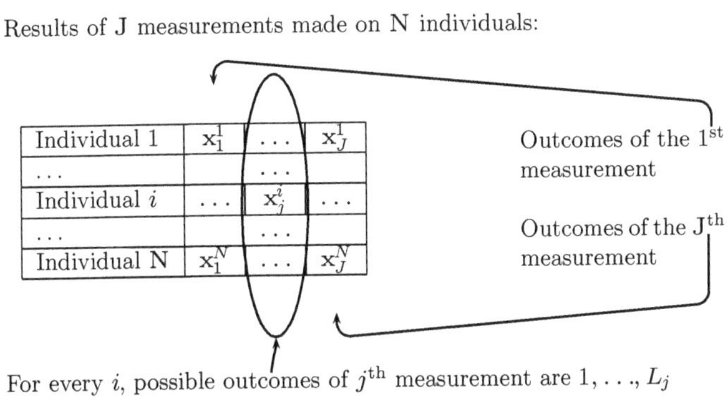


**Fig. 5.2** Representation of measurements for GoM and LLS analyses

is capable of predicting them and explaining the correlation structure between the random variables. It allows us to reduce the task of estimating billions of probabilities to the task of estimating several tens of model parameters that describe the population, and several individual parameters that reflect information contained in individual response patterns. As a result, instead of using rough estimates of these billion frequencies, we predict the individual response pattern using the model, compare it with the results of his/her measurements, construct the likelihood function or develop another strategy for parameter estimation, and then estimate parameters of the developed model.

### 5.4.1 GoM Model

The Grade of Membership (GoM) model is the first of two specific models of latent structure analysis considered here. This approach generalizes the definition of the state of the individuals to include states defined by "fuzzy" sets. If it is assumed that study populations can be represented as a set of $K$-specific compartments or population subgroups, then individual fuzzy state means that it can be represented by the degree of grade of membership in each of these $K$ comparators' sets. Formally, it means that for individuals with response pattern $\ell$, the set of the first-order probabilities is

$$\Pr(X_j = \ell_j) = \sum\nolimits_{k=1}^{K} g_{\ell k}\lambda_{j\ell_j}^{k}, \tag{5.4}$$

Thus, categorical data are described by two types of coefficients. The first is population structural parameters, $\lambda_{jl}^{k}$, characterizing the $k$th population subgroup. These parameters have the same structure as the first-order probabilities, i.e., $\sum_l \lambda_{jl}^{k} = 1$ and $\lambda_{jl}^{k} \geq 0$. An ideal person, who exactly belongs to the $k$th population subgroup, would have probabilities to answer the survey question exactly coinciding with $\lambda_{jl}^{k}$. The second is individual parameters, or GoM scores, $g_{\ell k}$ or $g_{ik}$, explicitly reflecting individual indices which are often more convenient. Two sets, $g_{\ell k}$ or $g_{ik}$, are equivalent, if keeping in mind that all $g_{ik}$ are equal for individuals with the same outcome pattern, i.e., individuals with the same set of outcomes are nondistinguishable. GoM scores represent a weight that describes how much the $k$th dimension contributes to the traits observed for the $i$th person, or in other words, a weight reflecting the degree in which the $i$th person has characteristics of $k$th ideal person from the $k$th population subgroup. In latent class analysis, $g_{ik}$ is, in effect, forced to take exactly the value of 0 or 1, to form the $K$ homogeneous discrete, or "crisp", classes in each of those models (Everitt, 1984). In GoM, $g_{ik}$'s are not probabilities, but mixing parameters, which may take any value between 0 and 1 to define $K$ fuzzy classes. The GoM scores are constructed in a convex space, where $\sum_k g_{ik} = 1$, $0 \leq g_{ik} \leq 1$.

Parameters are estimated using the equivalent of a multinominal type likelihood

$$L = \prod_i \prod_j \prod_l \left( \sum_{k=1}^{K} g_{ik} \lambda_{jl}^{k} \right)^{y_{ijl}}$$

by means of a variation of Newton–Raphson procedures suitably modified to ensure progress toward a solution with Kuhn–Tucker conditions, which is imposed for boundary conditions. Here, $y_{ijl}$ is an indicator variable, equaling to 1 if $i$th individual's response on $j$th question is $l$, and equaling to 0 otherwise. As the partial derivatives of the likelihood function for $g_{ik}$ are continuous (except at boundaries), the algorithm is (for a complex, nonlinear optimization problem) numerically relatively well behaved and solved fairly rapidly (Press et al., 1999).

A comparison of GoM with other methods of latent structure analyses, such as Rasch (or latent trait) and latent class models, was presented by Erosheva (2005). Manton and Land (2000) and Stallard (2007) generalized GoM for analysis of longitudinal data and applied the generalized version of the GoM analysis to trajectories of morbidity, disability, and mortality among the U.S. elderly population. An example of recent application of the GoM approach is described in Manton et al. (2004). In this study, the application of the methodology for analyses of complex genotype–phenotype relations was illustrated for apolipoprotein E (APOE) assessments made on 1805 people in the 1999 NLTCS (Manton et al., 2006).

### 5.4.2 LLS Model

The LLS analysis is the second method of latent structure analysis discussed here. Similar to other methods of latent structure analyses (LSA), the LLS analysis is designed to explain the mutual correlations between variables $X_1,\ldots,X_J$. Similar to GoM, LLS assumes equation (5.4) for relation between the first-order probabilities and model parameters, which are the same as in the GoM model. However, LLS analysis does not assume positivity of scores $g_{ik}$. Instead, it requires positivity of respective probabilities. The complete set of model restrictions in LLS analysis is

$$\sum_{l=1}^{L_J} \lambda_{jl}^{k} = 1, \quad \lambda_{jl}^{k} \geq 0, \quad \sum_{k=1}^{K} g_{ik} = 1 \quad \text{and} \quad \sum_{k} g_{ik} \lambda_{jl}^{k} \geq 0.$$

This set of model restrictions leads to the extension of the set of allowed parameter values comparing to GoM. One result of the generalization is that new scores (LLS scores also denoted as $g_{ik}$) are not restricted to $0 \leq g_i \leq 1$ and, therefore, are not interpreted as grades of membership. An important property of the LLS analysis is the existence of high-performance algorithms of parameter estimation (Akushevich et al. 2008). The properties of the new model (Kovtun et al., 2006, 2007; Akushevich et al., 2008) allowed for reducing the problem of estimating model parameters to a sequence of linear algebra problems. The algorithmic approach based on linear algebra methods assures a low computational complexity and an ability to handle data potentially involving thousands of variables. Preliminary studies demonstrated that the new numerical scheme is stable, and it will allow the researcher to include many more variables in an analysis than was possible using the GoM analysis and other latent-structure methods.

Basic steps of the algorithm include (1) determining the dimension of the investigated data; (2) identifying the linear subspace of the found dimensionality, which has the sense of the latent structure generating the data; (3) choosing a basis in the found subspace, using methods of cluster analysis and/or prior knowledge of the phenomenon of interest; (4) calculating empirical distributions of the LLS scores which reflect individual responses in the linear subspace; (5) investigating properties of the LLS score distribution to capture population and individual effects (e.g., heterogeneity); and (6) using components of the vectors of individual LLS scores, developing a scheme of prediction of individual lifespan and future changes in health.

Akushevich et al. (2008) performed detailed simulation studies to demonstrate the quality of reconstruction of the major components of the models. Results of these simulation studies proved the sufficient quality of reconstruction for typical sample size and demonstrated the potential of the methodology to analyze survey datasets with 1000 or more questions. This methodology has been applied to the 1999 NLTCS dataset, including 4791 individuals with responses on 57 questions on activities of daily living, instrumental activities of daily living, physical impairment, and self-reports of chronic diseases. It has been found that (1) the estimated dimensionality is three; (2) the components of the space correspond to healthy individuals, disabled (strongly disabled) individuals, and individuals with chronic diseases but without the evidence of disability; (3) empirical distribution of the LLS scores in the found subspace demonstrates heterogeneity of the population with respect to these characteristics; and (4) the components of the vectors of individual LLS scores can be used as predictors of individual life spans. To illustrate the last fact, the Medicare Vital Statistics Data from 1999 to 2003 has been linked to NLTCS.

LLS analysis can be naturally generalized to longitudinal data using the binomial quadratic hazard, or stochastic process models, where LLS scores play the role of covariates. After estimating the parameters, a scheme for the projection has to be developed, which, in general, can be based on microsimulation procedures.

An attractive feature of both LLS and GoM is their ability to convert information from individual response patterns into several continuous or discrete measures. If the method of latent structure analysis is chosen properly (i.e., if an assumption about the mathematical form of the latent structure corresponds well to the unobservable structure responsible for generating observable data), then such measures will absorb all available information and be largely free from stochasticity. These individual measures constitute a useful and convenient parameterization of population heterogeneity, which is originally hidden in a set of interrelated individual responses. Population measures, in the form of a set of individual scores, can be used to develop state-space models, the analysis of individual trajectories in this state space, and in joint analysis of data from different sources, e.g., in longitudinal analysis of surveys collected in different periods. The methods of GoM and LLS analyses also can be used in modeling for populations with heterogeneous mutational events and tumor growth rates (as it described in Chapter 2).

## 5.5 A Geo-epidemiological/Mapping Study of the U.S. Cancer Mortality Rates and Trends Based on an Empirical Bayes Approach

Mapping and the analysis of geographic patterns is not a new approach in epidemiological studies. John Snow (1813–1858), a British physician, in his classic study in 1854 mapped individual cases of a cholera epidemic in London, revealing a strong circular pattern centered on the Broad Street water pump, thus suggesting that cholera was a water-born disease. That was the first-known historic example of an epidemiologist analyzing geographic patterns of morbidity or mortality in the epidemiological investigation of a disease outbreak.

At present, the maps of cancer incidence and mortality are very useful instruments for various epidemiologic studies: these maps may be used to identify territories with high cancer incidence and mortality rates, to examine large shifts in the ranking of area risk levels over time, to investigate the possible causes and/or associations with various risk factors, etc. The maps may also be used to locate areas with unusual demographic, environmental, industrial characteristics, or employment patterns and to determine whether they exhibit elevated rates or unusual trends that might be attributed to these characteristics. Maps can be used for sex-specific analysis of cancer incidence and mortality: high rates for both sexes suggest a possible relation to an environmental exposure or other factors unrelated to sex, while high rates among men only, for example, might suggest occupational or other sex-related factors.

The mapping of age-standardized death rates for specific cancer sites is an important tool in assessing the environmental determinants of different types of cancer and in forming hypotheses about cancer risks for specific subpopulations (to be tested in specially designed epidemiologic population studies). Information about types of environmental exposure may point to genes that might modify cancer risk, so the identification of genes associated with risk could help to indict previously unrecognized environmental risk factors. Analysis of cancer mortality in the United States is a powerful instrument to describe the situation for the entire country. But it might not be, however, a very precise instrument to identify detailed associations. It is especially important for analysis of risk factors' association with cancer, when state or even state economic areas are too large for detailed analyses with wide variation in risk-factor presence and intensity. The finer the geographic detail (e.g., the smaller the area), the greater the capacity to identify the environmental causes of disease risks.

Geographic and temporal differences are useful in developing and examining hypotheses about the influence of various environmental factors on cancer risk. The consistency of such differences across race, sex, and age groups can provide insight into other possible causes. For local communities concerned about a specific cancer situation, these maps could provide both a spatial and temporal context in which to evaluate local conditions. Maps of county and state economic area (SEA) cancer death rates have produced important insights about etiologic factors for lung cancer (i.e., asbestos exposure), cancer of the oral cavity (i.e., "smokeless" tobacco), and several other cancer types (Blot et al., 1979; NCI, 1987). Several strategies have been discussed for using detailed maps of rates, in conjunction with demographic and industrial data at the county level, to develop etiologic hypotheses for detailed epidemiologic investigation (Blot et al., 1979).

Unfortunately, there is a difficulty in using observed rates for small areas to make such decisions, i.e., small areas tend to have small populations. The precision of a rate estimate is inversely related to the size of the local population and number of index events in the area. Use of the observed rates for small areas may introduce systematic errors in decision making, if it is necessary to identify rates that are truly extreme. This is especially important for rare cancer types, when unstable local area rate estimates, resulting from small population sizes, obscure the underlying spatial patterns of disease risk. If the observed rates are inadequate for mapping because of their large random fluctuations, the rate estimates that are more stable should be found to replace "fluctuating" rates. Rates for each area used in the study are assumed to be temporarily stable: e.g., in the analysis below, each area's data was pooled by decade. However, pooling data over a long period may cause temporal changes in risk to be missed. A decade was chosen as the longest period that could be substantially justified (Riggan et al., 1991). However, even decade-specific rates were still often unstable: thus, it was necessary to generalize the principle of averaging beyond the data which was available to a hypothetical case where the observed rate was one of an infinite number of outcomes that could have happened. This generalization is the underlying essence of the empirical Bayes model (Riggan et al., 1991).

Cancer mortality atlases were prepared by Mason with colleagues (1975, 1976) for the U.S. counties and SEAs for the 20-year period, 1950–1969. Two subsequent publications have independently updated these atlases by extending the period covered to 1950–1979 and by presenting decade-specific maps for three decades – 1950–1959, 1960–1969, and 1970–1979 (Pickle et al., 1987; Riggan et al., 1987). Analysis of the observed unadjusted mortality rates for small local area populations raises important statistical questions. With nonparametric methods, there are questions about the validity and reliability of the data used to construct the rates. With parametric methods, there are questions concerning the correct specification of the covariate function used to generate model-based estimates. Furthermore, the rates for local area populations are highly variable due to (1) small population sizes and (2) the rarity of events. Such statistical instability makes the identification of "extreme" rates difficult, and it tends to conceal patterns by breaking up clusters of counties with similar rates.

The model used to improve cancer rate estimates is a type of empirical Bayes procedure (Efron and Morris, 1973). The term "empirical Bayes" was introduced by Robbins (1955, 1964) to refer to decision problems in which the identical conditions are faced repeatedly. For each decision, new data are provided, and one wants to estimate the long-run average over repeated, identical experiments. This problem is often referred to as "nonparametric empirical Bayes". Nonparametric empirical Bayes is not directly applicable to the mapping problem because it requires multiple observed rates for each county, for each decade produced under identical conditions. This approach was developed for situations where the identical population health "experiment" could be repeated. We cannot repeat the identical conditions producing the set of cancer deaths in a county. However, we do have observations on multiple ($n = 3061$) U.S. counties. While the conditions in these counties are not identical, the decisions to be made for each county are the same. If the age-specific population counts were constant over counties, the decision problems would be identical (Riggan et al., 1991).

The term "parametric" empirical Bayes is used to refer to problems where the conditions producing each event (e.g., a county-cancer mortality rate) are similar in some respect, but not identical (Morris, 1983). By introducing a parametric distribution into an empirical Bayes model, the rate estimator for each county can be made dependent, through the parameters of the selected distribution, on the rate estimator for all other counties. In effect, this allows the information required to estimate one county's long-run average to be obtained from the rates in all other counties (i.e., an average is obtained for the entire set of counties).

Specifically, a negative binomial regression model, defining relationships between dependent and independent variables, utilizes the information in the total ensemble of subarea rates, to determine the best statistical weighting of individual subarea rates (Manton and Stallard, 1981; Manton et al., 1981). The statistical theory underlying empirical Bayes procedure (and a number of successful applications) is reviewed in Morris (1983). Much of this theory deals with the properties of conjugate prior distributions of key parameters. If the local area mortality count is Poisson distributed with an unknown mortality rate that is gamma distributed (the assumed form of the "prior" distribution), then (1) the observed mortality count will exhibit super-Poisson variability, which will be described by a negative binomial distribution and (2) the "posterior" (also called conditional) distribution of the unknown local area rate, given the observed mortality count, will also be a gamma distribution, but with the two parameters (i.e., mean and variance) modified to reflect the new information provided by the observed mortality count. The term "conjugate" indicates that under the Poisson model, both the prior and the posterior distributions of the mortality rates are members of the same parametric family of distributions (i.e., the gamma family). Morris (1983) showed that in the absence of information contradicting the gamma conjugate prior assumption, the empirical Bayes procedure is the method of choice for (1) identifying patterns of high- and low-risk areas in a large group of small area populations and (2) selecting specific high-risk areas by ranking a number of such areas. The composite rate estimate is superior for the uses for which the cancer maps and the analysis of detailed geographic patterns of cancer rates are usually applied.

Two distinct forms of empirical Bayes analysis exist. The first – the quintile model – is based on an extension of the standardized mortality ratio (SMR) model (Manton et al., 1987) to age-specific death rates. The second – the two-stage model – analyzes both total and age-specific death rates, but sequentially (Manton et al., 1989). In the first stage, a negative binomial regression model is used to assess the variability of the local area mortality rates and to produce an estimate of the rate for each local area. Second, estimates of the super-Poisson variability of local area rates derived from the negative binomial regression models are used to calculate local area composite rate estimates as weighted averages of the model-based and observed rates.

The variation of local area cancer rates is evaluated with a negative binomial regression model. The model is based on the assumption that the numbers of cancer deaths in each county population are negative binomially distributed. The negative binomial distribution is a generalization of the Poisson distribution, which is often employed in the analysis of health event rates (Manton et al., 1985). Negative binomial variance is larger than that expected for the Poisson distribution with the same aggregate mean rate. The additional variance in the negative binomial case, which is referred to as "super-Poisson variance", can be attributed to population heterogeneity in disease risk, e.g., as when the Poisson rate parameters for individuals in an area differ and are gamma distributed. The rate differential between areas can be attributed to the clustering of rate differentials between people, so one need not unrealistically assume all people in a given area to have the same cancer mortality risks.

If individual cancer rates are assumed to be independently and identically gamma distributed, then the average rate in each local area is also gamma distributed with the local area population size introduced as a parameter. Combining this with the standard assumption that the cancer mortality count is Poisson distributed, with an expected value given by the product of the population size, one obtains the marginal negative binomial distribution. The choice of the gamma distribution to represent the uncertainty in local area mortality rates is justifiable (1) as a physically meaningful model in a covariate regression model with heterogeneous subpopulations and (2) as a statistically optimal choice of a conjugate, prior distribution, in a composite estimation strategy based on empirical Bayes procedures.

There are several types of "observed death rate" that might be used in cancer risk analyses in small populations. One of them is the "crude death rate" (CDR), which is, however, unsatisfactory for comparing county rates, since two counties with identical age-specific death rates can yield different crude death rates because of differences in age-specific population counts. In the examples given below, we will use two other rates – the "direct age standardization" and "marginal age standardized" age-specific death rates (DASDR and MASDR, respectively) (see Manton et al. (1989) and Riggan et al. (1991) for exact definitions and further details).

The data employed in the analysis of cancer mortality were drawn from files of county-specific cancer deaths, prepared by the U.S. Environmental Protection Agency (EPA) from detailed microdata mortality files (which were prepared by the National Center for Health Statistics) and population estimates provided by the U.S. Bureau of the Census. Tabulations of death and population counts were prepared for 3061 counties (or county equivalents), for 31 cancer sites, 18 age groups of white and nonwhite populations.

The motivation for these cancer analyses was to obtain empirical Bayes rate estimates suitable for mapping. As it was discussed by Manton et al. (1989), the particular model structure, which is a two-stage application of empirical Bayes principles, has several advantages over other empirical Bayes models (Gaver and O'Muircheartaigh, 1987; Tsutakawa, 1988; Tsutakawa et al., 1985). Table 5.1 contains MASDRS estimates for white males and females. There are significant statistical variations between counties and between decades for all 15 cancer sites selected for mapping. The statistical variation of the marginal rates is negligible (the coefficient of variation was 0.1–1.0%), so one can be confident in these temporal changes.

The effects of using such rate estimators are illustrated in Fig. 5.3 for kidney cancer. Map A represents unadjusted rates for U.S. counties, and Map B

**Table 5.1** County-level MASDR estimates obtained under two-stage maximum likelihood: U.S. white population, site-specific cancers, 1950–1979 (data from Manton et al., 1989)

| | MASDR $\times 10^5$, male/female | | |
|---|---|---|---|
| Cancer site | 1950–1959 | 1960–1969 | 1970–1979 |
| Stomach | 20.2 / 10.7 | 13.2 / 6.7 | 9.0 / 4.3 |
| Large intestine | 17.1 / 18.0 | 18.3 / 17.1 | 20.0 / 16.4 |
| Rectum | 9.0 / 6.0 | 7.5 / 4.6 | 5.8 / 3.4 |
| Liver/gallbladder | 3.0 / 4.2 | 3.7 / 3.9 | 4.6 / 3.8 |
| Pancreas | 9.3 / 5.9 | 10.9 / 6.5 | 10.9 / 6.7 |
| Lung | 29.6 / 5.1 | 46.8 / 7.6 | 64.0 / 15.3 |
| Prostate | 20.7 | 19.7 | 20.3 |
| Breast (female) | 26.3 | 26.4 | 27.0 |
| Cervix uteri | 9.1 | 6.9 | 4.2 |
| Other uterus | 7.8 | 5.4 | 4.3 |
| Ovary | 8.6 | 8.9 | 8.8 |
| Kidney/ureter | 3.7 / 2.0 | 4.2 / 2.1 | 4.6 / 2.1 |
| Bladder | 7.4 / 2.9 | 7.3 / 2.4 | 7.3 / 2.1 |
| Brain | 4.0 / 2.6 | 4.4 / 2.9 | 4.9 / 3.3 |
| Total | 176.6 / 141.6 | 190.0 / 132.4 | 204.1 / 131.7 |

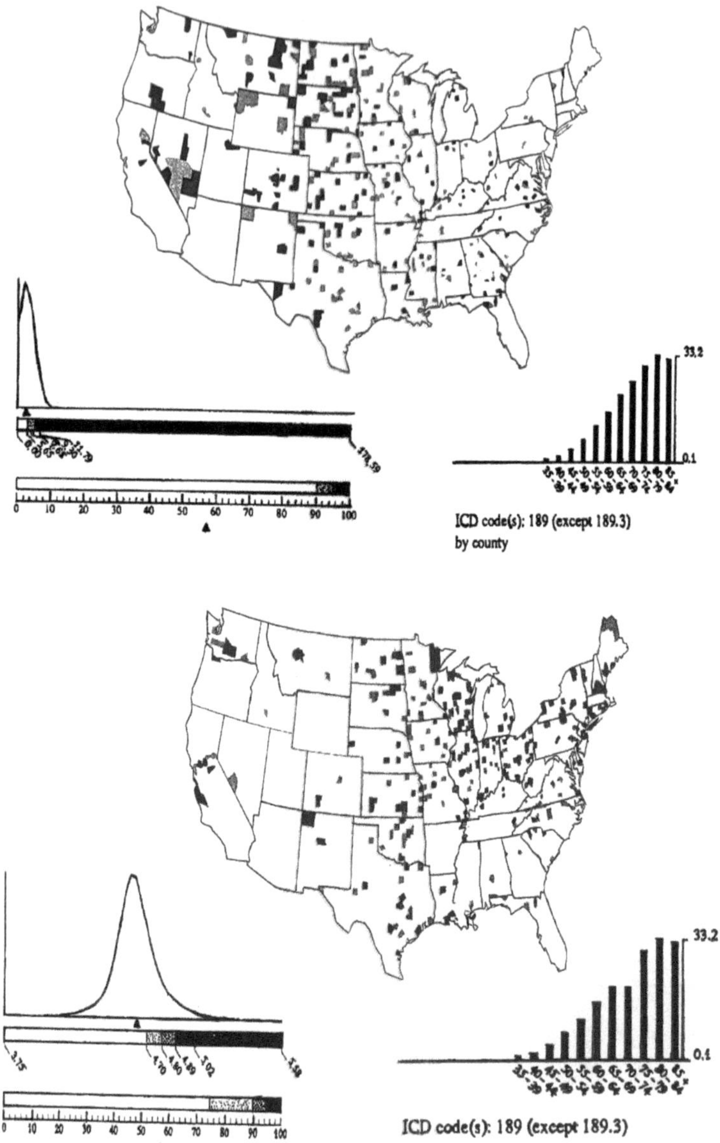


**Fig. 5.3 A and B** Direct age-standardized death rates (DASDRs). **A** Empirical Bayes age-standardized death rates (EBASDRs); **B** for cancer of kidney/ureter per 100,000 according to county: U.S. white males, 1970–1979 (from Manton et al., 1989). (The frequency function in the lower left of the figure is a graph of the unweighted frequencies of the 3061 county-specific DASDRs. The first tone bar below the graph indicates the range of the distribution (in units of $10^{-5}$) and the locations of the 75th, 90th, 95th, and 98th percentiles, as defined on the second tone bar. The *arrowheads* below the graph and below the second tone bar indicate the location of the national death rate (MASDR). The bar graph in the lower right is a graph of the age-specific death rates (in units of $10^{-5}$) for ages 35–39 to 85 years and older)

represents regression of adjusted rates. Apparently, elevated rates in very small counties no longer appear to be elevated. The shrinkage is illustrated for the total distribution in the three figures at the bottom of the maps. Given the MASDR of $204.1 \times 10^{-5}$ for total cancer, the entire set of high rates in the range from 11.79 to $178.59 \times 10^{-5}$ for kidney/ureter cancer is suspected because of the small number of death represented (i.e., kidney/ureter cancer accounted for about 2% of all cancer deaths for white males in those period). That is confirmed by the largest empirical Bayes rate ($5.58 \times 10^{-5}$), which is below the 75th percentile of the DASDR's ($5.65 \times 10^{-5}$). The first-stage $z(\beta)$ score for kidney/ureter cancer is 8.44, i.e., a strong evidence that the marginal rates do not fit the data for all 3061 counties. On the other hand, the second-stage $z(\alpha)$ score of 0.66 indicates that the residual extra-Poisson variation after fitting the first-stage model is nonsignificant.

Maps based on the stabilized rate estimates allow new spatial features of cancer mortality risks to be identified, which reflect absolute levels of risk by using composite estimators that weight the statistical evidence in several ways. The generalized empirical Bayes procedure is capable of incorporating a wide variety of area-specific covariates (Manton et al., 1987). This example illustrates three features of the empirical Bayes approach: (1) comparison of the frequency functions for the DASDR's versus empirical Bayes rates shows substantial reduction in the range and variance of the distribution – especially the upper tail of the distribution; (2) the empirical Bayes rates provide a good fit to the data, with only two parameters (one in each stage), with the second-stage test, showing that the hypothesis about the variation in rates can be described by a proportional hazards model that cannot be rejected; (3) a substantial reordering of the counties occurs when the DASDRs are replaced by the empirical Bayes rates.

There are several associations found when analyzing cancer mortality rates and trends over the counties, which are missed if the analysis is restricted to state or SEA:

- Analyses showed certain race associations between geographic mortality rate patterns of cancers of uterus/chorion and ovary/Fallopian tube: counties with highest mortality rate in whites were located in northeast, while in nonwhite populations the highest mortality rates were in central-east and southeast. That might be related to race-dependent susceptibility to risk factor, which is associated with place of residence. Some studies showed that cancers of the female reproductive system may be highly associated with geographical factors even after the short periods of exposure: e.g., Japanese immigrants to the United States, related to the place of residence, had risk factor-related mortality rates of ovarian cancer change within a relatively short period of time (Haenszel and Kurihara, 1968).
- Relative changes of cancer mortality rates over the 30-year period (1950–1979) per county were compared to the U.S. national rates. In some counties, the highest increases in cancer mortality were registered the same time in both sexes: i.e., cancer of large intestine (certain counties in Ohio),

rectum (counties in Ohio and Pennsylvania), liver/gallbladder/bile ducts (counties in Pennsylvania), and trachea/bronchus/lung (counties in Maryland). Some cancers showed the most significant relative changes in rates in white populations (i.e., cancer of esophagus – increase, and stomach cancer – decrease), when other had the highest increase in both white and nonwhite population (i.e., cancer of large intestine, rectum, and liver/gallbladder/bile ducts).
- Certain cancers showed associated trends in relative changes per county. The most common combinations were cancers of the 1) ovary and uterus, 2) liver and uterus, and 3) breast and uterus.

Mortality rates are generally more available, but they do not completely reflect the differences in risk for cancers for which early diagnosis and therapy substantially increases survival. However, analyses of cancer incidence rates could give a possibility for deeper insight in these associations, especially when data will be extended into the 2000s.

Thus, the empirical Bayes rate stabilization procedure has appropriate operating characteristics for accomplishing several types of scientific and environmental policy-related tasks. Such analyses could be extended by representing area-exposure factors to determine what factors were associated with spatial differences. Thus the two-stage regression and rate adjustment procedure represents a general strategy for analyzing geographic patterns of cancer mortality risks. The application of these procedures to death rates is not their only possible use. They could be used for analysis of any type of health- or health service- related event in small local populations, e.g., the risk of accidents or the use of renal dialysis. This can be accomplished without the necessity of making strong modeling or distributional assumptions about the spatial distribution of mortality or morbidity risks, using standard epidemiological measures of risk.

## 5.6 Summary

In this chapter, we considered several methods of analyses and modeling data associated with cancer research: (1) stochastic process model, generalizing approaches presented in Chapter 4; (2) microsimulation models with a broad spectrum of the potential application to forecasting and analyses of risk factors and diagnostic/therapeutic interventions; (3) methods of latent structure analyses, such as Grade of Membership and linear latent structure analyses; and (4) empirical Bayes approaches to perform the geo-epidemiological/mapping analysis of the U.S. cancer mortality rates and trends on the county level. Because of relative complexity, these methods are not broadly used yet. However, these innovative methods will certainly contribute to future progress in understanding of mechanisms of cancer initiation, promotion, and progression.

# References


Akushevich I., Kulminski A., Manton K.G., 2005. Life tables with covariates: life tables with covariates: dynamic model for nonlinear analysis of longitudinal data. Math Popul Stud, 12(2):51–80.

Akushevich I., Manton K.G., Kulminski A., Kovtun M., Kravchenko J., Yashin A., 2006. Population models for the health effects of ionizing radiation. Radiats Biol Radioecol 46(6):663–674.

Akushevich I., Kravchenko J. S., Manton K.G., 2007. Health based population forecasting: effects of smoking on mortality and fertility. Risk Anal 27(2):467–82.

Akushevich I., Kovtun M., Manton K.G., Yashin A.I., 2008. Linear latent structure analysis and modeling of multiple categorical variables. Comput Math Methods Med (in press)

Armitage P., Doll R., 1954. The age distribution of cancer and a multistage theory of carcinogenesis. Br J Cancer 8:12.

Armitage P., Doll R., 1961. Stochastic models for carcinogenesis. In: Proceedings of the Fourth Berkeley Symposium on Mathematical Statistics and Probability, Berkeley, CA: University of California Press.

Bartholomew D.J., Knott M., 1999. Latent Variable Models and Factor Analysis, 2nd edition. New York: Oxford University Press.

Berkman L., Singer B., Manton K.G., 1989. Black-white differences in health status and mortality among the elderly. Demography 26(4):661–678.

Blot W.J., Fraumeni J.F., Mason T.J., Hoover R., 1979. Developing clues to environmental cancer: a stepwise approach with the use of cancer mortality data. Environ Health Perspect 32:53–58.

Clogg, C.C., 1995. Latent Class Models. Handbook of Statistical Modeling for the Social and Behavioral Sciences. New York: Plenum Press, pp. 311–360.

Efron B., Morris C., 1973. Combining possibly related estimation problems. J R Statist Soc B 35:379–421.

Erosheva, E.A., 2005. Comparing Latent Structures of the Grade of Membership, Rasch, and Latent Class Models. Psychometrika 70:619–628.

Everitt B., 1984. An introduction to latent variable models. London, New York: Chapman and Hall.

Frank S., 2007. Dynamics of Cancer. Incidence, Inheritance, and Evolution. Princeton and Oxford: Princeton University Press 378 pages.

Gaver D.P., O'Muircheartaigh I.G., 1987. Robust empirical Bayes analyses of event rates. Technometrics 29:1–15.

Gavrilov L.A., Gavrilova N.S. 2006a Models of Systems Failure in Aging. In: Michael Conn P. (Editor): Handbook of Models for Human Aging, Burlington, MA: Elsevier Academic Press. 45–68.

Gavrilov L.A., Gavrilova N.S., 2006b. Reliability Theory of Aging and Longevity. In: Masoro E.J., Austad S.N. (Eds.): Handbook of the Biology of Aging, 6th edition. Academic Press. San Diego, CA, USA, 3–42.

Gilbert N., Troitszch K.G., 1999. Simulation for the Social Scientist. Buckingham: Open University Press.

Gompertz B., 1825. On the nature of the function expressive of the law of human mortality, and on a new mode of determining the value of life contingencies. Philos Trans R Soc Lond 115(1825):513–585.

Goodman L.A., 1974. Exploratory latent structure analysis using both identifiable and unidentifiable models. Biometrika 61:215–231.

Goodman L.A., 1978. Analyzing qualitative/categorical data: log-linear models and latent-structure analysis. Cambridge, MA: Abt Books.

Haenszel W., Kurihara M., 1968. Studies of Japanese migrants. I. Mortality from cancer and other diseases among Japanese in the United States. J Natl Cancer Inst 40:43–68.

Heinen T., 1996. Latent class and discrete latent trait models: similarities and differences. Thousand Oaks, CA: SAGE Publications.

Kossenko M.M., Thomas T.L., Akleyev A.V. et al., 2005. The Techa River Cohort: study design and follow-up methods. Radiat Res. 164(5):591–601.

Kovtun, M., Akushevich, I., Manton, K.G. et al., 2006. Grade of membership analysis: one possible approach to foundations. In: Focus on Probability Theory, pp. 1–26, Hauppauge, NY: Nova Science Publishers.

Kovtun, M., Akushevich, I., Manton, K.G., Tolley, H.D., 2007. Linear latent structure analysis: mixture distribution models with linear constrains. Stat Methodol 4:90–110.

Kravchenko J., Goldschmidt-Clermont P.J., Powell T. et al., 2005. Endothelial progenitor cell therapy for atherosclerosis: the philosopher's stone for an aging population ? Effect of therapy on human life span is predicted to be comparable to that caused by eliminating cancer. Sci Aging Knowl Environ 25:18.

Krestinina L.Y., Preston D.L., Ostroumova E.V. et al., 2005. Protracted radiation exposure and cancer mortality in the extended Techa river cohort. Radiat Res 164:602–611.

Langeheine, R., Rost, J. (Eds.), 1988. Latent Trait and Latent Class Models. New York, NY: Plenum Press.

Lazarsfeld P.F., Henry N.W., 1968. Latent Structure Analysis. Boston, MA: Houghton Mifflin Company.

Little M.P., 2003. Risks associated with ionizing radiation. Br Med Bull 68:259–275.

Manton K.G., Stallard E., 1981. Methods for the analysis of mortality risks across heterogeneous small populations: examination of space-time gradients in cancer mortality in NC counties. Demography 18:217–230.

Manton K.G., Stallard E., 1988. Chronic Disease Modeling: Measurement and Evaluation of the Risks of Chronic Disease Processes. London: Griffin.

Manton K.G., Land K.C., 2000 Active life expectancy estimates for the U.S. elderly population: a multidimensional continuous-mixture model of functional change applied to completed cohorts, 1982–1996. Demography 37(3):253–265.

Manton K.G., Yashin A.I., 2000. Mechanisms of Aging and Mortality: Searches for New Paradigms. Monographs on Population Aging, 7, Odense, Denmark: Odense University Press.

Manton K.G., Woodbury M.A., Stallard E., 1981. A variance components approach to categorical data models with heterogeneous cell populations: analysis of spatial gradients in county lung cancer mortality rates in North Carolina counties. Biometrics 37:259–269.

Manton K.G., Stallard E., Creason J P. et al., 1985. U.S. cancer mortality 1950–1978: a strategy for analyzing spatial and temporal patterns. Environ Health Perspect 60:369–380.

Manton K.G., Stallard E., Woodbury M.A. et al., 1987. Statistically adjusted estimates of geographic mortality profiles. J Natl Cancer Inst 78:805–815.

Manton K.G., Woodbury M.A., Stallard E. et al., 1989. Empirical Bayes procedures for stabilizing maps of U.S. cancer mortality rates. J Am Stat Assoc 84(407):637–650.

Manton K.G., Stallard E., Singer B.H., 1992. Projecting the future size and health status of the U.S. elderly population. Int J Forecast 8:433–458.

Manton K.G., Woodbury M.A., Tolley H.D., 1994. Statistical applications using fuzzy sets. New York: John Wiley and Sons.

Manton K.G., Gu X., Huang H. et al., 2004 Fuzzy set analyses of genetic determinants of health and disability status. Stat Methods Med Res. 13(5):395–408.

Manton K.G., Gu X., Lamb V.L. 2006. Change in chronic disability from 1982 to 2004/2005 as measured by long-term changes in function and health in the U.S. elderly population. Proc Natl Acad Sci USA. 103(48):18374–18379.

Manton K.G., Akushevich I., Kulminski A., 2008. Human mortality at extreme ages: data from the national long term care survey and linked medicare records, Math Popul Stud. 15(2):137–159.

Marcoulides G.A., Moustaki I. (Eds.), 2002. Latent Variable and Latent Structure Models. Methodology for Business and Management. Mahwah, NJ: Lawrence Erlbaum Associates.

Mason T.J., McKey F.W., Hoover R. et al., 1975. Atlas of cancer mortality for U.S. counties, 1950–1969. Department of Health, Education, and Welfare, publication 75–780, Washington, DC: U.S Government Printing Office.

Mason T.J., McKey F.W., Hoover R. et al., 1976. Atlas of cancer mortality among U.S. nonwhites: 1950–1969. Department of Health, Education, and Welfare, publication 76–1204, Washington, DC: U.S. Government Printing Office.

Mislevy R.J., 1984. Estimating latent distributions. Psychometrika 49:359–381.

Morris C.N., 1983. Parametric empirical Bayes inference: theory and applications. J Am Stat Assoc 78:47–55.

NCI (National Cancer Institute), 1987. Research contributions made possible by the NCI Cancer Atlases published in the 1970s. Office of Cancer Communications Report (Backgrounder series), June 9, Bethesda, MD.

Pickle L.W., Mason T.J., Howard N. et al., 1987. Atlas of U.S. cancer mortality among whites, 1950–1980. Department of Health and Human Services, Publication 87–2900, Washington, DC: U.S. Government Printing Office.

Press W.H., Flannery B.P., Teukolsky S.A. et al., 1999. Numerical recipes in FORTRAN, the art of scientific computing. Cambridge: Cambridge University Press.

Qu Y., Tan M., Kutner M.H., 1996 Random effects models in latent class analysis for evaluating accuracy of diagnostic tests. Biometrics 52:797–810.

Riggan W.B., Creason J.P., Nelson W.C. et al., 1987. U.S. cancer mortality rates and trends, 1950–1979. Volume IV: Maps. Environmental Protection Agency, Health Effects Research Laboratory, Publication 600/1-83/015e, Washington, DC: U.S. Government Printing Office.

Riggan W.B., Manton K.G., Creason J.P. et al., 1991. Assessment of spatial variation of risks in small populations. Environ Health Perspect 96:223–238.

Robbins H., 1955. The empirical Bayes approach to statistics. Proceedings of the Third Berkeley Symposium on Mathematical Statistics and Probability Volume 1. Berkely, CA: University of California Press. pp. 157–164.

Robbins H., 1964. The empirical Bayes approach to statistical decision problems. Ann Math Stat 35:49–68.

Sacher G.A., Trucco E., 1962 The stochastic theory of mortality. Ann NY Acad Sci 96:985–1007.

Singer, B. 1989. Grade of membership representations: concepts and problems. In: T.W. Andersen, K.B. Athreya, and D.L. Iglehart (Eds.), Probability, Statistics, and Mathematics: Papers in Honor of Samuel Karlin. New York: Academic Press, Inc. pp. 317–334.

Stallard E., 2007. Trajectories of disability and mortality among the U. S. elderly population: evidence from the 1984–1999 NLTCS. North Am Actuar J 11(3):16–53.

Strehler B.L., Mildvan A.S., 1960. General theory of mortality and aging. Science 132:14–21.

Tolley H.D., Manton K.G., 1992. Large sample properties of estimates of a discrete grade of membership model. Ann Inst Stat Math, 44:85–95.

Tsutakawa R.K., 1988. Mixed model for analyzing geographic variability in mortality rates. J Am Stat Assoc 83:37–42.

Tsutakawa R.K., Shoop G.L., Marienfeld C.J., 1985. Empirical Bayes estimation of cancer mortality rates. Stat Med 4:201–212.

Uebersax J.S., 1997. Analysis of student problem behaviors with latent trait, latent class, and related probit mixture models. In: Rost J., Langeheine R., (Eds). Applications of Latent Trait and Latent Class Models in the Social Sciences. New York, NY: Waxmann. pp. 188–195.

Uebersax, J.S., Grove, W.M. 1993. A latent trait finite mixture model for the analysis of rating agreement. Biometrics 49:823–835

UNSCEAR 2000, published 2001. United Nations Scientific Committee on the Effects of Atomic Radiation. Health Phys. 80(3):291.

Wachter K.W., 1999. Grade of membership models in low dimensions. Stat Papers 40:439–458.

Witteman J.C.M., Grobbee D.E., Valkenburg H.A. et al., 1994. J-shaped relation between change in diastolic blood pressure and aortic atherosclerosis. Lancet 343:504–507.

Woodbury M.A., Clive J., 1974. Clinical pure types as a fuzzy partition. J Cybernetics 4:111–121.

Woodbury M.A., Manton K.G. 1977. A random walk model of human mortality and aging. Theor Popul Biol 11:37–48.

Yashin A.I., Manton K.G., 1997: Effects of unobserved and partially observed covariate processes on system failure: a review of models and estimation strategies. Stat Sci 12(1):20–34.

Yashin A.I., Iachine I.A., Begun A.S., 2000. Mortality modeling: a review. Math Popul Stud 8(4):305–332.

# Chapter 6
# U.S. Cancer Morbidity and Mortality Trends

## 6.1 Introduction: Cancer Mortality and Morbidity Registration

The vital records and statistics system in the United States began at the local level in the eighteenth century, and then progressed to the state level in the nineteenth century. In the 1930s, the national vital statistics system was developed (U.S. Vital Statistics System, 1950). Population-based cancer mortality data in the U.S. began to be collected at the beginning of the twentieth century. From 1930 to 1998, mortality data included information on race, gender, year, and age at death, which were published annually in *Vital Statistics of the United States* (U.S. Department of Health, Education, and Welfare, 1950–1959; Federal Security Agency, 1945–1949; U.S. Department of Commerce, 1930–1936, 1937–1944). By 1930, 47 of the existing 48 states and the District of Columbia were included in the national vital statistics system, in 1933 – Texas, in 1959 – Alaska, and in 1960 – Hawaii (Hetzel, 1997; Wingo et al., 2003).

Cancer incidence data before 1973 were collected mostly by the periodic surveys conducted in different areas of the United States (1937–1939, 1947–1948, 1969–1971) (Dorn and Cutler, 1959; Cutler and Young, 1975). The End Results Program, conducted by NCI, collected data on survival of cancer patients based on hospitalization cases up to 1973 (Axtell et al., 1976). Since 1973, this program was succeeded by the Surveillance, Epidemiology, and End Results (SEER) register. It collects population-based data on newly diagnosed cancers in the United States through multiple cancer registries located in various U.S. states. It was intended to represent the U.S. population. This register initially captured approximately 14% of the U.S. population. Population coverage later has been increased to approximately 26%. As a complement to SEER Register data, there are files with population counts for 1-year intervals for calendar year and age, and for sex and race groups.

In this chapter we analyze key characteristics of the U.S. cancer epidemiology, such as mortality and incidence, for different cancer sites, such as lung and bronchus, breast, colon, rectum, prostate, corpus uteri, esophagus, stomach, kidney, ovarian, cervix uteri, pancreas, liver, thyroid, and testis. These are solid tumors with major contributions to U.S. cancer incidence and mortality.

K.G. Manton et al., *Cancer Mortality and Morbidity Patterns in the U.S. Population*, Statistics for Biology and Health, DOI 10.1007/978-0-387-78193-8_6,
© Springer Science+Business Media, LLC 2009

Further analyses of these cancers, such as of patterns/trends of histologic types, analysis of the selected cancer histotypes for the two-disease model (to check their homogeneity and discuss the factors that might cause heterogeneity), analysis of time trends of squamous cell carcinomas versus adenocarcinomas, and discussion of hereditary and environmental factors contributing to their genesis are presented in Chapter 7.

Using national death certificate data (NBER, 2007), we first analyzed mortality for specific cancer sites, studying age-, sex- and race-specific patterns and their time trends. For the analysis of the U.S. cancer mortality we used data based on death certificate reports for all individual decedents in the U.S. and U.S. territories, occurring during the 46 years (from 1959 to 2004) – 1.7–2.4 million deaths per year in last decade. Each death certificate record includes data on underlying and associated causes of death, date of death, geographic location (region, state, county, and division) of death, residence of the deceased (region, state, county, city, and population size), sex, race, age, marital status, state of birth, and origin of descent. The cause of death fields was coded using ICD-7 for 1958–1967, ICD-8 for 1968–1978, ICD-9 for 1979–1998, and ICD-10 for 1999 and later. The data are collected from death certificates filed in the vital statistics offices of each state and the District of Columbia. Totally there are 65 million individual death certificate records in these files. Several considerations need to be taken into account when interpreting mortality data: e.g., cancer mortality trends may reflect changes in population and individual cancer risk factors, medical technology advances, implementation of new screening programs and treatment protocols, adjustments to societal norms, as well as changes in death registration and classification of causes of death, data collection definitions and procedures, changes in legislation, and statistical methods used for analyses (Ries et al., 2002; Wingo et al., 2003).

Then we analyzed cancer incidence using NCI's SEER tumor registry data, which, in contrast to mortality data, contains information on tumor histotypes, from 1973 to 2003. The information collected about each cancer case includes the patient's demographic characteristics, date of diagnosis, data on up to 10 diagnosed cancer characteristics (including histology), type of surgical treatment and radiation therapy recommended or provided within 4 months of diagnosis, follow-up of vital status, and age and cause of death, if applicable.

## 6.2 U.S. Cancer Mortality

Codes for all cancers from International Disease Classifications (ICD) revisions 7th, 8th, 9th, and 10th were converted as shown in Table 6.1.

The death certificate files contain no information on survivors. Therefore, only the frequencies of cause-specific deaths relative to the total number of deaths for a group of individuals can be calculated using death certificate files. Population information is necessary to estimate cause-specific mortality rates.

**Table 6.1** Cancer codes from ICD-7, -8, -9, and -10 used for conversion of diagnosis for time trends analyses

| Cancer site | ICD-7*** 1958–1967 | ICD-8** 1968–1978 | ICD-9* 1979–1998 | ICD-10* 1999 + |
|---|---|---|---|---|
| Lung and bronchus | 162.1 + 162.8, + 163 | 162.1 | 162.2–162.5, 162.8–162.9 | C34 |
| Breast | 170 F + M | 174 F + M | 174 F 175 M | C50 |
| Prostate | 177 | 185 | 185 | C61 |
| Esophagus | 150 | 150 | 150 | C15 |
| Stomach | 151 | 151 | 151 | C16 |
| Thyroid | 194 | 193 | 193 | C73 |
| Colon | 153 | 153 | 153 | C18 |
| Testis | 178 | 186 | 186 | C62 |
| Rectum and rectosigmoid junction | 154 | 154.0 + 154.1 | 154.0 + 154.1 | C19 + C20 |
| Ovarian | 175.0 | 183.0 | 183.0 | C56 |
| Corpus uteri | 172 | 182.0 | 182 | C54.1-C54.3, C54.8, C54.9 |
| Cervix uteri | 171 | 180 | 180 | C53 |
| Pancreas | 157 | 157 | 157 | C25 |
| Liver, intrahepatic biliary ducts, gallbladder[1] | 155, 156 | 155 + 156 + 197.8 | 155, 156 | C22 + C23 + C24 |
| Kidney | 180 | 189.0 | 189.0 | C64 |

*NIH, NCI "ICD-9 to ICD-10 Neoplasms", Edited by C. Percy, 1995. At: seer.cancer.gov/tools/conversion/ICD9-10manual.pdf. **"Neoplasms ICD-9 to ICD-8". Conversion of neoplasm section. U.S. Department of Heath and Human Services, Public Health Service, NIH. Edited by Percy C.L., 1983. NIH Publication No. 83-2638. ***NIOSH-92 Death Categories and corresponding ICD codes, 1940–2004. (NIOSH is the National Institute for Occupational Safety and Health is the United States federal agency responsible for conducting research and making recommendations for the prevention of work-related injury and illness. NIOSH is part of the *Centers for Disease Control and Prevention* (CDC) within the US *Department of Health and Human Services*). Table III. At: //www.cdc.gov/niosh/LTAS/PDFs/NIOSH-92_table_2006-05.pdf and International Classification of Diseases, Seventh Revision. Neoplasms (140–239). At: www.health.nsw.gov.au/public-health/icd/icd7tabl.htm#140-239.
[1]Code 156 in ICD7 secondary and unspecified cancers cannot be separated. For this reason, death rates for 1950–1959 and 1970–1979 are not comparable and the comparison should not be made.

For population counts we used information from the Human Mortality Database (HMD, 2006). The database includes death rates and life tables calculated by age, time, and sex, along with all of the raw data (vital statistics, census counts, and population estimates) used in computing these quantities. The data were presented in a variety of formats with regard to age groups and time periods (see www.demog.berkeley.edu/wilmoth/mortality). Thus we consider two quantities characterizing mortality caused by specific cancers. The first is the relative frequency, $f_c(a, s, y) = N_c(a, s, y)/ N(a, s, y)$, where $N_c(a, s, y)$ is the number of deaths $c$ registered for sex $s$, age $a$, and calendar year $y$, and $N(a, s, y)$ is the total number of deaths. The second is the cause-specific mortality rate $m_c(a, s, y) = f_c(a, s, y)d(a, s, y)$, where $d(a, s, y)$ is death rate for the group from HMD. Only ratios of counts are used to define these quantities. This minimizes possible biases: e.g., different criteria for inclusion of certain records in different data sets. In this sense relative frequencies are more stable quantities. Furthermore, they used only death records, not population estimates, which are known with limited accuracy from census data, especially at older ages and, in addition, require assumptions for interpolation between censuses. For further discussion about possible uncertainties in estimating death records at later ages and methods of their estimations see Manton et al. (2008) and Chapter 4.

The age-specific rates of mortality from the most common cancers were analyzed for 1959–2004 period (see Fig. 6.1). Rates for different cancers are rescaled to use the same scale on all plots, to compare rates for different cancers, and to incidence rates (see Section 6.3 of this chapter). The real rate for a specific cancer is calculated by dividing values obtained from the plot to the rescaled factor. Rates for three-time periods (1959–1978, 1979–1992 and 1993–2003) for each cancer are presented to characterize cancer mortality time trends, and the time trends of age patterns of cancer mortality. Cancers of stomach, testis, rectum, corpus, and cervix uteri showed decreases in mortality over the 46 years of observation. The age-specific mortality of breast, esophagus (females), liver, pancreas, thyroid, and colon cancers did not change significantly over the observation period. Cancers of lung and bronchus, prostate, esophagus (male), ovarian, and kidney showed increases in age-specific mortality rates [see also Fig. 6.2 for trends of cancer-specific frequencies, averaged over age], which were especially pronounced in the last decade for lung cancer in females and esophageal carcinoma in males.

Analyses of certain cancer mortality trends might be influenced by factors, causing rates' under- or overestimation. For example, separating death rates for cancer of the colon from those of the rectum might be inappropriate because death rates for rectal cancer are underestimated: during death certification, colon cancer is often designated as the underlying cause of death, when the hospital diagnosis was rectal cancer. The impact of this misclassification has changed over time (Ries and Devesa, 2006; Percy et al., 1981; Chow and Devesa, 1992). The reported death rates for liver cancer might be also overestimated because some deaths attributed to liver cancer on the death certificate may be in reality due to metastasized cancers (Percy et al., 1990).

Some cancers showed age-related shifts to older ages in their peak mortality rates 1959–2003 (see Fig. 6.1), i.e., in females the age of the highest mortality for lung, kidney, ovarian, corpus uteri, and thyroid cancers increased (became "older") about 7 years. In males the age of the highest mortality for lung, stomach, and kidney cancers increased almost 10 years. Esophageal cancer in males seems to have a mortality peak which became "younger" (approximately 5 years).

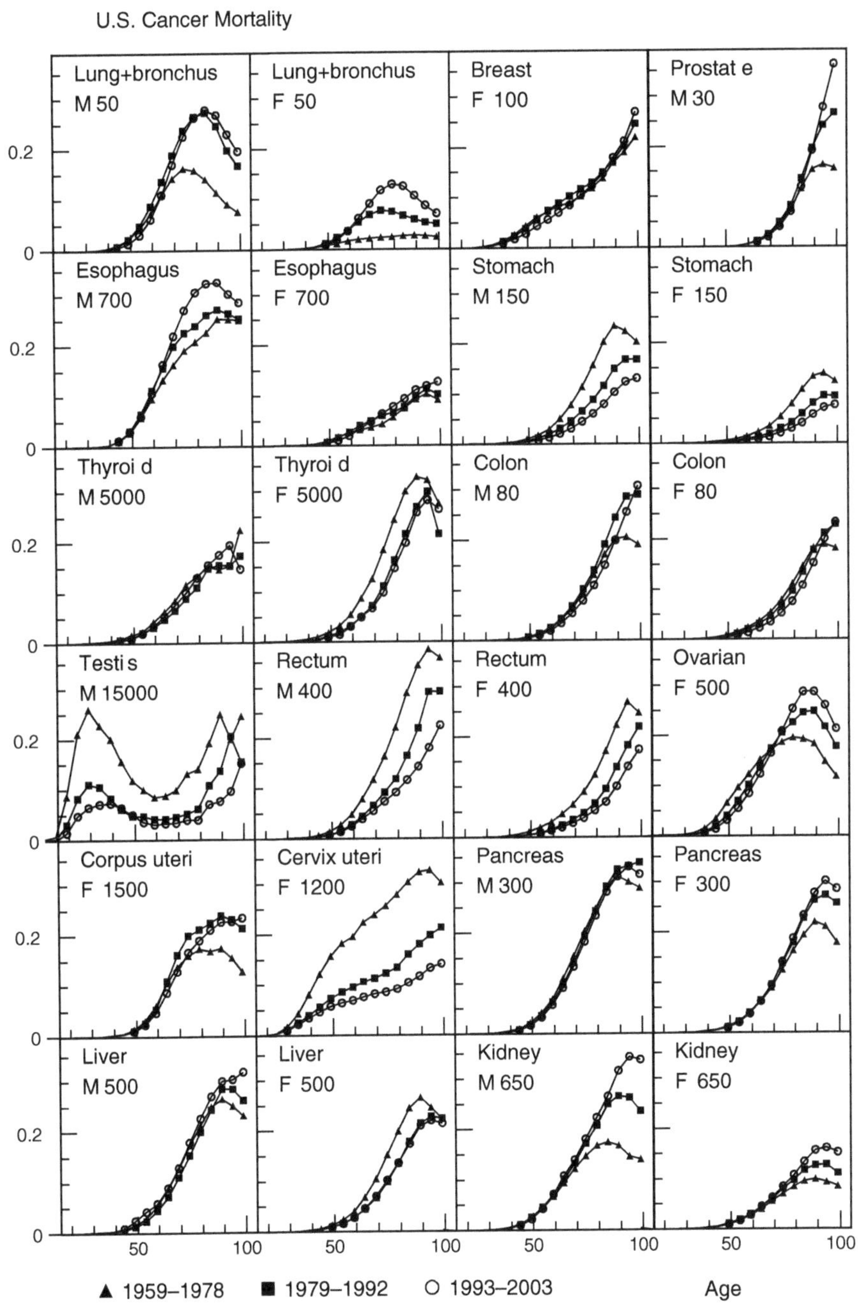


**Fig. 6.1** Age patterns of U.S. cancer mortality for males (M) and females (F) over the three-time periods (1959–1978, 1979–1992, and 1993–2003). The number at each plot is the rescaling factor (see text for explanation)

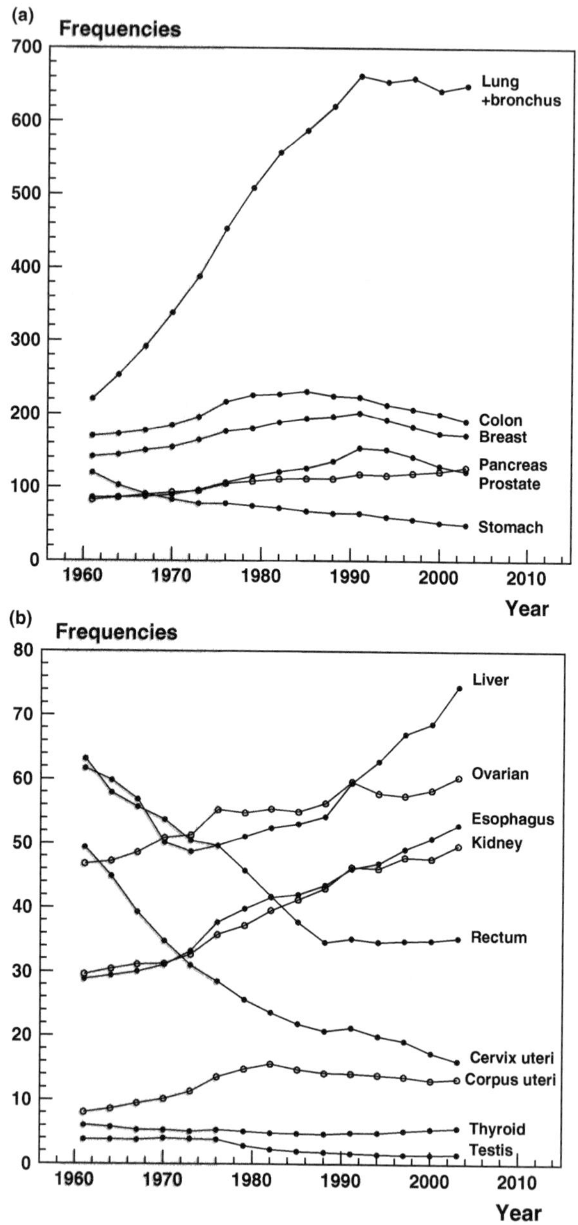


**Fig. 6.2** Time trends of frequencies of cancers as deaths causes in the United States, 1961–2003, cancers of high frequency rates (**a**), and cancers of low frequency rates (**b**)

Sex differences in age-specific cancer mortality were observed for some nonreproductive cancer sites (see Fig. 6.1): mortality from cancers of lung, esophagus, stomach, rectum, pancreas, liver, and kidney were higher in males. Thyroid cancer mortality was higher in females.

Further insight of time trends can be obtained from Fig. 6.2, where the calendar year dependence of the relative frequencies of deaths, caused by these cancers, is presented (relative frequencies shown in absolute units multiplied by $10^4$, i.e., per 10,000 death at a specific year). Lung cancer is by far the leading cause of cancer death in the United States, peaking at the end of the 1980s, with a slow decrease beginning in the 1990s. Cancers of colon, breast, and prostate, which trailed lung cancer, had their maximum frequency at the end of the 1970s – the beginning of the 1980s, with decreases starting in the 1990s. Cancers of the pancreas, liver, ovarian, esophagus, and kidney have significantly increased as causes of death over the 40-year period. Stomach, rectal, and cervix uteri cancers demonstrated the opposite, i.e., decreases in the frequency of deaths. Small decreases were also observed for thyroid and testicular cancers, which had the lowest death frequencies among 15 of the cancers studied. By definition, the number of death caused by a specific cancer is normalized by the total population. Therefore cancer frequencies, attributable to gender (e.g., prostate, corpus uteri, cervix uteri, ovarian, testis, and partly breast), appear underestimated. Sex-specific frequencies are obtained by taking into account the male/female ratios of total population and case counts.

## 6.3 U.S. Cancer Incidence

According to the annual report to the nation on the status of cancer for 1973–2003, based on data on cancer incidence from SEER registry (for invasive cancer), cancer incidence rates for all sites combined increased from the mid-1970s through 1992, and then decreased in 1992–1995, remaining stable in 1995–2000, when the delay-adjusted trend (delay in receiving cancer reports) showed an increase that had borderline statistical significance for all cancers combined and for the four leading cancers (Weir et al., 2003). In 1971, 3.0 million persons were living with cancer in th U.S. (1.5% of the U.S. population). This increased to 9.8 million (3.5% of the U.S. population) in 2001, with more female than male survivors, although more males than females were diagnosed with cancer (Rowland et al., 2004).

We analyzed the dynamics of various cancers incidence rates from 1973 to 2003 (see Fig. 6.3). It corresponds to Fig. 6.2 for cancer-specific mortality frequencies. The definition of the quantities presented in Fig. 6.3 is the number of cases of a specific cancer onset normalized to the total population. Cancers of lung and bronchus, breast, prostate, kidney, thyroid, and liver had the most prominent increase in incidence rates over 30 years of observation, when increases of esophageal and testicular cancers were less obvious. Cervical and gastric carcinomas demonstrated an apparent decrease in incidence over the study period. Slight fluctuations in incidence rates over time were observed for cancers of colon, rectum, ovary, corpus uteri, and pancreas.

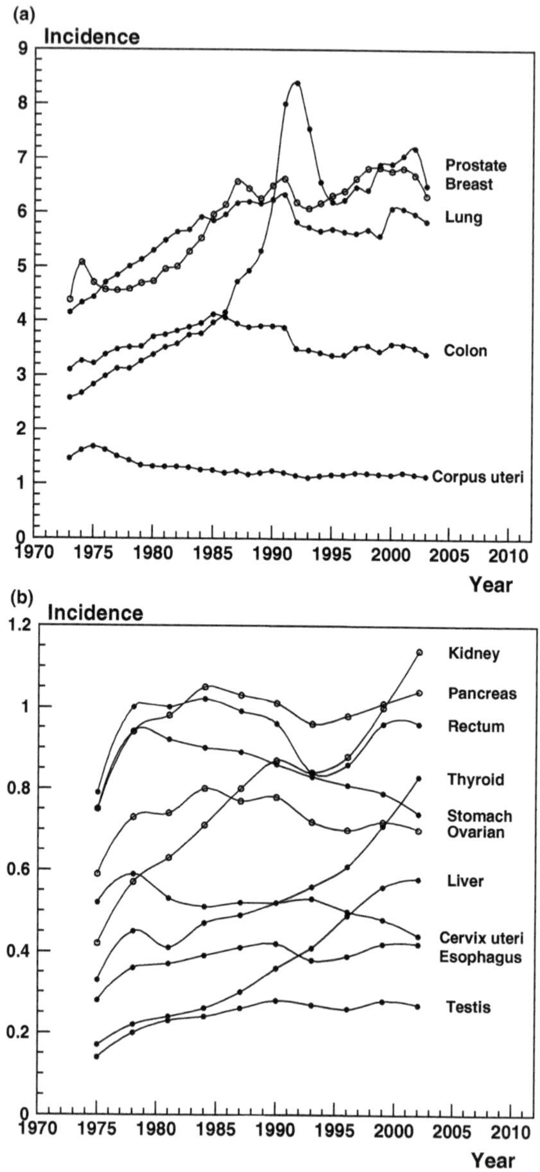


**Fig. 6.3** Time trends for incidence rates for high rates (**a**) (with 1-year intervals calculated), and low rates (**b**) (with 3-year intervals calculated)

We analyzed age-specific cancer incidence rates in white and black populations for 1973–2003 (see Fig. 6.4). Incidences of lung (male), prostate, esophagus, stomach, cervix uteri, pancreas, and liver cancers were higher in blacks [i.e., the U.S. black male have one of the highest incidence rate of prostate cancer in the world (Hsing and Devesa, 2001; Parkin et al., 2003)]. Whites had higher incidence rates of breast, thyroid, testicular, ovarian, and corpus uteri cancers. The incidence peaks of lung (males) and esophageal (both sexes) cancers were "younger" in blacks than in whites:

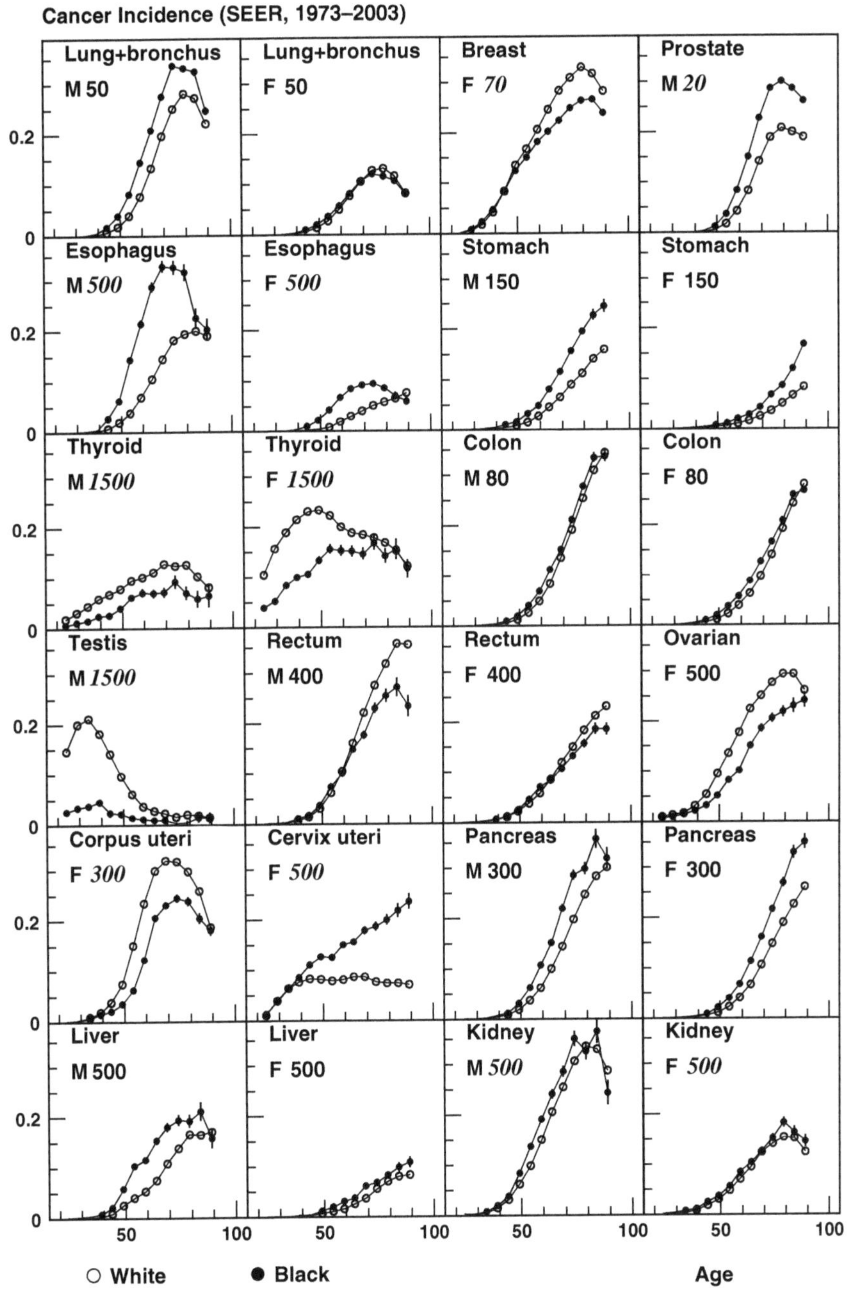


**Fig. 6.4** Cancer incidence rates for black and white U.S. population, females and males, 1973–2003. The number is the rescaling factor as in Fig. 6.1 (rescaling factors marked by an italic font mean that they differ from mortality rates rescaling factors for the same cancers)

i.e., about 5 and 15 years, respectively. The thyroid cancer incidence rate was higher in women, being the highest in whites around the age 45 – that was almost 10 years earlier than for blacks. Cervical cancer demonstrated equal trends for both races till age 40. Then rate among black women continued to increase, while in whites the rate was almost stable. For all cancers the tendencies for both sexes were similar within race (i.e., both, black males and black females, had higher incidence rates of the same cancer compared to whites), with the exception of lung cancer – black males had a significantly higher incidence rate compared to white males, while women of both races had nearly equal age-specific incidence rates.

Blacks had higher incidence for 5 out of 7 cancers of nonreproductive system (excluding two cancers – kidney and colon, which had no race difference). Whites had higher incidence rates of cancers of the reproductive system, compared to blacks, for 4 out of 6 cancers.

There was a significantly higher incidence for males in 8 out of 9 cancers of nonreproductive system. Only one nonreproductive system cancer – thyroid cancer – had a higher incidence in females.

Trends in age-specific incidence rates for 15 cancers in white and black males and females were analyzed for 1973–2003 (see Fig. 6.5). The main characteristics of these trends are in Table 6.2. and discussed in the subsequent Section 6.4 of this chapter by cancer site.

We analyzed the time trend (from 1973–1978 to 1999–2003) of the age at which cancer incidence rates began to rise, and the age of the highest incidence rate (peak). Based on these two characteristics, some of the cancers became 5–10 years "older" over the 30 years of observation, such as lung (in males), esophagus (i.e., in black females incidence peak became almost 15 years "older" – 50–60 years versus 65–75 years), stomach, testis, corpus uteri (in blacks), and pancreas. Cancers of prostate, colon and rectum (in males), liver (in females), and kidney (in females), in contrast, became 5–10 years "younger." Increases in incidence of hepatocellular carcinoma and cancer of intrahepatic bile duct began 13–14 years earlier in females (67 versus 53 years in whites, and 60 versus 47 years in blacks) in 1999–2003 compared to 1973–1978. The higher incidence of breast cancer in white women over blacks in 2003 started almost 10 years earlier (in 1973–1978 both races had equal incidence rates up to age 42, while in 1999–2003 they had equal rates up to age 52; only at age 52 + white women had a higher incidence). Cancer of the cervix uteri did not show a change in the age of incidence increase onset (23 years of age), but in 1973–1978 black women started to have a higher incidence rate compared to whites at age 27. In 1999–2003 that happened almost at age 57 (a shift associated with the dramatic decrease of cervical cancer incidence in black women in the mid-1990s).

When comparing the time trends of frequencies of incidence and deaths, cancers of lung, breast, prostate, colon, and rectum kept the leading positions over the several past decades for both of parameters. Prostate and thyroid cancers, despite having large increases in incidence, nevertheless, had almost stable death frequencies over time. Cancers of the liver and kidney showed increases in both incidence and deaths frequencies.

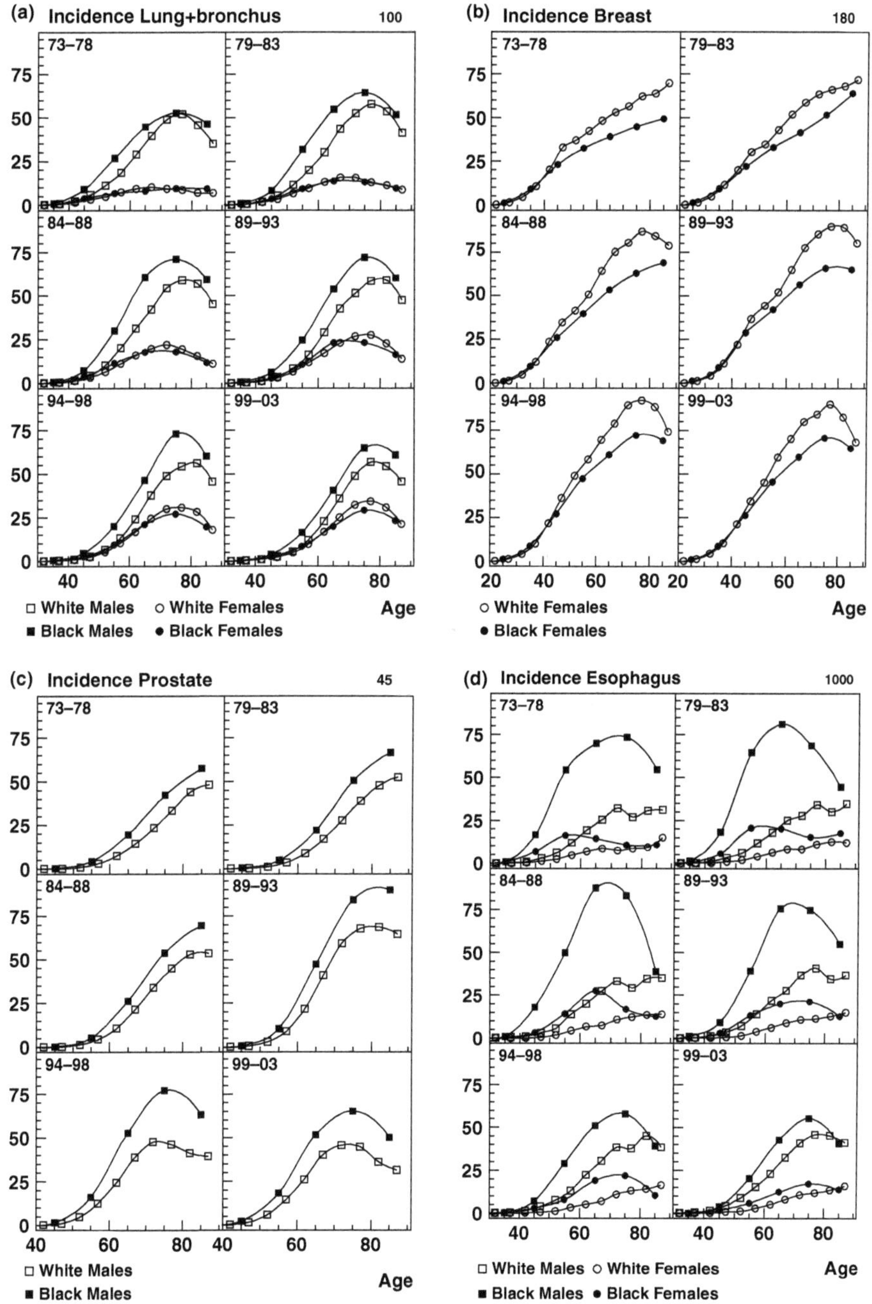


**Fig. 6.5** Trends of age-specific incidence rates (in %) of 15 cancers in white and black males and females in the United States, 1973–2003

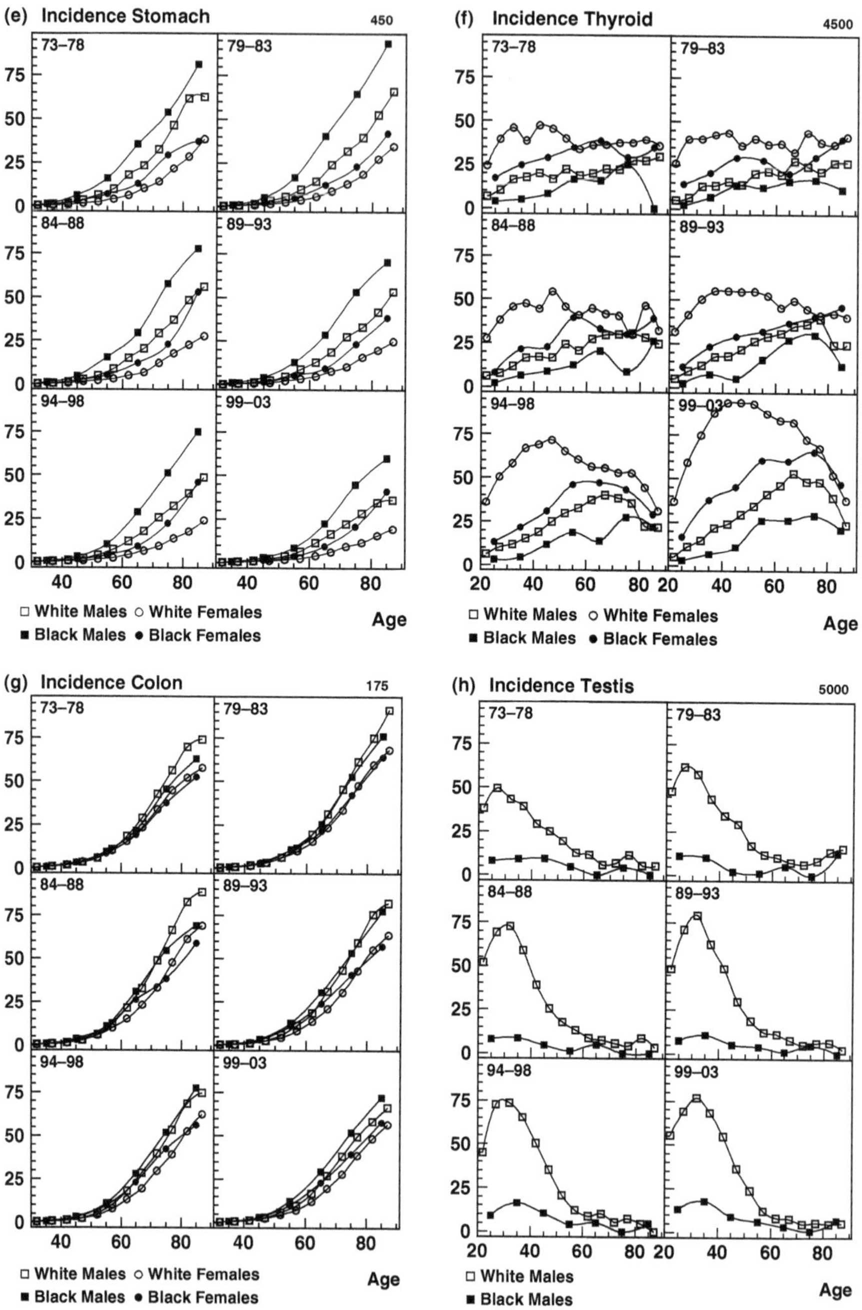


**Fig. 6.5** (continued)

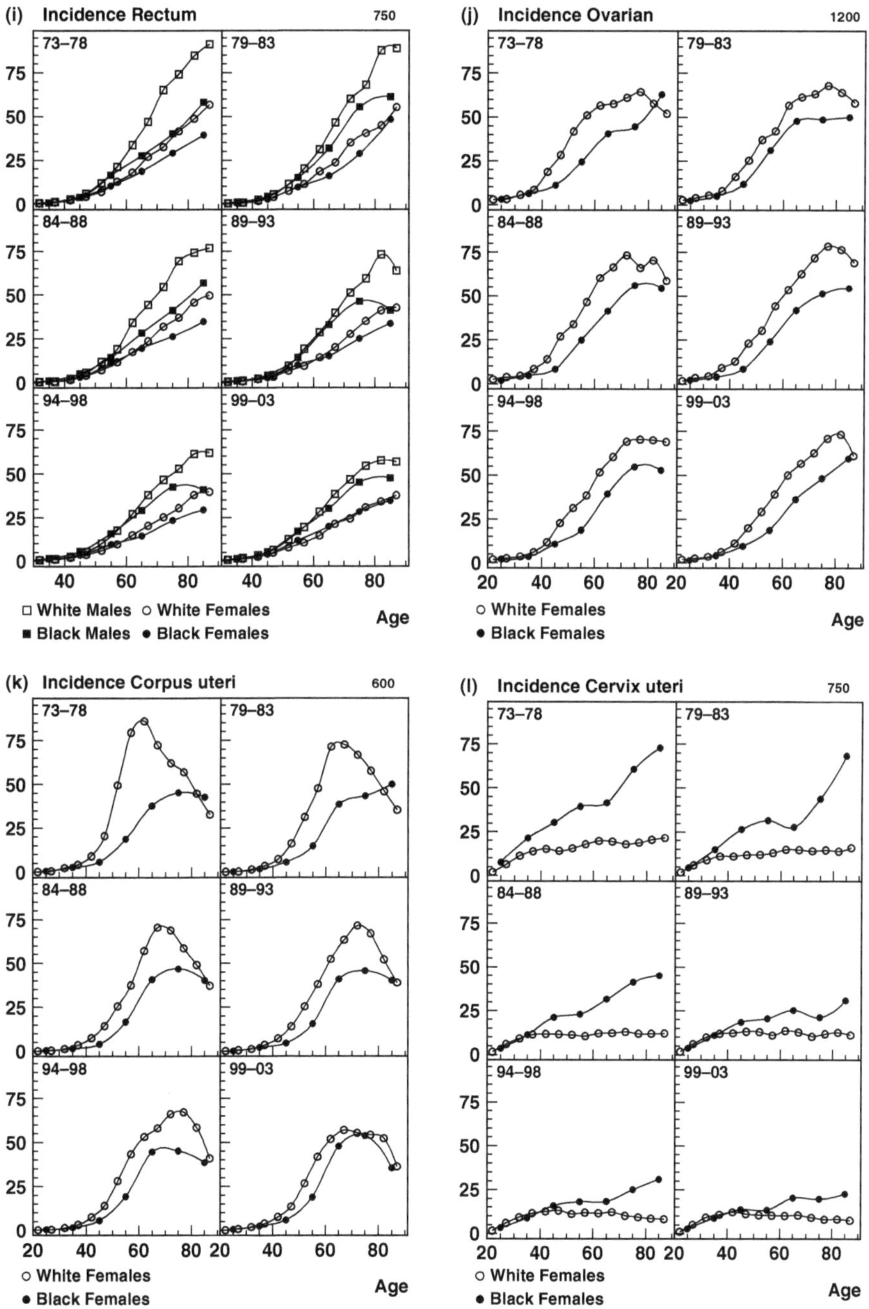


**Fig. 6.5** (continued)

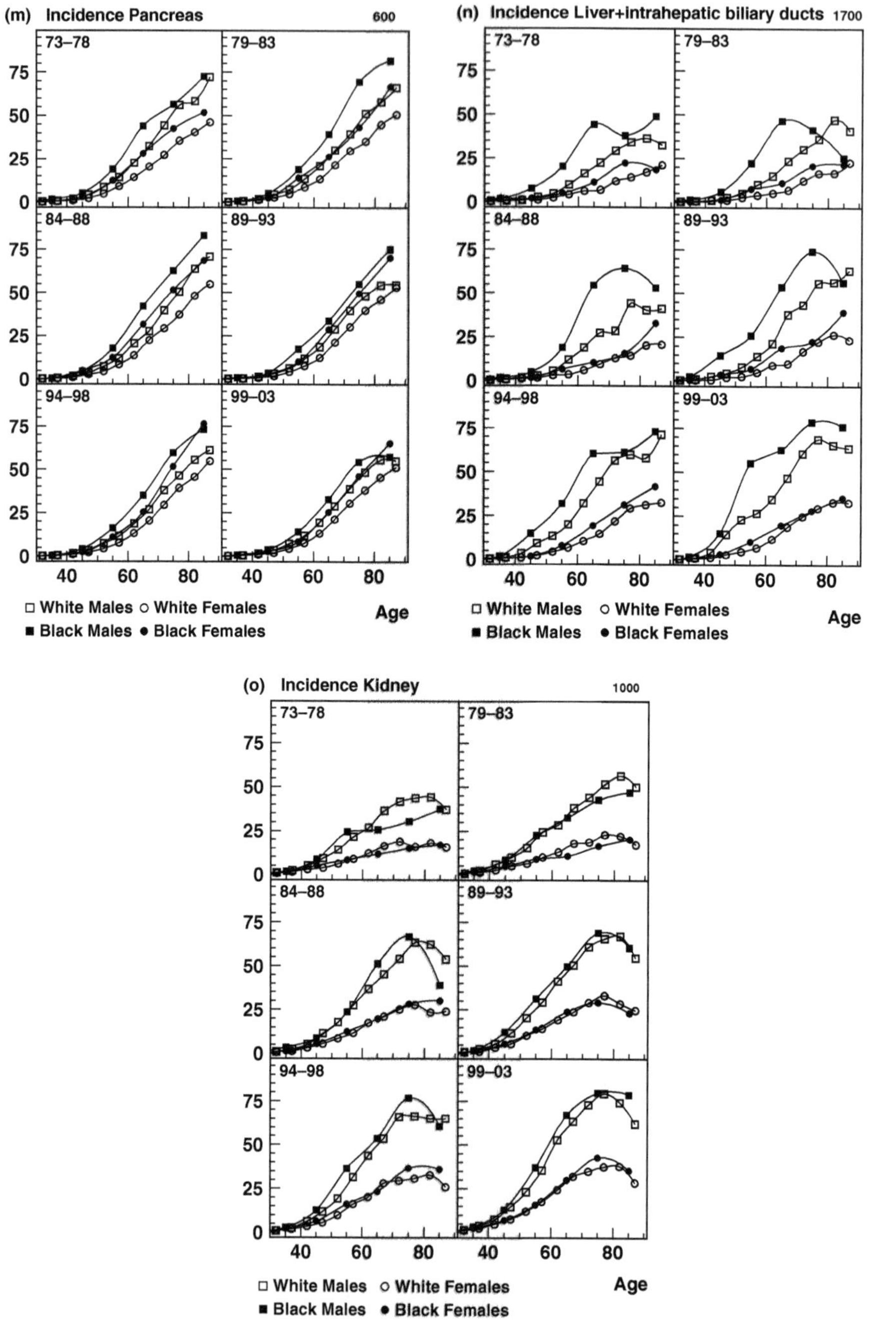


**Fig. 6.5** (continued)

**Table 6.2** Trends of age-specific incidence rates for white and black populations, United States, 1973–2003

| Cancer site | | Race differences: Over time race predominance | Race differences: Beginning of incidence increase in 1973 vs. 2003, in years of age | Sex differences: Predominance over time |
|---|---|---|---|---|
| Lung and bronchus | M | B+ | 40 for B and 45 for W vs. 55 in W and 50 in B | M>F |
| | F | B = W | 50, no dynamics | |
| Breast | F | W+ | W predominance started at 42 in 1973 vs. 52 in 2003 | – |
| Prostate | M | B++ | 53 in B and 57 in W vs. 45 in B and 50 in W | – |
| Esophagus | M | B+++ in 1973–1993, and B+ in 2003 | 37 in B and 47 in W vs. 47 in B and 50 in W | M>F |
| | F | B+ | 37 in B and 53 in W vs. 47 in B and 57 in W | |
| Stomach | M | B+ | 45 in B and 55 in W vs. 52 in B and 57 in W | M>F |
| | F | B+ | 53 in B and 63 in W vs. 65 in B and 75 in W | |
| Thyroid | M | W+ | N/A | F>M |
| | F | W+ in 1973–1993 to W++ in 1994–2003 | N/A | |
| Colon | M | B = W | 50 vs. 45 | M>F (slightly) |
| | F | B = W | 53, no dynamics | |
| Testis | M | W+++ | N/A | – |
| Rectum | M | W+ | 47 vs. 43 | M>F |
| | F | B = W | 47, no dynamics | |
| Ovarian | F | W+ | 37 in W and 47 in B vs. 40 in W and 47 in B | – |
| Corpus uteri | F | W++ in 1973–1998, and W+ in 1999–2003 | 37 in W and 40 in B vs. 37 yrs in W and 45 in B | – |
| Cervix uteri | F | B++ in 1973–1988, and B+ in 1989–2003 | 27 of B predominance vs. 57 | – |
| Pancreas | M | B+ slightly | 47 vs. 53 | M>F slightly |
| | F | B+ slightly | 47 vs. 53 | |
| Liver + IBDs[1] | M | B+ | 37 in B and 47 in W vs. 43 for W and B | M>F |
| | F | B = W | 60 in B and 67 in W vs. 47 in B and 53 in W | |
| Kidney | M | B = W | 37, no dynamics | M>F |
| | F | B = W | 47 vs. 37 | |

M, males; F, females; W, whites; B, blacks; +, light increase/predominance; ++, moderate increase/predominance; +++, high increase/predominance; N/A, no data estimated.
[1]IBDs – intrahepatic biliary ducts.

For cancers with poor survival, the death rates are only slightly lower than the incidence rates (Ries and Devesa, 2006). By comparing age patterns and time trends of mortality and incidence (taking into account the rescaling factors for every cancer site), cancers of lung and bronchus, stomach, ovary, pancreas, and liver had the poorest survival over the observation period. Based on the same criteria, the best survival was found for thyroid, uterine, cervical, and testicular cancers.

## 6.4 Morbidity and Mortality of Specific Cancer Sites

### *6.4.1 Lung Cancer*

Lung cancer keeps a leading position in the U.S. cancer mortality. Its changing dynamics over the past several decades in both, males and females, made it the focus point of numerous studies. Decline in lung cancer death rates in men since the beginning of the 1990s reflects a reduction in tobacco use that began in the 1960s, when the first Surgeon General Report on Smoking and Health was published (U.S. Public Health Service, 1964). In women, the recent stabilization in lung cancer incidence and a slowing rate of increase in death rate suggested the beginning of trend's reversal, following the reversal in male lung cancer mortality and incidence.

Time trends for age groups and birth cohorts for lung and bronchus cancer were analyzed. Age-specific analysis of lung cancer mortality (see Fig. 6.6) showed a mortality peak for male cohort aged 90+ in at the mid-1990s, compared to a peak in the mid-1980s for male cohort aged 50–59. In females a similar tendency was observed: though, mortality peaks have not yet been reached for cohorts older than 70, the peak for 50–59 age cohort was registered at the beginning of 1990s, and for the cohort of 60–69 – at the mid-1990s.

In Fig. 6.7 we present birth cohort analyses of age-specific lung cancer mortality in males (10-year birth cohorts) and females (5-year birth cohorts). Several 5-year cohorts for males gave similar results, so they were combined into 10-year cohorts. In males the slope of increasing lung cancer mortality and the age of the mortality peak shifted about 5 years (from age 85 to age 80) in 1911–1920 versus 1901–1910 birth cohorts, with subsequent trend to stabilize in 1921–1930 cohort, and shifting back to older ages (based on slope dynamic) in 1931–1935, and further – in 1936–1940 cohorts. In females, mortality rates almost doubled for the birth cohort born in 1921–1925 versus the cohort of 1901–1905. Both, female slope and mortality peak, shifted to younger ages with every succeeding birth cohort. Only in 1936–1940 birth cohort slope started reversing, shifting to mortality at older ages. If in males the average difference in age-related slope between 1901–1910 and 1936–1940 birth cohorts was approximately 5 years, in females this difference was more dramatic – about 20 years.

Fig. 6.6 Time trends of age patterns of death (in %) caused by lung cancer for males (a) and females (b)

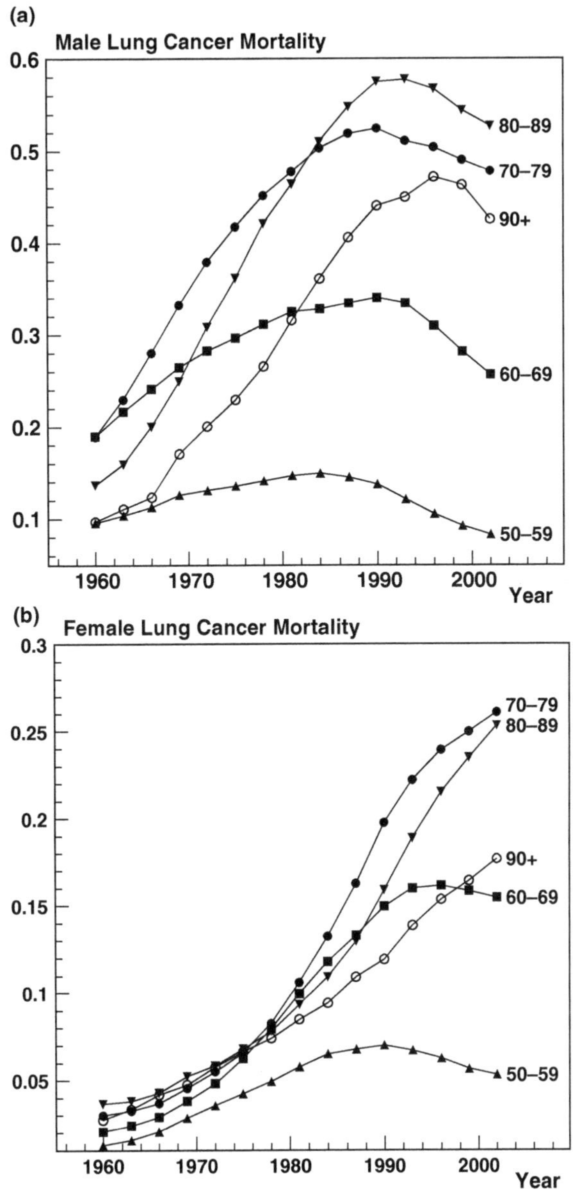


The incidence of lung cancer in males after many years of rising peaked during the mid-1980s and the beginning of the 1990s, and then decreased in both, whites and blacks. Incidence declined predominantly among the young and middle-aged males over the 30 years period, in both races. In females, incidence more than doubled in 1973–2003, with more rapid increase through the mid-1990s. This

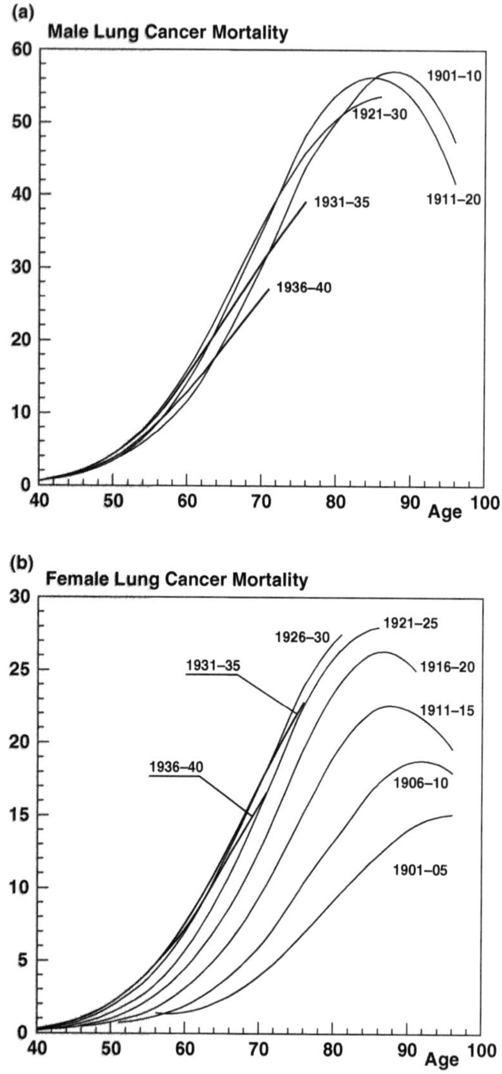


**Fig. 6.7** Male (**a**) and female (**b**) lung cancer mortality (per 10,000) for different birth cohorts

increase slowed and started decreasing at the beginning of the 2000s. The male/female ratio declined with time, but incidence rate remained higher in males (however, in whites this "gender gap" started vanishing).

Despite some encouraging observations, especially in males, tobacco-attributable cancers will not continue decreasing without sustained efforts to reduce the percentage of smoking adolescents and to increase the percentage of adults who successfully quit. Generally, the decrease of smoking prevalence among adults reflects the combined effects of both, smoking initiation and cessation. This is a less sensitive indicator of smoking trends than prevalence of this habit among teenagers/adolescents (reflects trend of smoking initiation) (Weir et al., 2003).

### 6.4.2 Breast Cancer

Female breast cancer incidence has varied since 1975. In 1974 incidence spiked when women reacted to highly publicized breast cancer diagnoses in two nationally prominent diagnostic testing programs (Ries and Devesa, 2006). Since the early 1980s, incidence rate increased dramatically, with a subsequent leveling since the early 1990s. These trends were observed in white and black women of both age groups – younger then 50 and 50 + . It has been suggested, that incidence increase was mostly due to the early-stage small tumors, due to the increased practice of mammography (Miller et al., 1993).

Breast cancer risk increases rapidly with age during child-bearing years (Brinton et al., 2002; Lacey et al., 2002). After menopause, risk continues to increase, but not so rapid. During childbearing period, the incidence rates were higher in black women than in white, at ages 40–44 years rates became equal, and in older ages rates were substantially higher in whites (Ries and Devesa, 2006). The reported differences in breast cancer incidence rates among the U.S. states may be explained not only by different demographic characteristics and coverage by screening program (e.g., mammography) but also by variations in modifiable risk factors, such as obesity, lack of physical activity, smoking, and use of hormone replacement therapy (Hulka and Stark, 1995; Collaborative Group on Hormonal Factors in Breast Cancer, 1997; Vainio et al., 2002; Weir et al., 2003).

The breast cancer death rate increased till the 1990s, with a subsequent slight decrease, more prominent in white than in black females. This decrease was more obvious in younger women. It has been suggested that the mortality decrease maybe due to mammography screening and dissemination of adjuvant chemotherapy including multiagent chemotherapy and tamoxifen (Wingo et al., 1998; Howe et al., 2001; Edwards et al., 2002; Mariotto et al., 2002; Harlan et al., 2002). More detailed discussion of breast cancer nonhomogeneity and risk factors is presented in Chapter 7.

### 6.4.3 Prostate Cancer

The trend of prostate cancer incidence is one of the most changeable over time. This is why it is important when analyzing trends to take into account delays in reporting cancer incidence to registries/data sets (this is less important for cancers with slow changes in incidence). Rates for both races increased during the late 1970s to the early 1990s, more prominently in blacks, with peaks at approximately 1995. There is indirect evidence that the increasing prostate cancer incidence over the 1973–1986 period may be due to increased detection of clinically asymptomatic cases associated with increasing rates of transurethral resection of the prostate (TURP) for benign prostatic hyperplasia (Potosky et al., 1990). Increases after 1987 do not seem to be associated with TURPs. During the late 1980s, increases in incidence were seen for all stages, except of distant (Ries and Devesa, 2006). With the introduction of widespread screening for prostate-specific antigen (PSA) in the early 1990s, the prostate cancer incidence rate increased substantially (Potosky et al., 1995).

Prostate cancer death rates increased from 1959 to early 1990s, then started decreasing (more prominently for younger males). The decline in prostate cancer mortality was, at least partly, due to decrease in late-stage cancers (Feuer et al., 1999). There were no currently recognized risk factors accounting for the decline in prostate cancer mortality, although the decrease might reflect the improvement in treatment combined with the PSA early diagnostic (Hsing et al., 2001; Boyle et al., 2003; Weir et al., 2003). Some studies have found that the decreasing prostate cancer mortality was unlikely to be due to only intensive screening (Etzioni et al., 1999; Oliver et al., 2001), and treatment advances may contribute in the mortality decline (Harris and Lohr, 2002; Meng et al., 2002). For further discussion of certain prostate cancer features see Chapter 7.

### *6.4.4 Colorectal Cancer*

Colon cancer represents more than 70% of colorectal cancer (with more than half of colon cancer localized in sigmoid) and less than 30% – cancer of rectum. The incidence rates of colon cancer appeared more stable over 1973–2003 period (with moderate increase in males in 1979–1988, succeeding gradual decrease, and with consistent decrease in female) for both races, while rectal cancer had more prominent dynamics over this period, decreasing substantially in males, especially in white males, and moderately decreasing in females.

According to some studies, trends in colorectal cancer incidence have varied by subsite, with declines in rectal cancer, especially among whites, and increases in proximal colon cancer rates among blacks (Devesa and Chow, 1993; Troisi et al., 1999). Screening and removal of precancerous lesions may be affecting rates, when individual at "average risk" usually receive a sigmoidoscopy, which can detect and remove distal adenomas, but not proximal adenomas. However, some other factors, such as difference in risk factors for proximal and distal colorectal cancers, may influence incidence rates (Wu et al., 2001). The black/white incidence ratio decreased, and the male/female incidence ratio increased, from the proximal to the distal colon (with race ratios of 1.31–1.25–1.17–1.05–1.00–0.92, and sex ratios of 1.2–1.2–1.2–1.5–1.6–1.6 for caecal, ascending, transverse, sigmoid, rectosigmoid, and rectal cancers, respectively) (Ries and Devesa, 2006). It is still unclear, why colorectal cancer location differs by sex. Differences in sex hormones, which may influence bile acid metabolism, fecal transit time and composition, may in part be responsible (McMichael and Potter, 1985; Lampe et al., 1993). Differences in sex hormones may also be the factor influencing the overall male-to-female ratio for colorectal cancer, which is lower at ages younger than 54 (i.e., for premenopausal women), than at ages 55+ (i.e., for postmenopausal women) (Bufill, 1990; Dornschneider et al., 1990; Lampe et al., 1993). A recent study showed that estrogens may decrease the risk of microsatellite instability-positive colon cancer (MSI +) (Slattery et al., 2001). In term of racial difference in distal-to-proximal colorectal cancer prevalence, it is still unclear whether other factors, additional to screening effect and precancerous lesions removal, may affect the race prevalence (Ries and Devesa, 2006).

Mortality rates of colorectal cancer started to decrease after the mid-1980s, predominantly because of decrease in rectal cancer death rates. Decreases in mortality rates were highest in white males, followed by white females, black males, and then black females.

### *6.4.5 Esophageal Cancer*

Esophageal cancer occurs about three times more frequently in males (both, white and black). Both, white males and females still have a lower incidence rate than blacks. Increasing mortality among the U.S. blacks was observed from the mid-50s [that supposed unlikely to be due to poorer medical care (Blot and Fraumeni, 1987)], while incidence started declining since the beginning of the 1990s. There is an opposite trend observed for white males, who have steady increase in incidence and death rates over the period of observation. There are no obvious changes in white female rates, while dynamic of rates in black women is similar to those in black males (but at considerably lower scale).

According to other studies, increasing trends of esophageal cancer mortality have been found since 1970 among females at age 65+, and in males of all ages (Qiu and Kaneko, 2005). The 5-year survival rate for esophageal cancer showed little improvement in the mid-1990s compared to the 1950s (Welch et al., 2000). Analyzing the histology-specific cancer incidences (see Chapter 7), it has been found that the highest incidence rate in both sexes and both races has a squamous cell carcinoma (SCC), especially in black males. However, SCC rates, supposed to be associated with alcohol consumption, smoking, nitrosamines in food, and zinc and vitamin A deficiency, were steadily declining during the more than 30 years, especially in black male, while esophageal adenocarcinoma (which may be associated with obesity and gastroesophageal reflux disease) was rising, especially fast among white males (more detailed analysis of the trends of esophageal cancer histotypes see in Chapter 7).

### *6.4.6 Stomach Cancer*

Stomach cancer is almost twice as prevalent in males as in females. It is also more prevalent in blacks than in whites. Its incidence and mortality have been steeply declining over the past several decades. Reduction of population H. pylori prevalence, as well as smoking reduction, and, probably, improved diet may explain this decrease (Brown and Devesa, 2002).

### *6.4.7 Cancer of Pancreas*

Cancer of pancreas is more frequent in males, especially black males. Pancreatic cancer incidence almost did not change since the early 1970s. The slight decline in incidence over the past 20 years was mainly due to males' rates. Given epidemiologic data that strongly implicate cigarettes as a risk factor for pancreatic cancer, trends in cigarette use might explain increase seen until the 1970s and the more recent decrease in males (Fontham and Correa, 1989; CDC, 2004). In contrast to high incidence rates in Afro-Americans, the incidence rates in Africans (based on limited data) are low (Parkin et al., 2002). Some evidence has been found to suggest that the disparity between rates in black and white Americans is due to differences in risk factor prevalence, and modifications of their effects by race (Silverman et al., 2003).

### *6.4.8 Cancer of Liver and Intrahepatic Duct*

This category includes hepatocellular carcinoma (HCC) and the considerably rarer cancer of the intrahepatic bile duct. This cancer had substantial incidence increase in both sexes and races. The aging of the population with chronic hepatitis C may, at least in part, explain the increase in HCC rates (Shaheen et al., 2006). HCC has an incidence rate more than twofold higher in males, especially in black males. The reasons why males are at increased risk are not completely understood but may be partly explained by the sex-specific prevalence of risk factors: e.g., it is supposed that males are more likely to be infected with HBV and HCV, consume alcohol, smoke cigarettes, and have increased iron stores (London and McGlynn, 2006). It remains unclear, whether androgenic hormones and/or increased genetic susceptibility influence gender differences in rates of liver cancer. Race/ethnic variation in rates almost certainly reflects the differences in the likelihood of infection with HBV and/or HCV, although genetic susceptibility and different patterns of exposure to other risk factors may also play a role.

Liver cancer showed substantial differences between males and females in the ages at which incidence started increasing. In 1973–1978, black women had 23 years later disease onset and white women had 20 years later onset, than males (at age 60 versus 37, and at age 67 versus 47, respectively). These differences were smaller in 1999–2003 – 4 and 10 years (at age 47 versus 43, and at age 53 versus 43, respectively). These changes were dramatic in black women, possibly because of rapidly increasing rates of HCV infection in this group, rather than intensity of alcohol consumption, which declined in black females over the 30 years [i.e., at the beginning of the 2000s, the death rates of alcohol-related liver cirrhosis in black women dropped considerably and met the rate for white females (Yoon and Yi, 2006)]. The detailed discussion of HCC rates, risk factors, and nonhomogeneity is presented in Chapter 7.

### *6.4.9 Cancer of Corpus Uteri*

The incidence of corpus uteri cancer increased in the mid-1970s, shortly after the use of postmenopausal estrogens gained wide acceptance. While many studies support the association between the decline in uterine cancer risk and the continuous use of combined postmenopausal hormonal therapy, the long-term risks associated with different dosage and duration of progestogen use still needs evaluation (Cook et al., 2006).

Use of oral hormonal contraceptives (widespread since 1960s) may have played a role in dynamic of uterine cancer incidence: combination (concurrent estrogen and progesterone) pills have been associated with the decreased cancer risk, depending on duration of use, doses of components, women body weight, and parity (Benshushan et al., 2001; Armstrong et al., 1988; Levi et al., 1991; The Centers for Disease Control, Cancer and Steroid Hormone Study, 1987; Stanford et al., 1993; Weiderpass et al., 1999). The decrease of uterine cancer incidence in women who used the depot medroxyprogesterone acetate has been reported (DMPA) (injected progesterones) (WHO Collaborative Study of Neoplasia and Steroid Contraceptives, 1991; Weiderpass et al., 1999). Started in the mid-1970s use of sequential oral contraceptive pills (estrogen only followed by a short course of estrogen plus progestogen) increased risk of endometrial cancer (Henderson et al., 1983; Silverberg et al., 1977; The Centers for Disease Control, Cancer and Steroid Hormone Study, 1983), and in 1976 sequential preparations were removed from the consumer market in the United States and Canada (Cook et al., 2006). Uterine cancer incidence rates stabilized in the mid-1990s. A significant gap in incidence between white and black women existed in the 1970s, with white females reaching a higher incidence peak at younger ages than in black females. At present, this gap is almost vanished, and both races have almost equal rates. However, incidence peak in white women is still "younger" – it occurs about 7 years earlier than in blacks.

Mortality due to cancer of the uterus declined since the end of the 1980s. It has been suggested that introduction of nonsequential oral contraceptives may have played a role in the decrease of mortality in subsequent decades (Grady and Ernster, 1996). The effect of the estrogen replacement therapy on mortality rate needs further investigations. The interpretation of incidence and mortality rates is rather complicated because of the lack of corrections made in reports from cancer registries and vital records agencies for the percentage of women with hysterectomies, who are no longer at risk for developing endometrial cancer (Cook et al.,

2006). The reported age-standardized mortality due to corpus cancer and uterine cancer not otherwise specified (NOS) decreased by 60% from 1950 to 1985 in the U.S. females (Division of Cancer Prevention and Control, National Cancer Institute, 1988), remaining relatively stable since that time. However, this decrease is an overestimate of a true dynamic in corpus uteri cancer mortality (Cook et al., 2006), due to the dramatic decrease in cervical cancer mortality over the same period (uterine cervix, corpus, and uterus NOS were combined for analysis of historical trend, because separating causes of death reported with respect to these sites was not possible in early decades) and rise of the hysterectomy rate (Wingo et al., 2003; Division of Cancer Prevention and Control, 1988; U.S. Department of Health and Human Services, 1981; Koepsell et al., 1980; Lyon, 1977). The prevalence of women with an intact uterus varies by age, race, geographical location, and time period. Thus the degree of underestimation of cancer risk also varies (Pokras and Hufnagel, 1987; CDC, 2002).

### *6.4.10 Cervical Cancer*

Incidence and mortality rates of cervical cancer significantly decreased over the past 30 years due, at least in part, to widespread use of the Papanicolaou smear test and advances in diagnostic techniques (Ries and Devesa, 2006). The largest decrease was for black females. Black women had a significantly higher incidence rates in the 1970s. By 2003, both races had almost the same incidences at younger ages. At ages 55 and older incidence rates are still higher in black females. An incidence rates decrease occurred mostly due to earlier diagnosis and treatment of squamous cell carcinoma, the predominant type of cervical cancer. The more difficult to detect (and likely associated with other risk factors) adenocarcinoma did not show incidence decline.

Cervical cancer mortality decreased significantly over the 30-year period, with a rapid decrease between 1978 and 1993 in both, white and black women.

Incidence and mortality trends for cervical SCC and AC, and how changes in cervical cancer registration (*in situ* and invasive), active screening strategy, and age-related hysterectomy may influence cervical cancer prevalence are described in Chapter 7.

### *6.4.11 Kidney Cancer*

Kidney cancer has both, increasing incidence and mortality rates over 30 years in both races and genders. Kidney cancer incidence is more than twice as high in males, without substantial race differences. Improved detection of presymptomatic kidney tumors by imaging procedures, such as ultrasonography, computed tomography, and magnetic resonance imaging does not fully explain the increasing incidence of renal-cell carcinoma, the dominant kidney cancer form

(Chow et al., 1999). Other factors may also contribute, such as increasing prevalence of obesity, hypertention, smoking, and dietary habits (Mathew et al., 2002). The incidence of kidney cancer increased substantially not only in the United States, but in many countries around the world (Ries and Devesa, 2006).

### 6.4.12 *Thyroid Cancer*

Thyroid cancer incidence increased during the past couple of decades in the United States. Only incidence of liver cancer increased more rapid than thyroid carcinoma. The acceleration of thyroid cancer incline was the first among other cancers in women and third – in men (Ron and Schneider, 2006). The incidence of thyroid cancer, especially in females, became higher, and peaked at younger ages in whites since the early 1980s. Thyroid cancer showed sex differences in the dynamic of incidence: over time it increased dramatically in females, compared to slight increases in males. The increased incidence of thyroid cancer among whites is confined to papillary carcinoma (Correa and Chen, 1995; Ries et al., 2004) and suggested that it may be related to better cancer screening. Furthermore, in a large national survey, whites had higher serum thyroid-stimulating hormone (TSH) concentrations than blacks (Hollowell et al., 2002), which might enhance their thyroid cancer risk. More details about thyroid cancer trends and risk factors are presented in Chapter 3.

## 6.5 Summary

In this chapter we analyzed mortality caused by 15 solid cancers and their incidence rates in the U.S. population, using an epidemiological approach. Special focus was on age patterns of these characteristics, their time trends, as well as on sex and race differences. We revealed the dynamic properties of the most prevalent and the most lethal cancers in the United States over the observed period through analyses of their mortality and incidence trends. Note that because of different classifications during observation period, and newly developed and recently used medical technologies, certain care is required to perform such analyses of trends.

Lung and bronchus cancer is still the leading cause of cancer death in the United States, followed by cancers of colon and breast. While having the tendency to improving mortality trends of some cancers over the more that half a century period (e.g., cancers of stomach, testis, rectum, cervix uteri), mortality trends for some other cancers are not so impressive. Cancers of lung, prostate, esophagus (in male), ovary, and kidney have the increased age-specific mortality rates over the observed period (however, some cancers, such as lung, kidney, ovarian, corpus uteri, thyroid, and stomach now have age-related shifts to about 7–10 years older ages in their mortality peaks).

Lung and bronchus, breast, prostate and colorectal cancers over the several past decades keep the leading positions for incidence rates. The most prominent increase in incidence rates over the 30 years were registered for cancers of lung and bronchus, breast, prostate, kidney, thyroid, and liver, while cervical and gastric carcinomas demonstrated obvious decreases in their incidences. The incidences of cancers of lung (in males), esophagus (in black females), stomach, testis, corpus uteri (in blacks), and pancreas became about 5–10 years "older", while cancers of prostate, colorectum (in males), and liver and kidney (in females), on the contrary, became about 5–10 years "younger".

Deeper analyses of cancer histotypes (such as SCC and AC) are performed in Chapter 7 to describe their time trends and to characterize their homogeneity/no-homogeneity. To do that, a modeling approach is used that let us hypothesize about the possible underlying biological mechanisms of carcinogenesis.

## References

Armstrong B.K., Ray R.M., Thomas D.B., 1988. Endometrial cancer and combined oral contraceptives. The WHO Collaborative Study of Neoplasia and Steroid Contraceptives. Int J Epidemiol 17:263–269.

Axtell L.M., Asire A.J., Myers M.H., 1976. Cancer Patient Survival: Report Number 5. NIH Publ. 81–992. Bethesda: National Cancer Institute.

Benshushan A., Paltiel O., Brzezinski A. et al., 2001. Ovulation induction and risk of endometrial cancer: a pilot study. Eur J Obstet Gynecol Reprod Biol 98:53–57.

Blot W.J., Fraumeni J.F., 1987. Trends in esophageal cancer mortality among US blacks and whites. Am J Public Health 77:296–298.

Boyle P., Severi G., Giles G.G. 2003. The epidemiology of prostate cancer. Urol Clin North Am 30:209–217.

Brinton L.A., Lacey J., Devesa S.S., 2002. Epidemiology of breast cancer. In: Donegan W.L., Spratt J.S. (Eds.) Cancer of the Breast. 5th edition. Orlando, FL: Saunders, pp. 111–132.

Brown L.M., Devesa S.S., 2002. Epidemiologic trends in esophageal and gastric cancer in the United States. Surg Oncol Clin N Am 11:235–256.

Bufill J.A., 1990. Colorectal cancer: evidence for distinct genetic categories based on proximal or distal tumor location. Ann Intern Med 113:779–788.

CDC, 2002. National Center for Chronic Disease Prevention and Health Promotion. Behavioral Risk Factor Surveillance System, Nationwide. 2002 Women's Health Prevalence Data. At: www.cdc.gov/

CDC, 2004. Cigarette smoking among adults – United States, 2002. MMWR 53:428–431.

Centers for Disease Control and Prevention, 1960–1998. National Center for Health Statistics. Multiple cause of death public use data files. Hyattsville, MD: National Center for Heath Statistics.

Chow W-H., Devesa S.S., 1992. Death certificate reporting of colon and rectal cancers [letter to the editors]. JAMA 267:3028.

Chow W-H., Devesa S.S., Warren J.L., Fraumeni J.F. Jr., 1999. Rising incidence of renal cell cancer in the United States. JAMA 281:1628–1631.

Collaborative Group on Hormonal Factors in Breast Cancer, 1997. Breast cancer and hormone replacement therapy: collaborative re-analysis of data from 51 epidemiological studies of 52,705 women with breast cancer and 108,411 women without breast cancer. Lancet 350:1047–1059.

Cook L.S., Weiss N.S., Doherty J.A., Chen C., 2006. Endometrial cancer. In: Schottenfeld D., Fraumeni J.F. Jr. (Eds.) Cancer Epidemiology and Prevention. 3rd edition. Oxford: University Press.

Correa P., Chen V.W., 1995. Endocrine gland cancer. Cancer 75:338–352.

Cutler S.J., Young J.L. (Eds.), 1975. Third National Cancer Survey: Incidence Data. National Cancer Institute Monograph 41. Washington, DC: U.S. Government Printing Office.

Devesa S.S., Chow W-H., 1993. Variation in colorectal cancer incidence in the United States by subsite of origin. Cancer 71:3819–3826.

Division of Cancer Prevention and Control., 1988. National Cancer Institute. 1987 Annual Cancer Statistics Review: including cancer trends 1950–1985. NIH Publication No. 88-2789.

Dorn H.F., Cutler S.J., 1959. Morbidity from cancer in the United States: Parts I and II. Public Health Monograph 56-1-207. Bethesda: DHEW.

Dornschneider G., Izbicki J.R., Wilker D.K. et al., 1990. The effects of sex steroids on colon carcinogenesis. Anticancer Drugs 1:15–21.

Edwards B.K., Howe H.L., Ries L.A. et al., 2002. Annual report to the Nation on the status of cancer, 1973–1999, featuring implications of age and aging on U.S. cancer burden. Cancer 94:2766–2792.

Etzioni R., Legler J.M., Feuer E.J. et al., 1999. Cancer surveillance series: interpreting trends in prostate cancer – part III: quantifying the link between population prostate-specific antigen testing and recent declines in prostate cancer mortality. J Natl Cancer Inst 91:1033–1039.

Federal Security Agency, 1945–1949. Public Health Service, National Office of Vital Statistics. Vital statistics of the United States. Part I. Washington: Government Printing Office.

Feuer E.J., Merrill R.M., Hankey B.F., 1999. Cancer surveillance series: interpreting trends in prostate cancer – part II: cancer of death misclassification and the recent rise and fall in prostate cancer mortality. J Natl Cancer Inst 91:1025–1032.

Fontham E.T.H., Correa P., 1989. Epidemiology of pancreatic cancer. Surg Clin N Am 69:551–567.

Grady D., Ernster V.L., 1996. Endometrial cancer. In: Schottenfeld D., Fraumeni J.F. Jr. (Eds). Cancer Epidemiology and Prevention. New York: Oxford University Press.

Harlan L.C., Abrams J., Warren J.L. et al., 2002. Adjuvant therapy for breast cancer: practice patterns of community physicians. J Clin Oncol 20:1809–1817.

Harris R., Lohr K.N., 2002. Screening for prostate cancer: an update of the evidence for the U.S. Preventive Services Task Force. Ann Intern Med 137:917–929.

Henderson B.E., Casagrande J.T., Pike M.C. et al., 1983. The epidemiology of endometrial cancer in young women. Br J Cancer 47:749–756.

Hetzel A.M., 1997. U.S. Vital Statistics System, Major Activities and Developments, 1950–1995. Hyattsville, MD: Centers for Disease Control and Prevention, National Center for Health Statistics.

Hollowell J.G., Staehling N.W., Flanders W.D., et al., 2002. Serum TSH, T4 and thyroid antibodies in the United States population (1988 to 1994): National Health and Nutrition Examination Survey (NHANES III). J Clin Endocrinol Metab 87:489–499.

Howe H.L., Wingo P.A., Thun M.J., Ries L.A., Rosenberg H.M., Feigal E.G., Edwards B K., 2001. Annual report to the nation on the status of cancer (1973 through 1998), featuring cancers with recent increasing trends. J. Natl. Cancer Inst. (Bethesda) 93:824–842.

Hsing A., Devesa S.S., 2001. Trends and patterns of prostate cancer: what do they suggest? Epidemiol Rev 23:3–13.

Hsing A., Nomura A., Isaacs W. et al., 2001. Epidemiologic reviews: prostate cancer. Baltimore (MD): Oxford University Press, Vol. 23, No. 1..

Hulka B.S., Stark A.T., 1995. Breast cancer: cause and prevention. Lancet 346:883–887.

Human Mortality Database. University of California, Berkeley (USA), and Max Planck Institute for Demographic Research (Germany). Available at www.mortality.org or www.humanmortality.de (data downloaded on January 2006).

Koepsell T.D., Weiss N.S., Thompson D.J. et al., 1980. Prevalence of prior hysterectomy in the Seattle-Tacoma area. Am J Public Health 70:40–47.
Lacey J.V., Devesa S.S., Brinton L.A., 2002. Recent trends in breast cancer incidence and mortality. Environ Mol Mutagen 39:82–88.
Lampe J.W., Fredstrom S.B., Slavin J.L. et al., 1993. Sex differences in colonic function: a randomized trial. Gut 34:531–536.
Levi F., La Vecchia C., Gulie C. et al., 1991. Oral contraceptives and the risk of endometrial cancer. Cancer Causes Control 2:99–103.
London W.T., McGlynn K.A., 2006. Liver cancer. In: Schottenfeld D., Fraumeni J.F. Jr. (Eds.) Cancer Epidemiology and Prevention 3rd edition. Oxford: Oxford University Press.
Lyon J.L., Gardner J.W., 1977. The rising frequency of hysterectomy: its effect on uterine cancer rates. Am J Epidemiol 105:439–443.
Manton K.G., Akushevich I., Kulminski A. 2008 Human mortality at extreme ages: data from the National Long Term Care Survey and Linked Medicare Records, Math Popul Stud 15:137–159.
Mariotto A., Feuer E., Harlan L. et al., 2002. Trends in use of adjuvant multi-agent chemotherapy and tamoxifen for breast cancer in the United States: 1975–1999. J Natl Cancer Inst 94:1626–1634.
Mathew A., Devesa S.S., Fraumeni J.F. Jr., Chow W-H., 2002. Global increases in kidney cancer incidence 1973–1992. Eur J Cancer Prev 11:171–178.
McMichael A.J., Potter J.D., 1985. Host factors in carcinogenesis: certain bile-acid metabolic profiles that selectively increase the risk of proximal colon cancer. J Natl Cancer Inst 75:185–191.
Meng M.V., Grossfeld G.D., Sadetsky N. et al., 2002. Contemporary patterns of androgen deprivation therapy use for newly diagnosed prostate cancer. Urology 60(3 Suppl 1):7–11.
Miller B.A., Feuer E.J., Hankey B.F., 1993. Recent incidence trends for breast cancer in women and the relevance of early detection: an update. CA Cancer J Clin 43:27–41.
NBER, 2007 Multiple Cause-of-Death Mortality Data from the National Vital Statistics System of the National Center for Health Statistics, 1959–2004, www.nber.org/data/multicause.html, access in 2007.
Oliver S.E., May M.T., Gunnell D., 2001. International trends in prostate-cancer mortality in the "PSA era". Int J Cancer 92:893–898.
Parkin D.M., Whelan S.L., Ferlay J. et al., 2002. Cancer Incidence in Five Continents. volume 8. Lyon, France: International Agency for Research on Cancer.
Parkin D.M., Whelan S.L., Ferlay J., Teppo L., Thomas D.B. (Eds.), 2003. Cancer Incidence in Five Continents volume 8. IARC Scientific Publ. No. 155 Lyon: IARC.
Percy C., Stanek E. III, Gloeckler L., 1981. Accuracy of cancer death certificates and its effect on cancer mortality statistics. Am J Public Health 71:242–250.
Percy C.L., Ries L.A.G., Van Holten V.D., 1990. The accuracy of liver cancer as the underlying cause of death on death certificates. Public Health Rep 105:361–368.
Pokras R., Hufnagel V.G., 1987. Hysterectomies in the United States. DHHS Publ. No. (PHS) 88–1753. Hyattsville, MD: National Center for Health Statistics.
Potosky A.L., Kessler L., Gridley G., Brown C.C., Horm J.W., 1990. Rise in prostatic cancer incidence associated with increased use of transurethral resection. J Natl Cancer Inst 82:1624–1628.
Potosky A.L., Miller B.A., Albertsen P.C., Kramer B.S., 1995. The role of increasing detection in the rising incidence of prostate cancer. JAMA 273:548–552.
Qiu D., Kaneko S., 2005. Comparison of esophageal cancer mortality in five countries: France, Italy, Japan, UK and USA from the WHO Mortality Database (1960–2000). Jpn J Clin Oncol 35:564–567.
Ries L.A.G., Devesa S.S., 2006. Cancer incidence, mortality and patient survival in the United States. In: Schottenfeld D., Fraumeni J.F. Jr. (Eds). Cancer Epidemiology and Prevention. 3rd edition. Oxford, University Press, pp. 139–173.

Ries L.A.G., Eisner M.P., Kosary C.L. et al., 2002. SEER Cancer Statistics Review, 1973–1999. Bethesda, MD: National Cancer Institute.

Ries L.A.G., Eisner M.P., Kosary C.L. (Eds.) et al., 2004. SEER Cancer Statistics Review, 1975–2001. Bethesda, MD: National Cancer Institute. At: seer.cancer.gov/csr/1975_2001/

Ron E., Schneider A.B., 2006. Thyroid cancer. In: Schottenfeld D., Fraumeni J.F. Jr. (Eds.) Cancer Epidemiology and Prevention. 3rd edition. Oxford: Oxford University Press.

Rowland J., Mariotto A., Aziz N. et al., 2004. Cancer Survivorship – United States, 1971–2001. MMWR 53(24):526–529.

Shaheen N.J., Hansen R.A., Morgan D.R. et al., 2006. The burden of gastrointestinal and liver diseases. Am J Gastroenterol 101:2128–2138.

Silverberg S.G., Makowski E.L., Roche W.D., 1977. Endometrial carcinoma in women under 40 years of age: comparison of cases in oral contraceptive users and non-users. Cancer 39:592–598.

Silverman D.T., Hoover R.N., Brown L.M. et al., 2003. Why do black Americans have a higher risk of pancreatic cancer than white Americans? Epidemiology 14:45–54.

Slattery M.L., Potter J.D., Curtin K. et al., 2001. Estrogens reduce and withdrawal of estrogens increase risk of microsatellite instability-positive colon cancer. Cancer Res 61:126–130.

Stanford J.L., Brinton L.A., Berman M.L. et al., 1993. Oral contraceptive and endometrial cancer: do other risk factors modify the association? Int J Cancer 54:243–248.

The Centers for Disease Control, Cancer and Steroid Hormone Study, 1983. Oral contraceptive use and the risk of endometrial cancer. JAMA 249:1600–1604.

The Centers for Disease Control, Cancer and Steroid Hormone Study, 1987. Combination oral contraceptive use and the risk of endometrial cancer. JAMA 257:796–800.

Troisi R., Freedman A., Devesa S., 1999. Incidence of colorectal carcinoma in the United States: an update of trends by gender, race, age, subsite, and stage, 1975–1994. Cancer 85:1670–1676.

U.S Public Health Service, 1964. Smoking and Health. Report of the Advisory Committee to the Surgeon General of the Public Health Service. U.S. Department of Health, Education and Welfare, Public Health Service Publication No. 1103. U.S. Government Printing Office: 1964.

U.S. Department of Commerce, 1930–1936. Bureau of the Census. Mortality Statistics. Washington, DC: U.S Government Printing Office.

U.S. Department of Commerce, 1937–1944. Bureau of the Census. Vital Statistics of the United States. Part I. Washington, DC: Government Printing Office.

U.S. Department of Health and Human Services, 1981. Hysterectomy in women aged 15–44. United States, 1970–1978. MMWR 30:173–176.

U.S. Department of Health and Human Services, 2000. Tracking Healthy People 2010. Washington, DC: U.S. Government Printing Office.

U.S. Department of Health, Education and Welfare, 1950–1959. Public Health Service, National Office of Vital Statistics. Vital Statistics of the United States. Volume 2. Washington, DC: Government Printing Office.

U.S. Vital Statistics System, 1950. History and Organization of the Vital Statistics System. Volume 1. Hyattsville, MD: U.S. Vital Statistics System 1950:1–19.

Vainio H., Bianchini F., 2002. IARC Handbook of Cancer Prevention. Volume 6: Weight Control and Physical Activity. Lyon, France: IARC Press.

Weiderpass E., Adami H.O., Baron J.A. et al., 1999. Use of oral contraceptives and endometrial cancer risk (Sweden). Cancer Cause Control 10:277–284.

Weir H.K., Thun M.J., Hankey B.F. et al., 2003. Annual report to the Nation on the status of cancer, 1975–2000, featuring the uses of surveillance data for cancer prevention and control. J Natl Cancer Inst 95(17):1276–1299.

Welch H.G., Schwartz L.M., Woloshin S., 2000. Are increasing 5-year survival rates evidence of success against cancer? JAMA 283(22):2975–2978.

WHO Collaborative Study of Neoplasia and Steroid Contraceptives, 1991. Depot medroxyprogesterone acetate (DMPA) and risk of endometrial cancer. Int J Cancer 49:186–190.
Wingo P.A., Ries L.A., Rosenberg H.M. et al., 1998. Cancer incidence and mortality, 1973–1995: a report card for the U.S. Cancer 82:1197–1207.
Wingo P.A., Cardinez C.J., Landis S.H. et al., 2003. Long-term trends in cancer mortality in the United States, 1030–1998. Cancer Supplement 97(12):3133–3275.
Wu X.C., Chen V.W., Steele B. et al., 2001. Subsite-specific incidence rate and stage of disease in colorectal cancer by race, gender, and age group in the United States, 1992–1997. Cancer 92:2547–2554.
Yoon Y-H., Yi H-Y., 2006. Liver cirrhosis mortality in the United States, 1970–2003. At NIAAA (National Institute on Alcohol Abuse and Alcoholism) site: pubs.niaaa.nih.gov/publications/surveillance75/images/fig8.gif

# Chapter 7
# U.S. Cancer Morbidity: Modeling Age-Patterns of Cancer Histotypes

## 7.1 Introduction

In this chapter, we use modeling to study the features of possible mechanisms of carcinogenesis that may underlie morbidity trends described in Chapter 6. Hypotheses considered include that (1) observed data represent a mixture of diseases, each of which can be described by different models or by a model with different combinations of parameters, and (2) the main differences are due to cancer histotypes contributing to observed cancer incidence patterns. Hence, the model represents disease heterogeneity by estimating histotype-specific incidence rates. Other factors, which may contribute to heterogeneity of cancer incidence, include unobserved factors, such as a genetic predisposition and other host-related features that modulated cancer risk when exposed to risk factors. Analytic approaches used in this chapter include the models of carcinogenesis reviewed in Chapter 2, as well as their generalizations, such as the two-disease model (see Chapter 4). Analysis of the contribution of specific histotypes is not only motivated by the desire to obtain more homogeneous groups of cancer cases for modeling but also because the different histological forms may differ by their time trajectories, by the spectrum of risk factors, and by carcinogenesis mechanisms.

Nowadays approaches for analyzing histology-specific cancer incidences (as well as other factor making the subgroups investigated more homogeneous) using concepts based on biological theories of carcinogenesis are not well established. In this chapter we will evaluate only several of many possible such tactics. Instead, we present the general direction of how these approaches can be developed. For detailed and comprehensive analyses, numerous sources of potential biases have to be kept in mind (e.g., effects of screening). Analysis of specific cancer histotypes using modeling allows to hypothesize about possible mechanisms underlying observed morbidity trends.

K.G. Manton et al., *Cancer Mortality and Morbidity Patterns in the U.S. Population*, Statistics for Biology and Health, DOI 10.1007/978-0-387-78193-8_7,
© Springer Science+Business Media, LLC 2009

## 7.2 Analyses of Trends of Cancer Histotypes in the U.S. Population

The most prevalent histotypes of cancers were analyzed for homogeneity of their age-specific trends. The ICD-O-2 SEER site/histology validation list (http://www.facs.org/cancer/ncdb/icdO2sitetype.pdf) was used to code histotypes (including morphological characteristics for certain cancers) of cancers of esophagus, stomach, rectum, colon, liver, pancreas, lung, breast, cervix uteri, corpus uteri, ovary, prostate, and kidney. Frequencies of incidences of these histotypes were estimated for males and females (see Table 7.1). Then the cancer histotypes were selected with incidence frequencies of more then 15% to be included in further analysis: esophageal SCC, esophageal AC, gastric AC, rectal AC and papillary AC of rectum, AC of colon, hepatocellular carcinoma, AC of pancreas, small-cell carcinoma of lung and bronchus, SCC of lung and bronchus, AC of lung and bronchus, breast ductal and lobular ACs, SCC of cervix uteri, AC of corpus uteri, endometrioid AC, papillary serous cystadenocarcinoma of ovary, AC of prostate, and clear-cell AC of kidney.

Age-specific incidence rates over a 30-year period (1973–2003) for selected histotypes for white and black males and females are shown in Fig. 7.1. The value in upper right corner of each graph denotes the histological code, e.g., code 807 corresponds to SCC and 814 to AC. The letters "M" and "F" correspond to male and female. The number next to these letters is the rescaling factor defined as in Figs. 6.1 and 6.4. The rate for a cancer is obtained by dividing values obtained from the plot by the rescaling factor. Apart from histology specific considerations, rates in Figs. 7.1 and 6.4 differ due to the contribution by cancers *in situ*, which are included in Fig. 7.1, but not in Fig. 6.4, where only the contributions of invasive cancers were considered. The rates in Fig. 7.1 were calculated by averaging over all SEER datasets from 1973 to 2003, so care is required in interpretation. Screening effects and the fact that *in situ* cervical cancer were not in the database since 1996 have to be kept in mind (we discuss those effects below).

Age-specific and histology-specific patterns for several time periods are presented in Figs. 7.2 and 7.3. Curves in these plots present one-disease and two-disease models. Details of the modeling procedure and interpretation of estimated parameters are discussed in Section 7.3. The time periods for cervix cancer differ from others to reflect dynamics of screening and the specifics of SEER registration of cases *in situ*.

When analyzing the time trends of specific cancer histotypes (see Figs. 7.2 and 7.3), it was shown that ACs have opposite directions in incidence dynamics with time compare to SCCs (for such cancer sites as esophagus, lung and bronchus, and cervix uteri).

### *7.2.1 Lung Cancer*

As shown in Fig. 7.1, lung SCC is the predominant histotype in males, with a higher incidence in black males. In females, a slight predominance of lung AC

**Table 7.1** Incidence frequencies of cancer histotypes, males/females (based on ICD-O-2 code) (frequencies of more than 3% are shown, in %)

| Histological code | Esophagus | Stomach | Rectum | Large intestine (excl. appendix) | Liver | Pancreas | Lung and bronchus | Breast | Cervix uteri | Corpus uteri | Ovary | Prostate | Kidney |
|---|---|---|---|---|---|---|---|---|---|---|---|---|---|
| C | 150–159 | 160–169 | 209 | 180, 182–189, 199 | 220 | 250–259 | 340–349 | 500–509 | 530–539 | 540–549 | 569 | 619 | 649 |
| 804 | <3.0 | <3.0 | <3.0 | <3.0 | <3.0 | <3.0 | 13.4/15.6 | <3.0 | <3.0 | <3.0 | <3.0 | <3.0 | <3.0 |
| 805 | <3.0 | <3.0 | <3.0 | <3.0 | <3.0 | <3.0 | <3.0 | <3.0 | <3.0 | <3.0 | <3.0 | <3.0 | <3.0 |
| 807 | 48.0/66.2 | <3.0 | <3.0 | <3.0 | <3.0 | <3.0 | 27.1/16.1 | <3.0 | -/64.0 | <3.0 | <3.0 | <3.0 | <3.0 |
| 814 | 37.0/18.1 | 66.0/55.7 | 64.2/59.3 | 63.8/63.0 | 3.1/5.4 | 55.6/51.7 | 22.5/27.8 | 7.6/3.4 | -/4.0 | -/51.0 | -/13.6 | 93.1/- | 3.9/3.6 |
| 816 | <3.0 | <3.0 | <3.0 | <3.0 | -/4.8 | <3.0 | <3.0 | <3.0 | <3.0 | <3.0 | <3.0 | <3.0 | <3.0 |
| 817 | <3.0 | <3.0 | <3.0 | <3.0 | 81.1/67.6 | <3.0 | <3.0 | <3.0 | <3.0 | <3.0 | <3.0 | <3.0 | <3.0 |
| 821 | <3.0 | <3.0 | 6.8/6.7 | 4.8/3.7 | <3.0 | <3.0 | <3.0 | <3.0 | <3.0 | <3.0 | <3.0 | <3.0 | <3.0 |
| 824 | <3.0 | <3.0 | 3.7/4.6 | <3.0 | <3.0 | <3.0 | <3.0 | <3.0 | <3.0 | <3.0 | <3.0 | <3.0 | <3.0 |
| 825 | <3.0 | <3.0 | <3.0 | <3.0 | <3.0 | <3.0 | -/4.1 | <3.0 | <3.0 | <3.0 | <3.0 | <3.0 | <3.0 |
| 826 | <3.0 | <3.0 | 13.7/16.1 | 10.8/9.9 | <3.0 | <3.0 | <3.0 | <3.0 | <3.0 | -/3.0 | -/6.5 | <3.0 | 3.2/- |
| 829 | <3.0 | <3.0 | <3.0 | <3.0 | <3.0 | <3.0 | <3.0 | <3.0 | <3.0 | <3.0 | <3.0 | <3.0 | <3.0 |
| 831 | <3.0 | <3.0 | <3.0 | <3.0 | <3.0 | <3.0 | <3.0 | <3.0 | <3.0 | <3.0 | -/3.5 | <3.0 | 82.3/82.6 |
| 833 | <3.0 | <3.0 | <3.0 | <3.0 | <3.0 | <3.0 | <3.0 | <3.0 | <3.0 | <3.0 | <3.0 | <3.0 | <3.0 |
| 834 | <3.0 | <3.0 | <3.0 | <3.0 | <3.0 | <3.0 | <3.0 | <3.0 | <3.0 | <3.0 | <3.0 | <3.0 | <3.0 |
| 838 | <3.0 | <3.0 | <3.0 | <3.0 | <3.0 | <3.0 | <3.0 | <3.0 | <3.0 | -/23.0 | -/9.3 | <3.0 | <3.0 |
| 844 | <3.0 | <3.0 | <3.0 | <3.0 | <3.0 | <3.0 | <3.0 | <3.0 | <3.0 | <3.0 | -/8.9 | <3.0 | <3.0 |
| 846 | <3.0 | <3.0 | <3.0 | <3.0 | <3.0 | <3.0 | <3.0 | <3.0 | <3.0 | <3.0 | -/26.7 | <3.0 | <3.0 |
| 847 | <3.0 | <3.0 | <3.0 | <3.0 | <3.0 | <3.0 | <3.0 | <3.0 | <3.0 | <3.0 | -/7.8 | <3.0 | <3.0 |
| 848 | <3.0 | 4.1/3.5 | 5.7/5.4 | 10.6/12.1 | <3.0 | 4.6/4.5 | <3.0 | <3.0 | <3.0 | <3.0 | -/3.4 | <3.0 | <3.0 |
| 849 | <3.0 | 8.9/12.6 | <3.0 | <3.0 | <3.0 | <3.0 | <3.0 | <3.0 | <3.0 | <3.0 | <3.0 | <3.0 | <3.0 |
| 850 | <3.0 | <3.0 | <3.0 | <3.0 | <3.0 | 4.2/4.3 | <3.0 | -/75.8 | <3.0 | <3.0 | <3.0 | <3.0 | <3.0 |

**Table 7.1** (continued)

| Histological code | Esophagus | Stomach | Rectum | Large intestine (excl. appendix) | Liver | Pancreas | Lung and bronchus | Breast | Cervix uteri | Corpus uteri | Ovary | Prostate | Kidney |
|---|---|---|---|---|---|---|---|---|---|---|---|---|---|
| C | 150–159 | 160–169 | 209 | 180, 182–189, 199 | 220 | 250–259 | 340–349 | 500–509 | 530–539 | 540–549 | 569 | 619 | 649 |
| 851 | <3.0 | <3.0 | <3.0 | <3.0 | <3.0 | <3.0 | <3.0 | <3.0 | <3.0 | <3.0 | <3.0 | <3.0 | <3.0 |
| 852 | <3.0 | <3.0 | <3.0 | <3.0 | <3.0 | <3.0 | <3.0 | -/13.8 | <3.0 | <3.0 | <3.0 | <3.0 | <3.0 |
| 857 | <3.0 | <3.0 | <3.0 | <3.0 | <3.0 | <3.0 | <3.0 | <3.0 | <3.0 | -/3.3 | <3.0 | <3.0 | <3.0 |
| 968 | <3.0 | 4.5/6.1 | <3.0 | <3.0 | <3.0 | <3.0 | <3.0 | <3.0 | <3.0 | <3.0 | <3.0 | <3.0 | <3.0 |

804 – small cell carcinoma, 807 – squamous cell carcinoma, 814 – adenocarcinoma, 817 – hepatocellular carcinoma, 826 – papillary adenocarcinoma, 850 – duct carcinoma, 852 – lobular carcinoma, 848 – mucinous adenocarcinoma, 846 – papillary serous cystadenocarcinoma, 838 – endometroid carcinoma, 831 – clear cell adenocarcinoma.

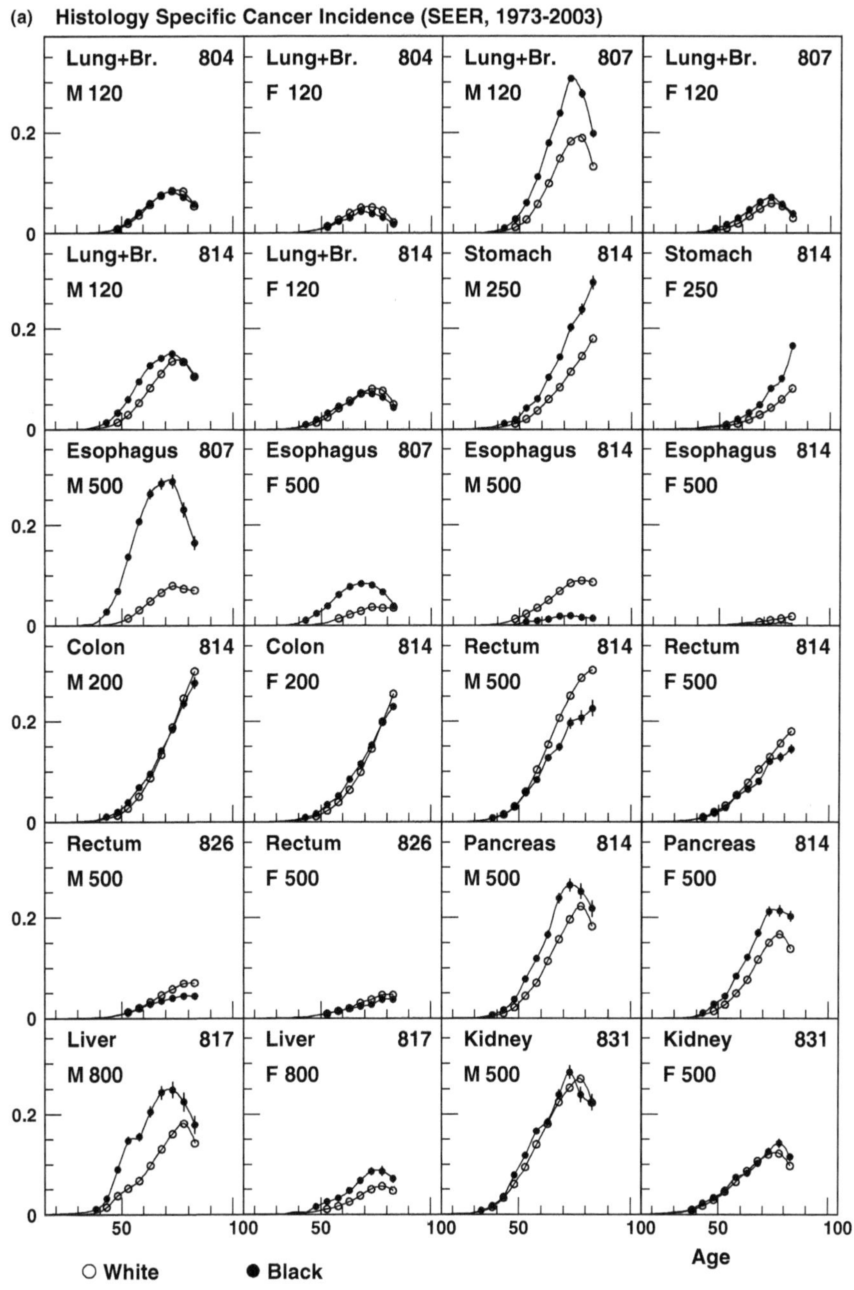


**Fig. 7.1** Age-specific patterns of the incidence of selected cancer histotypes for *white* and *black* males and females, 1973–2003

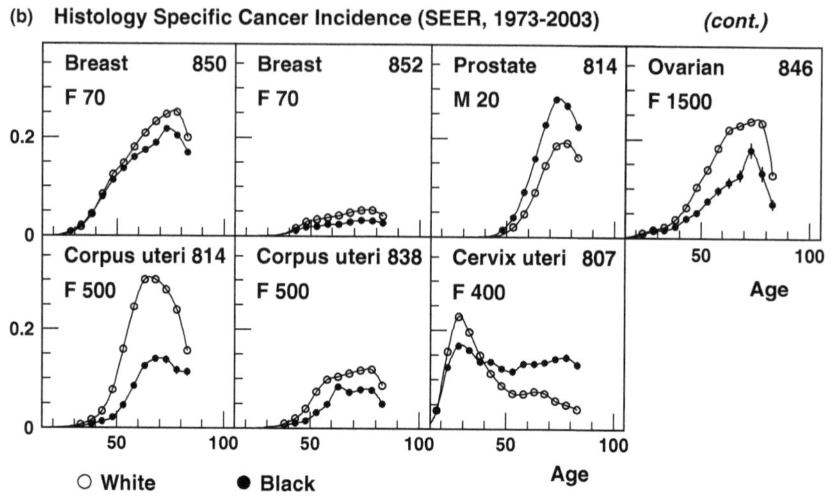


**Fig. 7.1** (continued)

over other histotypes was observed. The small-cell lung carcinoma was less common compared to SCC and AC. There were higher incidences of SCC, AC, and small-cell carcinoma of lung in males than in females. The highest incidence of SCC in black males was approximately 7 years earlier than in white males. Analyzing time trends (Fig. 7.2), the incidence of lung SCC decreased from 1973–1983 to 1994–2003 in black and white males, with an increase in white and black females. AC incidence increased in both sexes and both races, especially in females. In 1994–2003, lung AC became more common in white females than SCC, and almost equal to SCC in black females, while in males, SCC is still predominant.

There were many recent studies of lung cancer histotype prevalence. Trends in histological subtypes of lung cancer in the United States, and many other countries, have shifted over recent decades: AC became more common, especially in white females, SCC decreased (Travis et al., 1996; Li et al., 2001; Janssen-Heijnen and Coebergh, 2001; Harkness et al., 2002; Morita, 2002). It has been suggested that AC increase may represent, at least in part, the improved diagnostic techniques (Boffetta and Trichopoulos, 2002). This phenomenon needs further evaluation. Changes in classification and pathology techniques can account for only a fraction of this trend. The shift may also reflect changes in the type of cigarette smoked (Spitz et al., 2006). The average nicotine and tar delivery in cigarettes decreased by more than half from the 1950s through the 1990s. The tobacco in filter cigarettes, the percentage of which has increased since the 1950s, is richer in nitrates than that of nonfilter cigarettes, raising the yield of *N*-nitrosamines (Wynder and Muscat, 1995). It has been suggested that smoking of low-yield cigarettes has higher risks for AC, while smokers of high-yield cigarettes may be more likely to develop SCC of

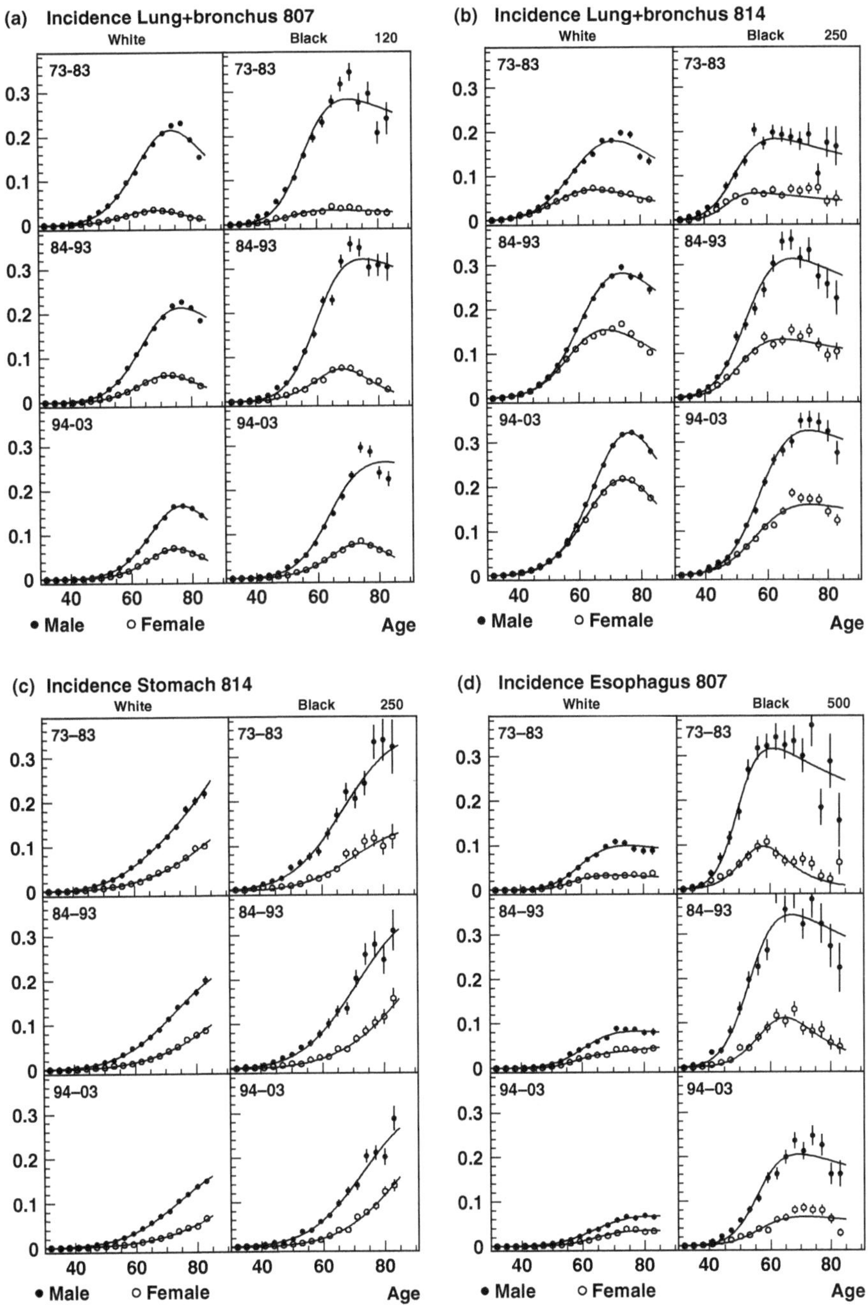


**Fig. 7.2** Age-specific incidence rates of selected cancer histotypes in *white* and *black* males and females, 1973–2003 (*points*) and the fit of the M0/M2 models (*lines*) (with parameters as in Table 7.3)

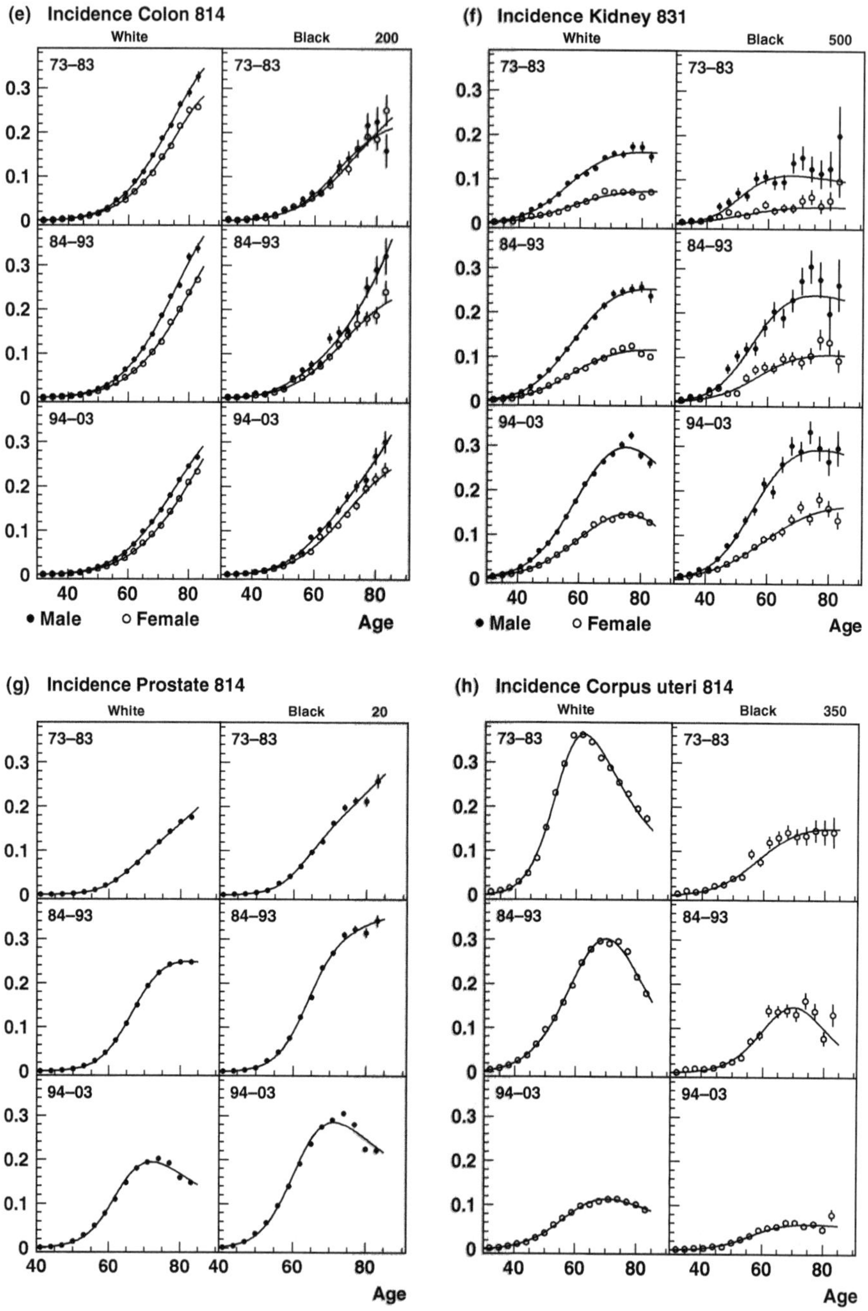


**Fig. 7.2** (Continued)

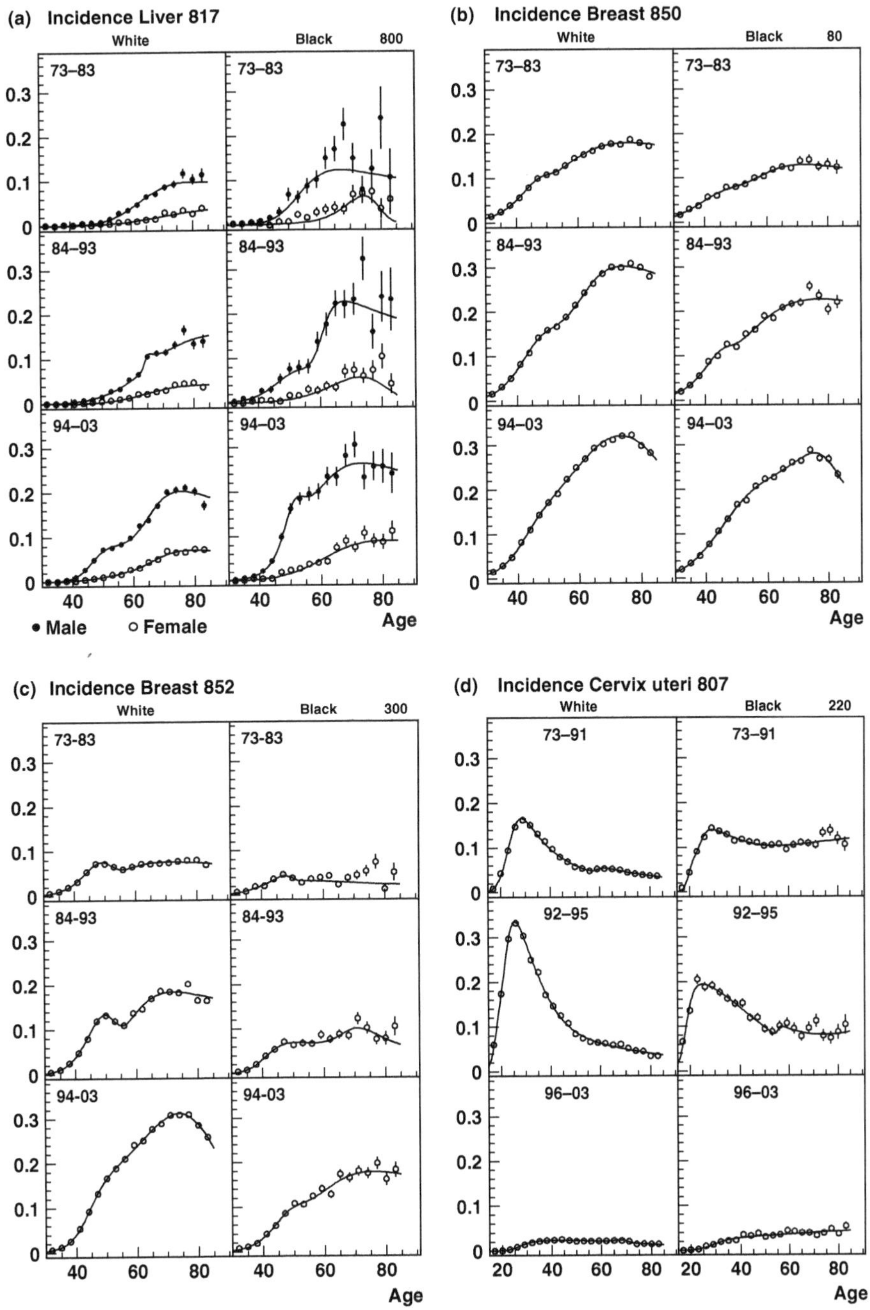


**Fig. 7.3** Age-specific incidence rates of selected cancer histotypes in *white* and *black* males and females, 1973–2003, and fit using the two-disease model

lung – the former are thought to smoke more intensely and inhale more deeply to satisfy the need for nicotine, resulting in greater exposure of bronchioalveolar regions and smaller bronchi to the organ-specific lung carcinogen, tobacco-specific nitrosamines 4-(methylnitrosamino)-1-(3-pyridyl)-1-butanone (NNK) (Stellman et al., 1997).

Recent studies show that lung cancer arising in persons younger than age 40 tends to occur more frequently among women and is characterized by a predominance of ACs (Skarin et al., 2001; Kreuzer et al., 1998, 1999; Liu et al., 2000). The reason for that gender predilection among young nonsmokers is not known. Approximately 79% of lung cancers in females were attributed to cigarette smoking 15 years ago (Surgeon General, 1989), when women smokers had a higher risk for lung cancer (Zang and Wynder, 1996; Risch et al., 1993). It is still unclear whether increased female lung cancer mortality suggests a higher gender-related risk (e.g., due to greater genetic susceptibility) or reflects changing smoking patterns (Jemel et al., 2003; Kreuzer et al., 2000; Khudar, 2001).

Each of the three major histologic subtypes of lung cancer – SCC, AC, and small-cell carcinoma – are associated with smoking and tobacco exposure (Women and smoking, 2001; Blot and Fraumeni, 1996). However, smoking-associated risk for AC is not believed to be as strong as for the two other histotypes (Lubin and Blot, 1984; Kabat, 1996; Prescott et al., 1998). Most reports come from case–control studies, sometimes with only a small amount of never smokers available for analysis (Khuder, 2001; Prescott et al., 1998). These studies mostly estimated relative, rather than excess, risk or population-attributable risk. Recently, it was suggested that AC in women might be more strongly associated with smoking than believed before: e.g., the multivariate-adjusted excess risk for women-heavy smokers, compared with never smokers, was higher for AC than for SCC and small-cell carcinoma (excess risk of 206/100,000 versus 122/100,000, and 104/100,000, respectively) (Yang et al., 2002). However, comparing relative risks as the measure of effect, a stronger association of tobacco has been found for small cell and squamous cell lung carcinomas, than for AC.

### 7.2.2 *Esophageal Cancer*

Esophageal cancer, with predominance of both histotypes in males, showed race differences in histotype prevalence: SCC was predominant in black males and females, and it had a "younger" age at manifestation and age at the highest incidence compared to whites. AC was found more often in whites. Over time, SCC incidence declined, while AC incidence rose.

In developing countries, AC is still relatively rare. In a number of developed countries, AC is equaled, or even exceeded the prevalence of SCC. In the United States, by 2000, almost 70% of esophageal carcinomas were ACs (Blot et al., 2006). Rates for SCC have recently declined, probably, related to decreases in cigarette smoking and hard alcohol consumption, while AC rates have been rising, especially among white males, likely associated with increasing obesity and gastro-esophageal reflux disease, and, probably, partly associated with the declining prevalence of *H. pylori* infection due to improvements in sanitation and antibiotic use (Brown and Devesa, 2002; Blaser, 1999; Henrik et al., 2001; Vieth et al., 2000). The rising incidence of esophageal AC may be partly related to how cancers are reported: i.e., early (pre-1970) coding rules called for tumors at the junction of the esophagus and stomach to be classified as stomach (cardia) cancers. Changes in diagnostic or recording practice could account for some of the increase in esophageal cancer incidence. However, there is a concomitant rise in the incidence of gastric cardia cancer, whereas a decline would have been expected if that was just due to a shift in classification (Blot et al., 1991). An improvement in endoscopic and imaging technology may have led to more precise histological, as well as anatomic, classification but the parallel trends for esophageal and gastric cardia ACs suggest that the increasing incidence of both tumors is real and reflects shared causal factors (Blot et al., 2006).

### *7.2.3 Stomach Cancer*

Adenocarcinoma, a malignant tumor of glandular epithelium, is the most common histotype of gastric cancer. Gastric AC had higher incidence rates in blacks (in both sexes), and it was found more often in males. Its incidence rate decreased gradually from 1973 to 2003 in both sexes and in blacks and whites. However, it is still more prevalent in blacks. Decrease in the noncardia gastric AC might be associated with decrease of the prevalence of *H. pylori* infection in the United States, improved dietary patterns (i.e., consumption of more fruits and vegetables and less salt), and reduced smoking over recent decades.

Gastric AC has been classified into two morphological subtypes (Lauren, 1965) – intestinal and diffuse, which differ by age, sex, and risk factors predominance and by their time trends (Correa and Chen, 1994; Hanai et al., 1982; Hamilton and Aaltonen, 2000). Distinct time trends for these two morphological subtypes of gastric AC have not yet been clearly described (Lundegardh et al., 1991).

### *7.2.4 Colorectal Cancer*

The predominant histotype of rectal cancer – AC – showed a higher incidence in males and higher rates in whites in both sexes. AC of the colon also was more prevalent in males, but with a considerably lower male/female ratio compared to rectal cancer. The incidences of colon AC decreased moderately in white males since the mid-1980s. In black males, it leveled after an increase at the beginning of the 1990s. Changes in incidence rates among white and black females were not substantial.

ACs are the predominant type of colorectal cancer. In most cases, they are preceded by adenomas or adenomatous polyps (Levin and Raijman, 1995; Giovannucci and Wu, 2006). Growing evidence suggests the existence of etiologic differences for different parts of the colorectum: e.g., cancer risk in proximal, distal parts of colon, and rectum may differ by association with sex, alcohol intake, physical activity, calcium intake, cholecystectomy, and genetic predisposition (Yoo et al., 1999; Lampe et al., 1993; Thune and Furberg, 2001; Knekt et al., 2000; Dornschneider et al., 1990; Wu et al., 2002; Todoroki et al., 1999; Soong et al., 2000; Lleonart et al., 1998).

### *7.2.5 Cancer of Pancreas*

AC of the pancreas had higher incidence in blacks of both sexes, than in whites. Incidence rates were higher in males than in females. The peak incidence in black men and women was about 10 and 8 years "younger", respectively.

It has been reported that the male/female ratio for the U.S. population depends on age: it is 1.54 for ages 20–54 and 1.13 for ages 75 and older (Anderson et al., 2006). Higher incidence rates in males decline with time (Muir et al., 1987; Parkin et al., 2002). That may be associated with increased smoking in women and decreased smoking in men, and with likely differences in diagnosis and treatment between sexes, as well as differences in reporting of pancreatic carcinoma over time (Anderson et al., 2006).

U.S. African-Americans have one of the highest incidence rates for pancreas cancer in the world (about 14.7 per 100,000 for males and 9.5 for females), i.e., 1.5–1.9 time higher than whites, and higher than the incidence rates in Africa (Parkin et al., 2002). This may be due to differences in risk factors (between whites and blacks, and between Afro-Americans and Africans), as well as due to different susceptibility to risk factors (between whites and blacks) (Silvermann et al., 2003).

### *7.2.6 Liver Cancer*

Hepatocellular carcinoma (HCC) has higher incidence in blacks, especially for black males. Males of both races had higher incidence, than females. Over time (1973–2003), cancer incidences rose in both sexes and both races. The largest increase was among black males.

HCC is the predominant type of liver cancer – 75–90% of all liver cancer, depending on the country (Okuda et al., 2002). It is highly correlated with age. Differences in age patterns are possibly related to the prevalence of HBV or/and HCV infection in the population and to the age at infection. It is not clear why males have higher incidence of HCC than females. It has been supposed that males may be more likely infected with HBV and HCV, smoking, consuming alcohol and have increased iron stores. The role of androgens or sex-related genetic susceptibility in higher prevalence of HCC in men is still not clear (London and McGlynn, 2006). The likelihood of infection with HBV and HCV, different patterns of exposure to other HCC risk factors, as well as differences in genetic susceptibility to risk factors may result in different incidences in whites and blacks (see Section 7.3).

### *7.2.7 Breast Cancer*

Duct and lobular carcinoma trends were analyzed. Both ductal and lobular carcinomas showed higher incidence in white women. Ductal carcinoma was more prevalent than lobular. Over 1973–2003, both types of breast carcinomas increased in both white and black women. Since the mid-1990s, breast ductal carcinoma is almost 4 times more prevalent then lobular in white females, and almost 7 times more prevalent then lobular in black females.

More than 95% of breast cancers are ACs. Breast cancers *in situ* are morphologically similar to invasive cancers. They may be limited by the ductal (i.e., duct carcinoma *in situ*) or by the lobule (i.e., lobular carcinoma *in situ*). Invasive breast cancer is a heterogeneous group of lesions, about 80% of which are infiltrating ductal carcinomas and about 10% infiltrating lobular carcinomas (Colditz et al., 2006).

Breast cancer incidence in the United States is one of the highest in the world. Increased use of mammography for screening since 1980s accounts for part of the increase (Chu et al., 1996; Miller et al., 1991). Breast cancer risk at younger ages is modestly higher in black women, while at older ages (50 + ) white women have higher incidence. More detailed discussion of breast cancer features is presented in Section 7.3.

### *7.2.8 Cancer of Uterus*

AC and endometrioid carcinoma of the corpus uteri had higher incidence in white, compared to black, females. Incidence of uterine AC was considerably higher than endometrioid carcinoma. Cases of uterine ACs were started diagnosed, and it had the highest incidence, almost 10 years earlier in both white and black women, than cases of endometrioid carcinoma. Trends demonstrated substantial decreases of uterine ACs incidence in 1973–2003, in both white and black females. White women had a rapid decrease from initially very high incidence rates; however, uterine AC is still higher among white than black females.

In the past, the incidence of uterine corpus cancer had no further specification of tumor location within the uterine body [endometrium (the inner mucosal layer), myometrium (the middle muscular layer), serosa (the external layer), and the lower portion of the uterus below the uterine body – cervix uteri]. The term "corpus uteri" has been suggested as a relatively good proximation for endometrial cancer (Cook et al., 2006).

### *7.2.9 Ovarian Cancer*

Ovarian papillary serous cystadenocarcinoma had higher incidence in white, compared to black, females. About 50% of invasive ovarian epithelial tumors are serous, 24% – endometrioid, 12% – mucinous, 8% – clear cell, and about 6% undifferentiated or other (Crum, 1999). Several studies demonstrated a significant increase of serous ovarian cancer in white females during the last 25 years, while rates of other subtypes (e.g., mucinous, papillary not otherwise specified, and other epithelial cancers) decreased. It has been suggested that differences in trends of ovarian cancer subtypes, being associated with different risk factors, may be at least partially due to changes in tumor classification (e.g., while not classified any more as "papillary NOS", tumors may demonstrate an increase in other – serous – cancers) (Mink et al., 2002).

### *7.2.10 Cancer of Cervix Uteri*

Incidences of cervical SCC in older ages (50+) are higher in black females when compared to white. Over time, "natural" trends of cervical cancer were substantially changed by screening. In general, there is a decrease in cervical SCC in both white and black females, with an increasing role of cervical AC. During active screening (1992–1995), there was a higher peak of SCCs, predominantly due to increased detection of carcinoma *in situ*. This peak was at ages 25–35 and higher in white women. Postscreening incidence decreased at older ages in both white and black females (partly due to early-stage tumors diagnosed by active screening).

Both *in situ* and invasive carcinomas were analyzed. Carcinoma *in situ* is no longer reported by SEER (since 1995) because the definitions of pre-cancer is not sufficiently uniform, e.g. carcinoma *in situ* compared to CIN3 (cervical intraepithelial neoplasia) or to CIN2 (Schiffman and Hildesheim, 2006). SCC showed predominance in black females at ages older than 40, while at younger ages, white females had higher incidence. This may be due to screening effects, when very active screening programs was applied in the beginning of the 1990s. As a consequence, cervical carcinoma *in situ* changed the "natural" history of cervical cancer incidence (see Section 7.3). The decline in the incidence of SCC may be attributed, at least partly, to the success of screening to detect preinvasive tumors.

Cervical cancer prescreening (i.e., "natural") trends and effects of screening were discussed in many publications. The high incidence of SCC *in situ* in young women might reflect a displacement of invasive carcinoma at older ages (Wang et al., 2004). If effective screening for black women was received more recently than for white women, as the delay in rising SCC *in situ* suggests, then a higher incidence of invasive SCC among black women would be expected. Although numerous surveys have reported similar Pap screening practices for white and black women (Hewitt et al., 2002; Swan et al., 2003), these rates still may be due to differences in the quality of screening and subsequent management of cases. However, race-specific differences in risk factors for SCC cannot be excluded (Wang et al., 2004; Devesa, 1984; Schairer et al., 1991). A number of recent studies demonstrated that rarer cervical ACs have been increasing, especially in young women born after the 1960s-sexual revolution (Zaino, 2002; Schwartz and Weiss, 1986).

### *7.2.11 Prostate Cancer*

AC of the prostate had higher incidence in black males. Prostate AC incidence increased in 1983–1993, with a subsequent recent decrease accompanied by the shift of peak incidence to younger ages in both races (20 and 25 years earlier in white and black men, respectively). Differences in the prevalence of prostate carcinoma by race still have no clear explanation. Differences in the prevalence of inherited predisposition and risk factors may play a role. In the early 1990s, PSA screening was used widely, so tumors could be detected at earlier stages. Incidence rates started declining from the peak in the mid-1990s to its previous rates at the beginning of 1970s. The age at diagnosis in the United States dropped slightly with the introduction of PSA screening (Platz and Giovannucci, 2006). For a discussion of prostate AC heterogeneity, see Section 7.3.

### *7.2.12 Cancer of Kidney*

Clear-cell AC of kidney does not show significant race differences in incidence. It was found more than twice often in males, than females, of both races. Its incidence increased in both races and sexes from 1973 to 2003. The most rapid increase was observed in black females. The rapid increase of renal cell cancer cannot be entirely explained by improved diagnostics, such as ultrasonography, because an increased incidence of late-stage cancer has been also observed (Mathew et al., 2002; Kosary and McLaughlin, 1993; Chow et al., 1999).

## 7.3 Analyses of Selected Cancer Histotypes for the Two-Disease Model

The age-adjusted time trends of selected cancer histotypes were studied for homogeneity of their age patterns and the quality of fit of various models. The carcinogenesis models discussed in Chapter 2 can describe the age- specific incidence of cancers, if these rates are calculated to be maximally homogeneous over all hidden characteristics of population groups and cancer diagnoses. We considered 1-year incidence rates for specific histological types of cancer sites, conditional on sex and race. To diminish the impact of new medical technologies on diagnostics, we calculated age patterns for three periods: 1973–1983, 1984–1993, and 1994–2003. Thus, 93 age patterns were obtained from the 31 sex-specific plots of 19 cancer histotypes (see Fig. 7.1), each calculated for white and black U.S. populations. Each of the 186 race-specific age patterns was fit to ten carcinogenesis models.

- M1: Armitage-Doll two-parameter model: $I_1(t) = ct^{m-1}$;
- M2–M4: models with hidden frailty with three distributions corresponding to gamma-(M2), inverse Gaussian (M3), and a new generalized frailty distribution (M4) (Manton et al., 2008). For all models, the baseline hazard was given by an Armitage-Doll (Weibull) function. These three-parameter models are

$$I_2(t) = \frac{ct^{m-1}}{1 + \gamma t^m},$$

$$I_3(t) = \frac{ct^{m-1}}{(1 + \gamma t^m)^{1/2}},$$

$$I_4(t) = \beta t^{(m-1)} \exp(-\gamma \beta t^m).$$

- M5: two-stage clonal expansion (TSCE) model is the most popular version of the Moolgavkar–Venzon–Knudson model:

$$I_5(t) = \frac{X(e^{(\gamma+2q)t} - 1)}{q(e^{(\gamma+2q)t} + 1) + \gamma},$$

- M6: gamma-model as in M2 with an additional constant as a fourth parameter:

$$I_6(t) = c_0 + \frac{ct^{m-1}}{1 + \gamma t^m}$$

- M7–M10: For completeness and diversity, we added four models, modifying the baseline hazard from the Armitage-Doll to the Gompertz in models M1–M4:

$$I_7(t) = \alpha e^{\theta t},$$

$$I_8(t) = \frac{\alpha e^{\theta t}}{1 + \gamma(e^{\theta t} - 1)},$$

$$I_9(t) = \frac{\alpha e^{\theta t}}{(1 + \gamma(e^{\theta t} - 1))^{1/2}},$$

$$I_{10}(t) = \alpha \exp(\theta t) \exp\left(-\frac{\alpha\gamma}{\theta}[\exp(\theta t) - 1]\right).$$

In total, we investigated 1860 models of sex-, race-, time period-, and histology-specific age patterns of incidence rates. For each pattern, the fit was validated, and the results obtained by different models were compared, using standard criteria for quality of the fits, such as $\chi^2$ and Fisher's criteria. NLP software from the SAS package was used for analyses. For each age pattern, we compared fits for all 10 models. Table 7.2 shows how many times each of the models was ranked first, second, etc., in the description of each age pattern.

The conclusions of these analyses can be briefly summarized. First, models with Gompertz baseline functions (M7–M10) do not work very well. In almost all cases, the models with an Armitage-Doll baseline function describe age patterns better. Second, an additional parameter in M6 did not significantly improve fit. This conclusion was made based on analyses of goodness-of-fit using $\chi^2$ and Akaike Information Criterion (AIC).[1] Therefore, in spite of the good scores for this model in Table 7.2, we do not use them in further analyses. Third, the best family of the model to describe the majority of age patterns is the three-parameter frailty models with Armitage-Doll baseline function (e.g., M2 and M4). Probably the best fits are provided by the model with the gamma-distribution of frailty, i.e., M2. Models with inverse Gaussian frailty distributions do not fit age patterns as well. Other models successfully describing patterns of specific cancer histotypes are the generalized frailty model with Armitage-Doll baseline function, M4; the classic two-parameter Armitage-

[1] Akaike's information criterion (AIC) has been developed by Hirotsugu Akaike as a measure of the goodness of fit of an estimated statistical model. It is grounded in the concept of entropy, in effect offering a relative measure of the information lost when a given model is used to describe reality and can be said to describe the tradeoff between bias and variance in model construction, or loosely speaking that of precision and complexity of the model. This is not a test of the model in the sense of hypothesis testing. Rather it is a tool for model selection. Given a dataset, several competing models may be ranked according to their AIC, with the one having the lowest AIC being the best.

**Table 7.2** Results of modeling of sex- and histotype-specific age patterns of cancer incidence rates by the models M1–M10. Numbers show how many times each of the models was ranked first (*line* marked by "1"), second (*line* marked by "2"), and so on in describing each age pattern

| Model | M1 | M2 | M3 | M4 | M5 | M6 | M7 | M8 | M9 | M10 |
|---|---|---|---|---|---|---|---|---|---|---|
| 1 | 4 | 80 | 0 | 43 | 3 | 52 | 1 | 2 | 0 | 1 |
| 2 | 3 | 63 | 1 | 15 | 11 | 74 | 2 | 16 | 0 | 1 |
| 3 | 18 | 24 | 0 | 14 | 54 | 30 | 1 | 43 | 1 | 1 |
| 4 | 38 | 4 | 5 | 11 | 49 | 11 | 2 | 58 | 8 | 0 |
| 5 | 10 | 6 | 20 | 50 | 12 | 16 | 14 | 45 | 8 | 5 |
| 6 | 58 | 2 | 25 | 10 | 18 | 1 | 15 | 2 | 8 | 47 |
| 7 | 51 | 4 | 62 | 10 | 24 | 0 | 15 | 6 | 12 | 2 |
| 8 | 4 | 2 | 52 | 14 | 13 | 1 | 19 | 10 | 62 | 9 |
| 9 | 0 | 0 | 20 | 12 | 2 | 1 | 67 | 3 | 42 | 39 |
| 10 | 0 | 1 | 1 | 7 | 0 | 0 | 50 | 1 | 45 | 81 |

Doll baseline function, M1; and two-stage clonal expansion model, M5. Finally, there are several histotypes for which no one model provides a satisfactory description, e.g., breast, hepatocellular, prostate, and cervical cancers.

Because the family of frailty models with the Armitage-Doll baseline function best describes age-specific patterns of incidence rates, we can formulate the notions of homogeneity and latent heterogeneity. By homogeneous age patterns (patterns without hidden heterogeneity), we mean age patterns well described by one of the models from the family. It corresponds to the multistage hypothesis of carcinogenesis (in the form of Armitage-Doll mutation principles), with individual predispositions modeled by a frailty distribution as in M2–M4. Properties of these models, and specifically frailty distributions, are discussed by Manton et al. (1986, 2008) and reviewed in Chapter 4.

All of the models from this family can be described by (with redefined $\gamma$):

$$I(t) = \frac{ct^{m-1}}{(1 + n\gamma t^m/m)^{\frac{1}{n}}}.$$

M2, M3, and M4 correspond to $n = 1, n = 2$, and $n \to 0$. We will refer to this four-parameter model as M0. For further analyses, we keep M0 and M2. Analytical expressions for hazard rates can be specified in terms of interpretable parameters as

$$I_0(t; c, m, \sigma, n) = \frac{(t/c)^{m-1}}{c(1 + n\sigma^2 m^{-1}(t/c)^m)^{1/n}},$$

$$I_2(t; t_{\mathrm{p}}, m, \sigma) = \frac{m(m-1)(t/t_{\mathrm{p}})^{m-1}}{\sigma^2 t_{\mathrm{p}}(1 + (m-1)(t/t_{\mathrm{p}})^m)}.$$

The first is the scale parameter of the dimension of age. This is $c$ in M0 and $t_{\mathrm{p}}$ in M2. Different symbols for this parameter in M0 and M2 are used because in M2, it has the meaning of the age of maximal hazard (i.e., risk of incidence). In M0, age of maximal hazard does not exist for $n > 1$. Therefore we kept the simplest scale parameter. For $n \to 1$, M0 tends to M2. In this case, $c$ and $t_{\mathrm{p}}$ are related as

$$c = t_{\mathrm{p}} \left( \frac{\sigma^2}{m(m-1)} \right)^{\frac{1}{m}}$$

The second corresponds to the Armitage-Doll shape parameter $m$, describing the number of possible stages from cancer initiation (e.g., by exposure to carcinogen), following its promotion and progression, and up to the moment when tumor was manifested and diagnosed (registered as a cancer case in SEER Register). Strictly speaking, this definition of $m$ does not exactly correspond to those used in Armitage-Doll, MVK, TSCE, and other related models of carcinogenesis because it includes the stage of latency. Later, when a lag is explicitly included in the model, we will have an exact correspondence. The third parameter, $\sigma$, describes standard deviation of frailty distribution. For M0, the fourth parameter, $n$, describes the shape of frailty distribution (see Fig. 4.8).

Table 7.3A, B presents the results of fittings for all 186 age patterns. Only the best fit found using M0 is presented. When the additional fourth parameter $n$ is not significantly different from 1 (corresponds to gamma model M2), the results for M2 are presented. In the latter case, the cells corresponding to $n$ are empty, and estimate of $t_p$ is used for the scale parameter.

Many cancer histotype age patterns are well described by these models. This is proved by the value of $\chi^2/d.o.f.$ and by analyses of residuals for each fit. For a proper description/fit, all residuals fluctuate randomly around zero, there are no residuals with abnormally large values, no age periods with regular (i.e., nonstochastic) behavior, no large periods with residuals of the same sign, the distribution of values of residuals is approximately normal, with zero mean and unit variance.

A poor description of the age patterns of some cancers is interpreted as due to latent heterogeneity. This is in agreement with the above analyses, results of which are shown in Table 7.3A, B. One possible explanation of the poor description can be the presence of several groups (or subcohorts/ subpopulations) in a study population. These effects are analyzed below by considering a mixture of two models. Since even in the case of these cancers, the gamma-distributed frailty model provided a better description, we used this model as the components in the mixture models. Each component might be interpreted as describing a group with a genetic predisposition and/or having age-related periods of increased susceptibility to specific risk factor exposures. We came to the same result (i.e., the same models are mixed to describe the same cancer) as formulated and solved by Manton and Stallard (1980) in their paper on two-disease models. Explicitly, the

**Table 7.3** Modeling results (i.e., fitted parameters with SE) for selected cancer histotypes model fitting, white U.S. population, for three time periods: 1973–1983, 1984–1993, and 1994–2003 (see definition of symbols used in headline in text)

| Cancer site by histotype and sex | Time period | Initial age (years) | $X^2$ / *d.o.f* | c/$t_p$ (years) | *m* | σ | *n* |
|---|---|---|---|---|---|---|---|
| (A) | | | | | | | |
| Lung and bronchus 804 Male | 1973–1983 | 30 | 2.89 | 73.4 ± 2.3 | 11.0±0.9 | 14.7±1.2 | |
| | 1984–1993 | | 1.54 | 81.2±3.6 | 10.6±0.9 | 11.2±2.2 | 0.84±0.22 |
| | 1994–1903 | | 2.02 | 85.4±3.0 | 10.7±0.7 | 11.8±1.9 | 0.75±0.21 |
| Lung and bronchus 804 Female | 1973–1983 | 30 | 2.6 | 83.0±5.3 | 10.8±1.2 | 19.2±3.3 | 0.64±0.18 |
| | 1984–1993 | | 1.93 | 82.5±4.0 | 10.9±1 | 13.9±2.2 | 0.69±0.18 |
| | 1994–1903 | | 1.24 | 88.5±3.1 | 10.1±0.7 | 11.6±1.4 | 0.48±0.18 |
| Lung and bronchus 807 Male | 1973–1983 | 30 | 3.07 | 79.0±2.3 | 9.8±0.6 | 6.4±1 | 0.66±0.25 |
| | 1984–1993 | | 1.97 | 78.4±2.3 | 10.9±0.7 | 7.5±1.3 | 0.85±0.22 |
| | 1994–1903 | | 1.36 | 83.2±1.8 | 10.9±0.5 | 7.6±1 | 0.64±0.22 |
| Lung and bronchus 807 Female | 1973–1983 | 30 | 1.12 | 95.6±7.1 | 8.6±1 | 14.2±2.8 | 0.38±0.32 |
| | 1984–1993 | | 1.93 | 88.9±4.2 | 10.3±0.9 | 11.8±2.2 | 0.51±0.29 |
| | 1994–1903 | | 1.64 | 88.9±2.9 | 11.0±0.8 | 11.2±1.5 | 0.5±0.22 |
| Lung and bronchus 814 Male | 1973–1983 | 30 | 1.82 | 85.4±5.5 | 8.6±0.9 | 10.2±2.5 | 0.76±0.33 |
| | 1984–1993 | | 1.49 | 81.1±3.4 | 9.5±0.8 | 8.8±1.6 | 0.83±0.24 |
| | 1994–1903 | | 1.27 | 85.2±2.2 | 9.4±0.5 | 7.1±1 | 0.59±0.25 |
| Lung and bronchus 814 Female | 1973–1983 | 30 | 1.63 | 88.6±9.1 | 8.3±1.2 | 17±4.1 | 0.77±0.25 |
| | 1984–1993 | | 2.25 | 82.7±4.6 | 9.2±0.9 | 11.7±2 | 0.77±0.19 |
| | 1994–1903 | | 1.51 | 88.0±2.7 | 8.8±0.5 | 8.5±1.1 | 0.58±0.22 |
| Stomach 814 Male | 1973–1983 | 30 | 1.13 | 97.5±16.5 | 8.2±1.7 | 13.8±14.6 | 2.1±1.19 |
| | 1984–1993 | | 0.78 | 104.6±12.0 | 7.4±0.7 | 7.8±2.3 | |
| | 1994–1903 | | 0.92 | 106.2±10.1 | 7.4±0.6 | 8.6±2.1 | |
| Stomach 814 Female | 1973–1983 | 30 | 1.31 | 132.6±56.3 | 6.7±1 | 6.2±7.2 | |
| | 1984–1993 | | 0.93 | 136.4±67.6 | 6.8±1 | 6.1±8.7 | |
| | 1994–1903 | | 1.45 | 592.7±0.0 | 6.4±0.4 | 0.1±0 | |

**Table 7.3** (continued)

| Cancer site by histotype and sex | Time period | Initial age (years) | $X^2$ / *d.o.f* | c/$t_p$ (years) | $m$ | $\sigma$ | $n$ |
|---|---|---|---|---|---|---|---|
| Esophagus 807 Male | 1973–1983 | 30 | 0.76 | 75.1±4.3 | 12.3±2.1 | 27.4±4.5 | |
| | 1984–1993 | | 1.13 | 77.7±5.6 | 11.1±2 | 28.1±5.4 | |
| | 1994–1903 | | 1.09 | 83.6±5.8 | 10.0±1.3 | 28.3±4.9 | |
| Esophagus 807 Female | 1973–1983 | 30 | 0.97 | 69.9±5.6 | 14.2±4.3 | 54±12.7 | |
| | 1984–1993 | | 0.94 | 82.5±19.5 | 15.7±7.6 | 64.5±42.6 | 1.16±0.19 |
| | 1994–1903 | | 1.50 | 81.3±5.7 | 12.0±2.1 | 44.1±8.4 | |
| Esophagus 814 Male | 1973–1983 | 30 | 1.18 | 91.1±51.6 | 11.5±8.6 | 89.2±111.1 | 1.28±0.37 |
| | 1984–1993 | | 1.18 | 80.5±7.3 | 9.2±1.6 | 27.3±5.9 | |
| | 1994–1903 | | 1.04 | 82.9±4.6 | 8.7±0.8 | 19.5±2.5 | |
| Esophagus 814 Female | 1973–1983 | 30 | 0.75 | 129.2±50.8 | 11.8±7.4 | 82.3±24.8 | 0±0 |
| | 1984–1993 | | 1.02 | 87.9±28.2 | 12.4±9.6 | 74.7±74 | |
| | 1994–1903 | | 0.83 | 90.4±15.9 | 8.8±2.5 | 45.6±19.5 | |
| Colon 814 Male | 1973–1983 | 30 | 1.19 | 102.0±8.1 | 8.1±0.6 | 5.8±1.2 | |
| | 1984–1993 | | 1.19 | 104.3±7.9 | 8.0±0.5 | 5.5±1.1 | |
| | 1994–1903 | | 1.48 | 105.2±6.4 | 7.8±0.4 | 6±1 | |
| Colon 814 Female | 1973–1983 | 30 | 1.3 | 103.7±3.8 | 7.4±0.5 | 4.6±0.8 | 0±0 |
| | 1984–1993 | | 0.87 | 114.5±12.5 | 7.5±0.5 | 4.9±1.5 | |
| | 1994–1903 | | 0.86 | 117.3±11.2 | 7.4±0.4 | 5±1.3 | |
| Colon 826 Male | 1973–1983 | 30 | 0.83 | 89.9±5.9 | 8.2±0.8 | 9.9±1.6 | |
| | 1984–1993 | | 1.41 | 87.2±5.4 | 8.7±0.8 | 11.6±1.7 | |
| | 1994–1903 | | 1.33 | 89.9±4.8 | 7.7±0.5 | 11.8±1.4 | |
| Colon 826 Female | 1973–1983 | 30 | 0.87 | 93.5±8.6 | 7.5±0.9 | 12.3±2.6 | |
| | 1984–1993 | | 1.24 | 104.5±24.5 | 8.1±2.1 | 20.9±21.9 | 1.79±0.91 |
| | 1994–1903 | | 1.09 | 101.5±11.0 | 6.5±0.6 | 12.5±2.9 | |
| Rectum 814 Male | 1973–1983 | 30 | 0.78 | 86.1±11.4 | 10.1±2.4 | 29.5±10.6 | |
| | 1984–1993 | | 1.11 | 90.6±11.8 | 9.0±1.7 | 22.1±7.4 | |
| | 1994–1903 | | 1.05 | 85.0±6.6 | 8.9±1.1 | 26.5±4.9 | |

**Table 7.3** (continued)

| Cancer site by histotype and sex | Time period | Initial age (years) | $X^2$ / *d.o.f* | c/$t_p$ (years) | *m* | $\sigma$ | *n* |
|---|---|---|---|---|---|---|---|
| Rectum 814 Female | 1973–1983 | 30 | 1.45 | 86.0±31.1 | 14.9±9.8 | 77.4±81.8 | 1.27±0.28 |
| | 1984–1993 | | 1.47 | 94.4±15.6 | 7.9±1.7 | 24.4±9.5 | |
| | 1994–1903 | | 0.89 | 153.6±0.3 | 6.5±0.1 | 17.5±2 | 0±0 |
| Pancreas 814 Male | 1973–1983 | 30 | 0.92 | 84.2±5.8 | 8.6±1 | 14.8±2.5 | |
| | 1984–1993 | | 1.29 | 87.5±5.9 | 8.5±0.9 | 13.7±2.2 | |
| | 1994–1903 | | 1.59 | 86.9±3.9 | 8.7±0.6 | 13.9±1.5 | |
| Pancreas 814 Female | 1973–1983 | 30 | 1.32 | 82.1±5.2 | 9.5±1.2 | 19.5±3 | |
| | 1984–1993 | | 0.91 | 86.9±5.4 | 9.2±1 | 16.1±2.5 | |
| | 1994–1903 | | 0.94 | 106.9±7.0 | 8.4±0.8 | 11.4±3.5 | 0.29±0.84 |
| Liver 817 Male | 1973–1983 | 30 | 1.57 | 80.8±8.6 | 10.9±2.5 | 31.5±9.1 | |
| | 1984–1993 | | 1.29 | 87.5±9.5 | 8.5±1.5 | 21.1±5.5 | |
| | 1994–1903 | | 3.84 | 63.4±7.7 | 16.8±4.8 | 62.6±18.7 | 1.24±0.07 |
| Liver 817 Female | 1973–1983 | 30 | 1.31 | 108.0±53.3 | 6.7±2.6 | 30±34.3 | |
| | 1984–1993 | | 1.73 | 86.9±14.5 | 9.1±2.8 | 41.6±17 | |
| | 1994–1903 | | 1.37 | 89.0±8.8 | 8.1±1.3 | 28.9±6.5 | |
| Kidney 831 Male | 1973–1983 | 30 | 1.29 | 80.2±6.4 | 7.2±0.9 | 15.6±2.6 | |
| | 1984–1993 | | 0.96 | 82.3±5.2 | 7.4±0.7 | 12.3±1.6 | |
| | 1994–1903 | | 1.41 | 98.0±6.0 | 7.0±0.5 | 9.4±2.1 | 0.66±0.42 |
| Kidney 831 Female | 1973–1983 | 30 | 1.27 | 84.0±10.2 | 6.6±1.2 | 21.4±5.1 | |
| | 1984–1993 | | 1.22 | 83.0±7.1 | 7.0±0.9 | 17.6±3 | |
| | 1994–1903 | | 1.02 | 120.6±11.6 | 6.1±0.6 | 10.8±3 | 0.26±0.75 |
| Breast 850 Female | 1973–1983 | 30 | 3.15 | 61.8±4.3 | 7.7±0.9 | 9.3±1.7 | 1.27±0.05 |
| | 1984–1993 | | 9.96 | 57.6±2.6 | 8.4±0.8 | 8.2±1.2 | 1.31±0.03 |
| | 1994–1903 | | 13.3 | 62.1±1.5 | 7.4±0.3 | 5.7±0.5 | 1.17±0.04 |
| Breast 852 Female | 1973–1983 | 30 | 2.24 | 60.4±6.5 | 12.1±2.7 | 34.1±7 | 1.1±0.04 |
| | 1984–1993 | | 5.57 | 57.4±3.9 | 13.4±2.2 | 27.2±4.4 | 1.15±0.03 |
| | 1994–1903 | | 5.9 | 64.9±2.8 | 10.4±0.9 | 15.4±1.8 | 1.16±0.03 |

**Table 7.3** (continued)

| Cancer site by histotype and sex | Time period | Initial age (years) | $X^2$ / $d.o.f$ | c/$t_p$ (years) | $m$ | $\sigma$ | $n$ |
|---|---|---|---|---|---|---|---|
| Prostate 814 | 1973–1983 | 40 | 2.36 | 71.2±1.1 | 14.5±0.8 | 5.8±0.9 | 1.38±0.11 |
| | 1984–1993 | | 3.61 | 82.2±0.6 | 14.2±0.3 | 3.6±0.1 | |
| | 1994–1903 | | 27.2 | 64.6±0.3 | 13.3±0.2 | 3.7±0.1 | 0.85±0.03 |
| Ovary 846 | 1973–1983 | 30 | 1.20 | 122.3±37.8 | 6.1±1.7 | 24±9.9 | 0.58±0.56 |
| | 1984–1993 | | 1.26 | 146±15.2 | 5.0±0.5 | 12.7±0.8 | 0±0 |
| | 1994–1903 | | 1.02 | 129.7±16.2 | 5.6±0.6 | 12.4±3.2 | 0.15±0.65 |
| Corpus uteri 814 | 1973–1983 | 30 | 3.45 | 70.2±2.3 | 10.6±0.9 | 10±0.9 | 0.75±0.08 |
| | 1984–1993 | | 1.3 | 87.9±3.8 | 7.6±0.5 | 7.6±1 | 0.41±0.24 |
| | 1994–1903 | | 1.32 | 95.7±7.1 | 7.8±0.8 | 14.4±2.7 | 0.74±0.24 |
| Corpus uteri 838 | 1973–1983 | 30 | 0.16 | 794.9±0.0 | 1.5±123.4 | 2224.1±0 | |
| | 1984–1993 | | 1.38 | 138.0±36.5 | 7.0±1.7 | 24.6±14.6 | 0.06±1.44 |
| | 1994–1903 | | 2.73 | 85.4±4.6 | 8.3±0.7 | 13.3±1.8 | 0.83±0.13 |
| Cervix uteri 807 | 1973–1991 | 15 | 6.75 | 33.2±1.2 | 11.0±0.8 | 20.9±1.2 | 0.94±0.01 |
| | 1992–1995 | | 2.10 | 29.2±1.4 | 10.0±1.1 | 13.8±1.0 | 0.89±0.02 |
| | 1996–1903 | | 2.24 | 49.6±6.6 | 10.6±1.9 | 51.2±9.1 | 1.05±0.03 |
| (B) | | | | | | | |
| Lung and bronchus 804 Male | 1973–1983 | 30 | 0.91 | 81.2±13.5 | 10±3.1 | 10.8±6.9 | 0.55±0.87 |
| | 1984–1993 | | 1.01 | 71.4±6.2 | 11.8±2.8 | 14.6±3.3 | |
| | 1994–1903 | | 1.02 | 76.4±5.7 | 11.5±2 | 15.2±3 | |
| Lung and bronchus 804 Female | 1973–1983 | 30 | 1.32 | 89.9±28.5 | 9.8±4.6 | 18.6±12.8 | 0.27±1.01 |
| | 1984–1993 | | 1.27 | 85.4±12.3 | 10.4±2.9 | 14±5.5 | 0.44±0.54 |
| | 1994–1903 | | 1.51 | 89.6±11.6 | 10.3±2.3 | 15±7 | 0.63±0.64 |
| Lung and bronchus 807 Male | 1973–1983 | 30 | 2.13 | 70.3±3.6 | 11.9±1.6 | 8.1±1.1 | |
| | 1984–1993 | | 2.07 | 75.3±3.6 | 12.1±1.4 | 7.4±1 | |
| | 1994–1903 | | 2.55 | 81.9±3.6 | 11±0.9 | 7.5±0.9 | |
| Lung and bronchus 807 Female | 1973–1983 | 30 | 0.63 | 66.2±9.9 | 10.5±4.2 | 23.1±8.2 | |
| | 1984–1993 | | 0.98 | 82.9±8.4 | 10.3±2.1 | 10.9±3.8 | 0.46±0.51 |
| | 1994–1903 | | 0.92 | 88.6±7.7 | 10.1±1.6 | 10.1±3.9 | 0.48±0.68 |

**Table 7.3** (continued)

| Cancer site by histotype and sex | Time period | Initial age (years) | $X^2$ / *d.o.f* | c/$t_p$ (years) | *m* | *σ* | *n* |
|---|---|---|---|---|---|---|---|
| Lung and bronchus 814 Male | 1973–1983 | 30 | 1.46 | 63±5.1 | 11.7±3.1 | 15.2±3.2 | |
| | 1984–1993 | | 1.08 | 68.4±4.5 | 10.7±1.7 | 10.7±1.7 | |
| | 1994–1903 | | 1.75 | 73.7±3.4 | 10.7±1.1 | 10.1±1.2 | |
| Lung and bronchus 814 Female | 1973–1983 | 30 | 1.12 | 57.2±8.4 | 14.5±7.1 | 31.2±11.7 | |
| | 1984–1993 | | 1.08 | 65.2±5.4 | 11.2±2.5 | 17.2±3.3 | |
| | 1994–1903 | | 1.67 | 74.1±4.7 | 9.5±1.3 | 13.4±2 | |
| Stomach 814 Male | 1973–1983 | 30 | 0.8 | 95.2±22.5 | 7.4±1.8 | 7±3.9 | |
| | 1984–1993 | | 1.11 | 100.5±26.2 | 7.5±1.7 | 6.7±4.4 | |
| | 1994–1903 | | 0.95 | 102.3±20.7 | 7.7±1.3 | 7.3±3.8 | |
| Stomach 814 Female | 1973–1983 | 30 | 0.99 | 97.3±34 | 8.1±3.4 | 11.4±10.1 | |
| | 1984–1993 | | 1.04 | 128.9±140 | 7.2±2.4 | 5.7±18.4 | |
| | 1994–1903 | | 1.06 | 118.2±54.5 | 8.1±1.9 | 7.1±10.2 | |
| Esophagus 807 Male | 1973–1983 | 30 | 1.17 | 61±5 | 13.7±4.4 | 18.2±4.2 | |
| | 1984–1993 | | 1.03 | 67.1±5.6 | 11.8±2.7 | 15.3±3.2 | |
| | 1994–1903 | | 1.7 | 69.4±4.8 | 12.8±2.3 | 20.4±3.8 | |
| Esophagus 807 Female | 1973–1983 | 30 | 0.97 | 76.2±17.4 | 11.6±5.1 | 23.4±11.2 | 0.59±0.45 |
| | 1984–1993 | | 0.69 | 83.6±19.1 | 11.2±4.6 | 21.1±12.7 | 0.67±0.61 |
| | 1994–1903 | | 1.53 | 71.5±8.1 | 11.3±3.3 | 33.4±9.6 | |
| Esophagus 814 Male | 1973–1983 | 30 | 0.22 | 101±1916 | 7±161.2 | 112.6±4801 | |
| | 1984–1993 | | 0.53 | 68±76.8 | 12.9±64.4 | 158.4±559.1 | |
| | 1994–1903 | | 1.14 | 115.1±31.8 | 10.4±4.7 | 33.4±10.3 | 0±0 |
| Esophagus 814 Female | 1973–1983 | 30 | 0.03 | 69.5±2359 | 2.3±0 | 422.7±19584 | |
| | 1984–1993 | | 0.15 | 119.8±0 | 2.4±0 | 142.8±0 | |
| | 1994–1903 | | 0.73 | 61.9±18.1 | 20±0 | 330.3±328.1 | |
| Colon 814 Male | 1973–1983 | 30 | 1.06 | 93.6±21.2 | 8.2±2.2 | 8.4±4.9 | |
| | 1984–1993 | | 1.15 | 74.2±36.2 | 11.1±7.9 | 28.7±65.7 | 1.91±0.86 |
| | 1994–1903 | | 0.91 | 87.2±16.9 | 8.8±2.5 | 12.2±13.7 | 1.83±1.05 |

**Table 7.3** (continued)

| Cancer site by histotype and sex | Time period | Initial age (years) | $X^2$ / $d.o.f$ | c/$t_p$ (years) | $m$ | $\sigma$ | $n$ |
|---|---|---|---|---|---|---|---|
| Colon 814 Female | 1973–1983 | 30 | 0.72 | 105.8±37.3 | 7.3±1.9 | 6.4±5.7 | |
| | 1984–1993 | | 0.77 | 98.9±21.5 | 7.4±1.6 | 7.3±3.8 | |
| | 1994–1903 | | 1.04 | 100.8±15.7 | 7.4±1 | 6.8±2.5 | |
| Colon 826 Male | 1973–1983 | 30 | 1.27 | 76.5±16.3 | 10.4±4.8 | 17.5±9.6 | |
| | 1984–1993 | | 1.05 | 80.8±14.5 | 8.9±3 | 16.6±7.1 | |
| | 1994–1903 | | 1.31 | 77.6±37.7 | 10.9±7.4 | 41.4±66.7 | 1.58±0.43 |
| Colon 826 Female | 1973–1983 | 30 | 0.81 | 86.6±35.8 | 8.1±4.4 | 17.2±16.6 | |
| | 1984–1993 | | 0.94 | 87.3±23.7 | 9.1±3.7 | 18.1±12.3 | |
| | 1994–1903 | | 1.04 | 76.1±44.6 | 12.6±11.4 | 66.2±106.1 | 1.53±0.6 |
| Rectum 814 Male | 1973–1983 | 30 | 0.77 | 94.5±16.4 | 20±0 | 44.2±44.5 | 0±0 |
| | 1984–1993 | | 1.11 | 93.8±16.4 | 14.7±7.5 | 26±8.4 | 0±0 |
| | 1994–1903 | | 1.08 | 73.5±15.1 | 10.9±5.7 | 46.3±24.7 | |
| Rectum 814 Female | 1973–1983 | 30 | 0.66 | 188.3±11320 | 10.5±11.6 | 1.5±469.3 | |
| | 1984–1993 | | 0.82 | 119.1±40 | 9.3±4.9 | 27.4±9.3 | 0±0 |
| | 1994–1903 | | 0.87 | 70.8±5.8 | 20±0 | 133.7±80 | 1.22±0.22 |
| Pancreas 814 Male | 1973–1983 | 30 | 1.13 | 73.8±10.8 | 10.6±3.7 | 16.7±6.3 | |
| | 1984–1993 | | 1.01 | 77.8±10.3 | 10±2.7 | 14.8±4.9 | |
| | 1994–1903 | | 1.59 | 78.4±7.7 | 9.9±1.8 | 15.3±3.7 | |
| Pancreas 814 Female | 1973–1983 | 30 | 1.1 | 66.6±24.2 | 18.3±19.2 | 44.8±50.4 | 1.18±0.22 |
| | 1984–1993 | | 0.97 | 85.7±15.8 | 8.8±2.7 | 13.4±5.9 | |
| | 1994–1903 | | 2.37 | 78.5±6 | 11.6±2 | 18.1±3.7 | |
| Liver 817 Male | 1973–1983 | 30 | 1.3 | 66.8±13.2 | 12.7±8.7 | 33.7±18.7 | |
| | 1984–1993 | | 0.83 | 81.2±18.8 | 7.5±2.8 | 16.8±8.3 | |
| | 1994–1903 | | 0.92 | 65.5±12.4 | 14.3±6.4 | 31.3±15.4 | 1.1±0.11 |
| Liver 817 Female | 1973–1983 | 30 | 0.88 | 107±41.1 | 11.8±9.5 | 28.3±11.9 | 0±0 |
| | 1984–1993 | | 1 | 122.1±38.4 | 8.5±3.8 | 24.5±7.7 | 0±0 |
| | 1994–1903 | | 1.21 | 81.9±17.2 | 8.7±3.1 | 28.8±13.8 | |

**Table 7.3** (continued)

| Cancer site by histotype and sex | Time period | Initial age (years) | $X^2$ / $d.o.f$ | c/$t_p$ (years) | $m$ | $\sigma$ | $n$ |
|---|---|---|---|---|---|---|---|
| Kidney 831 Male | 1973–1983 | 30 | 0.98 | 67.5±17.6 | 9.2±5.9 | 23.5±14 | |
| | 1984–1993 | | 1.11 | 75.2±12 | 8.3±2.2 | 14.2±4.9 | |
| | 1994–1903 | | 1.13 | 76.9±7.8 | 7.9±1.3 | 12.3±2.7 | |
| Kidney 831 Female | 1973–1983 | 30 | 0.95 | 78±61.4 | 6±6 | 28.9±41.4 | |
| | 1984–1993 | | 1.47 | 79.8±20.3 | 7.5±3 | 19.6±10.3 | |
| | 1994–1903 | | 0.94 | 89.4±19.1 | 6.1±1.5 | 13.1±5.3 | |
| Breast 850 Female | 1973–1983 | 30 | 1.2 | 68.3±32.2 | 6.3±4 | 9.5±10.2 | 1.32±0.32 |
| | 1984–1993 | | 1.72 | 58.6±10.7 | 7.7±2.5 | 8.9±4.4 | 1.31±0.11 |
| | 1994–1903 | | 1.82 | 63.9±4.9 | 6.8±0.9 | 5.8±1.4 | 1.16±0.15 |
| Breast 852 Female | 1973–1983 | 30 | 1.36 | 51±11.1 | 10±8.3 | 38.9±21 | |
| | 1984–1993 | | 1.08 | 55.9±16.9 | 14.7±10.8 | 39.3±25.6 | 1.11±0.1 |
| | 1994–1903 | | 1.52 | 65.5±11.3 | 10.6±3.3 | 23±9.7 | 1.21±0.1 |
| Prostate 814 | 1973–1983 | 40 | 1.66 | 66.6±2.4 | 15.4±2.5 | 5.4±2 | 1.3±0.19 |
| | 1984–1993 | | 1.59 | 64.4±1.1 | 14.9±1.3 | 3.7±0.7 | 1.1±0.13 |
| | 1994–1903 | | 5.08 | 61.3±0.6 | 12.7±0.6 | 3±0.3 | 0.88±0.08 |
| Ovary 846 | 1973–1983 | 30 | 1.05 | 198.5±228.6 | 4.4±3.5 | 26.6±9.7 | 0±0 |
| | 1984–1993 | | 1.64 | 145.3±0.3 | 6.3±0.6 | 19.6±5.8 | 0±0 |
| | 1994–1903 | | 1.12 | 155.8±43.7 | 5.3±1.3 | 16.9±2.7 | 0±0 |
| Corpus uteri 814 | 1973–1983 | 30 | 0.97 | 82±19.5 | 7.6±3.2 | 13.7±7.1 | |
| | 1984–1993 | | 1.29 | 91±17.1 | 9.1±2.9 | 12.1±6.4 | 0.45±0.86 |
| | 1994–1903 | | 1.16 | 75.2±11.5 | 9±3.4 | 25.6±8.9 | |
| Corpus uteri 838 | 1973–1983 | 30 | 0 | 59.6±0 | 1±0.1 | 15.7±0 | |
| | 1984–1993 | | 0.73 | 101.6±36.6 | 13.1±10.6 | 42.3±18 | 0±0 |
| | 1994–1903 | | 1.4 | 107.2±21 | 7.2±1.8 | 11.5±4.9 | 0.05±0.97 |
| Cervix uteri 807 | 1973–1991 | 15 | 1.69 | 28.9±3.0 | 14.6±3.6 | 31.6±6.4 | 1.05±0.02 |
| | 1992–1995 | | 1.30 | 24.6±2.3 | 10.5±4.0 | 20.5±4.4 | |
| | 1996–1903 | | 1.43 | 51.5±24.2 | 10.2±6.4 | 51.9±36.5 | 1.18±0.11 |

mixture for hazards is presented in the form of the sum for conditional hazards:

$$I_2^{\text{mix}}(t) = \frac{wS(t;t_{\text{p1}},m_1,\sigma_1)}{S(t)} I_2(t;t_{\text{p1}},m_1,\sigma_1) + \frac{(1-w)S(t;t_{\text{p2}},m_2,\sigma_2)}{S(t)} I_2(t;t_{\text{p2}},m_2,\sigma_2);$$
$$S(t) = wS(t;t_{\text{p1}},m_1,\sigma_1) + (1-w)S(t;t_{\text{p2}},m_2,\sigma_2);$$

Moreover, we will consider the mixture model based on M0. In this case, the formula is appropriately generalized.

Our data did not show significant differences in the number of $m$ stages between races, sexes, and three time periods for every cancer histotype analyzed – it stayed constant and depended only on cancer site and histotype. That provided us with an opportunity to analyze the total cancer incidence for every cancer site and histotype, summarizing all cases observed for every cancer histotype for whites and blacks, males and females, for all, over the 30-year period. Table 7.4 shows the main characteristics obtained.

We calculated differences in the number of $m$ stages (as defined above) between various cancer sites/histotypes. Results are shown in Table 7.5. For lung cancer, there was a difference in the number of $m$ stages between SCC and

**Table 7.4** Results of model fitting (presented as fitted parameters ± SE) for selected cancer histotypes in the U.S. population in 1973–2003: parameters are summarized for both sexes and both races (see descriptions of symbols used in headline in text)

| Cancer site and histotype | Initial age (years) | $X^2$/ *d.o.f* | *c* (years) | *M* | $\sigma$ | *n* |
|---|---|---|---|---|---|---|
| Lung and bronchus 804 | 30 | 3.34 | 86.8±1.5 | 10±0.3 | 12.2±0.8 | 0.63±0.09 |
| Lung and bronchus 807 | 30 | 2.66 | 85.3±1 | 10±0.2 | 8.6±0.5 | 0.55±0.09 |
| Lung and bronchus 814 | 30 | 4.51 | 87.4±1.4 | 8.6±0.2 | 9±0.6 | 0.69±0.11 |
| Stomach 814 | 30 | 1.55 | 120.3±2.8 | 6.9±0.2 | 7.4±0.6 | 0±0 |
| Esophagus 807 | 30 | 1.52 | 97±5.2 | 10.1±0.8 | 29.6±4.6 | 0.91±0.16 |
| Esophagus 814 | 30 | 1.39 | 115.3±13.7 | 8.6±1.1 | 36.6±13.3 | 1.11±0.41 |
| Colon 814 | 30 | 2.03 | 103.1±2.4 | 7.5±0.2 | 5.1±1.4 | 0.45±1.39 |
| Colon 826 | 30 | 1.22 | 104.5±4.6 | 7.7±0.4 | 13.5±2.5 | 1.11±0.31 |
| Rectum 814 | 30 | 1.53 | 129.8±13.5 | 7.6±0.8 | 22.1±9.7 | 0.72±1.09 |
| Pancreas 814 | 30 | 1.92 | 103.6±3.5 | 8.2±0.4 | 12.3±1.9 | 0.63±0.33 |
| Liver 817 | 30 | 5.38 | 114.6±12.2 | 7.4±0.8 | 22.1±7.5 | 0.97±0.47 |
| Kidney 831 | 30 | 2.05 | 108.9±4.9 | 6.6±0.3 | 11.9±1.7 | 0.66±0.28 |
| Breast 850 | 30 | 22.32 | 59.4±1.2 | 8±0.3 | 7.7±0.5 | 1.27±0.02 |
| Breast 852 | 30 | 8.8 | 61±2.1 | 11.8±0.9 | 23.3±1.9 | 1.18±0.02 |
| Prostate 814 | 40 | 38.78 | 66.9±0.2 | 12.8±0.2 | 3.7±0.1 | 0.95±0.03 |
| Ovary 846 | 30 | 1.65 | 142.1±16.3 | 5.3±0.5 | 13.9±3.2 | 0.15±0.58 |
| Corpus uteri 814 | 30 | 4.82 | 84.3±2.2 | 8.5±0.4 | 11.2±0.8 | 0.65±0.08 |
| Corpus uteri 838 | 30 | 2.41 | 91.5±6 | 8.2±0.7 | 21.2±3.1 | 0.99±0.11 |
| Cervix uteri 807 | 15 | 15.05 | 32.6±0.9 | 10.9±0.6 | 24.6±1 | 0.96±0.01 |

Table 7.5 Differences in number of $m$ stages between various cancers/histotypes, 1973–2003

| Cancer histotype | Lung&brochus 804 | Lung&bronchus 807 | Lung&bronchus 814 | Stomach 814 | Esophagus 807 | Esophagus 814 | Colon 814 | Colon 826 | Rectum 814 | Pancreas 814 | Liver 817 | Kidney 831 | Breast 850 | Breast 852 | Prostate 814 | Ovary 846 | Corpus uteri 814 | Corpus uteri 838 | Cervix uteri 807 |
|---|---|---|---|---|---|---|---|---|---|---|---|---|---|---|---|---|---|---|---|
| Lung&bronch 804 | | | 1.4 | 3.1 | | 1.5 | 2.6 | 2.4 | 2.4 | 1.8 | 2.6 | 3.4 | 2.0 | 1.8 | 2.7 | 4.7 | 1.5 | 1.8 | * |
| Lung&bronch 807 | | | 1.4 | 3.1 | | 1.5 | 2.5 | 2.4 | 2.4 | 1.8 | 2.6 | 3.4 | 2.0 | 1.8 | 2.8 | 4.7 | 1.5 | 1.8 | * |
| Lung&bronch 814 | | | | 1.7 | 1.5 | | 1.1 | 0.9 | * | | 1.2 | 2 | 0.6 | 3.2 | 4.2 | 3.3 | | | 2.3 |
| Stomach 814 | | | | | 3.2 | 1.7 | 0.6 | 0.8 | | 1.4 | | | 1.1 | 4.9 | 5.9 | 1.6 | 1.6 | 1.3 | 4.0 |
| Esophagus 807 | | | | | | * | 2.6 | 2.5 | 2.5 | 1.9 | 2.7 | 3.5 | 2.1 | 1.7 | 2.6 | 4.8 | 1.6 | 1.9 | |
| Esophagus 814 | | | | | | | * | | | | * | 1.9 | | 3.2 | 4.2 | 3.3 | | | 2.3 |
| Colon 814 | | | | | | | | | | 0.8 | | 0.9 | * | 4.3 | 5.3 | 2.2 | 1.0 | * | 3.4 |
| Colon 826 | | | | | | | | | | * | | 1.0 | | 4.1 | 5.1 | 2.4 | 0.8 | | 3.2 |
| Rectum 814 | | | | | | | | | | | | * | | 4.2 | 5.2 | 2.3 | * | | 3.2 |
| Pancreas 814 | | | | | | | | | | | * | 1.6 | | 3.6 | 4.5 | 3 | | | 2.6 |

**Table 7.5** (continued)

| Cancer histotype | Lung&brochus 804 | Lung&bronchus 807 | Lung&bronchus 814 | Stomach 814 | Esophagus 807 | Esophagus 814 | Colon 814 | Colon 826 | Rectum 814 | Pancreas 814 | Liver 817 | Kidney 831 | Breast 850 | Breast 852 | Prostate 814 | Ovary 846 | Corpus uteri 814 | Corpus uteri 838 | Cervix uteri 807 |
|---|---|---|---|---|---|---|---|---|---|---|---|---|---|---|---|---|---|---|---|
| **Liver 817** | | | | | | | | | | | | | | 4.4 | 5.4 | 2.1 | * | * | 3.5 |
| **Kidney 831** | | | | | | | | | | | | | 1.4 | 5.2 | 6.2 | 1.3 | 1.9 | 1.6 | 4.2 |
| **Breast 850** | | | | | | | | | | | | | | 3.8 | 4.8 | 2.7 | * | | 2.9 |
| **Breast 852** | | | | | | | | | | | | | | | * | 6.5 | 3.3 | 3.6 | * |
| **Prostate 814** | | | | | | | | | | | | | | | | 7.5 | 4.3 | 4.6 | 1.9 |
| **Ovary 846** | | | | | | | | | | | | | | | | | 3.2 | 2.9 | 5.6 |
| **Corpus uteri 814** | | | | | | | | | | | | | | | | | | | 2.3 |
| **Corpus uteri 838** | | | | | | | | | | | | | | | | | | | 2.6 |
| **Cervix uteri 807** | | | | | | | | | | | | | | | | | | | |

Values represent the number of stages m from the initiation of carcinogenesis (e.g., by exposure to carcinogen) and up to clinical manifestation of tumor (as when it has been registered in SEER Registry as cancer case ).
Numbers represent the Δ – the difference between number of *m* stages; *dark-grey boxes* show no differences observed between numbers of *m* stages (i.e., differences of estimated values are less then SE); * in *light-grey boxes* show the tendency for having differences in number of *m* stages, but not strong enough to reach the proved value of statistical significance.

small-cell carcinoma lung cancers, and lung AC: the lung AC had 1.4 stages fewer compared with two other lung cancer histotypes. Our results (see section 7.2.1.), and results of other studies showed that lung AC differs from lung SCC and small-cell lung carcinoma by sex prevalence (it is more frequent in women), age prevalence (it is occurs at younger ages), and, probably, by the contribution of certain risk factors (smoking related and not related to smoking). It is possible that lung AC might in certain extent differ by its carcinogenesis pathway from SCC and small-cell carcinoma of lung.

Differences were also found between duct and lobular breast carcinomas: ductal carcinoma had 3.8 fewer stages than lobular carcinoma (taking into account the possible coding misclassifications in infiltrative form, such as infiltrating duct and lobular). That might suggest differences in carcinogenesis pathways between two tumor types. It has been shown in several studies that these two cancer types may, *in situ*, differ by their risks of development of invasive breast carcinomas (Schnitt and Morrow, 1999; Collins et al., 2005; Going, 2003) and by their age patterns (Li et al., 2005). Genetic predisposition to duct and lobular carcinomas is still unclear and needs further study, as well as associated with these lesions the risk of invasive breast carcinoma.

The estimates of $m$ had no, or small, differences between ACs of various sites (from 0 stages to a maximum of 2), such as lung, stomach, esophagus, colon, rectum, pancreas, kidney, corpus uteri, and breast ductal cancers. The AC of prostate, ovarian papillary serous cystadenocarcinoma, and breast lobular AC had more stages compared to other ACs, and differed by the number of stages ($m$) between each other. The SCCs of lung, esophagus, and cervix showed no differences in the number of stages, but showed a higher number of stages compared to ACs – with a difference from 1.5 to 4.8 stages (exceptions were breast lobular and prostate AC; this finding needs further study).

The estimates of $m$ may be compared over organs and systems of organs. The difference in number of stages between esophageal SCC and esophageal AC was fewer than the difference between esophageal SCC and other, nonesophageal, ACs, or between esophageal AC and other, nonesophageal SCCs. A similar tendency was observed for lung AC and SCC: differences between lung AC and SCC were smaller, than between lung SCC and nonlung ACs, and than between lung AC and nonlung SCCs. That might be, at least partly, because not only the same histotypes (e.g., SCC of different organs, such as lung and cervix uteri) might have some shared features related histotypically to carcinogenesis, but when histotypically different tumors (e.g., SCCs and ACs) develop in the same organ/cancer site (not on individual, but on populational level), the "shared" organ might play specific role in defining carcinogenesis.

The absence of or only small differences were observed between $m$ for stomach, colon, and rectum cancers and cancers of the parenchymal organs of the gastrointestinal tract, such as liver and pancreas. Being ACs by their histomorphological characteristics, however, these tumors had the smallest differences in the number of $m$ stages between each other compared with all other ACs analyzed. It might be supposed that not only cell-originated types (e.g., AC originates in the epithelial cells of glandular tissue, and SCC originates from the squamous cells which are found in the tissue that forms the surface of the skin, the lining of the hollow organs of the body, and the passages of the respiratory and digestive tracts), and not only the site by organ of tumor origination (as described in previous paragraph for lung and esophageal cancers) may be associated with determining the number of $m$ stages from tumor initiation to its manifestation, but some other tumor origin characteristics may also play roles. From the embryogenetic perspective, liver and pancreas developed as evaginations of the endoderm of the primitive gut and retained connections to the digestive tube by the way of ducts, such as pancreatic, hepatic, and common bile ducts. It is possible that organs which share embryogenesis might share (at least partly) certain features of their carcinogenesis. This "embryogenetic" feature may contribute to "histotype" and "organ" features (described above) by determining carcinogenesis stages. To prove this hypothesis and to estimate the degree of the possible contribution of every specific feature in carcinogenesis, it requires detailed analyses.

Cancers of reproductive organs demonstrated very nonhomogenous patterns of $m$, with the most stages for prostate cancer (see Table 7.4). It should be considered that some cancers are registered in SEER as *in situ*, increasing due to active screening, e.g., prostate cancer, cervical cancer, breast cancer, etc. (analysis of possible components, including the number of stages of tumorigenesis related to cancer stages, is the next step in developing our approach). Other factors are also of importance when analyzing and modeling incidence, such as hysterectomy prevalence for cervical and uterine cancers analyses.

*Estimation of lag period.* We analyzed the model fit with a lag for selected cancer histotypes generalizing M0 as

$$I_0(t; c, m, \sigma, n, l) = \frac{(t-l)^{m-1}}{c^m\left(1 + n\sigma^2 m^{-1} c^{-m} (t-l)^m\right)^{1/n}},$$

where $l$ is the lag in years. The results of model fitting are shown in Table 7.6. The parameter $m$ changes its biological meaning after the implementation of a lag: now $m$ includes stages since tumor initiation to the moment the first cancer cell appearance. The shortest lag among the analyzed cancers was for cervical cancer (9.5 years). The longest lag – 32.3 years – was for prostate cancer. Lags for other cancers were around 15–25 years. Most of the cancers in Table 7.6 are invasive. Because of active screening strategies, cancers of prostate, breast, and cervix include not only invasive, but also considerable proportion of *in situ*, cancers for certain periods of observation (e.g., absence of registered in SEER Registry cervical carcinomas *in situ* after 1992-1995 active screening period, but not before 1995). To obtain results specific to invasive cancers, separate analysis of invasive tumors is required. One can expect the quality of fit, represented by $\chi^2$/d.o.f., to improve for these histotypes.

To investigate nonhomogeneous cervical, liver, prostate, and breast cancers, we used a two-disease model. Statistical analyses based on the two-disease models demonstrated an improved quality of fit for cancers not described by a single-component model. The results of the fit for breast ductal carcinoma, breast lobular carcinoma, hepatocellular carcinoma, prostate AC, and cervical

**Table 7.6** Results of model fitting with estimated lag period for selected cancer histotypes in the U.S. population, 1973–2003, summarized for both sexes and both races (see descriptions of symbols used in headline in text)

| Cancer | $\chi^2/d.o.f$ | $C$ (years) | $m$ | $\sigma$ | $N$ | Lag (years) |
|---|---|---|---|---|---|---|
| Lung 804 | 1.43 | 79.6±2.2 | 6.2±1.1 | 9.9±1.1 | 0.35±0.2 | 19.8±5.8 |
| Lung 807 | 1.3 | 76.8±2.7 | 7.2±1.1 | 7.5±0.7 | 0.3±0.2 | 15.3±6.2 |
| Lung 814 | 1.31 | 83.3±2.4 | 5.3±0.7 | 6.7±0.7 | 0.2±0.3 | 19.3±4.0 |
| Pancreas 814 | 1.45 | 103.2±5.8 | 5.9±1.5 | 9.6±2.5 | 0.1±0.8 | 14.8±9.4 |
| Liver 817 | 4.42 | 147.6±0.7 | 4.1±0.2 | 13.1±0.4 | 0 | 21.2±2.3 |
| Kidney 831 | 1.57 | 118.6±15.5 | 4.5±1.2 | 8.9±2.5 | 0.1±0.8 | 15.4±8.5 |
| Breast 850 | 19.19 | 63.7±9.6 | 3.1±0.6 | 3.8±1.3 | 1.3±0.3 | 24.2±2.9 |
| Prostate 814 | 9.27 | 42.6±0.72 | 5.3±0.3 | 2.4±0.1 | 0.5±0.1 | 32.3±1.4 |
| Cervix uteri 807 | 13.78 | 29.3±1.9 | 5.8±1.7 | 21.6±1.9 | 1.0±0.01 | 9.5±3.2 |

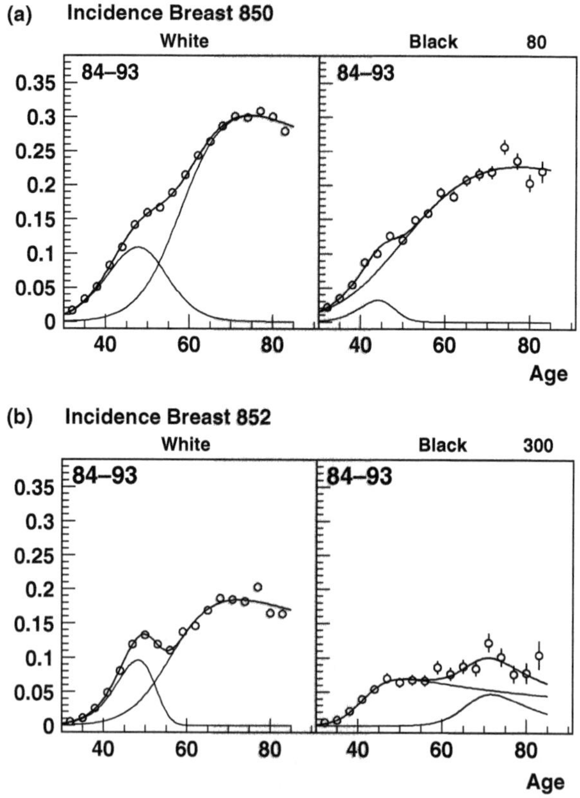


**Fig. 7.4** Age-specific incidence rates of nonhomogeneous cancer histotypes for selected time intervals. *Lines* correspond to the total two-disease model and its two components

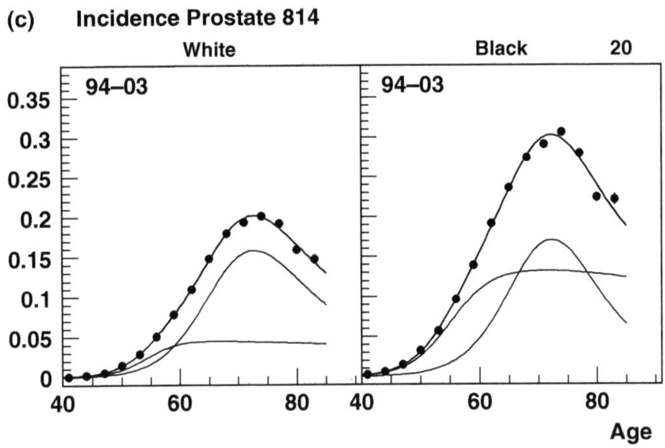


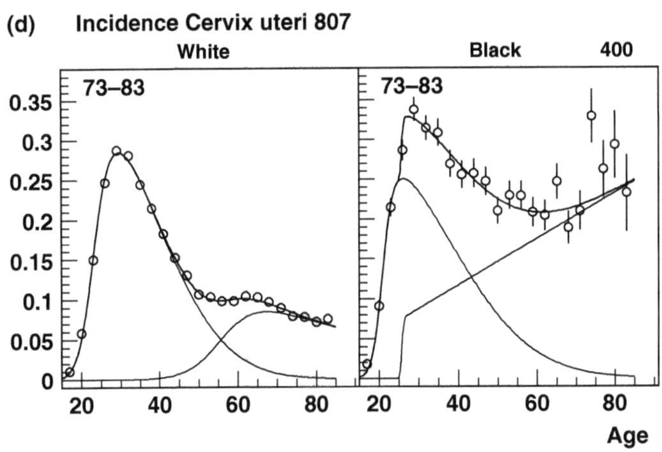


**Fig. 7.4** (continued)

SCC are shown in Figs. 7.3 and 7.4 (the latter also shows two components of the models). Below we discuss the results for specific cancer histotypes and suppose the interpretations of components of the two-disease model.

## *7.3.1 Breast Cancer: If Genetic Background May Result in Two Forms of Disease?*

At the initiation of menses, the growth and branching of the ducts in the mammary gland increases, and the terminal end buds of the ducts start giving rise to lobules. Lobule formation continues from menarche to age 25 (on average). It is supposed that only through pregnancy and lactation does complete differentiation of lobular tissue take place (Colditz et al., 2006).

More than 95% of breast cancers originate from the epithelial elements of the mammary gland and are ACs. Invasive cancers are a heterogeneous group of lesions characterized by tumor cells that invade breast stroma. Two breast cancer forms – duct and lobular carcinomas – are the most common, with a significant predominance of ductal carcinoma (about 80–85%); infiltrating lobular carcinoma represents around 5–10% of cases (Azzopardi et al., 1982). Other forms like mucinous, medullary, papillary, tubular, adenoid cystic carcinomas, are rare. In general clinical practice, the distinction between different histologic types has limited implications for either diagnosis or treatment (Hankinson and Hunter, 2002). Duct and lobular carcinomas *in situ* may differ by the average age at diagnosis (late fifties versus late forties), relation to menopausal status (70% postmenopausal versus 70% premenopausal), risks of subsequent carcinoma (30–50% at 10–18 years versus 25–30% at 15–20 years), and sites of subsequent breast carcinoma (99% in the same breast versus 50–60%) (Page et al., 1995). Studies of age-specific breast cancer incidence rates in women of Los Angeles County (1972–1998) showed that both duct and lobular carcinomas *in situ* increased from age 40, with a large predominance of ductal over lobular carcinoma. The highest incidence at age 75 was for duct carcinoma and at age 50 for lobular carcinoma (Bernstein, 2002).

Invasive ductal carcinoma has no specific histological features. Invasive lobular carcinoma is characterized microscopically by single-filing of small, regular epithelial cells that grow around ducts and lobules (Abeloff et al., 2004). Invasive lobular carcinoma *in situ*, which is a microscopic diagnosis, is prevalent in premenopausal women, and it is very rare in women older than 75 (Haagensen et al., 1978; Swain, 1989; Page and Jopaze, 1991). No association has been noted with use of exogenous estrogen (Abeloff et al., 2004).

Modeling of breast cancer started more than 25 years ago. Pike and colleagues (1983) proposed a model of tissue aging, to account for the relation between reproductive risk factors and breast cancer incidence. This model was based on the observed age-incidence curve and the known relations of ages at menarche, first birth, and menopause to the risk of breast cancer. The original Pike model, based on work by Moolgavkar and Knudson (1981), however, did not include terms for the second or subsequent births and for the spacing of pregnancies, nor did it easily accommodate pregnancies after 40 years of age (Colditz et al., 2006). An extension of the Pike model added a term to summarize birth spacing (Rosner et al., 1994; Colditz and Rosner, 2000). In the original Pike model, factors associated with reduced risk of breast cancer were considered to slow the rate of "breast tissue aging", which correlates with the accumulation of molecular damage in the pathway to breast cancer. In the extended model, the rate of tissue aging was highest between menarche and first birth. This is in accordance with hypothesis about the period of the highest vulnerability of breast to mutagenesis.

Rosner and Colditz expanded the Pike model of breast cancer incidence to include additional reproductive events: subsequent births after the first, type of menopause in addition to age at menopause, and the premenarche period

(Rosner et al., 1994; Rosner and Colditz, 1996). A further approach by Rosner was to follow Nunney (1999), who assumes that the number of cell divisions and incidence at time $t$ is proportional to the number of breast cell divisions accumulated up to age $t$, or Pikes "breast tissue age".

Parmigiani and colleagues (1998) used a Bayesian model to evaluate the probabilities that women is a carrier of a mutation of BRCA1 and BRCA2, based on breast and ovarian cancer history of first- and second-degree relatives as predictors. Efforts to combine both lifestyle and genetic factors have been limited, partly due to the divergent mathematical backgrounds of the approaches in the two areas (Colditz and Rosner, 2006).

Manton and Stallard (1980) used an approach that differed from those described above, by fitting the complex U.S. female breast cancer mortality curves (as described in Chapter 4). They described two components with different latencies: "early" breast cancer had a shorter latency of about 7 years, compared to a "later" one – about 20 years. The familial, "early", form seems to involve predetermined mutations passed through the germ cell line, with little dependence on duration of estrogen exposure, likely because of the early age of tumor initiation. This form of breast cancer might be very aggressive and rapidly proliferating.

Our study showed a high level of nonhomogeneity in both duct and lobular carcinomas of the breast. This nonhomogeneity was more prominent in white women compared to black. In incidence curves, both cancers demonstrated "steps" at pre-menopausal ages of 45–55 (see Fig. 7.3). These "steps" were registered for the 1973–1983 period, when mammography was not widely used, and most cancer cases at diagnosis were infiltrative or invasive rather than *in situ*. These "steps" did not disappear after the wide adoption of screening, when the number of early diagnosed cases increased, resulting in the decrease of the number of advanced tumors in older age groups. The two-disease model had better fits for both cancer subtypes, in both white and black females (see Fig. 7.4). It supposed the presence of at least two processes related to breast carcinogenesis. One might be characterized by early disease onset (peak around 50–55 years). It involves a smaller fraction of the female population, and it is "sharper" (more defined) in white, than in black, females. The second process is slower developing (with a peak around age 70). It involves the majority of females (with the exception of lobular cancer in black females, where the two processes have almost the same prevalence). One of the causes associated with breast cancer nonhomogeneity might be the influence of components of inherited susceptibility.

The differences in disease epidemiology may indicate a difference in the interaction between genetic and environmental carcinogenic factors across populations. Age, race, reproductive history, diet, and use of oral contraceptives have been identified as breast cancer risk factors. Patients with early breast cancer onset had higher prevalence of mutation carriers. As inherited cancers tend to develop earlier than sporadic ones, BRCA1 mutations were found in up to 15% of patients with early cancer onsets (Turchetti et al., 2000; FitzGerald et al., 1996).

Genetic predisposition to breast cancer is conferred by two categories of genes. A minority of patients inherit mutations in high-penetrance genes, which carry a high (up to 80%) lifetime risk of breast cancer (Claus et al., 1998). BRCA1 and BRCA2 genes belong to this category. Carriers of these mutations have more than a 10-fold higher risk of breast cancer. Mutations in BRCA1/BRCA2 are inherited in most cases of familial breast cancer. These tumors tend to have an early onset. The proportions of breast cancers due to BRCA1 and BRCA2 mutations are very different in different populations (Neuhausen, 1999). BRCA2 tumors tend to be estrogen receptor positive, whereas BRCA1 tumors are mostly estrogen receptor negative and highly proliferative. But not everyone who inherits a mutation in BRCA genes develops cancer. Other genetic and environmental factors may affect penetrance. Some of these factors may be modifying genes inherited by different populations at different rates: i.e., hormonal/reproductive factors, response to DNA damage risks, and smoking. Several studies indicated that smoking might reduce breast cancer risk in BRCA carriers. This finding, however, does not suggest that these women should smoke, but that there are issues of carcinogen metabolism and other enzyme activity to be considered (Wilson et al., 2002). Several other high penetrance cancer predisposition genes also may increase breast cancer risk, such as p53 (Li-Fraumeni syndrome), gene-expressing protein phosphatase and tensin homolog (PTEN, in Cowden syndrome), ataxia-teleangiectasia mutated gene (ATM, in ataxia-teleangiectasia syndrome).

Other genes have a lower penetrance, so they are believed to have a modest increase in cancer risk (Kelsey and Wiencke, 1998). These genes encode proteins responsible for degradation or activation of carcinogens. Carriers of these mutations have 2–3-fold higher risk of developing breast cancer. These alleles are prevalent; therefore, they are likely involved in the pathogenesis of breast cancer in a larger proportion of patients than the high-penetrance genes. Mutations in genes expressing enzyme glutathione-*S*-transferase, as well as CYP1A and CYP17, are associated with a 2–3-fold increase of breast cancer risk (Feigelson et al., 1997; Helzlsouer et al., 1998).

The reasons of breast cancer nonhomogeneity are still under discussion. Different mechanisms were proposed, but the possibility of existence of two forms of breast cancer is accepted in several studies. It is under discussion, if inherited mutations in BRCA1 are associated with "early" form of disease, or these are environmental carcinogens leading to increased rates of carcinoma not through mutagenic effect, but rather through mitogenic properties and support of the telomere crisis hypothesis of epithelial carcinogenesis (Frieboes and Brody, 2005). Future studies are required comparing carcinogenesis pathways and cancer "hallmarks" involved in each of breast cancer forms, and the roles of genetic and epigenetic mechanisms in both forms initiation, promotion, and progression (including metastatic potential).

Variation in transcriptional programs accounts for much of the biological diversity in human cancers, including breast carcinoma, which is diverse in its natural history and responsiveness to therapy (Tavassoli and Schnitt, 1992). Recently, it has been suggested to classify breast carcinoma into subtypes, distinguished by pervasive differences in their gene expression patterns (Perou et al., 2000). With an increasing ability to incorporate breast cancer risk factors into prediction models, it will be necessary to refine the ability to relate risk to individual women, identifying those who will benefit from traditional chemo-preventive measures and those who will not.

### 7.3.2 Cervical Cancer: Age Periods of Increased Susceptibility and Cancer Risk

Analyzing trends of cervical cancer incidence in white and black female U.S. populations from 1975 to 2005, based on SEER data, two peaks were observed in cervical SCC incidence (first, aged 25–30 years, and second, less prominent, aged 60–65 years). The first cases of SCC were diagnosed at age 17, while the active-screening period of 1992–1995 – even earlier, i.e., at age 13. To minimize effects of widely introduced screening for cervical cancer in 1992–1995, we also analyzed age-related patterns of cervical SCC and AC from 1973 to 1985 (see Fig. 7.5). During this period, two peaks were observed for SCC incidence: the first, at age 25–32, for both white and black females and the second, at age 58–68 in whites and age 70 in blacks. Patterns of cervical AC differed from those of SCC. Incidence of cervical AC was considerably lower than SCC. However, it was possible to suppose the existence of the first increase at age 35–45 in white and age 38–52 in black, females. The second peak was at age 75 in whites, and 70 in black, females. Inclusion of hysterectomy rates in cervical cancer analysis

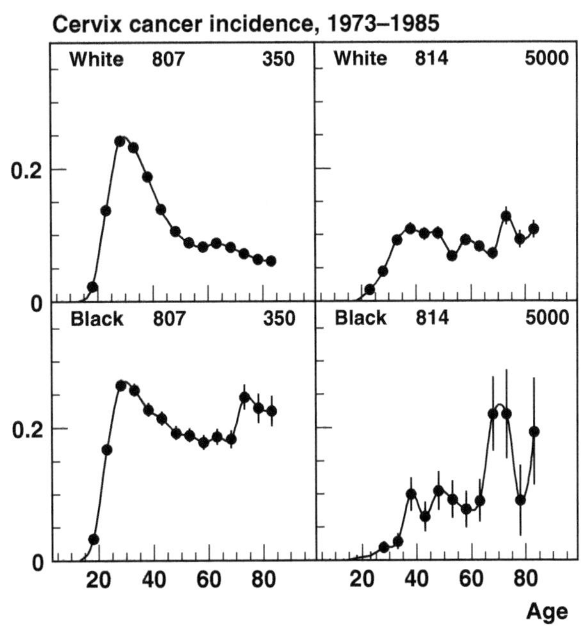


**Fig. 7.5** Incidence rates of cervical squamous-cell carcinoma (SCC) and adenocarcinoma (AC) in the U.S. population, 1973–1985, for *white* and *black* females

influences cervical SCC and AC incidence rates at ages 45 + , making the second ("older") peak more prominent.

Age, being the primary and most predictive host-dependent risk factor for most cancers, may modify effects of other risk factors: i.e., at certain "high-vulnerability" periods, susceptibility to cancer risk factors might increase. Mechanisms behind this increased "vulnerability" are not clear. They may be complex and include age-dependent features of host immune response, sex hormone levels, and changes in anatomo-physiological characteristics of organs and tissues. A cancer with one recognized predominant risk factor may be a good model for analyzing age-related patterns of cancer risks. To test the hypothesis about specific age intervals of increased susceptibility to carcinogenic exposure, we selected cervical cancer as a malignancy with one known predominant carcinogen – human papillomavirus (HPV). We hypothesized that women might be at increased risk of developing cervical cancer during certain age periods (e.g., puberty and perimenopause) due to increased susceptibility to HPV exposure.

Using modeling to study the age-related patterns of cervical cancer incidence, we applied the two-disease model (see Fig. 7.4). We used time intervals of 1973–1991, 1992–1995, and 1996–2003 to take into account changes in cervical cancer registration in SEER Register, when cervical cancer *in situ* was not reported to SEER just after a period of active screening in 1992–1995. When analyzing 1973–1991, in both races, the first incidence peaks were at ages 23–30. The second peaks were detected at ages 60–65 in whites and 70–80 in black, females. As consequences of active screening, the incidence of invasive cervical carcinoma decreased in both races. To evaluate the existence of age-related incidence peaks, further analyses may model screening effects (with an increased detection of cancers at early stages and decreased invasive cancer rates in later consequent age groups).

When a two-disease model was applied to age patterns of 1973–1983 cervical cancer incidence, two components were detected. The first, characterized by a high peak at age 30 in both races, had almost similar shapes in white and black females. The second process differed significantly between the two races: i.e., it peaked at age 65–70 in whites (it was considerably smaller than in black women). Interestingly, in black women, the second slope did not have a peak at all; it just continuously increased with age. Based on these results, one could hypothesize that two ages of increased cervical cancer risk may exist.

The importance of age-associated risk factors was proved by various research groups, when two peaks were observed for HPV prevalence (first at age 16, with a subsequent decrease to 35, and the second at age 65, but not as prominent as the first), for CIN3/carcinoma *in situ* incidence (high rates of CIN3 were observed in the young age groups, followed by second increase at age 50 + ), and also for invasive cervical cancer incidence (with a steep increase at ages 20–35, and a second less-prominent increase at ages 55–60) (Sherman et al., 2003; Herrero et al., 2005; Gustafsson et al., 1997b; Bosch and de Sanjose, 2003). Interpretation of these peaks remains uncertain.

The effects of screening on cervical cancer age patterns have been discussed in many publications. It has been shown that effective screening produced age-restricted declines in cervical cancer incidence confined primarily to women aged 30–70 years, leaving rates among younger and older women relatively unchanged (Gustafsson et al., 1997a). It has been supposed, based on results from cancer registries in developing countries, that in the absence of screening, cervical cancer tends to follow a linear relationship with age, which is the pattern of the majority of the epithelial tumors. Thus, it has been suggested that the shape of the age-specific incidence curve is highly dependent on screening (Bosch and de Sanjose, 2003). Myers et al. (2000) showed that as screening intervals decrease, the proportion of early-stage diseases increases, as does the proportion of cases among younger women. In younger women, "rapid-onset" cervical cancer tends to be an early-stage disease, thus screening contributes significantly to the first "wave" of incidence (Hildesheim et al., 1999). However, the screening effect is not able to explain the presence of the second, older incidence peak.

In Canada, two incidence peaks have emerged since the mid-1980s: one in women aged 35–44 and another in women aged 75 +, with the first peak shifting to younger ages (Duarte-Franco and Franco, 2004). In some countries, a second mode in the incidence of cervical cancer among women aged 55–60 is prominent, but its interpretation remains uncertain. Decrease in screening coverage or lower sensitivity of cytology in the old-age groups were suggested as explanations, but immunosuppression or acquisition of new HPV infections during middle age cannot be ruled out. It can be hypothesized that after a certain latency period, the presence of recently acquired viral DNA in the middle-aged group, with or without the concurrent effect of cofactors, is capable of generating a second increase in the incidence of CIN3 and of invasive cancer. The predicted height of the second mode was attenuated by both screening in these age groups and competing causes of death (Bosch and de Sanjose, 2003).

Two approaches may be used to analyze this age pattern of cervical cancer. The first one is to develop more complex models, which will include both screening effects on "early stage" of cancer and effects of screening on rates of "late stages" of cancer. The second approach is to analyze age-related patterns, not of cervical cancer, but of its main and well-recognized risk factor – HPV, to examine the hypothesis that two age periods in females are especially prone to HPV infection, which, persisting in women's cervix for years without being cleared by immune response, increases dramatically the risk of cervical cancer developing more than decade later.

Taking into account that the lag for cervical cancer was estimated to be 9–10 years, the ages when certain events might act as triggers for cervical carcinogenesis are 15–18 and 50–60. These intervals are close to the reported ages of puberty/early postpuberty for female cohorts in the United States (13–14 years, in average, for 1950–1970 birth cohort) and to the reported ages of menopause/perimenopause (49–50.5 years, in average, for women born in

1915 and 1939, respectively) (Nichols et al., 2006; McDowell et al., 2003; Chumlea et al., 2003). The role of additional factors besides HPV, such as host immune response, sex-hormone-associated cytokine shifts, and anatomo-physiological changes in cervix uteri, might be suggested by analyzing age-specific relative risks, particularly in puberty and premenopause periods. Various age trends in HPV prevalence might be related to different sexual practices, immunologic senescence, and/or cohort effects (Castle et al., 2005; Lajous et al., 2005). Some of these factors are important, possibly regulating the integration of HPV in host genome, and some of them as promoting cervical carcinogenesis.

The relationship of HPV and host seems different at different ages. Some of the observed phenomena still have no clear explanation, e.g., following a rapid accumulation of HPV infection after onset of sexual activity (women younger than 20 years), there is a transition of this balance in favor of virus clearance soon after age 25. That might explain (in most part) the constantly declining age-specific prevalence of HPV infections until menopause. Failure to eradicate the virus at postmenopause is not uncommon, explaining the second peak in HPV prevalence that was recently reported in many different populations. Data on the two periods of life, such as childhood and postmenopause, are still too fragmentary to enable the creation of a comprehensive view about how HPV infection behaves in these periods and what makes many women incapable of clearing virus postmenopausally. It is still unknown why some women over age 55 are likely to remain HR-HPV carriers, while most of them successfully clear their infection well before menopause (Syrjanen, 2007). Early detection of cervical cancer precursors among elderly HR-HPV positive women, who are usually out of the age frame of organized screening, remains a challenge.

Integration of HPV into the host genome is not a normal event and likely represents an important step in the progression to invasive disease. What induces this integration is unknown, but it may involve factors affecting host immune control (Stuver and Adami, 2002). It is proposed that the role of HLA type in cervical cancer occurrence involves the host immune response to HPV infection and the ability of the virus to persist and induce carcinogenesis. Unlike many other cancers, somatic mutations of tumor suppressor genes or cellular oncogenes are not commonly observed for cervical cancer (Southern and Herrington, 1998). Invasion seems to be a stochastic process requiring additional, genotoxic events with few prominent risk factors appreciable by conventional epidemiology (Schiffman and Hildesheim, 2006).

Some studies hypothesized that the early ages at first intercourse might represent a "vulnerable period" for cervix, when the carcinogenic effect of HPV is greatest, while some attempts to associate risk with the number of different sexual partners at early ages have not yielded consistent support for this hypothesis (Herrero et al., 1990; Peters et al., 1986). The host immune competency at the time of infection is important in either clearing the infection or generating an adequate cytotoxic T-lymphocyte response for setting a favorable host response to control latent infections, such as HPV. Factors that influence the initial host–agent interaction include age, route of infection, the presence of coinfections and other comorbidity (Lajous et al., 2005). The process of HPV acquisition and clearance supposedly produces the characteristic age distributions as infections are transmitted sexually when women have new partners and then infection is cleared by female's immune system. Persistence tends to increase at older ages (Castle et al., 2005).

Early age at first intercourse, while related to lifetime number of partners, is considered to be an independent risk factor for cervical SCC (Herrero et al., 1990; Cuzick et al., 1996; Deacon et al., 2000). Another study (Green et al., 2003) found that age at first intercourse was a strong independent risk factor for both AC and SCC of cervix (previous studies of AC of the cervix have generally found no association between age at first intercourse and cancer risk in analyses adjusted for the number of sexual partners) (Brinton et al., 1987a, b, 1993; Ursin et al., 1996; Chichareon et al., 1998; Ngelangel et al., 1998; Altekruse et al., 2003). Studies of adolescents, representing a unique opportunity to examine HPV infections in HIV-infected individuals with a more limited history of sexual activity and relatively recent HIV (before evidence of major immunosuppression) and HPV infections, theorized that infection with HIV may enhance HPV proliferation through mechanisms other than CD4 immunosuppression, particularly early in the course of HIV infection (Moscicki et al., 2000).

Cervix in adolescent seems to be especially vulnerable to carcinogenic exposures due to specific biological changes occurring in this organ during this period of life (Broso, 1994), as well because of specific features of host immune response. The general effect of the age of infection influences the age-incidence patterns of most infection-associated tumors, including cervical cancer. A cytokine-mediated immune response may be associated with this effect (Mosmann and Moore, 1991; Romagnani, 1994). There is growing evidence linking imbalanced immune responses with a variety of diseases, including malignancies (Lucey et al., 1996; Simpson et al., 1997).

The role that HPV infection plays in cervical carcinogenesis remains unclear: e.g., what increases cervical cancer risk more substantially, contact with HPV at age(s) of the highest "vulnerability" (such as puberty and perimenopause), or persistent, long-term HPV infection which is supposed to cause a second cervical cancer incidence peak at older ages (Liaw et al., 1999; Schlecht et al., 2001).

Results from studies of other than cervical cancers also observed specific age-related incidence trends. Related results have been obtained from our earlier study of thyroid cancer risk (see Chapter 3) in a population exposed to internal $^{131}$I irradiation: the excess relative risk (ERR) revealed two peaks in females at ages of menarche and of perimenopause. The hypothesis about the role of age-related (sex hormone associated) immune responses suggested that the thyroid gland might express a secondary neoplastic event (important to thyroid cancer onset) after being exposed to $^{131}$I irradiation at specific ages, such as puberty or perimenopause, when exposure was coincident with a hormonal shift (which occurs during puberty or perimenopause).

In puberty/early postpuberty periods, the female organism changes dramatically due to sex hormone shifts. That shift influences other systems of the organism. Immune system functioning is known to be strongly dependent on sex hormones (Pacifici et al., 1989; Deswal et al., 2001; Nguyen et al., 2003). During periods of age-associated sex hormone fluctuation, a so-called cytokine storm may occur, causing an increased risk of immune-associated disorders. Studies in phenotypically normal individuals indicate that serum cytokine levels are dynamic and are likely to be influenced by many factors, including age (Pawelec et al., 2002). Kingsmore and Patel (2003) analyzed 78 cytokines in 60 individuals ranging in age from birth to 21 years, revealing complex cytokine expression patterns: many cytokines are strongly expressed immediately before adolescence.

Some studies of puberty-related risk factors of breast proved that the age when women reached their maximum height (an important landmark of puberty/early postpuberty) was a significant risk factor for breast cancer (Li et al., 1997). Previous studies reported that age at menarche is an important determinant of risk. The age of maximum height may be an even more important parameter related to breast cancer risk. The physiologic basis for this may lie in the influence of exposure to growth hormones on breast development and insulin-like growth factor during puberty, believed to affect a woman's risk of breast cancer (Li et al., 1997). Several recent studies have shown that early-life events may also influence benign breast disease risk. Greater body fatness during childhood and adolescence may reduce the incidence of proliferative benign breast disease among premenopausal women (Baer et al., 2005). An inverse association has been found between body mass index in girls at age 18 and their risk of proliferative benign breast disease later in life (Baer et al., 2005). Adolescent diet (e.g., fats intakes and micronutrients) has been suggested to dramatically influence the risk of proliferative benign breast lesions (Baer et al., 2003). The significantly greater breast cancer susceptibility of teenage and prepubescent girls compared with adults exposed to ionizing radiation has been found among atomic bomb survivors (Tokunaga et al., 1994) and among patients radiologically treated for cancer (Bhatia et al., 1998) and ankylosing spondilitis (Nekolla et al., 1999). Recent evidence emerged on the increased risk of gastric cancer later in life in persons exposed to *H. pylori* infection in childhood/adolescence (Blaser et al., 2007; Imrie et al., 2001) and on increased melanoma risk associated with moderate sun exposure in childhood/adolescence (Nikolaou et al., 2008). Chronic HBV infection in childhood/adolescence may increase future hepatocellular carcinoma risk more dramatic than HBV infection acquired by adults (Munoz et al., 1989; Hsieh et al., 1992; Kuper et al., 2000). The perimenopausal period has not gotten a lot of researchers' attention in specifically studying its possible association with increased cancer risk.

In addition to host immune response factors and factors determining sexual behavior in adolescents and in perimenopaused females (with supposed higher average numbers of sexual contacts/partners compared with females of other ages), anatomo-physiological age-related changes in cervix uteri may also contribute to increased risk of HPV infection, increasing the cervical carcinoma risk. The normal cervix is covered by a nonkeratinizing squamous epithelium.

With increasing age, the squamocolumnar junction migrates from the exocervix into the distal endocervical canal, with a region between the original and subsequent locations – the transformation zone. This area has a unique sensitivity for neoplastic events (Jacobson et al., 1999) (there are several other transformation zones in human organism, e.g., the anus and oropharynx, which are also prone to HPV carcinogenesis). Reasons for this increased susceptibility of transformation zones are unknown. At puberty, in pregnancy (especially the first pregnancy), and in some steroid contraceptive users, changes in the size and shape of the cervix result in the squamocolumnar junction being moved out to the anatomical ectocervix, thus exposing tissues, previously found in the lower endocervical canal, to vagina (Arends et al., 1998). This is a physiological process, when exposed tissue forms the "cervical ectopia". This ectopia epithelium is where most CINs and invasive cervical carcinomas develop. When columnar epithelium in cervical ectopia undergoes transformation into a stratified squamous epithelium (e.g., under physiological conditions such as puberty or pregnancy), this part of the cervix becomes particularly vulnerable to HPV infection (and likely other damaging factors). Chronic HPV infection in this area may result in CIN (Arends et al., 1998), which may cause cervical carcinoma 8-10 years later. Women with an increased area of squamous metaplasia, such as diethylbestrol-exposed young females, may be at higher risk of cervical carcinoma (Robboy et al., 1984). The anatomical site of cervix uteri, where the majority of cervical cancers are diagnosed (in the anterior and posterior parts of the transformation zone), is also the most-exposed site to sexual trauma, being more susceptible because of the slow completion of squamous transformation, or might be immunologically compromised because of blood flow specifics (Guido et al., 2005).

Different approaches were used to study various aspects of cervical cancer features. Several models have been published over the past two decades to identify the best clinical practices and public policies for developing strategies of cervical cancer prevention and control (Mandelblatt and Fahs, 1988; Eddy, 1990; Fahs et al., 1992; Brown and Garber, 1999; Goldie et al., 1999; McCrory et al., 1999; Goldie et al., 2001; Kim et al., 2002; Mandelblatt et al., 2002). New interventions for cervical precancer, such as fluorescence spectroscopy, also can be evaluated using mathematical modeling (Cantor et al., 1998). The ultimate outcome of decision-analytic and cost-effectiveness models is to improve women's health at an economically reasonable cost. The effectiveness of human papillomavirus vaccine was modeled to predict the impact of HPV-6/11/16/18 vaccination, using a cohort model and measuring parameter uncertainty (Van de Velde et al., 2007).

Most often modeling is used for cost-effectiveness analyses of cervical cancer screening and vaccination preventive strategies. A Markov model was used to attempt to describe a natural history of HPV infection (Myers et al., 2000) and the lifetime costs and life expectancy of a hypothetical cohort of women screened for cervical cancer in the United States (Kulasingam and Meyers, 2003). The initial goal in constructing this model was to analyze the cost effectiveness of new technologies for improving the sensitivity of cervical smears (McCrory et al., 1999). In 2000, this model was used for the approximation of reported patterns of age-specific incidence and prevalence of HPV infection and cervical carcinoma in the United States (Myers et al., 2000).

A stochastic microsimulation of cervical cancer that distinguishes different HPV types by their incidence, clearance, persistence, and progression was developed to predict the expected benefits, costs, and cost effectiveness of different policies in the United States (Goldhaber-Fiebert et al., 2007), when age-specific prevalence of HPV by type, age-specific prevalence of CIN, HPV-type distribution within CIN and cancer, and age-specific cancer incidence were established through systematic reviews. The plausible ranges of the probability of HPV transmission per coital act among newly forming couples were estimated using stochastic simulation and empirical data from a cohort study of young women in Montreal (Burchell et al., 2006).

Monte Carlo simulation has been used to analyze the sensitivity of spectral measurements to a range of changes in epithelial and stromal optical properties that occur as cervical cancer dysplasia develops (Arifler et al., 2006). This study provided a framework to interpret optical signals obtained from epithelial tissues and to optimize design of optical sensors for *in vivo* measurements for precancer detection.

Cox proportional hazard modeling was used for all-cause and cervical cancer-specific survival analyses of women with cervical cancer aged 65 or older (SEER – Medicare linked data) in relevance to their socioeconomic status (Coker et al., 2006).

A difficulty exists in estimating the risk of cancer in unscreened populations when most available data represents both screened and unscreened populations. Some models (Oortmarssen and Habbema, 1991; IARC, 1986) used estimates from case–control or cohort studies. The consistency of the shape of the curve for age-specific incidence in unscreened women across populations (Gustafsson et al., 1997a, b) facilitates calibration of the model (Myers et al., 2000; Gustafsson and Adami, 1992).

At present, few studies have been conducted on mechanisms of cervical carcinogenesis, i.e., mechanisms of cancer initiation, promotion, and progression due to HPV infection. However, it is admitted that interdisciplinary teams could more successfully describe the pathogenesis of cervical cancer in molecular terms (e.g., persistent oncogenic HPV infection with genomic integration), related to the interplay of viral and host biomarkers, rather than using the traditional microscopic or macroscopic terms (Schiffman and Hildesheim, 2006).

### *7.3.3 Hepatocellular Carcinoma: Behavioral Risks and Viral Hepatitis Infections*

Liver cancer, especially hepatocellular carcinoma, is one of the few malignancies for which major etiologic factors have been clearly identified. They include hepatitis C virus (HCV), hepatitis B virus (HBV), alcohol, and dietary aflatoxins. A chronic process of liver damage and regeneration appears to play an important role, possibly through the stochastic accumulation of mutations over a long latent period (Tabor, 1998). HCV and HBV infections account for over three-quarters of all liver cancers (Stuver and Trichopoulos, 2002). In developed countries, the main etiologic factors for HCC include alcohol consumption and HCV infection, with both increasing in prevalence. North America, which is still categorized as having a low incidence of HCC, has had a dramatic increase in its incidence, primarily due to an increasing incidence of HCV infection (Weber et al., 2004). El-Serag (2002) reported a 3-fold increase in the age-adjusted rates for HCC associated with HCV between 1993 and 1998 in the U.S. Veterans Administration hospitals. HCV has been designated as carcinogenic to humans by the IARC (1994).

Convincing evidence has been provided for a causal role of chronic HBV infection in hepatocellular carcinogenesis. It has been shown that an early establishment of chronic HBV infection during childhood appears to increase future risks of carcinoma of liver (Munoz et al., 1989; Hsieh et al., 1992; Kuper et al., 2000). In populations from low- and intermediate-risk areas, HCC is rarely observed in persons younger than 40 years old, in contrast to populations with a high incidence of HCC, in which it is much more frequent in younger age groups (Bosch, 1997). It has been reported that in developed countries, about 23% of liver cancers are likely due to HBV infection, with around 9% in North America (Parkin et al., 1999a, b). In the United States, the number of newly infected cases of HBV declined almost one-third from 2001 to 2005 (CDC, 2007).

In 1988, IARC classified alcohol as a human carcinogen and specifically implicated alcohol consumption in liver cancer etiology (IARC, 1988). However, alcohol intake might be a liver carcinogen mainly by being involved in the development of liver cirrhosis (Adami et al., 1992). A large Swedish cohort study found a relative risk of 22.4 for HCC among patients hospitalized for alcoholism with liver cirrhosis, but only a 2.4-fold increased risk among patients hospitalized for alcoholism alone (Kuper et al., 2001). Although not specifically examined with respect to the temporal trends of liver carcinoma, an increased risk for this cancer in patients with nonalcoholic steatohepatitis and those with diabetes is supported by growing evidence (El-Serag et al., 2003; Adami et al., 1996; Hassan et al., 2002).

Liver cancer has a long latency. Hepatocarcinogenesis likely involves a multistep process representing an accumulation of genetic changes. Cirrhosis, observed in the majority of HCC cases, may play an important role in those mutation events through the associated cycles of liver necrosis and regeneration (IARC, 1994; Tabor, 1998). Allelic loss has been reported in conjunction with liver carcinoma (Chen et al., 1997; IARC, 1994), and somatic changes in several cellular genes have been related to the development of this cancer.

Our results agree with results of other studies that showed men have a higher risk of HCC than women (Stuver and Trichopoulos, 1994). The consistently higher incidence of HCC in men, particularly in developing countries, where liver cancer rates are high and chronic viral hepatitis infections endemic, suggests a role for testosterone in liver carcinogenesis (Yu and Chen, 1993).

Nonhomogeneity in HCC incidence rates was more significant in males than in females, with the predominance of black males over whites. Recently, it has been shown that the higher rates of HBV, HCV, concurrent HBV and HCV infections, and viral hepatitis associated with diabetes might explain the greater burden of liver carcinoma in African Americans (Yu et al., 2006). Nonhomogeneity in HCC rates increased with time, reaching their highest in the last decade (see Fig. 7.3). The "step" in the slope appeared in males since the mid-1980s. It became more prominent in 1994–2003. The age of the "step" is around 50. The highest incidence rates were registered in 1994–2003 at ages 70–80 in white and at ages 65–80 in black, males. One of the possible explanations for the "step" might be a dramatic increase of drug use, especially injected, associated with increased risk of HBV and HCV infection transmission. The appearance of a "younger" peak is in agreement with recent studies, which reported that concomitant with rising rates of liver carcinoma, there was a shift of incidence from, typically for HCC, elderly to younger patients – aged 40–60 years (El-Serag, 2002; El-Serag et al., 2003). A study from the Centers for Disease Control and Prevention estimated the HCV epidemic started in the 1960s and peaked in the 1980s (Armstrong et al., 2000). Risk factors for transmitting HCV were rampant during this period (e.g., injection drug use, needle sharing, and transfusion of unscreened blood and blood products). The HCV first appeared in the United States around 1910, with its widespread dissemination in the 1960s (Tanaka et al., 2002). This study suggests that HCC in the United States will continue to increase in the near future. According to the data from the Bureau of Justice Statistics, there was a dramatic increase in drug users (data based on drug offenders prison admissions) in the United States beginning from 1980s with a 10-fold increase to the 1990s. For drug users, male rates were 10 times higher than female rates, and black rates were 5.2 times higher than for whites (Harrison and Beck, 2006). According to estimates of the Office of National Drug Control Policy, in 1999 almost 977,000 persons were heroin users, and while cocaine use has fallen dramatically since the early 1980s, the number of hardcore heroin users remained virtually unchanged (National Drug Control Strategy, 2000). Recent reports have found HCV prevalence from 27 to 39% among injection drug users younger than 30 years of age (Garfein et al., 1998; Hahn et al., 2002; Thorpe et al., 2002).

The "step" on HCC incidence slope might be related to high-risk behavior males who are more likely to engage for acquiring HCV and coinfections with HBV and HIV, which may also increase risks of HCC (London and McGlynn, 2006). This "risky" behavior, such as injected drug use, rose dramatically over the recent two decades, especially among young males. The lag for HCC is about 21 years (see Table 7.6), assuming the persons had the HCC initiation at age about 29. It has been reported that HCV-associated HCC affects older patients than HBV-associated carcinoma (Shiratori et al., 1995). A prolonged period of liver damage progressing from chronic hepatitis to cirrhosis and then to carcinoma may be required for development of HCV-related malignancy (Kiyosawa et al., 1990). In several studies, an older age (>40) at infection of HCV was associated with an increased risk of HCC, in contrast to younger age at risk for HBV (Poynard et al., 1997; Niederau et al., 1998; Colombo, 1998). If considering the reports documenting decreases in HCV and HBV prevalences among drug users in the United States (Des Jarlais et al., 2005; Villiano et al., 1997; Levine et al., 1996; Burt et al., 2007), the "step" associated with HBV and HCV infections in drug users might be expected to be smoother during the next decades.

The second peak of increase of liver carcinoma incidence occurred at older ages. Increases in HCV prevalence during the recent two decades and the finding that being infected at older ages may increase the risk of developing HCV-associated HCC may be related to the increase of HCC incidence in elderly. Some studies reported the association with increased incidence of both HCV and HBV infections may explain a substantial proportion of the reported increase in HCC incidence during recent years in the U.S. population aged 65 and older (Davila et al., 2004). Nonalcoholic steatohepatitis and diabetes also may play roles in increase of HCC incidence in elderly population.

### *7.3.4 Prostate Cancer: Screening Effects, Genetic Predisposition, or Something Else?*

Traditionally prostate has been described as gland with homogeneous tissue. However, it is still unclear whether foci of prostate cancer develop from separate initiation or promotion events, or whether they represent the same original tumor cells that have traveled to other sites within the prostate via the ductal system (Platz and Giovannucci, 2006). The cells at risk for neoplastic transformation are not known, but are hypothesized to have a phenotype that is intermediate between a stem cell and an epithelial cell (van Leenders et al., 2003). Unlike most other solid tumors, prostate cancer tends to be multifocal (Wise et al., 2002).

Our analysis showed that incidence of prostate cancer has a better fit for the two-disease model, than for the one-disease model, with two components/peaks (see Fig. 7.4). The first is a sharp increase with a plateau at age 55–58 in white, and age 60–65 in black, men. This plateau was "higher" in black males. The second is a prominent peak at around age 75 in both races. The total incidence of prostatic AC was substantially higher in black population. The estimated lag for prostate AC was 32.3 years.

Several factors might be associated with observed nonhomogeneity of prostatic carcinoma in our model. The first is a screening effect. Active prostate cancer screening began in the United States in the early and mid-1990s, at age of 50, and was recommended to be performed annually. Its effects are noticeable in prostate cancer incidence dynamics (see Figs. 6.5 and 7.2): the increase in cancer incidences in 1989–1993, for both races was followed by decrease in subsequent years, with forming of "younger" incidence peaks in both whites and blacks. We may suppose that the first component of our modeling might reflect the screening dynamics.

Screening for PSA is highly adopted in the United States. In 2001, about 75% of American man aged 50 or older reported having a PSA test, and 54% of men aged 50–69 reported that they had a PSA test recently (Sirovich et al., 2003). However, screening may be not the only factor contributing to prostate carcinoma nonhomogeneity (further analyses might reveal the better fit of three- (or even more) disease models for this cancer). There is evidence from recent studies that suggest the importance of genetic factors in prostate cancer age and race patterns. So, the notable variation in prostate cancer incidence among black and white men in the United States may be due to that fact that inherent and modifiable factors determining epithelial cells growth in prostate are heterogeneous (Platz and Giovannucci, 2006). Prostate cancer could progress through one or more precursor lesions. By the third to fourth decades of life, a significant proportion of men already have prostatic intraepithelial neoplasia (PIN) (Sakr et al., 1993), while focal atrophy, another lesion that has received attention recently for its possible relation to prostate cancer, is common in the prostate of older men (Platz and Giovannucci, 2006). The evolution of focal histological cancer is not clear, but it takes several decades before clinically detected carcinoma develops. The initiating events leading to clinically relevant prostatic carcinoma likely occur at young ages (Sakr et al., 1993). Different prevalences of high-grade PIN has been found in African-American and Caucasian males: it was more prevalent in black males and appeared earlier in life (for almost a decade), than among whites (Sakr et al., 1995, 1996). Other lesion that may give rise to carcinoma is the focal atrophy. It is not related to decrease in androgen level while males age, and likely is derived from the proximate inflammatory cells (De Marzo et al., 1999). Prostatic cells in proliferative inflammatory atrophy lesions express high levels of glutathione-S-transferase (GSTP1) thus defending against oxidative genome damage. When this mechanism is damaged, cells may transform into PIN and prostatic cancer cells (Nelson et al., 2001).

Development of multigenic models of cancer susceptibility will be an important approach to cancer prediction, prevention, and diagnosis. For example, pedigree analysis suggests a genetic component for some individuals with prostate cancer; however, the majority of prostate cancer cannot be explained by a single-gene model, suggesting multigenic etiology. The international and racial–ethnic variations in prostate cancer incidence, combined with the effects of migration on risk patterns, suggest genetic factors are likely to play a central role in determining prostate cancer risk (Wilson et al., 2002). In a study of male health professionals, risk of prostate cancer remained elevated even after adjusting for dietary and lifestyle risk factors (Platz et al., 2000).

Androgens influence maturation of the prostate and are believed to contribute to the development and progression of prostate cancer (Hsing, 2001). The growth promotion of already initiated luminal cells via stimulation by hormonal (e.g., androgens) and growth factor (e.g., IGF-1, insulin) systems, that become excessive because of Western dietary and lifestyle patterns, was recently described (Hsing and Devesa, 2001). It has been suggested that the effects of high animal fat diets may be influential *in utero* and in childhood, as well as in adulthood, by altering androgen set points, the duration of androgen exposure, along with higher androgen levels (Ross and Henderson, 1994). These later-acting factors may help to explain differences in the incidence of clinically overt prostate cancer between high and low prostate cancer risk regions and may also explain the increased risk of prostate cancer in men who move from low to high prostate cancer risk areas. With continued genetic and epigenetic events (e.g., loss of 8p, perturbations in methylation at CpG sites), continued growth and metastasis advantage may occur (Platz and Giovannucci, 2006).

So, nonhomogeneity of prostatic carcinoma age-specific incidence rates, as well as reasons for race difference in components of the two-disease model, might be explained by a complex of factors. Among them are screening effects, genetic predisposition and associated precursors of prostatic AC, greater prevalence of modifiable risk factors in African-American men (e.g., diet, physical activity), androgens/testosterone level, and others, not recognized yet. Further research is needed to analyze contribution of these factors to prostate carcinoma developing and to define the key causes of its nonhomogeneity. Based on these detailed information, preventive strategies could be modified to be effective in various age groups and races.

## 7.4 Squamous Cell Carcinomas and Adenocarcinomas: The Time Trends

We analyzed the time trends of lung, esophageal, and cervical squamous-cell carcinomas (SCCs) and adenocarcinomas (ACs) from 1973 to 2004. The incidence rates over this period for all races, sexes, and ages in total are shown in Fig. 7.6. Incidence rates of SCCs of all three cancer sites decreased during observation, while rates of ACs increased (with stabilization for cervical AC).

Analyses of incidence rates demonstrated that lung AC shifted in 1988 and became the predominant lung cancer histotype. Shift for esophageal AC predominance was registered 10 years later than for lung cancer – in 1998. Both lung and esophageal cancers showed certain common features in their histotype trends: shifts occurred 10–20 years ago, and ACs incidences increased more dramatically in whites.

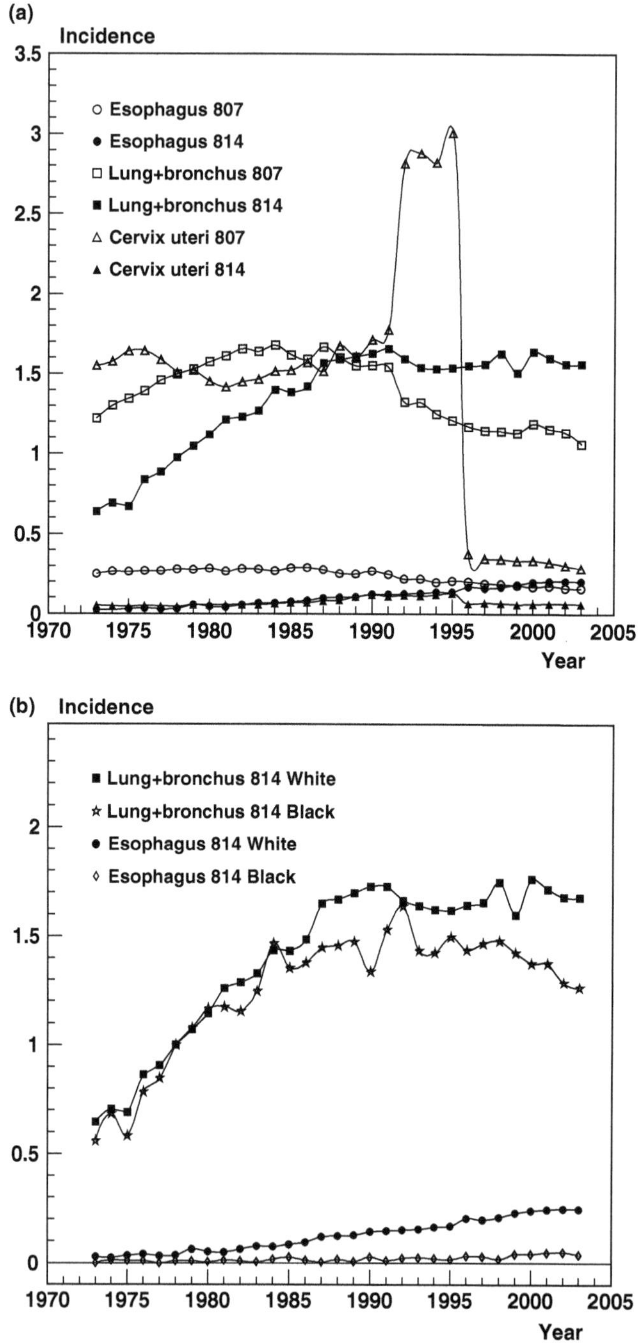


**Fig. 7.6** Time trends of SCCs and ACs of esophagus, lung, and cervix uteri, 1973–2003: (**a**) general incidence, both sexes, both races, (**b**) race-specific patterns of lung and esophageal ACs

Cervical cancer histotypes have not shifted yet, but beginning from 1996 they are moving closer to convergence since 1996 (especially in white females), with SCC incidence decreasing since 1973 (dramatic decrease in 1995 was due to changes in reporting cervical cancer *in situ* to SEER Register). The incidence of AC started to slightly increase in 1987 in white females, but in 1995 it dropped to the pre-1987 rate (supposed due to screening effects). These results are in accordance with other studies (Wang et al., 2004; Zheng et al., 1996), confirming the continuing decline in cervical SCC in U.S. women, attributable to screening. Dramatic increases of SCC incidence in both white and black women were observed in 1991–1995. They were due to a culmination of events, including the CDC nationwide National Breast and Cervical Cancer Early Detection Program (Henson et al., 1996; Reynolds, 1992), nomenclature and classification introduction to The Bethesda System in 1988–1989 and its subsequent revision in 1991 (The Bethesda System, 1993), the introduction of the Clinical Laboratory Improvement Act of 1988 (CLIA 1988), and the widespread introduction and use of the loop electrosurgical excision procedure (LEEP). Increases in AC incidence around the same time likely reflected the increasing rate of both AC *in situ* and invasive AC, predominantly in young white women (Wang et al., 2004). Possible reasons for that rise include increased recognition and awareness of AC, which lead to an increased number of referrals, and also the better description of cytomorphology of AC *in situ* in late 1980s–1990s (nomenclature changes that may have affected reporting of SCC *in situ* are not likely to impact AC in the same manner).

Taking into account not only cancers which at the same time have both SCCs and ACs histotypes but cancers of other organs which are SCCs/ACs, it is likely that the tendencies of the trends described above might be in some extent generalized to other cancers. All analyzed SCCs (lung, esophageal, and cervical) had decreased incidence rates during the past 30 years, while incidence rates of 7 of 12 ACs (lung, esophageal, hepatic, renal, breast, prostate, and cervical) increased (some of them dramatically) during the same period, and 3 of 12 ACs (colorectal, pancreatic, and ovarian) had almost stable rates. Only two ACs (uterine and gastric) had decreased incidences.

Risk factors for SCCs are mostly well known, such as smoking, HPV infection, alcohol use. They predominantly act directly at the spot of cancer onset (like epithelium of lung, cervix uteri or esophagus). Risk factors of many of ACs are still not clearly established (excluding infectious factors for gastric cancer – *H. pylori*, and for hepatocellular carcinoma – HBV and HCV): e.g., obesity, diet, reproductive factors and sex hormones, gastroesophageal reflux disease, physical activity, arterial hypertension. These factors act predominantly not by directly targeting the spot of cancer development, but through various "indirect" endogenous pathways/mechanisms. This makes it more difficult to forecast ACs trends by risk factor intervention, to estimate the groups of highest risk and to develop effective preventive strategies to reduce ACs prevalence. Recent study suggested that the role of smoking in lung AC in women might be underestimated (Yang et al., 2002). This finding needs further observation – not only for lung but also for other ACs.

SCCs and ACs probably differ by the mechanisms involved in carcinogenesis, the number of stages of carcinogenesis, as well as by response to therapy, 5-year survival, and the risk of the tumor recurring. So, prognosis was better for patients with esophageal AC than SCC, while SCC had a higher percentage of tumor recurrence and a high risk of second primary tumors (Goseki et al., 1992; Gayet et al., 1990; Yoshinaka et al., 1991; Holscher et al., 1995).

All of these facts should be taken into account while analyzing SCCs and ACs time trends to make a more realistic forecast, and for planning future preventive measures, effective screening, and treatment strategies.

## 7.5 Summary

In this chapter we extended analyses of cancer morbidity and mortality trends in the U.S. by site (see Chapter 6) to analyses by cancer histotype. Analysis of histotype-specific age patterns is the natural step in strategy of separation of observed effects into contributions of the most homogeneous components and apply the biological models of carcinogenesis to them. Analyzing the frequencies of all possible histotypes, we selected a set of the most frequent of them, and evaluated the basic characteristics of their age patterns and time trends. Detail was devoted to analyses of the trends of squamous-cell carcinomas and adenocarcinomas: the incidence rates of SCCs decreased during the past 30 years, while the incidence rates of most of ACs increased, especially in whites. These results allowed us to conclude that great attention to identifying adenocarcinoma risk factors would, probably, be paid in near future.

Not all of the important conclusions about the nature of carcinogenesis can be made from empirical analysis. The step we made was to apply mathematical models describing specific biological mechanisms of carcinogenesis (reviewed in Chapter 2). The benefit of this approach is in the possibility to uncover the underlying causes and biological mechanisms of carcinogenesis and to use the obtained models for studying interventions and for developing population forecasting. Specific advantage of using the two-disease model approach is in uncovering the possible unobserved structure of the age patterns and in describing each component (or each form of disease) by the biological model of carcinogenesis. First we analyzed all selected histology age patterns by a series of models assuming a lack of hidden structure. From analyses of the quality of fits, we concluded that gamma-frailty models with baseline hazard described by Weibull functions, which in original form represented a family of multihit models, was the best for describing these patterns. The notion of a lag and the number of latent stages were incorporated in these models. Application of the two-disease model for histotype-specific age patterns allowed us to investigate the effects of unobserved heterogeneity, which might be attributable to the genetic predisposition or to the environmental factors. Based on the results of the analyses, we hypothesized that (1) the stages of carcinogenesis and tumor progression to clinical manifestation might be likely associated with cancer histotype and (2) there exists an association detected in the estimates of model parameters which can be explained in terms of embryogenetic background shared by organs. Both hypotheses need further investigation (e.g., how these factors contribute to carcinogenesis mechanisms, whether they might influence cancer developing independently, or can they have certain synergistic effects).

Several sources for uncertainties are important while analyzing cancer patterns or making certain assumptions concerning the possible mechanisms of carcinogenesis. They include (1) adjusting population by the main risk factors (smoking, alcohol consumption, obesity, etc.); (2) nonhomogeneity of the white population in the statistical databases in the United States (i.e., racial classifications are often a social, but not a biological construct, which is important to remember when studying, for example, gene–environment interactions); (3) difficulties of some cancer histotypes subclassification (e.g., certain ACs might be highly heterogeneous histologically); and (4) necessity to take into account the dynamics of medical technology/diagnostic procedures, especially screening with increased possibility to diagnose cancer at earlier stages, thus changing (reducing) the fraction of unobserved cases in population.

Screening-related uncertainties, probably, are the most contributing in changing the "natural" history of cancer, thus obscuring the underlying patterns and associated risks. By definition, screening for cancer is the testing of apparently healthy volunteers from the general population for the purpose of separating them into high and low probabilities of having a given cancer (Prorok et al., 1999). Intervention at this point will lead to treatment of detected abnormalities (such as cancer precursor or cancer *in situ*), improved outcome for individual because of early treatment of early-stage disease, thus increasing survival and, in some cases, decreasing mortality. As to incidence patterns, screening usually has two effects. One is directly related to the screening procedure, when after implementation of screening the incidence of the tumor, to which screening has been applied, increases, often dramatically, with the peak at the age at which the screening strategy was directed to. The second effect is associated with the subsequent decrease of advanced cancer incidence (e.g., invasive cancer) decades, or more, later, due to "pre-selection" of cancers at their early stage, thus not giving the tumor the chance to mature. Thus, the screening results in redistribution of newly registered cases among cancer stages/precursors. When the data are collected in groups with strong latent distortion due to screening, this effect has to be either measured or modeled, or, at least, a respective contribution to uncertainty due to screening effect has to be estimated. Recently, modeling of screening effects using microsimulation became a promising method, which led to the reconstruction of the "natural history" of cancers distorted by specific widely adopted screening strategies (e.g., cervical, prostate, breast cancers) (Berkhof et al., 2005; Etzioni et al., 1999).

Evidently, only part of all real cases of cancer onset is detected and appears in cancer registers, and another part of the cases escapes the registration. The unobserved fraction will be larger for earlier disease stages. As we discussed in Chapter 2, one of the possible explanations of the leveling-off and the decline in age-specific incidence rate at advanced ages is that a larger fraction of all real cases has avoided registration. If the part of unobserved cases does not change during the period when data are collected, then this effect can be simply kept in mind until the stage of interpretation of results and comparison to theories predicting incidence rate in population. Methods exist to clarify this latent fraction of unobserved cases. One approach could be based on the comparison of information provided by SEER Register (incidence plus forthcoming survival) and related cause-specific rates provided by Multiple Cause of Death data. Comparison of the age and cause-specific death rates extracted by these two ways can shed a light on the unobserved fraction.

Modeling methods presented in this chapter are important in describing age patterns of cancer incidence using biological models of carcinogenesis. However, they have limited power for tasks where a dynamic component has to be investigated, e.g., for cohort design. Examples of respective tasks include forecasting and approaches to analyzing results of hypothetical medical interventions directed to cancer prevention. In Chapter 8, we will describe how such tasks can be investigated using methods of microsimulation and the quadratic hazard models presented in Chapters 4 and 5.

## References

Abeloff M.D., Wolff A.C., Wood W.C. et al., 2004. Cancer of the breast. In: Abeloff M.D., Armitage J.O., Niederhuber J.E., Kastan M.B., McKenna W.G. (eds.). Clinical Oncology. 3rd edition. Amsterdam: Elsevier.

Adami H.O., Chow W.H., Nyren O. et al., 1996. Excess risk of primary liver cancer in patients with diabetes mellitus [see comments]). J Natl Cancer Inst 88:1472–1477.

Adami H.O., Hsing A.W., McLaughlin J.K. et al., 1992. Alcoholism and liver cirrhosis in the etiology of primary liver cancer. Int J Cancer 51:898–902.

Altekruse S.F., Lacey J.V. Jr., Brinton L.A. et al., 2003. Comparison of human papillomavirus genotypes, sexual, and reproductive risk factors of cervical adenocarcinoma and squamous cell carcinoma: Northeastern United States. Am J Obstet Gynecol 188:657–663.

Anderson K.E., Mack T.M., Silverman D.T., 2006. Cancer of the pancreas. In: Schottenfeld D., Fraumeni J.F. Jr. (eds.). Cancer Epidemiology and Prevention. 3rd edition. New York: Oxford University Press, pp. 721–762.

Arends M.J., Buckley C.H., Wells M., 1998. Aetiology, pathogenesis, and pathology of cervical neoplasia. J Clin Pathol 51:96–103.

Arifler D., MacAulay C., Follen M. et al., 2006. Spatially resolved reflectance spectroscopy for diagnosis of cervical pre-cancer: Monte Carlo modeling and comparison to clinical measurements. J Biomed Opt 11(6):064027.

Armstrong G.L., Alter M.J., McQuillan G.M., Margolis H.S., 2000. The past incidence of hepatitis C virus infection: implications for the future burden of chronic liver disease in the United States. Hepatology 31:777–782.

Azzopardi J.G., Chepik O.F., Hartman W.H. et al., 1982. The World Health Organization histological typing of breast tumors. 2nd edition. Am J Clin Pathol 78:806.

Baer H.J., Schnitt S.J., Connoly J.L et al., 2003. Adolescent diet and uincidence of proliferative benign breast disease. Cancer Epidemiol Biomarkers Prev 12:1159–1167.

Baer H.J., Schnitt S.J., Connolly J.L. et al., 2005. Early life factors and incidence of proliferative benign breast disease. Cancer Epidemiol Biomarkers Prev 14:2889–2897.

Berkhof J., de Bruijne M.C., Zielinski G.D. et al., 2005. Natural history and screening model for high-risk humanpapillomavirus infection, neoplasia and cervical cancer in the Netherlands. Epidemiology 115(2):268–275.

Bernstein L., 2002. The epidemiology of breast cancer in situ. In: Silverstain M. (ed.). Ductal Carcinomas In Situ of the Breast. 2nd edition. Philadelphia, PA: Lippincott Williams & Wilkins, pp. 22–34.

Bhatia S., Meadows A.T., Robison L.L., 1998. Second cancers after pediatric Hodgkin's disease. J Clin Oncol 16(7):2570–2572.

Blaser M.J., 1999. Hypothesis: The changing relationship of *Helicobacter pylori* and humans: implications for health and disease. J Infect Dis 179:1523–1530.

Blaser M.J., Nomura A., Lee J. et al., 2007. Early-life structure and microbially induced cancer risk. PLoS Med 4(1):e7.

Blot W.J., Devesa S.S., Kneller R.W., Fraumeni J.F. Jr., 1991. Rising incidence of adenocarcinoma of the esophagus and gastric cardia. JAMA 265:1287–1289.

Blot W.J., Fraumeni J.F. Jr., 1996. Cancers of the lung and pleura. In: Schottenfeld D., Fraumeni J.F. Jr. (eds.). Cancer Epidemiology and Prevention. New York: Oxford University Press, Inc, pp. 637–665.

Blot W.J., MacLaughlin K., Fraumeni J.F. Jr., 2006. Esophageal cancer. In: Schottenfeld D., Fraumeni J.F. Jr. (eds.). Cancer Epidemiology and Prevention. 3rd edition. New York: Oxford University Press.

Boffetta P., Trichopoulos D., 2002. Cancer of the lung, larynx, and pleura. In: Adami H-O., Hunter D., Trichopoulos D. (eds.). Textbook of Cancer Epidemiology. New York: Oxford University Press.

Bosch F.X., 1997. Global epidemiology of hepatocellular carcinoma. In: Okuda K., Tabor E. (eds.). Liver Cancer. New York, Churchill Livingstone, pp.13–28.

Bosch F.X., de Sanjose S., 2003. Human papillomavirus and cervical cancer – burden and assessment of casuality. Chapter I. Journal of the National Cancer Institute Monographs No. 31. New York: Oxford University Press, pp. 3–13.

Brinton L.A., Hamman R.F., Huggins G.R. et al., 1987a. Sexual and reproductive risk factors for invasive squamous cell cervical cancer. J Natl Cancer Inst 79:23–30.

Brinton L.A., Herrero R., Reeves W.C. et al., 1993. Risk factors for cervical cancer by histology. Gynecol Oncol 51(3):301–306.

Brinton L.A., Tashima K.T., Lehman H.F. et al., 1987b. Epidemiology of cervical cancer by cell type. Cancer Res 47:1706–1711.

Broso P.R., 1994. The age for beginning and the frequency for performing the Pap test. Minerva Ginecol 46(10):565–569.

Brown L.M., Devesa S.S., 2002. Epidemiologic trends in esophageal and gastric cancer in the United States. Surg Oncol Clin N Am 11:235–256.

Brown A., Garber A., 1999. Cost-effectiveness of 3 methods to improve the sensitivity of Papanicolaou screening. JAMA 281:347–353.

Burchell A.N., Richardson H., Mahmud S.M. et al., 2006. Modeling the sexual transmissibility of human papillomavirus infection using stochastic computer simulation and empirical data from a cohort study of young women in Montreal, Canada. Am J Epidemiol 163(6):534–543.

Burt R.D., Hagan H., Garfein R.S. et al., 2007. Trends in hepatitis B virus, hepatitis C virus, and human immunodeficiency virus prevalence, risk behaviors, and preventive measures among Seattle injection drug users aged 18–30 years, 1994–2004. J Urban Health Bull NY Acad Med 84(3):436–454.

Cantor S.B., Mitchell M.F., Tortolero-Luna G. et al., 1998. Cost-effectiveness analysis of diagnosis and management of cervical squamous intraepithelial lesions. Obstet Gynecol 91:270–277.

Castle P.E., Schiffman M., Herrero R. et al., 2005. A prospective study of age trends in cervical human papillomavirus acquisition and persistence in Guanacaste, Costa Rica. J Infect Dis 191:1808–1816.

CDC, 2007. Disease burden from hepatitis A, B, and C in the United States. At: http://www.cdc.gov/ncidod/diseases/hepatitis/resource/PDFs/disease_burden.pdf

Chen C-J., Yu M-W., Liaw Y-F., 1997. Epidemiological characteristics and risk factors of hepatocellular carcinoma. J Gastroenterol Hepatol 12(Suppl):S294–S308.

Chichareon S., Herrero R., Munoz N. et al., 1998. Risk factors for cervical cancer in Thailand: a case–control study. J Natl Cancer Instit 90:50–57.

Chow W-H., Devesa S.S., Warren J.L., Fraumeni J.F. Jr., 1999. Rising incidence of renal cell cancer in the United States. JAMA 281:1628–1631.

Chu K., Tarone R., Kessler L., 1996. Recent trends in U.S. breast cancer incidence, survival, and mortality rates. J Natl Cancer Inst 88:1571–1579.

Chumlea W.C., Schubert C.M., Roche A.F. et al., 2003. Age at menarche and racial comparisons in U.S. girls. Pediatrics 11(1):110–113.

Claus E.B., Schildkraut J., Iversen E.S. Jr., Berry D. et al., 1998. Effect of BRCA1 and BRCA2 on the association between breast cancer risk and family history. J Natl Cancer Inst 90:1824–1829.

CLIA, 1988. Medicare, medicaid and CLIA programs: regulations implementing the clinical laboratory improvement amendments of 1988 (CLIA-88); proposed rule. 1990. Fed Regist 55:20896–20959.

Coker A.L., Du X.L., Fang S. et al., 2006. Socioeconomic status and cervical cancer survival among older women: findings from the SEER-Medicare linked data cohorts. Gynecol Oncol 102(2):278–284.

Colditz G.A., Baer H.J., Tamimi R.M., 2006. Breast cancer. In: Schottenfeld D., Fraumeni J. F. Jr. (eds.). Cancer Epidemiology and Prevention. 3rd edition. New York: Oxford University Press.

Colditz G.A., Rosner B.A., 2000. Cumulative risk of breast cancer to age 70 years according to risk factor status: data from the Nurses' Health Study. Am J Epidemiol 152(10):950–964.

Colditz G.A., Rosner B.A., 2006. What can be learnt from models of incidence rates? Breast Cancer Research 8:208. doi:10.1186/bcr1414

Collins L.C., Tamimi R.M., Baer H.J. et al., 2005. Outcome of patients with ductal carcinoma in situ untreated after diagnostic biopsy: results from the Nurses' Health Study. Cancer 103(9):1778–1784.

Colombo M., 1998. The role of hepatitis C virus in hepatocellular carcinoma. Recent Results Cancer Res 154:337–344.

Cook L.S., Weiss N.S., Doherty J.A., Chen C., 2006. Endometrial cancer. In: Schottenfeld D., Fraumeni J.F. Jr. (eds.). Cancer Epidemiology and Prevention. 3rd edition. New York: Oxford University Press.

Correa P., Chen V.W., 1994. Gastric cancer. In: Doll R., Fraumeni J.F. Jr., Muir C.S. (eds.). Trends in Cancer Incidence and Mortality. Plainview, NY: Cold Spring Harbor Laboratory Press, pp. 55–75.

Crum C.P., 1999. Ovaries. In: Cotran R.S., Tucker C. (eds.). Pathologic Basis of Disease. Philadelphia, PA: W.B. Saunders Company.

Cuzick J., Sasieni P., Singer A., 1996. Risk factors for invasive cervix cancer in young women. Eur J Cancer 32A:836–841.

Davila J.A., Morgan R.O., Shaib Y. et al., 2004. Hepatitis C infection and the increasing incidence of hepatocellular carcinoma: a population-based study. Gastroenterology 127(5):1372–1380.

De Marzo A.M., Marchi V.L., Epstein J.I. et al., 1999. Proliferative inflammatory atrophy of the prostate. Am J Pathol 155(6):1985–1992.

Deacon J., Evans C.D., Yule R. et al., 2000. Sexual behaviour and smoking as determinants of cervical HPV infection and of CIN3 among those infected: a case–control study nested within the Manchester cohort. Br J Cancer 83:1565–1572.

Des Jarlais D.C., Perlis T., Arasteh K. et al., 2005. Reductions in hepatitis C virus and HIV infections among injecting drug users in New-York City, 1990–2001. AIDS 19 (Suppl 3):S20–S25.

Deswal A., Petersen N.J., Feldman A.M., Young J.B., White B.G., Mann D.L., 2001. Cytokines and cytokine receptors in advanced heart failure. An analysis of the cytokine database from the Vesnarinone Trial (VEST). Circulation 24:2055–2059.

Devesa S.S., 1984. Descriptive epidemiology of cancer of the uterine cervix. Obstet Gynecol 63:605–612.

Dornschneider G., Izbicki J.R., Wilker D.K. et al., 1990. The effects of sex steroids on colon carcinogenesis. Anticancer Drugs 1:15–21.

Duarte-Franco E., Franco E.L., 2004. Cancer of the uerine cervix. BMC Women Health 4(Suppl 1):S13.

Eddy D.M., 1990. Screening for cervical cancer. Ann Intern Med 113:214–226.

El-Serag H.B., 2002. Hepatocellular carcinoma: an epidemiologic view. Yale University Workshop on hepatocellular carcinoma. J Clin Gastroenterol 35(5) Suppl:S72–S78.

El-Serag H.B., Davila J.A., Petersen N.J., McGlynn K.A., 2003. The continuing increase in the incidence of hepatocellular carcinoma in the United States: an update. Ann Intern Med 139:817–823.

Etzioni R., Legler J.M., Feuer E.J. et al., 1999. Cancer surveillance series: interpreting trends in prostate cancer – Part III: quantifying the link between population prostate-specific antigen testing and recent declines in prostatecancer mortality. J Natl Cancer Instit 91(12):1033–1039.

Fahs M.C., Mandelblatt J.S., Schechter C., Muller C., 1992. Cost-effectiveness of cervical cancer screening for the elderly. Ann Intern Med 117:520–527.

Feigelson H.S., Coetzec G.A., Kolonel L.N. et al., 1997. A polymorphism in the CYP17 gene increases the risk of breast cancer. Cancer Res 57:1063–1065.

Fitzgerald M.G., MacDonald D.J., Krainer M. et al., 1996. Germ-line BRCA1 mutations in Jewish and non-Jewish women with early-onset breast cancer. N Engl J Med 334:143–149.

Frieboes H.B., Brody J.P., 2005. Telomere loss limits the rate of human epithelial tumor formation. Arxiv preprint q-bio.TO/0502004 – arxiv.org

Garfein R.S., Doherty M.C., Monterroso E.R. et al., 1998. Prevalence and incidence of hepatitis C virus infection among young adult injection drug users. J Acquir Immune Defic Syndr Hum Retrovirol 18(Suppl 1):S11–S19.

Gayet B., Vons C., Molas G. et al., 1990. Superficial squamous cell carcinoma of the esophagus: an early lesion of good prognosis? A report of 58 patients resected. In: Ferguson N.K., Little A.G., Skinner D.B. (eds.). Diseases of the Esophagus. Vol. 1: Malignant Diseases. Mount Kisko, NY: Futura Publishing Inc., pp. 161–169.

Giovannucci E., Wu K., 2006. Cancers of the colon and rectum. In: Schottenfeld D., Fraumeni J.F. Jr. (eds.). Cancer Epidemiology and Prevention. 3rd edition. New York: Oxford University Press. pp. 809–829.

Going J.J., 2003. Stages on the way to breast cancer. J Pathol 199(1):1–3.

Goldhaber-Fiebert J.D., Stout N.K., Ortendahl J. et al., 2007. Modeling human papillomavirus and cervical cancer in the United States for analyses of screening and vaccination. Popul Health Metr 5:11.

Goldie S.J., Kuhn L., Denny L. et al., 2001. Policy analysis of cervical cancer screening strategies in low-resource settings: clinical benefits and cost-effectiveness. JAMA 285:3107–3115.

Goldie S.J., Weinstein M.C., Kuntz K.M., Freedberg K.A., 1999. The costs, clinical benefits, and cast effectiveness of screening for cervical cancer in HIV-infected women. Ann Intern Med 130:97–107.

Goseki N., Koike M., Yoshida M., 1992. Hismiddleathologic characteristics of early stage esophageal carcinoma: a comparative study with gastric carcinoma. Cancer 69:1088–1093.

Green J., Berrington de Gonzalez A., Sweetland S. et al., 2003. Risk factors for adenocarcinoma and squamous cell carcinoma of the cervix in women aged 20–44 years: the UK National Case–Control Study of Cervical Cancer. Br J Cancer 89:2078–2086.

Guido R.S., Jeronimo J., Schiffman M. et al., 2005. The distribution of neoplasia arising on the cervix: results from the ALTS trial. Am J Obstet Gynecol 193:1331–1337.

Gustafsson L., Adami H-O., 1992. Optimization of cervical cancer screening. Cancer Causes Control 3(2):125–136.

Gustafsson L., Ponten J., Bergstrom R. et al., 1997b. International incidence rates of invasive cervical cancer before cytological screening. Int J Cancer 71:159–165.

Gustafsson L., Ponten J., Zack M. et al., 1997a. International incidence rates of invasive cervical cancer after introduction of cytological screening. Cancer Causes Control 8:755–763.

Haagensen C.D., Lane N., Lattes N. et al., 1978. Lobular neoplasia (so-called lobular carcinoma in situ) of the breast. Cancer 42:737.

Hahn J.A., Page-Shafer K., Lum P.J. et al., 2002. Hepatitis C virus seroconversion among young injection drug users: relationships and risks. J Infect Dis 186:1558–1564.

Hamilton S.R., Aaltonen L.A., 2000. Tumours of the stomach. In: Hamilton S.R., Aaltonen L.A. (eds.). World Health Organization Classification of Tumours: Pathology of Tumours of the Digestive System. Lyon: IARC Press, pp. 37–66.

Hanai A., Fujimoto I., Taniguchi H., 1982. Trends of stomach cancer incidence and histologic types in Osaka. In: Magnus K. (ed.). Trends in Cancer Incidence: Causes and Practical Implications. Washington, DC: Hemisphere Publishing Corporation, pp. 143–154.

Hankinson S., Hunter D., 2002. Breast cancer. In: Adami H-O., Hunter D., Trichopoulos D. (eds.). Textbook of Cancer Epidemiology. New York: Oxford University Press.

Harkness E.F., Brewster D.H., Kerr K.M. et al., 2002. Changing trends in incidence of lung cancer by histologic type in Scotland. Int J Cancer 102:179–183.

Harrison P.M., Beck A.J., 2006. Bureau of Justice Statistics, Prison and Jail Inmates at Midyear 2005. Washington, DC: U.S. Department of Justice, May 2006. NCJ213133, p. 10.

Hassan M.M., Hwang L., Hatten C.J. et al., 2002. Risk factors for hepatocellular carcinoma: synergism of alcohol with viral hepatitis and diabetes mellitus. Hepatology 36:1206–1213.

Helzlsouer K.J., Selmin O., Huang H.Y. et al., 1998. Association between glutathione-*S*-transferase M1, P1, and T1 genetic polymorphisms and development of breast cancer. J Natl Cancer Inst 90:512–518.

Henrik S.J., Forsgren A., Berglund G., Floren C.H., 2001. *Helicobacter pylori* infection is associated with a decreased risk of developing oesophageal neoplasms. Helicobacter 6:310–316.

Henson R.M., Wyatt S.W., Lee N.C., 1996. The National Breast and Cervical Cancer Early Detection Program: a comprehensive public health response to two major health issues for women. J Public Health Manag Pract 2:36–47.

Herrero R., Brinton L.A., Reeves W.C. et al., 1990. Sexual behavior, veneral diseases, hygiene practices, and invasive cervical cancer in a high-risk population. Cancer 65:380–386.

Herrero R., Castle P.E., Schiffman M. et al., 2005. Epidemiologic profile of type specific human papillomavirus infection and cervical neoplasia in Guanacaste, Costa Rica. J Infect Dis 191:1796–1807.
Hewitt M., Devesa S., Breen N., 2002. Papanicolaou test use among reproductive-age women at high risk for cervical cancer: analyses of the 1995 National Survey of Family Growth. Am J Public Health 92:666–669.
Hildesheim A., Hadjimichael O., Schwartz P.E. et al., 1999. Risk factors for rapid-onset cervical cancer. Am J Obstet Gynecol 180:571–577.
Holscher A.H., Bollschweiler E., Schneider P.M., Siewert J.R., 1995. Prognosis of early esophageal cancer. Comparison between adeno- and squamous cell carcinoma. Cancer 76(2):178–186.
Hsieh C-C., Tzonou A., Zavitsanos X. et al., 1992. Age at first establishment of chronic hepatitis B infection and hepatocellular carcinoma risk: a birth order study. Am J Epidemiol 136:1115–1121.
Hsing A.W., 2001. Hormones and prostate cancer: What's next? Epidemiol Rev 23:42–58.
Hsing A., Devesa S.S., 2001. Trends and patterns of prostate cancer: What do they suggest? Epidemiol Rev 23:3–13.
IARC, 1986. Screening for squamous cervical cancer: duration of low risk after negative results of cervical cytology and its implications for screening policies. IARC Working Group on evaluation of cervical cancer screening programmes. BMJ 293:659–664.
IARC, 1988. Alcohol Drinking. IARC Monogr Eval Carcinog Risk Humans. Vol. 44. Lyon: IARC.
IARC, 1994. Hepatitis Viruses. IARC Monogr Eval Carcinog Risk Humans. Vol. 59. Lyon: IARC.
Imrie C., Rowland M., Bourke B. et al., 2001. Is *Helicobacter pylori* infection in childhood a risk factor for gastric cancer? Pediatrics 107(2):373–380.
Jacobson D.L., Peralta L., Farmer M. et al., 1999. Cervical ecmiddley and the transformation zone measured by computerized planimetry in adolescents. Int J Gynaecol Obstet 66:7–17.
Janssen-Heijnen M.L.G., Coebergh J.W.W., 2001. Trends in incidence and prognosis of the histological subtypes of lung cancer in North America, Australia, New Zealand and Europe. Lung Cancer 31:123–137.
Jemel A., Travis W.D., Tarone R.E. et al., 2003. Lung cancer rates convergence in young men and women in the United States. Analysis by birth cohort and hystologic type. Int J Cancer 105:1001.
Kabat G.C., 1996. Aspects of the epidemiology of lung cancer in smokers and nonsmokers in the United States. Lung Cancer 15:1–20.
Kelsey K.T., Wiencke J.K., 1998. Growing pains for the environmental genetics of breast cancer: observations on a study of the glutathione-*S*-transferases. J Natl Cancer Inst 90: 484–485.
Khuder S.A., 2001. Effect of cigarette smoking on major histological types of lung cancer: a meta-analysis. Lung Cancer 31:139–148.
Kim J.J., Wright T.C., Goldie S.J., 2002. Cost-effectiveness of alternative triage strategies for atypical squamous cells of undertermined significance. JAMA 287:2382–2390.
Kingsmore S.F., Patel D.D., 2003. Multiplexed protein profiling on antibody-based microarrays by rolling circle amplification. Curr Opin Biotechnol 14:74–81.
Kiyosawa K., Sodeyama T., Tanaka E. et al., 1990. Interrelationship of blood transfusion, non-A, non-B hepatitis and hepatocellular carcinoma: analysis by detection of antibody to hepatitis C virus. Hepatology 12:671–675.
Knekt P., Hakulinen T., Leino A. et al., 2000. Serum albumin and colorectal cancer risk. Eur J Clin Nutr 54:460–462.

Kosary C.L., McLaughlin J.K., 1993. Kidney and renal pelvis. In: Miller B.A., Ries L.A.G., Hankey B.E. et al. (eds.). Cancer Statistics Review: 1973–1990. Bethesda, MD: National Cancer Institute. NIH Pub. No. 93-2789, pp. 1–22.

Kreuzer M., Bofferra P., Whitley E. et al., 2000. Gender differences in lung cancer risk by smoking: a multicenter case–control study in Germany and Italy. Br J Cancer 82:227.

Kreuzer M., Kreienbrock L., Gerken M. et al., 1998. Risk factors for lung cancer in young adults. Am J Epidemiol 147:1028.

Kreuzer M., Kreienbrock L., Muller K.M. et al., 1999. Histologic types of lung carcinoma and age at onset. Cancer 85:1958.

Kulasingam S.I., Meyers E.R., 2003. Potential health and economic impact of adding a human papillomavirus vaccine to screening programs. JAMA 290(6):781–789.

Kuper H., Hsieh C.C., Stuver S.O. et al., 2000. Birth order, as a proxy for age at infection, in the etiology of hepatocellular carcinoma. Epidemiology 11:680–683.

Kuper H., Ye W., Broome U. et al., 2001. The risk of liver and bile duct cancer in patients with chronic viral hepatitis, alcoholism, or cirrhosis. Hepatology 34:714–718.

Lajous M., Mueller N., Cruz-Valdéz A.. et al., 2005. Determinants of prevalence, acquisition, and persistence of human papillomavirus in healthy Mexican military men. Cancer Epidemiol Biomarkers Prev 14(7):1710–1716.

Lampe J.W., Fredstrom S.B., Slavin J.L. et al., 1993. Sex differences in colonic function: a randomized trial. Gut 34:531–536.

Lauren P., 1965. The two histological main types of gastric carcinoma: diffuse and so-called intestinal-type carcinoma. Acta Path et Microbiol Scandinav 64:31–49.

Levin B., Raijman I., 1995. Malignant tumors of the colon and rectum. In: Haubrich W.S., Schaffner F., Berk J. (eds.). Bockus Gastroenterology. 5th edition. Philadelphia, PA: W.B. Saunders, pp. 1744–1772.

Levine O.S., Vlahov D., Brookmeyer R. et al., 1996. Differences in the incidence of hepatitis B and human immunodeficiency virus infections among injecting drug users. J Infect Dis 173:579–583.

Li C.I., Daling J.R., Malone K.E., 2005. Age-specific incidence rates of in situ breast carcinomas by histologic type, 1980 to 2001. Cancer Epidemiol Biomarkers Prev 14(4):1008–1011.

Li C.I., Malone K.E., White E., Daling J.R., 1997. Age when maximum height is reached as a risk factor for breast cancer among young U.S. women. Epidemiology 8(5):559–565.

Li X., Mutanen P., Hemminki K., 2001. Gender-specific incidence trends in lung cancer by histological type in Sweden. Eur J Cancer Prev 10:227–235.

Liaw K.L., Glass A.G., Manos M.M. et al., 1999. Detection of human papillomavirus DNA in cytologically normal women and subsequent cervical squamous cervical squamous intraepithelial lesions. J Natl Cancer Inst 91:954–960.

Liu N.S., Spitz M.R., Kemp B.L. et al., 2000. Adenocarcinoma of the lung in young patients: the MD Anderson experience. Cancer 88:1837.

Lleonart M.E., Garcia-Foncillas J., Sanchez-Prieto R. et al., 1998. Microsatellite instability and p53 mutations in sporadic right and left colon carcinoma: different clinical and molecular implications. Cancer 83:889–895.

London W.T., McGlynn K.A., 2006. Liver cancer. In: Schottenfeld D., Fraumeni J.F. Jr. (eds.). Cancer Epidemiology and Prevention. 3rd edition. New York: Oxford University Press, pp. 763–786.

Lubin J.H., Blot W.J., 1984. Assessment of lung cancer risk factors by histologic category. J Natl Cancer Inst 73:383–389.

Lucey D.R., Clerici M., Shearer G.M., 1996. Type 1 and type 2 cytokine dysregulation in human infectious, neoplastic and inflammatory diseases. Clin Microbiol Rev 9:532–562.

Lundegardh G., Lindgren A., Rohul A. et al., 1991. Intestinal and diffuse types of gastric cancer: secular trends in Sweden since 1951. Br J Cancer 64:1182–1186.

Mandelblatt J.S., Fahs M.C., 1988. The cost effectiveness of cervical cancer screening for low income elderly women. JAMA 259:2409–2413.
Mandelblatt J.S., Lawrence W.F., Womack S.M. et al., 2002. Benefits and costs of using HPV testing to screen for cervical cancer. JAMA 287:2372–2381.
Manton K.G., Akushevich I., Kulminski A., 2008. Human mortality at extreme ages: data from the national long term care survey and linked medicare records. Mathe Popul Stud 15:137–159.
Manton K.G., Stallard E., 1980. A two-disease model of female breast cancer: mortality in 1969 among white females in the United States. J NCI 64(1):9–16.
Manton K.G., Stallard E., Vaupel J.W., 1986. Alternative models for the heterogeneity of mortality risks among the aged. J Am Stat Assoc 81(395):635–644.
Mathew A., Devesa S.S., Fraumeni J.F. Jr., Chow W-H., 2002. Global increases in kidney cancer incidence 1973–1992. Eur J Cancer Prev 11:171–178.
McCrory D.C., Matchar D.B., Bastian L. et al., 1999. Evaluation of cervical cytology. Evidence report/technology assessment no. 5. (Prepared by Duke University under contract no. 290-97-0014). Rockwill, MD: Agency for Health Care Policy and Research. (AHCPR publication no. 99-E010).
McDowell M., Brody D., Hughes J., 2003. Has age at menarche changed? Results from the National Health and Nutrition Survey (NHANES), 1999–2004. J Adolesc Health 40(3):227–231.
Miller B.A., Feuer E.J., Hankey B.F., 1991. The increasing incidence of breast cancer since 1982: relevance of early detection. Cancer Causes Control 2:67–74.
Mink P.J., Sherman M.E., Devesa S.S., 2002. Incidence patterns of invasive and borderline ovarian tumors among white women and black women in the United States. Results from the SEER Program, 1978–1998. Cancer 95:2380–2389.
Moolgavkar S., Knudson A. Jr., 1981. Mutation and cancer: a model for human carcinogenesis. J Natl Cancer Inst 66:1037–1052.
Morita T., 2002. A statistical study of lung cancer in the annual of pathological aumiddlesy cases in Japan, from 1958 to 1997, with reference to time trends of lung cancer in the world. Jpn J Cancer Res 93:15–23.
Moscicki A-B., Ellenberg J.H., Vermund S.H. et al., 2000. Prevalence of and risks for cervical human papillomavirus infection and squamous intraepithelial lesions in adolescent girls. Impact of infection with Human Immunodeficiency Virus. Arch Pediatr Adolesc Med 154:127–134.
Mosmann T.R., Moore K.W., 1991. The role of IL-10 in crossregulation of Thl and Th2 responses. Immunol Today 12:A49–A53.
Muir C., Waterhouse J., Mack T. et al., 1987. Cancer Incidence in Five Continents. Vol. V. Lyon, France: International Agency for Research on Cancer.
Munoz N., Lingao A., Lao J et al., 1989. Patterns of familial transmission of HBV and the risk of developing liver cancer: a case–control study in the Philippines. Int J Cancer 44:981–984.
Myers E.R., McCrory D.C., Nanda K. et al., 2000. Mathematical model for the natural history of human papillomavirus infection and cervical carcinogenesis. Am J Epidemiol 151(12):1158–1171.
National Drug Control Strategy, 2000. Annual Report. Table 3, Appendix: Drug Related Data. At: http://www.ncjrs.gov/ondcppubs/publications/policy/ndcs00/strategy2000.pdf
Nekolla E.A., Kellerer A.M., Kuse-Isingschulte M., Eder E., Spiess H., 1999. Malignancies in patients treated with high doses of radium-224. Radiat Res 152(Suppl 6):S3–S7.
Nelson W.G., De Marzo A.M., Deweese T.L. et al., 2001. An opportunity for prostate cancer prevention. Ann NY Acad Sci 952:135–144.
Neuhausen S.L., 1999. Ethnic differences in cancer risk resulting from genetic variation. Cancer 86:2575–2582.

Ngelangel C., Munoz N., Bosch F.X. et al., 1998. Causes of cervical cancer in the Philippines: a case–control study. J Natl Cancer Inst 90:43–49.
Nguyen L.T., Ramanathan M., Weinstock-Guttman B., Baier M., Brownscheidle C., Jacobs L.D., 2003. Sex differences in vitro pro-inflammatory cytokine production from peripheral blood of multiple sclerosis patients. J Neurol Sci 209(1–2):93–99.
Nichols H.B., Trentham-Dietz A., Hampton J.M., et al., 2006. From menarche to menopause: trend among U.S. women born from 1912 to 1969. Am J Epidemiol 164(10):1003–1011.
Niederau C., Lange S., Heintges T. et al., 1998. Prognosis of chronic hepatitis C: results of a large prospective cohort study. Hepatology 28:1687–1695.
Nikolaou V.A., Sypsa V., Stefanaki I. et al., 2008. Risk associations of melanoma in a Southern European population: results of a case/control study. Cancer Causes Control 19:671–679.
Nunney L., 1999. Lineage selection and the evolution of multistage carcinogenesis. Proc Biol Sci 266:493–498.
Okuda K., Nakanuma Y., Miyazaki M., 2002. Cholangiocarcinoma: recent progress. Part I: epidemiology and etiology. J Gastroenterol Hepatol 17:1049–1055.
Oortmarssen G.V., Habbema J., 1991. Epidemiological evidence for age-dependent regression of pre-invasive cervical cancer. Br J Cancer 64:559–565.
Pacifici R., Rifas L., McCraken R., Vered I., McMurfy C., Avioli L.V., Peck W.A., 1989. Ovarian steroid treatment blocks a postmenopausal increase in blood monocyte Il-1 release. Proc Natl Acad Sci USA 86(7):2398–2402.
Page D.L., Jopaze H., 1991. Non infiltrating (in situ) carcinoma. In: Bland K.I., Copeland E.M. (eds.). The Breast, Comprehensive Management of Benign and Malignant Diseases. Philadelphia, PA: W.B. Saunders, p. 169.
Page D.L., Steel C.M., Dixon J.M., 1995. ABC of breast cancer: carcinoma in situ and patients at high risk of breast cancer. BMJ 310:39–42.
Parkin D.M., Pisani P., Munoz N. et al., 1999a. The global health burden of infection associated cancers. In: Newton R., Beral V., Weiss R.A. (eds.). Infections and Human Cancer. Plain-view, NY: Cold Spring Harbor Laboratory Press, pp. 5–33.
Parkin D.M., Pisani P., Ferlay J., 1999b. Estimates of the worldwide incidence of 25 major cancers in 1990. Int J Cancer 80:827–841.
Parkin D.M., Whelan S.L., Ferlay J et al., 2002. Cancer Incidence in Five Continents. Vol. VIII. Lyon, France: International Agency for Research on Cancer.
Parmigiani G., Berry D., Aquilar O., 1998. Determining carrier probabilities for breast cancer-susceptibility genes BRCA1 and BRCA2. Am J Hum Genet 62:145–158.
Pawelec G., Barnett Y., Forsey R., Frasca D., Globerson A., McLeod J., Caruso C., Franceschi C., Fulop T., Gupta S. et al., 2002. T cells and aging. Front Biosci 7:1056–1183.
Perou C.M., Sorlie T., Elsen M.B. et al., 2000. Molecular portraits of human breast tumours. Nature 406:747–752.
Peters P.K., Chao A., Mack T.M. et al., 1986. Increased frequency of adenocarcinoma of the uterine cervix in young women in Los Angeles County. J Natl Cancer Inst 76:423–428.
Pike M.C., Krailo M.D., Henderson B.E. et al., 1983. "Hormonal" risk factors, "breast tissue age" and the age-incidence of breast cancer. Nature 303:767–770.
Platz E.A., Giovannucci E., 2006. Prostate cancer. In: Schottenfeld D., Fraumeni J.F. (eds.). Cancer Epidemiology and Prevention. 3rd edition. New York: Oxford University Press, pp. 1128–1150.
Platz E.A., Rimm E.B., Willett W.C. et al., 2000. Racial variation in prostate cancer incidence and in hormonal system markers among male health professionals. J Natl Cancer Inst 92:2009–2017.
Poynard T., Bedossa P., Opolon P., 1997. For the OBSVIRC, METAVIR, CLINIVIR, and DOSIVIR groups. Natural history of liver fibrosis progression in patients with chronic hepatitis C. Lancet 349:825–832.

Prescott E., Osler M., Hein H.O. et al., 1998. Gender and smokingrelated risk of lung cancer. The Copenhagen Center for Prospective Population Studies. Epidemiology 9:79–83.
Prorok P.C., Kramer B.S. Gohagan J.K., 1999. Screening theory and study design. In: Kramer B.S., Gohagan J.K., Prorok P.C. (eds.). Cancer Screening. New York: M. Dekker, pp. 29–53.
Reynolds T., 1992. States begin CDC-sponsored breast and cervical cancer screening. J Natl Cancer Inst 84:7–9.
Risch H.A., Howe G.R., Jain M. et al., 1993. Are female smokers at higher risk for lung cancer than male smokers? Am J Epidemiol 138:281.
Robboy S.J., Noller K.L., O'Brien P. et al., 1984. Increased incidence of cervical and vaginal dysplasia in 3,980 diethylstilbestrol-exposed young women. Experience of the National Collaborative Diethtylbestrol Adenosis Project. JAMA 252:2979–2983.
Romagnani S., 1994. Lymphokine production by human T cells in disease states. Annu Rev Immunol 1994.:227–257.
Rosner B., Colditz G., 1996. Nurses' health study: log-incidence mathematical model of breast cancer incidence. J Natl Cancer Inst 88:359–364.
Rosner B., Colditz G.A., Willett W.C., 1994. Reproductive risk factors in a prospective study of breast cancer: the Nurses' Health Study. Am J Epidemiol 139:819–835.
Ross R.K., Henderson B.E., 1994. Do diet and androgens alter prostate cancer risk via a common etiologic pathway? J Natl Cancer Inst 86:252–254.
Sakr W.A., Grignon D.J., Haas G.P. et al., 1995. Epidemiology of high grade prostatic intraepitheluial neoplasia. Pathol Res Pract 191(9):838–841.
Sakr W.A., Grignon D.J., Haas G.P. et al., 1996. Age and racial distribution of prostatic intraepithelial neoplasia. Eur Urol 30(2):138–144.
Sakr W.A., Haas G.P., Cassin B.F. et al., 1993. The frequency of carcinoma and intraepithelial neoplasia of the prostate in young male patients. J Urol 150:379–385.
Schairer C., Btrinton L.A., Devesa S.S. et al., 1991. Racial differences in the risk of invasive squamous cell cervical cancer. Cancer Causes Control 2:283–290.
Schiffman M.H., Hildesheim A., 2006. Cervical cancer. In: Schottenfeld D., Fraumeni J.F. Jr. (eds.). Cancer Epidemiology and Prevention. 3rd edition. New York: Oxford University Press.
Schlecht N.F., Kulaga S., Robitaille J. et al., 2001. Persistent human papillomavirus infection as a predictor of cervical intraepithelial neoplasia. JAMA 286:3106–3114.
Schnitt S.J., Morrow M., 1999. Lobular carcinoma in situ: current concepts and controversies. Semin Diagn Pathol 16(3):209–223.
Schwartz S.M., Weiss N.S., 1986. Increased incidence of adenocarcinoma of the cervix in young women in the United States. Am J Epidemiol 124:1045–1047.
Sherman M.E., Lorincz A.T., Scott D.R. et al., 2003. Baseline cytology, human papillomavirus testing, and risk for cervical neoplasia: a 10-year cohort analysis. J Natl Cancer Inst 95(1):46–52.
Shiratori Y., Shiina S., Imamura M. et al., 1995. Characteristic difference of hepatocellular carcinoma between hepatitis B- and C-viral infection in Japan. Hepatology 22:1027–1033.
Silvermann D.T., Hoover R.N., Brown L.M. et al., 2003. Why do black Americans have a higher risk of pancreatic cancer than white Americans? Epidemiology 14:45–54.
Simpson R.J., Hammacher A., Smith D.K., Matthews J.M., Ward L.D., 1997. Interleukin-6: structure–function relationships. Protein Sci 6(5):929–955.
Sirovich B.E., Schwartz L.M., Woloshin S., 2003. Screening men for prostate and colorectal cancer in the United States: Does practice reflect the evidence? JAMA 289:1414–1420.
Skarin A.T., Herbst R.S., Leong T.L. et al., 2001. Lung cancer in patients under 40. Lung Cancer 32:255.
Soong R., Powell B., Elsaleh H. et al., 2000. Prognostic significance of TP53 gene mutation in 995 cases of colorectal carcinoma. Influence of tumor site, stage, adjuvant chemotherapy and type of mutation. Eur J Cancer 36:2053–2060.

Southern S.A., Herrington C.S., 1998. Molecular events in uterine cervical cancer. Sex Transm Inf 74:101–109.
Spitz M.R., Wu X., Wilkinson A., Wei Q., 2006. Cancer of the lung. In: Schottenfeld D., Fraumeni J.F. Jr. (eds.). Cancer Epidemiology and Prevention. 3rd edition. New York: Oxford University Press, 2006.
Stellman S.D., Muscat J.E., Thompson S. et al., 1997. Risk of squamous cell carcinoma and adenocarcinoma of the lung in relation to lifetime filter cigarette smoking. Cancer 80:382–388.
Stuver S., Adami H-O., 2002. Cervical cancer. In: Adami H-O., Hunter D., Trichopoulos D. (eds.). Textbook of Cancer Epidemiology. New York: Oxford University Press, p. 347.
Stuver S.O., Trichopoulos D., 1994. Liver cancer. In: Doll R., Fraumeni J., Muir C. (eds.). Trends in Cancer Incidence and Mortality. Cancer Surveys. Vol. 19/20. London: Imperial Cancer Fund, pp. 99–124.
Stuver S., Trichopoulos D., 2002. Cancer of the liver and biliary tract. In: Adami H-O., Hunter D., Trichopoulos D. (eds.). Textbook of Cancer Epidemiology. New York: Oxford University Press, pp. 212–232.
Surgeon General, 1989. Reducing the health consequences of smoking: 25 years of progress. Washington, DC: U.S. Government Printing Office.
Swain S.M., 1989. Non-invasive breast cancer. Lobular carcinoma in situ. Incidence, presentation, guidelines to treatment. Oncology 3:35.
Swan J., Breen N., Coates R.J., Rimer B.K., Lee N.C., 2003. Progress in cancer screening practices in the United States: results from the 2000 National Health Interview Survey. Cancer 97:1528–1540.
Syrjanen K., 2007. Mechanisms and predictors of high-risk human papillomavirus (HPV) clearance in the uterine cervix. Eur J Gynaecol Oncol 28(5):337–351.
Tabor E., 1998. Viral hepatitis and liver cancer. In: Goldin R.D., Thomas H.C., Gerber M.A. (eds.). Pathology of Viral Hepatitis. London: Arnold, pp. 161–177.
Tanaka Y., Hanada K., Mizokami M. et al., 2002. A comparison of the molecular clock of hepatitis C virus in the United States and Japan predicts that hepatocellular carcinoma incidence in the United States will increase over the next two decades. Proc Natl Acad Sci USA 99(24):15584–15589.
Tavassoli F.A., Schnitt S.J., 1992. Pathology of the Breast. New York: Elsevier.
The Bethesda System, 1993. The Bethesda System for reporting cervical/vaginal cytologic diagnosis: revised after the second National Cancer Institute Workshop, April 29–30, 1991. Acta Cytol 37:115–124.
Thorpe L.E., Ouellet L.J., Hershow R. et al., 2002. Risk of hepatitis C virus infection among young adult injection drug users who share injection equipment. Am J Epidemiol 155:645–653.
Thune I., Furberg A.S., 2001. Physical activity and cancer risk: dose-response and cancer, all sites and site-specific. Med Sci Sports Exerc 33(6 Suppl):S530–S550.
Todoroki I., Friedman G.D., Slattery M.L. et al., 1999. Cholecystectomy and the risk of colon cancer. Am J Gastroenterol 94:41–46.
Tokunaga M., Land C.E., Tokuoka S., Nishimori I., Soda M., Akiba S., 1994. Incidence of female breast cancer among atomic bomb survivors. Hiroshima and Nagasaki, 1950–1985. Radiat Res 138:209–223.
Travis W.D., Lubin J., Ries L. et al., 1996. United States lung carcinoma incidence trends: declining for most histologic types among males, increasing among females. Cancer 77:2464–2470.
Turchetti D., Cortesi L., Federico M. et al., 2000. BRCA1 mutations and clinicopathological features in a sample of Italian women with early-onset breast cancer. Eur J Cancer 36:2083–2089.

Ursin G., Pike M.C., Preston-Martin S. et al., 1996. Sexual, reproductive, and other risk factors for adenocarcinoma of the cervix: results from a population-based case–control study (California, United States). Cancer Causes Control 7:391–401.

Van de Velde N., Brisson M., Boily M.C., 2007. Modeling human papillomavirus vaccine effectiveness: quantifying the impact of parameter uncertainty. Am J Epidemiol 165(7):762–775.

van Leenders G.J., Gage W.R., Hick J.L. et al., 2003. Intermediate cells in human prostatic epithelium are enriched in proliferative inflammatory atrophy. Am J Pathol 162:1529–1537.

Vieth M., Masoud B., Meining A., Stolte M., 2000. *Helicobacter pylori* infection: protection against Barrett's mucosa and neoplasia? Digestion 62:225–231.

Villiano S.A., Vlahov D., Nelson K.E. et al., 1997. Incidence and risk factors for hepatitis C among injection drug users in Baltimore, Maryland. J Clin Microbiol 35:3274–3277.

Wang S.S., Sherman M.E., Lacey J.V. Jr., Devesa S., 2004. Cervical adenocarcinoma and squamous cell carcinoma incidence trends among white women and black women in the United States for 1976–2000. Cancer 100(5):1035–1044.

Weber S., O'Reilly E.M., Abou-Alfa G.K. et al., 2004. Liver and bile duct cancer. In: Abeloff M.D., Armitage J.O., Niederhuber J.E., Kastan M.B., McKenna W.G. (eds.). Clinical Oncology. 3rd edition. London: Elsevier Inc. Churchill, Livingstone.

Wilson S., Jones L., Coussens C. et al. (eds.), 2002. Cancer and the Environment. Gene-Environment Interaction. Washington, DC: Institute of Medicine, National Academy Press.

Wise A.M., Stamney T.A., McNeal J.E., Clayton J.L., 2002. Morphologic and clinical significance of multifocal prostate cancers in radical prostatectomy specimens. Urology 60:264–269.

Women and Smoking: A Report of the Surgeon General, 2001. Rockville, MD: Office of the Surgeon General, Public Health Service, U.S. Department of Health and Human Services, pp. 193–212.

Wu K., Willett W.C., Fuchs C.S. et al., 2002. Calcium intake and risk of colon cancer in women and men. J Natl Cancer Inst 94:437–446.

Wynder E.L., Muscat J.E., 1995. The changing epidemiology of smoking and lung cancer histology. Environ Health Perspect 103(Suppl):143–8.

Yang P., Cerhan J.R., Vierkant R.A., Olson J.E., Vachon C.M., Limburg P.J., Parker A.S., Anderson K.E., Sellers T.A., 2002. Adenocarcinoma of the lung is strongly associated with cigarette smoking: further evidence from a prospective study of women. Am J Epidemiol 156(12):1114–1122.

Yoo K.Y., Tajima K., Inoue M. et al., 1999. Reproductive factors related to the risk of colorectal cancer by subsite: a case–control analysis. Br J Cancer 79:1901–1906.

Yoshinaka H., Shimazu H., Fukumoto T. et al., 1991. Superficial esophageal carcinoma: a clinicopathological review of 59 cases. Am J Gastroenterol 86(10):413–417.

Yu M-W., Chen C-J., 1993. Elevated serum testosterone levels and risk of hepatocellular carcinoma. Cancer Res 53:790–794.

Yu L., Sloane D.A., Guo C. et al., 2006. Risk factors for primary hepatocellular adenocarcinoma in black and white Americans in 2000. Clin Gastroenterol Hepatol 4:355–360.

Zaino R.J., 2002. Symposium, part I: Adenocarcinoma in situ, glandular dysplasia, and early invasive adenocarcinoma of the uterine cervix. Int J Gynecol Pathol 21:314–326.

Zang E.A., Wynder E.L., 1996. Differences in lung cancer risk between men and women: examination of the evidence. J Natl Cancer Inst 88:183.

Zheng T., Holford T.R., Ma Z. et al., 1996. The continuing increase in adenocarcinoma of the uterine cervix: a birth cohort phenomenon. Int J Epidemiol 25(2):252–258.

# Chapter 8
# Risk Factors Intervention

## 8.1 Environmental Risk Factor Contribution to Cancer and Noncancer Diseases

Age–incidence relationships and experimental evidence suggest that cancer is a polygenic multifactorial disease, where the environmental components are important causes of most cancer types (see Chapter 3). A recent study of 90,000 twins (Lichtenstein et al., 2000) showed that the nonshared random environmental effect was the largest factor for all cancers, accounting for 58–82% of the total variation. Various environmental factors, sometimes isolated, but most often in combinations, influence human health, contributing to both cancer and noncancer deaths, including such leading death causes as cardio- and cerebrovascular diseases, cancer, and chronic obstructive pulmonary disease.

Risk factor intervention analysis is used for investigating how systematic changes of age-specific values of parameters, featuring one or group of risk factors, impact life expectancy, mortality, morbidity, and other demographic and population health characteristics. An advantage of the binomial quadratic hazard model (described in Chapter 4) is its capacity to represent a wide spectrum of potential interventions by modifying selected combinations of model parameters. Manton and Stallard (1988) modeled a series of interventions related to the dynamic control of specific risk factors and their variances. The effect was roughly 5% of the baseline value of life expectancy when one risk factor is modeled in the interventions and can exceed 10% (especially at advanced ages) when dynamic control of multiple risk factors is considered (see Table 4.9.1 of Manton and Stallard, 1988). Furthermore, they estimated changes in direct and indirect costs under different interventions. However, as we discussed in Chapter 5, assumptions for which the binomial quadratic hazard model was constructed, are not always valid. Moreover, only limited sets of intervention scenarios can be reduced to analyses of means and variances of the risk factors. Most interventions require individual specific interventions and/or interventions dependent on a current

K.G. Manton et al., *Cancer Mortality and Morbidity Patterns in the U.S. Population*, Statistics for Biology and Health, DOI 10.1007/978-0-387-78193-8_8,
© Springer Science+Business Media, LLC 2009

value of a risk factor(s). This requires more flexibility than the standard binomial quadratic hazard model. One approach with potential for performing such individual-oriented interventions is based on microsimulation (see Chapter 5). The ability to estimate standard errors of projected characteristics and to perform sensitivity and error propagation analysis to estimate the impact of various factors, which are not modeled in the basic scenario, are the advantages of microsimulation. Another is the potential to incorporate various individual characteristics, performing projections for population groups, and modeling latent heterogeneity in cases when data sources are limited.

Therefore, for health-based population forecasting of risk factors' effects on human health, we used dynamic microsimulation, which has become popular in social sciences (Gilbert and Troitszch, 1999; Wolf, 2001). It is employed in such health effects studies as analysis of risk factor patterns, including smoking (Wolfson, 1994; Evans et al., 1995), in cancer screening (e.g., van den Akker-van Marle et al., 2002, Feuer et al., 2004), and in the evaluation of interventions (Vanness et al., 2005, Akushevich et al., 2005, Kravchenko et al., 2005) and cost effectiveness (Ramsey et al., 2000). One fruitful approach is based on individual event histories. A "simplified" life of an individual includes essential life events like birth, pregnancy/childbirth, and death. These events (both in real life and in the model) occur stochastically at rates which can be age-, sex-, and race-specific. Implementation of this strategy in a microsimulation model allows one to project population characteristics under specific temporal scenarios. We extended the multidimensional stochastic process model used for projecting population changes under simulated temporal changes in the distribution of major risk factors in population (Manton and Stallard, 1988; Manton et al., 1992) to make fertility a function of changes in simulated parameters (Manton and Akushevich, 2003) and to include the nonlinear dynamics of risk factors (Akushevich et al., 2005).

Based on this model, we developed a microsimulation strategy to forecast the effects of risk factors on individual health-related parameters changes and on mortality risk, thus providing a flexible platform for interventional studies. Generalizations of these forecasting models to describe large populations, while including the effects of risk factors on fertility, are often hampered by the lack of the necessary longitudinal data. Simplifying assumptions are thus required to make forecasts for large population, such as the U.S. population: (1) all model parameters and rates have to be defined by observations or obtained by nonparametric estimations; (2) the model has to be biomedically motivated; (3) the model has to be flexible enough to study specific, "precise" interventions; and (4) the model should be able to simulate the effects on an individual basis. A model constructed using such principles will be a flexible tool for modeling and forecasting, while allowing sensitivity analyses to estimate both, statistical uncertainty and uncertainties concerning specific choices of model parameters.

### *8.1.1 Health-Based Population Forecasting Effects of Smoking on Mortality from Cancer and Noncancer Diseases*

We used a one risk factor (cigarette smoking) version of the model to project future changes in the structure/health effects of the U.S. population assuming the importance of this environmental risk factor for human health:

- Smoking is the most widespread of human addictions – one in three adults in the world are regular smokers (Gajalakshmi et al., 2000). In 2005, approximately 20.9% (45.1 million) of U.S. adults were current cigarette smokers, with 80.8% of those (36.5 million) who were smoking every day (MMWR, 2006).
- Smoking-associated diseases are still the leading causes of death among diseases associated with preventable risk factors (18.1% of all deaths in 2000), poor diet and physical inactivity share the second position (16.6% of all deaths), with the third position held by mortality from diseases associated with alcohol consumption (3.5% of all deaths) (Mokdad et al., 2004).
- Smoking is associated with a broad spectrum of cancer and noncancer diseases, including those which are the leading causes of human mortality. The average annual number of deaths attributable to cigarette smoking in the United States in 1997–2001 was 259,494 in men and 178,408 in women, with 39.8% of those deaths from cancer, 34.7% from cardiovascular diseases and 25.5% from respiratory diseases.
- Lung cancer is the leading cause of smoking-attributable death (123,836 deaths per year), followed by chronic obstructive pulmonary disease (90,582 deaths per year), and ischemic heart disease (86,801 deaths per year) (MMWR, 2005) (plus an estimated 38,112 lung cancer and heart disease deaths are attributed annually to exposure to secondhand smoking). In the United States, during the last decade, the death rate from lung cancer continued to decrease in males; in females the rate of increase has recently slowed, thus reflecting cohort based reductions in tobacco use in both sexes (JAMA, 2002, and discussion in Chapter 6). Recently it has been concluded by the IARC (2004) that smoking is associated with an increased risk of cancers of the lung and bronchus, nasal cavity and nasal sinuses, oral cavity, pharynx, esophagus, stomach, pancreas, liver, kidney (renal-cell carcinoma), urinary bladder and renal pelvis, cervix uteri, and myeloid leukemia. The IARC theorized that the positive findings might be attributable to bias in case of prostate cancer, and to confounding in colorectal cancer, and the evidence showed a negative relationship with smoking and risk for endometrial cancer. The evidences indicated no relationship with smoking and breast cancer in females and were inconsistent or sparse for cancers of the salivary gland, small intestine, gallbladder and bile ducts, soft-tissue sarcoma, adrenals, melanoma, other skin cancers, cancers of the ovary, testis, CNS, thyroid, Hodgkin's lymphoma, other lymphomas, myeloma, and lymphatic leukemia. Data from a large follow-up of a cohort of male British doctors (50 years of observation) were generally in accordance with IARC conclusions (Doll et al., 2005).

- It has been found in series of case–control studies that cigarette smoking was associated with an increased risk of both primary and secondary infertility in women (Dorfman, 2008; Sepaniak et al., 2006). The adjusted primary infertility odds ratio, estimated for current female smokers at ages from 15 to 49, was 1.9 (95% CI 1.5–2.3) (Joesoef et al., 1993). The relative risk of secondary infertility in female ever-smokers was 2.6 (95% CI 1.2–6.0) (Tzonou et al., 1993). Recent studies demonstrated that smoking is hazardous to the female fetus not only in the short term but may affect her future ability to conceive (the fecundability odds ratio for women who were exposed to smoking in utero was 0.65, 95% CI 0.47–0.91) (Jensen et al., 2006). In males, decreases in sperm quality, alteration of male sex hormones, and erectile dysfunction were found in smokers compared with nonsmokers, influencing male smokers' fertility (Robbins et al., 2005; Sepaniak et al., 2005; Peate, 2005; Arabi and Moshtaghi, 2005; Mostafa et al., 2006; Ramlau-Hansen et al., 2007) – however, the exact pathological mechanisms of smoking effects on male fertility have not yet been verified.
- An important Federal Health Objective for 2010 is to reduce the prevalence of cigarette smoking among adults by half, i.e., to 12% prevalence in the U.D. adult population (U.S. DHHS, 2000, objective 27.1a). The current recent decline in cigarette smoking prevalence in the adults, however, is not fast enough to meet the 2010 objectives: during the period 1998–2005, cigarette smoking decreased only from 24.0 to 20.9%, so it is unlikely the rate will drop to 12.0% by 2010 (MMWR, 2006).

Smoking has effects on multiple health outcome variables, i.e., disease and mortality, as well as fertility. The core of the model is the parametric linkage of mortality caused by specific diseases, and female fertility rates with a risk factor, through the relative risks of infertility or disease incidence in exposed groups (i.e., individuals who smoke). We did not include male fertility changes due to smoking in our current model because of the much smaller effect on male, compared with female, fertility.

The basic scenario in our projection is formulated with (1) time-independent age patterns of fertility, cause-specific mortality, and smoking rates, (2) fertility, cause-specific mortality, and smoking rates equal to those observed in a base year, and (3) age- and time-independent relative risks of infertility and incidence of smoking-attributable diseases. In spite of the simplifying assumptions, the basic scenario is still realistic. Implementation of distinct effects, (e.g., a 30-year lag period for smoking effects), does not require additional assumptions, because the rates are constructed to be time independent. However, although posterior characteristics are defined as ratios, and, therefore, are expected to be stable with respect to various corrections, the contributions of these effects are important and are analyzed in a sensitivity analysis.

It was not this projection's purpose to describe the "essential" changes in smoking prevalence in long-term projections (which may be, to a great extent, under the influence of various factors that may increase smoking prevalence, such as stress-related behavior in the response of the individual to social, economic, and political changes with time–events difficult to forecast), but to forecast the consequences for a population of active "nonessential" interventions in smoking prevalence (including medical and social countermeasures to prevent the onset of smoking behavior and by encouraging smoking cessation). Our current model has several other limitations on model parameters, because it does not include characteristics of smoking behavior/patterns (e.g., how many cigarettes an individual smokes per day, the age of a person's smoking habit initiation, duration of smoking, persons being under passive smoking influence, such as only or in addition to the person being a smoker her or himself), as well as individuals who quit smoking (age of quitting and the duration and intensity of the habit before quitting). The type of tobacco used such as cigarettes, pipes, cigars, or smokeless tobacco can also be important, because different ways of tobacco consumption have different risks for smoking-associated diseases. These limitations illustrate a simpler type of proposed intervention – a sort of "skeleton" covered with the "muscles" of key parameters for easier demonstration, but which may be covered step by step with all the necessary "flesh and skin", that will improve the simulation of real populations with interferences and feedback of various risk factors and variations in risk factor intensity, duration, subtype, and other traits. If the data sets containing this information become available, it is possible for this model to be modified to include all of these parameters for future applications.

In our projection, we used the white population to reflect the average U.S. smoking tendencies and population health effects, because that is the population group for which we have the most detailed data. Data sources for population characteristic estimation included National Vital Statistics Reports (NVSR), Morbidity and Mortality Weekly Reports (MMWR), and consists of basic population characteristics (Health, United States, 2002; Martin et al., 2002a), death rates in population (Minino et al., 2002; Anderson, 2002; GMWKIII, 1998; Giovino et al., 1994; Fellows et al., 2002), and fertility characteristics in a population (Martin et al., 2002b; Ventura et al., 2003).

The recalculation of age-specific mortality rates for specific risk factor scenarios was performed using information on the smoking-attributed relative risks of disease incidences ($RR_d$) and disease-specific mortality rates ($\mu_d$). A straightforward calculation for mortality rates for age, sex, and race groups is

$$\bar{\mu} = \mu \left[ 1 + \sum_d \frac{(RR_d - 1)(\bar{r}_s - r_s)}{RR_d r_s + 1 - r_s} \frac{\mu_d}{\mu} \right], \tag{8.1}$$

where $r_3$ are the age-, sex-, and race-specific smoking rates observed in the studied population; $\mu$ is the observed mortality rate; and $\bar{\mu}$ is the mortality rate corresponding to the hypothetical smoking level $\bar{r}_s$. Index $d$ runs over all diseases of interest. A disease contributes to a change in a mortality rate when the smoking-attributed relative risk $RR_d$ deviates significantly from one, and when the disease-specific mortality rate $\mu_d$ is significant.

Based on these criteria, we selected the following smoking-associated diseases to be included in model:

- cancers of lip, oral cavity, pharynx, esophagus, pancreas, larynx, trachea, lung, bronchus, cervix uteri, urinary bladder, kidney and other urinary;
- cardio- and cerebrovascular diseases: hypertensive disease, ischemic heart disease, other forms of heart disease, cerebrovascular disease, atherosclerosis, aortic aneurysm, and other arterial diseases;
- diseases of the respiratory tract: pneumonia and influenza, bronchitis and emphysema, chronic airway obstruction, other respiratory diseases;
- infant's diseases: respiratory distress syndrome, sudden infant death syndrome.

The observed mortality rate ($\mu_d$) due to a disease includes contributions from smoking ($\mu_s^d$) and nonsmoking ($\mu_n^d$) individuals:$\mu^d = \mu_s^d + \mu_n^d$. The relative risk of the incidence of a smoking-attributable disease $d$ is

$$RR_d = \frac{\mu_s^d/p_s}{\mu_n^d/p_n} = \frac{\mu_s^d(1-r_s)}{\mu_n^d r_s},$$

where $p_s$ and $p_x$ are prevalence of smokers and nonsmokers, and $p_s$ is equal to the smoking rate $r_s = p_s$. These equations are sufficient to express $\mu_s^d$ and $\mu_n^d$ in terms of observed quantities:

$$\mu_n^d = \frac{1-r_s}{RR_d r_s + 1 - r_s}\mu^d, \mu_s^d = \frac{RR_d r_s}{RR_d r_s + 1 - r_s}\mu^d.$$

An important question is how this would change, if another smoking rate $\bar{r}_s$ is observed. Formally, we have the same set of equations with $r_s \to \bar{r}_s$. An equation, that connects equations with $r_s$ and $\bar{r}_s$, is obtained by observing that the mortality of nonsmokers can change only because of a change in the number of nonsmokers, i.e., the ratios $\mu_n^d/p_n$ and $\bar{\mu}_n^d/\bar{p}_n$ are equal, being independent of smoking rate. This implies

$$\bar{\mu}^d = \frac{RR_d\bar{r}_s + 1 - \bar{r}_s}{RR_d r_s + 1 - r_s}\mu^d.$$

Summing over all diseases produces the recalculated total mortality rate in equation (8.1).

This calculation admits the inclusion of former smokers, i.e., three groups of individuals are considered: smokers, nonsmokers, and former smokers. The result is,

$$\bar{\mu} = \mu\left[1 + \sum_d \left(\frac{(RR_d^s - 1)(\bar{r}_s - r_s) + (RR_d^f - 1)(\bar{r}_f - r_f)}{1 - r_s - r_f + RR_d^s r_s + RR_d^f r_f}\right)\frac{\mu_d}{\mu}\right], \quad (8.2)$$

Equation (8.2) leads to equation (8.1) for $r_f \rightarrow 0$.

For infertility, we consider three groups of women: fertile, infertile with primary, and infertile with secondary infertility. Observed quantities are (1) the birth rate $b = Cp_f$ (where $p_f$ is the prevalence of fertile women, and coefficient $C$ reflects the fraction of women desiring to be pregnant), (2) prevalences of women with primary and secondary infertility ($p_{1,2}$), and (3) infertility relative risks $RR_{1,2}$ of women with primary and secondary infertility:

$$RR_{1,2} = \frac{p_{1,2}^s / p^s}{p_{1,2}^n / p^n},$$

The calculation, where two infertility types are considered, instead of a group of smoking-associated diseases, produces equation (8.3).

The age-specific birth rate corresponding to the new smoking rate $\bar{r}_s$, and the observed primary ($p_1$) and secondary ($p_2$) infertility prevalence, is

$$\bar{b} = b\left[1 - \sum_i \frac{(RR_i - 1)(\bar{r}_s - r_s)}{RR_i r_s + 1 - r_s} \frac{p_i}{1 - \sum_{i'} p_{i'}}\right], \quad (8.3)$$

where $i$, $t' = 1,2$; $RR_{1,2}$ are the primary and secondary infertility relative risks, and $b$ is the age-specific birth rate observed in a base year. Equations (8.1) and (8.3) reproduce the observed mortality and fertility rates, when new and currently observable smoking rates are equal, i.e., when $\bar{r}_s$ tends to $r_s$.

The population is projected by ageing individuals and having them (1) giving birth to new persons according to equation (8.3) and (2) dying according to equation (8.1). Age is the only attribute assigned to individuals which changes annually. All quantities are derived from the projected population size of the group$P_g^{y_i,y_f}(r)$, where $g = m,f$ – denotes gender; $y_{i,f}$ – initial and final ages of the population subgroup, $r$ – correction factors to the current smoking rate ($r = 1/2$ means hypothetical 50% reduction in smoking prevalence, $r = 1$ – current smoking prevalence, $r = 2$ – hypothetical doubled smoking prevalence).

Population losses are defined as follows:

$$PL_g^{y_i,y_f}(r_b, r_a) = \left[\frac{P_g^{y_i,y_f}(r_b)}{P_g^{y_i,y_f}(r_a)} - 1\right] \times 100\%$$

The complete list of projected parameters includes initial 1-year population distribution for white population in 2000; age- and sex-specific smoking rates; age- and sex-specific death rates; sex-specific relative risks of death caused by smoking-attributable diseases; relative risks of primary and secondary female infertility; birth probability of males versus females; age-specific birth rate for 2000 (see Akushevich et al., 2007). The results we present below include forecasting the contribution of cancer and noncancer diseases to human mortality, population loses, and life expectancy changes (other results see in Akushevich et al., 2007).

The first component of the microsimulation is the initial population age distribution (i.e., the age distribution of the U.S. population as recorded in the 2000 Census). This age distribution is presented in Fig. 8.1a. It describes the changes in the population age distributions: e.g., smoothing of the "baby boom" peak and population aging. Figure 8.1b gives the age distribution of the U.S. population projected 100 years.

Using the simulated age distribution $P_g^{y_i,y_f}(r)$ values, the *population losses* dependent on smoking prevalence were calculated for males and females (see Table 8.1). These are calculated for three-age intervals (0–18, 19–67, and 68+ years old, as children/adolescents, working population, and retired persons, respectively). We used age 67 because that is the new normal retirement age for

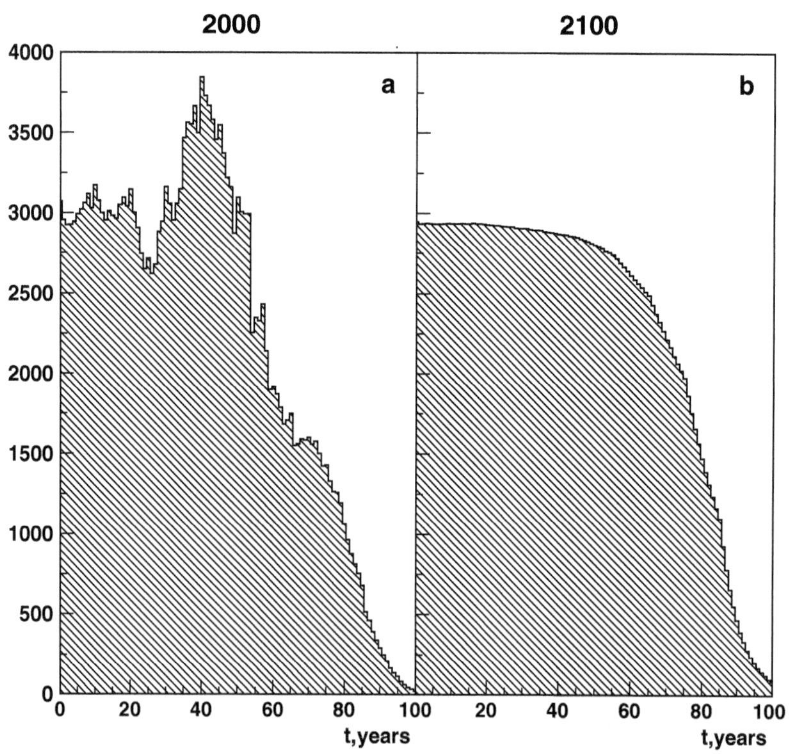


**Fig. 8.1** (**a, b**) Initial (**a**) and 100 years projected age distribution of white American population with current smoking prevalence, years 2000 and 2100, respectively

**Table 8. 1** Population losses (%) in population in 20- and 100-year projections, depending on smoking prevalence (as ratio of populations with three different smoking prevalences: $R_{1/2}$ – 50% reduced smoking prevalence, $R_1$ – current smoking prevalence, and $R_2$ – doubled smoking prevalence)

| Ratio of smoking | Male (age groups, years) | | | Female (age groups) | | |
|---|---|---|---|---|---|---|
| prevalence | 0–18 | 19–67 | >68 | 0–18 | 19–67 | >68 |
| $R_1/R_{1/2}$: | | | | | | |
| in 20 years | 0.79±0.03 | 0.44±0.00 | 4.87±0.02 | 0.81±0.03 | 0.22±0.00 | 2.92±0.02 |
| in 100 years | 2.96±0.05 | 2.43±0.03 | 7.56±0.04 | 2.99±0.05 | 2.22±0.04 | 4.88±0.04 |
| $R_2/R_1$: | | | | | | |
| in 20 years | 1.63±0.03 | 0.87±0.00 | 9.78±0.03 | 1.63±0.03 | 0.46±0.00 | 5.77±0.02 |
| in 100 years | 5.95±0.05 | 4.93±0.03 | 15.38±0.04 | 6.00±0.05 | 4.46±0.04 | 9.72±0.04 |
| $R_2/R_{1/2}$: | | | | | | |
| in 20 years | 2.39±0.03 | 1.32±0.00 | 15.20±0.03 | 2.41±0.03 | 0.68±0.00 | 8.88±0.02 |
| in 100 years | 9.04±0.05 | 7.43±0.03 | 24.06±0.04 | 9.09±0.05 | 6.74±0.04 | 14.98±0.04 |

Note: Data are presented as mean ± standard error.

Social Security (being 65 for many years, age at retirement gradually increases beginning with people born in 1938 or later, until it reaches 67 for people born after 1959). Losses were the highest in the population at age 68+. Figure 8.2 presents population losses as a function of projection time. Male losses were

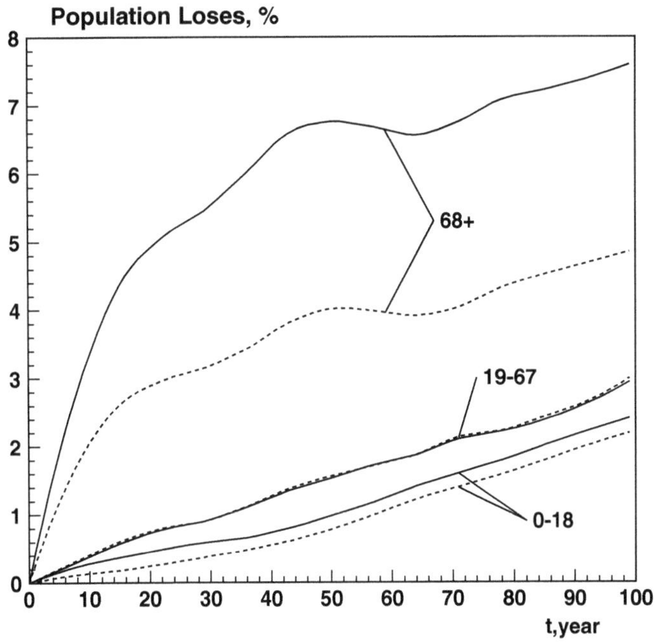


**Fig. 8.2** Population losses for male (*solid lines*) and female (*dashed lines*) versus projected year for three age groups (50% – less smoking versus current smoking population)

higher than for females at ages 68+ due to the higher male smoking rate and higher male mortality for most of smoking-attributable diseases.

#### 8.1.1.1 Life Expectancy

Life expectancy is determined by death rates. This quantity is the same for all projection times, though different for different smoking prevalences (see Table 8.2). Reducing smoking prevalence by 50% increases life expectancy almost 1 year for males and more than a half year for females. The largest difference in life expectancy is for age 30.

#### 8.1.1.2 Cancer and Noncancer Disease Contributions to Mortality

Using the data from the "Annual deaths, smoking-attributable mortality and years of potential life lost" (Fellows et al., 2002) and "Relative risk attributable to smoking- and estimated smoking-attributable mortality" (Giovino et al., 1994), we analyzed the effects of smoking-attributable cancer and smoking-attributable noncancer diseases, including cardio- and cerebrovascular diseases and diseases of respiratory tract, on male and female life expectancy and population losses at various ages. Male life expectancy is reduced more significantly by smoking-attributable cancers compared with noncancer diseases: in the doubled-smoking ($R_2$) population male life expectancy at birth is reduced by 0.92 year for cancer versus 0.75 year for noncancer diseases, and life expectancy at age 65 is reduced by 0.60 year for cancer versus 0.54 year for noncancer diseases. In females, smoking-associated noncancer diseases have a stronger influence on life expectancy: life expectancy at birth is reduced by 0.56 year for cancer versus 0.66 year for noncancer diseases, and life expectancy at age 65 is reduced by 0.34 year for cancer versus 0.56 year for noncancer diseases. In 100 years of projection, the size of the $R_2$ population compared with the $R_1$ population is reduced 0.14% because of deaths from smoking-attributable cancers in the age group of younger than 18, in age group of 19–67 – by 0.46%, and in age group of older than 65 – by 4.5%.

Cancer and noncancer contributions to losses in the $R_2$ population are shown in Fig. 8.3. A comparison of cancer versus noncancer effect on population losses (comparing the extreme, i.e., $R_2$ and $R_{1/2}$, populations over 100 years) shows 6.76% (versus 6.57%) of the population will be "lost" in

**Table 8.2** Life expectancy at birth and ages 30 and 65 depending on smoking prevalence

| | 0 year | | 30 years | | 65 years | |
|---|---|---|---|---|---|---|
| Smoking prevalence | Male | Female | Male | Female | Male | Female |
| $\overset{o}{e}(\mathbf{R}_{1/2})$ | **75.41** | **80.13** | **46.94** | **51.12** | **16.56** | **19.19** |
| $\overset{o}{e}(\mathbf{R}_{1/2})-\overset{o}{e}(\mathbf{R}_1)$ | 0.99 | 0.64 | 0.91 | 0.65 | 0.62 | 0.48 |
| $\overset{o}{e}(\mathbf{R}_1)$ | **74.42** | **79.49** | **46.03** | **50.47** | **15.94** | **18.71** |
| $\overset{o}{e}(\mathbf{R}_1)-\overset{o}{e}(\mathbf{R}_2)$ | 1.61 | 1.18 | 1.64 | 1.20 | 1.10 | 0.87 |
| $\overset{o}{e}(\mathbf{R}_2)$ | **72.81** | **78.29** | **44.39** | **49.27** | **14.84** | **17.84** |

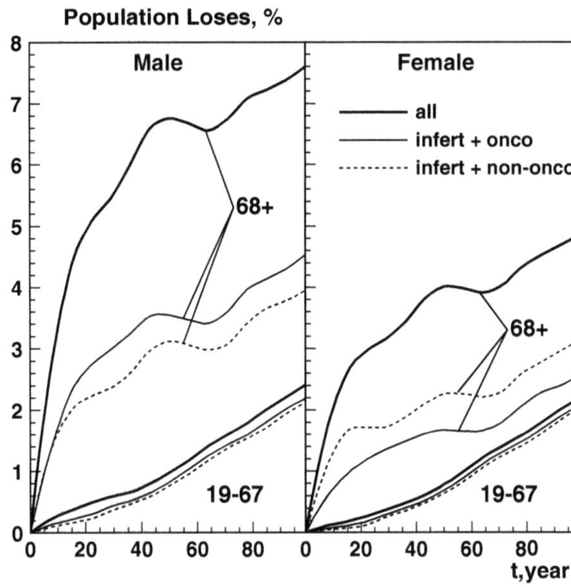


**Fig. 8.3** Cancer and noncancer contribution to population losses (%) versus projection year for two age groups (50% – smoking versus current smoking population)

males aged 19–67, and 14.03% (versus 12.20%) in males aged 68 + . In females, the proportions are 6.38% (versus 6.27%) at ages 19–67 and 7.91% (versus 9.54%) at age 68 + . Taking into account that smoking prevalence in males and females during the last 50 years has tended to become more equal [but still with a higher prevalence in males: 23.4 versus 18.5% in 2004 (TIPS, 2005)], the observed sex differences in population loses at older ages (68 + ), when male losses from smoking-associated diseases were higher than for females, may be explained by the different contributions of smoking-related cancer and smoking-related noncancer diseases in male and female mortality.

#### 8.1.1.3 Sensitivity Analysis

We checked the sensitivity of our results with respect to effects, which influence on the projected characteristics is expected to be important (e.g., Smoking and Tobacco Control Monograph, 1997; DHHS, 2004). The assumptions of the basic scenario chosen for testing in sensitivity analyses include (1) the specific choice of the base year for mortality and birth rates, (2) ignoring time trends of mortality and birth rates, (3) ignoring time trends of smoking rates, (4) ignoring the standard error of relative risk estimates, and (5) ignoring the effects of former smokers. The impact of each of the basic scenario assumptions was analyzed using the same procedure, i.e., by modeling the investigated effect, using microsimulation of 10 samples with the effect included, and comparing the average of these sample outcomes with the base scenario. Results of these analyses are shown in Table 8.3. The quantities characterizing the relative and absolute uncertainties are defined as follows:

**Table 8.3** Estimates of sensitivity of WRR due to different potential sources of bias

| Source of bias | 20-year projections | | | | | 100-year projections | | | | |
|---|---|---|---|---|---|---|---|---|---|---|
| | $\delta_{1/12}$ | $\delta_1$ | $\delta_2$ | $\Delta_{1/2}$ | $\Delta_2$ | $\delta_{1/2}$ | $\delta_1$ | $\delta_2$ | $\Delta_{1/2}$ | $\Delta_2$ |
| No intervention | 0 | 0 | 0 | −3.31 | 6.79 | 0 | 0 | 0 | −3.57 | 7.09 |
| Birth rate base year: 1980 | 0.31 | 0.31 | 0.32 | −3.31 | 6.78 | 18.66 | 18.54 | 18.32 | −3.72 | 7.38 |
| 1990 | 0.07 | 0.08 | 0.08 | −3.31 | 6.78 | 4.48 | 4.47 | 4.39 | −3.58 | 7.18 |
| 2000 | −0.05 | −0.05 | −0.05 | −3.31 | 6.79 | −3.06 | −3.03 | −3.02 | −3.54 | 7.08 |
| Total mortality trend | −0.31 | 0 | 0.58 | −3.00 | 6.18 | −0.24 | 0 | 0.35 | −3.32 | 6.74 |
| Trend of SAM[1] rate | 0.13 | 0 | −0.29 | −3.45 | 7.09 | 0.03 | 0 | −0.05 | −3.60 | 7.14 |
| Trend of smoking rate | 0.95 | 0 | −1.83 | −4.24 | 8.75 | −0.32 | 0 | 0.48 | −3.26 | 6.58 |
| Uncertainty in RR | −0.02 | 0 | 0.02 | −3.30 | 6.77 | −0.02 | 0 | 0.01 | −3.55 | 7.08 |
| Another set of RR | 1.03 | 0 | −1.96 | −4.31 | 8.87 | 1.39 | 0 | −2.52 | −4.90 | 9.80 |
| Former smokers | 0.53 | 0 | −1.01 | −3.82 | 7.87 | 0.72 | 0 | −1.29 | −4.26 | 8.49 |

[1]Smoking-associated mortality.

$$\delta_{R'} = \frac{\mathrm{WRR}(\mathrm{R}=\mathrm{R}') - \mathrm{WRR}_0(\mathrm{R}=\mathrm{R}')}{\mathrm{WRR}(\mathrm{R}=\mathrm{R}')}; \quad \mathrm{R}' = 1/2, 1, \text{ or } 2,$$

and

$$\Delta_{R'} = \frac{\mathrm{WRR}(\mathrm{R}=\mathrm{R}') - \mathrm{WRR}(\mathrm{R}=1)}{\mathrm{WRR}(\mathrm{R}=1)}; \quad \mathrm{R}' = 1/2 \text{ or } 2.$$

## Base Year and Time Trend of Mortality and Birth Rates

Mortality is known with better accuracy, and its time trend is more stable, than fertility. There is no evidence on long-term trends (e.g., 25 years) in birth rates (Martin et al., 2001). However, birth rates can have short-term, year-to-year, fluctuations. Results calculated for the fertility rates observed in 1960, 1980, 1999, and 2000 were compared. The highest fertility was in 1960. The minimum was in 1980. In this situation, it is reasonable to model time trends for mortality rates and use different base years for the birth rate. Relative corrections $\delta_R$ and $\Delta_R$ for 1980, 1990, and 2000 are shown in Table 8.3. The relative correction $\delta_R$ is less than 1% for 20 year projections, and much larger for 100-year projections. The correction $\Delta_R$ is stable: no effects on 20-year projections and variations not exceeding 4% for 100-year projections. The difference in population losses due to change in base years for the birth rate calculation is found to be almost "random" (i.e., the size of the difference is 12 standard deviations with random signs).

To analyze bias due to mortality trends, we use age-adjusted mortality rate trend from 1990 to 1999 (Hoyert and Anderson, 2001; Shi et al., 2005) and apply this trend to death rates (see $\Delta_R$ and $\delta_R$ in Table 8.3). Another intervention experiment took into account trends in smoking-associated mortality (SAM) rates. The trend of SAM rates comprises trends in mortality and in the prevalence of smokers, which operate through smoking-associated factors. To estimate the trend, we compared estimates of SAM for 1984 and 1997–2001 (CDC, 1997, 2005) using population estimates from the Human Mortality database (HMD, 2006). The trend is negative for males and positive for females. We assumed this trend holds for 15 years. The results on $\Delta_R$ and $\delta_R$ are shown in Table 8.3. Both, $\delta_{1/2}$ and $\delta_2$, are within a range of several percent. The two tendencies have opposite impacts on smoking effects, and, as expected, $\delta_{1/2}$ and $\delta_2$ have opposite signs.

## Time Trends of Smoking Rates

Time trends of smoking rates were recently analyzed by DHHS, 2004 (Table 8.3). We used those data to estimate time trends in age-specific smoking rates. For the nonconstant smoking rate, we lagged the health effects of smoking for 30 years. Due to this lag, the current situation has higher mortality, because the smoking rate was higher 30 years ago. This is reflected in the 20-year projections ($\delta_{1/2} > 0$ and $\delta_2 < 0$). The situation is different for the 100-year projections. Trends in smoking rates influence population losses most in the 20-year projections. The increase of population losses for this case can exceed 20%.

Uncertainties of Relative Risks

Relative risks of smoking-associated diseases are known with imperfect accuracy. This can cause biases. To examine this, we designed two experiments. First, we assumed the relative risks are random, normally distributed variables (parameters for the distribution are from Chapter 4 of "Smoking and Tobacco Control" Monograph No. 8 published by NIH, 1997). A similar approach was used for infertility rates, for which standard errors were provided above. Both, mortality and fertility rates, are simulated using these distributions for each individual for each year. The results are compared with the basic scheme. No important effects emerged.

Second, we examined the impact of shifts in terms of relative risks. We used relative risks calculated for the Cancer Prevention Study-II (DHHS, 2004). The relative risk values differed considerably between the two sources. This produced a relatively large effect.

Former Smokers

DHHS (2004) provided information on age-specific rates of former smokers and relative risks of death caused by smoking-attributable diseases. Updated formulas for mortality rate recalculation were used for projections, taking into account the contribution of former smokers to mortality risk. The corrections were several percent.

#### 8.1.1.4 Summary

We presented a microsimulation model to forecast short- and long-term population changes reflecting the effects of smoking on mortality and fertility. Population changes result from the aggregation of changes in individual event histories. The events in the model include pregnancy, birth, disease, and death. These events occur stochastically and are stratified by age, sex, and race using rates extracted from official reports (Center for Disease Control and Prevention, MMWR, NVSR, U.S. Department of Health and Human Services and U.S. Environmental Protection Agency). Thus, our forecasts were conducted by constructing an individual's event history and simulating projections on a "year-by-year" and a "individual-by-individual" basis. One advantage of the approach is that the recalculation of birth and mortality rates can be performed analytically. This allows us to control several key quantities (e.g., life expectancy) before simulations. Several assumptions made for the basic scenario could be questioned, e.g., assumptions of time-independent mortality trends or absence of the health risks of former smokers. These were assumed, because of insufficiency of reliable data. Since the characteristics of interest are expressed as ratios of projected population groups, the bias in analysis of impact of smoking on the projected population is minimized. Moreover, the effects of different corrections were evaluated in sensitivity analysis and found to be small. Considering this example, we demonstrated that the methods of microsimulation are a powerful tool for constructing the projections based on realistic or presumptive scenarios, playing "what-if" scenarios, and analyses of the effects of interventions and uncertainties of different sources. Further discussion of microsimulation methods in developing the cancer prevention models is presented in Chapter 9.

## 8.2 Atherosclerosis and Cancer: Anything in Common?

In previous section of this chapter, we compared the contribution of cancer to noncancer diseases (which included cardio- and cerebrovascular diseases, and diseases of respiratory tract) related to one of the most widespread environmental risk factors – smoking. In this section, we discuss the relationship between cancer and atherosclerosis, whose similarity might be based not only on shared risk factors but also, as recent studies showed, on some biological mechanisms of disease development and progression.

Heart disease, stroke, and cancer are the leading causes of death in the United States, accounting for more than one-and-a-half million of deaths annually. Cancer and atherosclerosis for many years were considered to have completely unrelated pathogenesis and disease progression pathways. Recently, a series of molecular markers and gene pathways have been associated with disease development common in both, atherosclerosis and cancer, such as genetic predisposition, oxidative stress, sex hormones, environmental risk factors (diet, smoking, toxins and other), infections and inflammation, DNA instability, nonspecific injury, cell proliferation, and clonal expansion (including apoptosis and telomerase shortening) (Ross et al., 2001). Some examples of these common features are presented in Table 8.4.

Cancer and atherosclerosis have some common molecular pathways in their progression, including transforming growth factor beta (TGF-β) pathway, other peptide growth factors, cell adhesion molecules and the β-catenin pathway, nuclear factor kappa B (NFkB) and the proteasome, matrix digestion, proteases and tissue protease inhibitors, and angiogenesis and angiogenesis inhibitors (Ross et al., 2001). The expression of TGF-β, a member of a polypeptide growth factor family, is an inhibitor of both, atherosclerosis and early stage of cancer (Blobe et al., 2000). In addition to TGF-β, several other peptide factors, including epidermal growth factor receptor (EGFR), fibroblast growth factor 2 (FGF-2), granulocyte macrophage colony-stimulating factor (GMCSF), platelet-derived growth factors (PDGF), and insulin growth factors

**Table 8.4** Some examples of the common features of atherosclerosis and cancer development

| Pathways | Cancer | Atherosclerosis |
|---|---|---|
| Genetic predisposition | Inherited defects of homocysteine metabolism associated with high blood levels of homocysteine may also be associated with neoplasia (McCully, 1994). | Aberrant homocysteine metabolism accelerates atherosclerosis (Chambers et al., 2001; Thambyrajah and Townsend, 2000) |
| Oxidative stress | Oxidative stress is one of the major contributing factor for cancer development by both, causing genetic mutations and preventing the correction of mutations in affected cells. Oxidative stress is linked to DNA instability, hypermethylation, mutations in DNA repair genes, heterozygosity loss, point mutations in DNA microsatellites. Also it is associated with cell cycle deregulation and a shortening of the cell cycle (Shackelford et al., 2000; Bartsch, 2000; Marnett, 2000) | Reactive oxygen species (ROS) are associated with leukocyte chemotaxis during early atherosclerosis lesion formation and can directly damage both, endothelial cells and smooth muscle cells of the vascular wall (Aviram, 2000; Tardif, 2000; Irani, 2000; Chisolm and Chai, 2000; Shackelford et al., 2000). Oxidative stress is associated with hyperlipidemia, hypertension, cigarette smoking, endothelial dysfunction, platelet aggregation, loss of vasodilatation capability, local inflammation, growth of smooth muscles, and linked to the production of mitogenes and growth factors stimulating cellular proliferation at the site of an early atheromatous lesion (Zalba et al., 2000; Cai and Harrison, 2000; Ruef et al., 1999; Jeremy et al., 1999; Zettler and Pierce, 2000b) |
| Smoking | Smoking is associated with lung, larynx, oral cavity, tongue, pharynx, esophagus, pancreas, urinary bladder, and kidney cancers, and probably increases the risk of cancers of lip, liver, cervix uteri, stomach, and leukemia (Smith and Glynn, 2000; IARC, 1990) | Cigarette smoking is a recognized risk factor for CVD, and smoking cessation is associated with reduction in risk and clinical reversal of atherosclerotic lesions (Wood, 2001; Misra, 2000; Cerami et al., 1997). Smoking-related generation of ROS is associated with cell proliferation in early atherosclerotic lesions (Wang, 2000) |
| Sex hormones | Estrogens have been implicated in breast, endometrial, and probably cervical cancer development, and testosterone has been accepted as a promoter of prostate cancer (Clemons and Goss, 2001; Slater and Oliver, | Estrogens slow or prevent the development of atherosclerosis in women, when in men androgens have been implicated in atherosclerosis development (Maxwell, 1998; Hermann and Berger, 1999) |

**Table 8.4** (continued)

| Pathways | Cancer | Atherosclerosis |
|---|---|---|
| | 2000). Incidences of some of the cancers of nonreproductive organs (e.g., thyroid carcinoma) differ between males and females during females' reproductive period, but almost similar in childhood and in postmenopause (Ron et al., 1987; Harach and Williams, 1995; Manole et al., 2001; Ivanov et al., 2005) | |
| Diet | Dietary fat intake increased risk of atherosclerosis and certain cancers (colorectal, breast). Dietary intake of omega-3-fatty acids is associated with decreased incidence of both atherosclerosis and cancer (Jump and Clarke, 1999) | |
| Inflammation | Inflammation has been considered a major precursor for the development of cancer in both, infectious and noninfectious conditions. So, an inflammatory bowel disease (IBD) is associated with chemokine signaling, increased local cell proliferation, shortened cell cycles, accumulation of mutations, and inadequate DNA repair, which are precursors of malignancy (Ardestani et al., 1999; Wang et al., 1998) | Chemokines and cytokines are associated with increased levels of growth factors and cell proliferation in endothelial cells, mononuclears, macrophages, and smooth muscle cells during the early phases of development of the atherosclerotic plaque (Gerszten et al., 2000; Sasayama et al., 2000) |
| Infection | Infection with *Chlamydia pneumoniae* may be associated with an increased risk of lung cancer (Littman et al., 2004; Koyi et al., 1999).<br>*Helicobacter pylori* is classified by WHO as a class I carcinogen for stomach cancer | *Chlamydia pneumoniae* has been linked to the atherosclerosis development as potential causative agent (Morre et al., 2000; Leinonen and Saikku 2000).<br>*Helicobacter pylori* has been proposed, but not widely confirmed, as a potential cause or co-factor for development of coronary atherosclerosis (Markle, 1997) |
| Nonspecific injury | Chronic local injury has been associated with an increased risk of neoplasia. The incidence of malignancy is increased in scars of the lung associated with infections and tissue infarcts, surgical procedures, and cutaneous thermal burn (Moore and Tsuda, 1998) | Mechanical and arterial shear stress are associated with atherosclerosis development (Li and Xu, 2000; Bakker and Gans, 2000) |

**Table 8.4** (continued)

| Pathways | Cancer | Atherosclerosis |
|---|---|---|
| Genomic and microsatellite instability | Impaired DNA repair and microsatellite instability have been associated with the development of several cancers (e.g., colorectal and endometrial cancers). In addition to germ-line mutations in DNA repair genes, gene silencing by promoter gene CpG island hypermethylation as a cause of microsatellite instability has been associated with cancer development (Parsons et al., 1995; Coleman and Tsongalis, 1999). Microsatellite instability is also linked to impaired function of the transforming growth factor beta (TGF-β) type 2 receptor in colorectal cancer (Parsons et al., 1995) | Both genomic instability (associated with loss of heterozygosity) and mutations in DNA microsatellites may play an important role in the development of atherosclerotic plaque at early stages (Andreassi et al., 2000; Hatzistamou et al., 1996). Microsatellite instability has been associated with atherosclerosis (Parsons et al., 1995) |
| Clonality, cell cycle regulation and cell proliferation | Studies of cancer cell cycles have shown dysregulation of cell cycle checkpoints, particularly the G1-S transition, as a major cause of rapid cell growth and accumulated mutations (Israels E.D. and Israels L.G., 2000) | Molecular assays of atherosclerotic plaque have demonstrated monoclonal cell proliferation and cell cycle dysregulation linked to cell proliferation in early atherosclerotic plaques (Schwartz and Murry, 1998; Zettler and Pierce, 2000a). Clonal proliferation of myointimal cells in vascular walls has been found when studying coronary restenosis in patients, who have undergone angioplasty or stent procedures (O'Brien et al., 2000) |
| Apoptosis | Apoptosis has been associated with cell proliferation and clonal expansion of malignant neoplasms, but the apoptosis-mediated mechanism of cancer and atherosclerosis progression may differ (Reed, 2000) | Generation of ROS is associated with oxidation of low-density lipoproteins and apoptotic cell death of endothelial cells and smooth muscle cells in vascular wall (Dimmeler and Zeiher, 2000; Okura et al., 2000) |
| Telomeres | Telomerase activity has been confirmed as an important factor in the development of many forms of cancer (Aragona et al., 2000) | Studies of telomeres and telomerase in patients with atherosclerosis were focused on aging processes and indirectly implicated endothelial and smooth muscle cell longevity in the development of vascular senility (Banks and Fossel, 1997) |

(IGF) have been considered as participating in cell cycle regulation in atherosclerosis, as well being involved in regulation of primary tumor growth, invasion, and metastasis (Newby and Zaltsman, 2000; Nugent and Iozzo, 2000; Ciardiello, 2000; Yancopoulos et al., 2000). Thus, it is supposed that the growth factor receptor-targeted therapies might slow the progression of cancer and also might reduce the incidence of restenosis of atherosclerotic lesions subsequent to angioplasty and stent placement procedures (Ross et al., 2001). Dysregulation of matrix metalloproteases is associated with early and advanced atherosclerosis and cancer (Curran and Murray, 2000; George, 2000). The nuclear factor kappa B (NFkB)-signaling pathway, a transcription factor involved in the regulation of expression of chemokines, cytokines, growth factors, cell adhesion molecules, and cell cycle regulatory proteins plays an important role in atherosclerosis and cancer progression and appears to be a potential therapeutic target for both diseases (Collins and Cybulsky, 2001; Mayo and Baldwin, 2000). Angiogenesis plays an important role in both atherosclerosis and cancer progression: in atherosclerosis, the development of angiogenesis has beneficial, as well negative, effects, whereas enhanced angiogenesis may be a favorable sign in the ischemic tissue's healing after myocardial infarction, the progressive angiogenesis in a primary atherosclerotic lesion might cause plaque expansion, complicated by its rupture and vascular thrombosis; in cancer patients, tumor angiogenesis has a negative prognostic value (Isner, 1999; Kerbel, 2000).

In Table 8.5, we summarized the presence of proven atherosclerosis risk factors in various cancers. The major shared risk factor for both diseases is smoking. Smoking not only increases CVD mortality as an independent risk factor but also influences CVD mortality by increasing arterial blood pressure, increasing the risk of diabetes, and decreasing BMI (Pekkanen et al., 1992). The other proposed risk factors which have both atherogenic and carcinogenic effects include diet, obesity, low physical activity, and endogenously and exogenously derived sex hormones. Most cancers are ageing-associated diseases, which is common with atherosclerosis.

There are several risk factors for atherosclerosis, i.e., arterial hypertension, hyperglyceridemia, and diabetes mellitus, which might be of importance for some cancer risks, but their role has not yet been clearly proven. So, the striking similarity of lifestyle risk factors for colorectal cancer and insulin resistance suggests that hypertriglyceridemia, as well as hyperinsulinemia and hyperglycemia associated with insulin resistance, might support the development of colon cancer (Bruce et al., 2000; Yamada et al., 1998; Tabuchi et al., 2006). Several animal studies showed direct evidence, suggesting a positive effect of triglycerides on the development of aberrant crypt foci (Koohestani et al., 1998) and adenomatous polyps (Niho et al., 2003a, b), suggesting an overlapping mechanism that increases the risk of colon cancer and hyperlipidemia (however, the true association between serum lipid levels and colonic cancer has not been clearly determined in humans).

**Table 8.5** Presence of atherosclerosis risk factors in cancer (risk factors marked with gray color prove their role in increasing cancer risk in meta-analyses)

| Cancer site | Age | Sex hormones | Smoking | Alcohol | Obesity | High fat diet | Physical inactivity | Arterial hypertension | Hypertriglyceridemia | Diabetes mellitus | IR |
|---|---|---|---|---|---|---|---|---|---|---|---|
| Colorectal | + | + | + | + | | + | + | | + (?) | + | |
| Breast | | | + | + | + | + | | | | | + |
| Lung | + | | + | + | | + | | | | | |
| Esophageal carcinoma | | | + | + | + | + | | | | | |
| Stomach | | | + | | + | | | | | | |
| Corpus uteri | | | | | + | | | + | | + | |
| Urinary bladder | + | + | + | + | | | | | | | |
| Thyroid | | | | | | | | | | | + |
| Pancreas | | | + | | + | | | | | + | |
| Hepatocellular carcinoma | | | + | + | | | | | | + | |
| Adult leukemia | | | + | | | | | | | | |
| Oral/oropharyngeal carcinoma | | | + | + | | | | | | | |
| Cervix uteri | | | + | | + | | | | | | |
| Brain | | | + | + | | | | | | | |
| Kidney | | | + | | + | | + | + | | | |
| Acute lymphoblastic leukemia | | | | | + | | | | | | |
| Acute myeloid leukemia | | | | | + | | | | | | |
| Prostate | | | | | + | + | | | | | |
| Ovarian | | | | | + | | | | | | |
| Gallbladder | | | | | + | | | | | | |

Ionizing radiation is a shared risk factor for both cancer (e.g., proven for breast and thyroid) and atherosclerosis especially for cerebrovascular disease (IR as risk factor is described in details in Chapter 3).

If atherosclerosis and cancer are linked in certain degree by common risk factors, a person with either condition should be at elevated risk for the other. The results obtained from some studies support this hypothesis, while others do not demonstrate an obvious correlation. So, a cross-sectional study (Henderson et al., 1974) and a case–control study (Neugut et al., 1998) both reported an association between a history of coronary heart disease and prostate cancer (no association was reported in two earlier case–control studies by Thompson et al., 1988; Checkoway et al., 1987). Several case– control studies examined the association between the self-reported ischemic heart disease and stroke and colorectal cancer, but the findings were contradictory (Neugut et al., 1998, 1995; Kune et al., 1988). Dreyer and Olsen (1999) did not find any association between atherosclerosis and cancers of the colon, rectum, or breast. The incidence of cancers non-associated with smoking in patients with extracoronary manifestations of atherosclerosis did not differ from general population for any specific cancer sites.

Lifestyle factors, such as smoking and presumably diet, may be changed by hospitalization and recommendations for atherosclerosis treatment, which could also decrease the risk for a subsequent cancer. The risk for smoking-associated cancers, however, was higher in the long- than the short-term follow-up, indicating that a healthier lifestyle after discharge may not have reduced smoking-related cancer risk to expected levels among patients with atherosclerosis during the 17 years of follow-up (Dreyer and Olsen, 1999). The overall risk for cancers of the brain and nervous system in patients discharged with occlusion of precerebral arteries, or atherosclerosis of the cerebral arteries, appeared to be increased (40%), but the excess was restricted to the initial 1–3 years from date of discharge – the most likely explanation is that the symptoms of preclinical brain cancers were misinterpreted. The nonincreased risk of brain cancer observed among patients with atherosclerosis does not support the hypothesis of an association between high cholesterol concentration and subsequent brain tumors (Dreyer and Olsen, 1999).

Even presently there are no absolutely clearly proven impacts on cancer incidence for the "classic" atherosclerosis risk factors, however, it is clear that the existence of some shared features of atherosclerosis and cancer might be significant in emerging therapeutic strategies, such as new antiinflammatory agents, proteasome inhibitors, and cell cycle and angiogenesis regulators, which may be effective in blocking the progression of both atherosclerosis and cancer (Ross et al., 2001).

## 8.3 Making Projections of Cardiovascular Disease Risk

We modeled the health effects of cardiovascular disease (CVD) risk factor intervention in human, compared with projection of the effects effects of progenitor cell therapy (Kravchenko et al., 2005). We analyzed the consequences of these interventions on age-specific CVD mortality and life expectancy, compared with the effects of the virtual elimination of cancer as a cause of death. It is well known that such a "classic" CVD risk factors as cigarette smoking, hypertension, dyslipidemia (abnormal serum concentrations of lipoproteins and triglycerides), diabetes mellitus, obesity, and other can promote a multistage inflammatory process in atherosclerosis. Recently it has been supposed that atherosclerosis may begin in childhood, and risk factors for its clinical syndromes appear to determine, to a large degree, its rate of progression rather than its presence (Strong et al., 1999). The evidences have been accumulated that factors other than conventional risk factors may contribute to the development of atherosclerosis, and conventional risk factors likely predict less than half of future cardiovascular events. "Newer" risk factors may play an important role, i.e., homocysteine, fibrinogen, impaired fibrinolysis, increased platelet reactivity, hypercoagulation, lipoproteins, and inflammatory-infectious markers (Kullo et al., 2000).

In a hypothetical situation, when "classic" risk factors are eliminated completely, there remains a substantial residual risk of CVD mortality related to age. Specifically, if assuming the levels of hypertension, smoking, diabetes, dyslipidemia, and sedentary behavior remain constant from age 20 to age 60, the risk of a coronary event at age 60 would still be 10–100 times higher than risk at age 20. It is believed that much of the extra risk at age 60 is attributable to age-related declines in the capacity of precursor cells to repair damage in the arterial endothelium (Goldschmidt-Clermont and Peterson, 2003; Edelberg, 2002). Bone marrow appears to contain endothelial progenitor cells (EPCs) that help to repair areas of vascular senescence, a function that, if lost as a result of aging and/or risk factor exposure, would lead to the acceleration of atherogenesis with age (Rauscher et al., 2003).

We modeled the health effects of progenitor cell therapy using data from the 46-year follow-up of the Framingham Heart Study. In 1950, 2336 males and 2873 females aged 29–62 were enrolled in this study. From 1950 to 1996, 23 exams of each volunteer were conducted, once every 2 years. The parameters followed were age, diastolic blood pressure, pulse pressure (the difference between the systolic and diastolic pressure), serum cholesterol, vital capacity index (the amount of air that can be exhaled after a maximum inhalation), blood glucose, BMI, hematocrit, smoking, left ventricular hypertrophy, and pulse rate. The projection scenario was to control "classic" CVD risk factors over the individual's lifetime, and then to project mortality, assuming that progenitor cell therapy for atherosclerosis is used at age 30 without changing the observed dynamics of conventional risk factors. In our projections, we used parameters that characterized the individuals' initial health status, 2-year changes in risk factors, and the age-dependent hazard function (the probability of the event, e.g., mortality, occurring at a given time point) for CVD.

The National Heart, Lung, and Blood Institute (NHLBI) Atherosclerosis Risk in Communities (ARCIC) surveillance study (1987–2000) found that the annual rate of first heart attack in males and females starts to increase at age 35–40 (Heart Disease and Stroke Statistics, 2005). The Framingham Study includes persons aged 29 years and older, so we were able to study the effects of interventions on population health at age 30 and older (i.e., when the depletion of bone marrow cells' ability to repair arterial endothelium appears to be first manifest), but before the most of individuals have clinical signs of CVD. At present, no clinical information is available about the effective duration of a course of progenitor cell therapy in humans. We made an assumption about this duration on the finding that patients who have experienced myocardial infarction exhibit shorter telomeres than do healthy controls: the difference in mean leukocyte telomere-restriction fragment (TRF) length (a measure of telomere length) between patients who had myocardial infarction and controls represents a biological age gap of more than 11 years (Brouilette et al., 2003). We made our projections assuming that the effective period for progenitor cell therapy might be close to 10 years.

The logistic regression model, which is often used for longitudinal analyses, has no mechanism to describe changes in risk factor values that ordinarily occur in longitudinal studies (Woodbury et al., 1981). Simple nonlinear regression models of changes in risk factor values might be considered for this purpose, but they do not describe the health effects (such as CVD mortality) of the age dynamics of risk factors. The model employed herein to calculate the population effect of progenitor cell therapy is constructed to describe risk factor dynamics and mortality as linked stochastic processes (Woodbury and Manton, 1977, 1983a, b; Manton and Stallard, 1988; Manton et al., 1994; Witteman et al., 1994; Manton et al., 1992; Manton and Yashin, 2000; Akushevich et al., 2005). In this model, age projections are constructed using the microsimulation technique, which is based on the simulation of trajectories (the serial values of physiological parameters that define a health state of an individual in his or her life). Two laws govern this process: (1) how new values of physiological parameters are defined by a set of prior values, and (2) under what conditions this life trajectory stops because of death. These two laws are probabilistic, i.e., one can predict only distributions of changes in physiological parameters or times of death, but not their exact values for individuals. Using data from human studies, parameters of these laws are estimated such that an artificial simulated cohort would likely reproduce these data. The probabilistic property of these laws is fundamental. It reflects the individual risk heterogeneity and competing disease risks. An intervention is performed by making changes in the two laws of the model to reflect the properties of a study. Comparison of simulation results with and without an intervention defines the effect at a population level.

A key quantity in this modeling approach is the mortality rate $\mu(x,t)$, which is defined as a sum of three competing risks of mortality due to cancer, CVD, and "other" causes. The competitive risk model for mortality rate, $\mu(y,s)$ where $t$ is age and $x$ is a set of 11 risk factors/parameters (defined in the text) measured or modeled at age $t$, is,

$$\mu(x,t) = \mu_{can}(x,t) + \mu_{cvd}(x,t) + \mu_{res}(x,t)$$

We considered three types of interventions in our analysis. The first is designed so that "classic" CVD risk factors (serum cholesterol, glucose, pulse and diastolic blood pressures, BMI, and cigarette smoking) are kept within selected limits to model current clinical recommendations. The second is designed to describe the effects of progenitor cell therapy. It is assumed that after therapy the entire cardiovascular system is "rejuvenated" by 10 years, i.e., the CVD mortality component is modified as $\mu_{cvd}(x,t) \rightarrow \mu_{cvd}(x,t-10)$. The third type of intervention is the simple elimination of the cancer competing risk term, reflecting the hypothetical situation in which cancer is completely "beaten." In our projections, we assumed that it was possible to control "classic" CVD risk factors during an individual's lifetime and compared these effects with the simulated effects of progenitor cell therapy and to observed age-dependent, sex-specific mortality rates. The results are presented in Fig. 8.4, where mortality, $\mu$, is plotted against age.

The continuous thick curve shows the age-specific CVD mortality rate without intervention. The continuous thin curve represents age-specific CVD mortality when a progenitor cell therapy is performed at age 30 (with the effect assumed to be a 10-year delay in age-related atherosclerosis progression; the observed risk factor dynamics are used in this example). The discontinuous (dashed) curve represents the CVD mortality rate obtained when risk factors

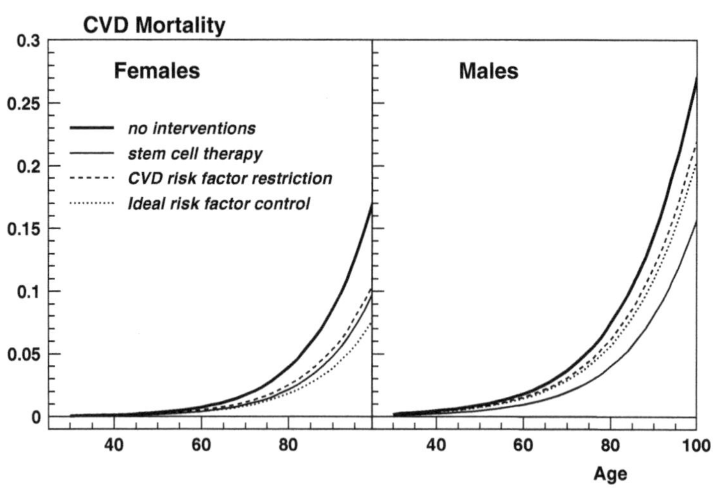


**Fig. 8.4** Age-specific CVD mortality for four scenarios: without intervention (*continuous thick line*), progenitor cell therapy (*continuous thin line*), restriction of contemporary risk factors for CVD (*discontinuous [dashed] line*), and restrictions of risk factors to clinically "ideal" values (*dotted line*)

are minimized for the selected time frames, but not to their "ideal" normal values. In this situation, it is presumed that medicines to control hypertension, serum lipid concentrations, and glucose levels are administered, and that diet and physical activity are controlled over the person's lifetime, but the result of the joint efforts of the physician and patient do not produce clinically "ideal" values for the patient, resulting in pulse pressure of 30–55 mm Hg, diastolic blood pressure of 90–70 mm Hg, BMI of 18.5–29.9 kg/m$^2$, serum cholesterol level of 180–240 mg/dl, serum glucose level of 70–124 mg/dl, cigarette smoking at rate of 0–10 cigarettes/day, and pulse rate of 72–90 beats/minute. Finally, the dotted curve represents the CVD mortality rate if CVD risk factors are restricted to "ideal" ranges, simulating a situation in which antihypertensive, lipid lowering, and glucose control therapy, as well as diet and physical activity, kept individuals in "ideal shape," such that they display pulse pressure of 30–40 mm Hg, diastolic blood pressure of 70–85 mm Hg, BMI of 18.5–24.9 kg/m$^2$, cholesterol level of 180–200 mg/dl, glucose level of 70–100 mg/dl, a complete lack of smoking, and pulse rates of 72–84 beats/minute.

Conventional CVD risk factors were used by many researches to explain mortality differences between populations: their roles are well studies and appropriately modeled (Pekkanen et al., 1992). Among "classic" CVD risk factors, the high serum cholesterol level is one of the most discussed as factor, controlling which might substantially decrease CVD mortality. Chen et al. (2004) showed that the number of EPCs is inversely correlated with total cholesterol and low-density lipoprotein cholesterol levels. It has been previously demonstrated that levels of circulating epithelial progenitor cells (EPCs) were a better predictor of vascular reactivity than the presence or absence of conventional risk factors (Hill et al., 2003). In humans, a strong correlation was observed between the number of circulating EPCs and the subjects' combined Framingham risk factor score (Hill et al., 2003).

Our results show that even "ideal" lifetime control of all conventional CVD risk factors, including cholesterol level, had effects on CVD mortality that were comparable to (in females) or smaller than (in males) the effect of progenitor cell therapy (see Fig. 8.4). Males received progenitor cell therapy had the lowest projected CVD mortality rate (probability of death within 2 years), compared with those who have got other interventions. The most effective intervention for females was the "ideal" control of risk factors. These effects are more striking at older ages. In females, we observe the possibility of estrogen-associated differences in the effectiveness of progenitor cell therapy: the effectiveness of progenitor cell therapy was about 11% higher in women at ages 36–44 compared with those at age 50 and older; postmenopausal women display decreases in CVD mortality similar to males.

A country's health status has historically been measured by life expectancy. Life expectancies for females and males at age 30, with and without intervention, are presented in Table 8.6. According to our calculations, progenitor cell therapy performed at age 30 might add 3.67 years to a female's life and

**Table 8.6** Life expectancy in females and males at age 30 when various interventions are applied

| | Life expectancy (years) | |
|---|---|---|
| Type of intervention | Female | Male |
| No intervention | 49.75 | 43.64 |
| Progenitor cell therapy | 53.42 EYL* = +3.67* | 49.58 EYL = +5.94 |
| "Nonideal" control over conventional risk factors | 51.62 EYL = +1.87 | 45.11 EYL = +1.47 |
| Elimination of CVD as a cause of death | 54.26 EYL = +4.51 | 50.84 EYL = +7.20 |
| Elimination of all cancers as a cause of death | 53.12 EYL = +3.37 | 46.50 EYL = +2.86 |

Note: * – EYL (extra years of life) is the number of years that life expectancy is increased as a result of different interventions.

5.94 years to a male's life, which is less than the modeled 10-year delay in CVD progression. That might be due to the non-CVD diseases/conditions intervene causing the death of treated with progenitor cells people before the full 10-year period had elapsed. Virtual elimination of all cancers as a cause of deaths is comparable with the projected effect of progenitor cell therapy in females. CVD as a factor influencing human life expectancy is stronger in males, including the projected effect from progenitor cell therapy. Note that difference in the base life expectancies presented in Tables 8.6 and 8.2 are due to different underlying populations.

## 8.4 Discussion and Comparison of the Results of the Two Intervention Studies

We have paralleled the results obtained from two interventions described in Sections 8.1 and 8.3 of this chapter and compared our results with the results of several published studies. Almost 25 years ago, it had been suggested that different rates of cigarette smoking might explain the differences in male and female longevity (Miller and Gerstein, 1983). This study immediately aroused a great deal of controversy (Holden, 1983; Enstrom, 1984; Feinleib and Luoto, 1984), with a variety of arguments that, although smoking makes a major contribution to sex differences in mortality, other factors may also be important.

According to some published studies, contribution of smoking in both ischemic heart disease/CVD (Powles and Sanz, 1996) and lung cancer mortality was higher in males than in females, with the percent of the sex differences in mortality attributable to smoking [calculated as $100\left(1 - (r^{mal}_{nsm} - r^{fem}_{nsm})/(r^{mal}_{tot} - r^{fem}_{tot})\right)$, where $r$: means death rates for certain population groups defined by Superscripts *mal* (males) and *fem* (females) and Subscripts *nsm* (nonsmokers) and *tot* (total)] about 30–40% (from 18 to 60%, depending on results of various studies) in smoking-attributable ischemic heart disease mortality, and about 90% (from 81 to 95% in various studies) in smoking-attributable lung cancer mortality (Waldron, 1986). Results of the smoking forecasting model presented in Section 8.1 also demonstrated the higher importance of smoking for males as a factor contributing in both cancer and noncancer diseases: e.g., if smoking prevalence in the population is doubled, then males life expectancy at birth would decrease by 0.92 years due to smoking-related cancer diseases, and by 0.75 years due to smoking-related noncancer diseases (including CVD and respiratory diseases), compared with 0.56 and 0.66 years in females, respectively (see Section 8.1).

Analyzing the MONICA data [MONItoring CArdiovascular disease project established by the W.H.O. to monitor trends in cardiovascular diseases and to relate these to risk factor changes in the population over a 10-year period (Tunstall-Pedoe, 2003)], it has been found that prevalence of four risk factors (namely, systolic blood pressure, BMI, total cholesterol and smoking) may explain around 30% of coronary heart disease (CHD) mortality variations in men, and around 45% in women, and that smoking made the largest contribution to increase in age-adjusted male CHD mortality (calculated after controlling for the four risk factors between different medical centers, participating in the study), while in females this effect was not as prominent (Marmot and Elliott, 2005). Thereafter, smoking played a major role in CHD mortality in men, while arterial blood pressure, BMI, and total cholesterol level were more important contributors to CHD mortality in women. This is in agreement with the results of our modeling of the effects of progenitor cell therapy (Section 8.3), which demonstrated the higher importance of the "classic" CVD risk factors for controlling CVD mortality in females than in males: the control of these factors in their virtually "ideal" frames was the most effective intervention to reduce CVD mortality in females, even more effective than the forecasted efficacy of progenitor cell therapy, especially at older ages (while dependence of males mortality from "classic" CVD risk factors was weaker, its efficacy being somewhere in the middle between the complete absence of intervention and the projected higher efficacy of progenitor cell therapy) (see Section 8.3).

When analyzing two of our studies in parallel, the effects of various interventions made in both studies might be compared (with keeping in mind all the possible biases – see below) in both sexes. Generally, males had shorter LE than females in both scenarios, without interventions, and in all interventions that have been performed (see Sections 8.1 and 8.3). In two interventions – elimination of all of the cancers as a cause of death, and control over "classic" CVD risk factors (both interventions are from progenitor cell therapy projection), women gained more extra years of life (EYL) than men, benefiting from these interventions. However, men gained more EYL than women resulting from such interventions as elimination of CVD as a cause of death, application of progenitor cell therapy for atherosclerosis, and reduction of smoking prevalence in the population up to 50%. Also males showed more obvious effects from modeling intervention of virtual doubling the smoking prevalence in the population: years of life lost by males exceeded those lost by females (see Section 8.1). Therefore, based on our two forecasts, males and females might differ by the contribution of certain risk factors in cancer and noncancer mortality: (1) in males smoking plays a considerably more important role in both cancer and noncancer (including CVD) mortality trends than in females, (2) smoking-associated cancers in males have a higher mortality than in females (at present, lung cancer is one of the dominant contributors here), (3) CVD-associated mortality is more important factor for effecting LE in males than in females, (4) progenitor cell therapy is supposed to be more effective in males than in females at least based on the decrease of CVD mortality and on increasing LE, and (5) in females the elimination of all cancers as a cause of death, and control over the "classic" CVD risk factors might be the most "beneficial" interventions, in which women gained more EYL than men. It needs to be noted that in our model smoking was included as risk factor in group of "classic" CVD risk factors. Taking into account that the contribution of smoking in male's CVD mortality is higher than in females and that the more pronounced effect from intervening females "classic" CVD risk factors was forecasted, it might be hypothesized that role of nonsmoking CVD risk factors (e.g., serum lipids level, obesity and other) in females is even more substantial than it is appeared in our models (thus, more substantial sex-related differences might be suggested for CVD risk factors).

When analyzing the contributions to cancer and noncancer diseases projected in each of these two studies, and comparing them with results of other published studies, several important observations should be taken into account, such as the difference in smoking prevalence in males and females, difference in relative risks of disease age-specific incidences and mortality rates in males and females, and specific characteristics of populations included in different studies. When interpreting results one should be concerned about other factors that can cause biases: e.g., the effects of sex differences in mortality increase that is caused by a given history of smoking habits, the absence of controls for other factors that may co-vary with smoking (e.g., alcohol consumption), interactions between smoking and risk factors that differ between males and females due to physiological differences (e.g., sex hormones related) or due to differences in environmental exposures (e.g., occupational exposures to asbestos, radon or due to uranium mining). One more problem may be related to the assumptions about the temporal relationship (i.e., latency period) between smoking and smoking-associated noncancer diseases that might be similar to that for lung cancer, or might differ.

The microsimulation approach is quite general and depends on the specifics of underlying biological processes, so the approach has the potential to describe both, CVD and cancer. This is possible because in many respects this allows us to reduce the specifics of underlying disease mechanisms to certain observable quantities (like relative risks), so the difference between mechanisms appears only in the different values of observed quantities. Although some differences can appear when considering effects (e.g., different dose–response functions, different lag periods, different forms of risk dependence on risk factors), the overall projection scheme based on a microsimulation approach remains general and applicable to a wide class of analyses. The approach developed for analysis of medical interventions can be applied as well to cancer prevention. What is necessary for that are estimates of the direct or indirect reduction of cancer-specific mortality (or incidence rates, if "healthy life" is analyzed). Several examples of specification of respective risk factor properties approach to formulating corresponding tasks for intervention analyses and discussion of modeling methods will be considered in Chapter 9.

## 8.5 Summary

Combinations of environmental risk factor exposures may obscure exposure–disease relationships. Additionally, the crucial exposures usually have happened years before a tumor is diagnosed – for many cancers (e.g., gastric cancer, melanoma) in childhood/teenage years. For some cancers, such as colon and breast, it is not clear whether the increased cancer risk in obese people is due to extra weight, a high-fat high-calorie diet, a lack of physical activity, or a combination of factors. That is why the IARC (Vainio and Bianchini, 2002) made recommendation for future trials: (1) conduct long-term intervention studies on the effect of dietary changes on weight gain and cancer risk; (2) conduct long-term intervention studies on the effect of patterns of physical activity (the intensity, frequency, and duration of various physical activities) in relation to weight gain and cancer risk; (3) conduct long-term intervention studies on the combined effects of changes in diet and physical activity on obesity and cancer risk; and (4) conduct community intervention studies to prevent weight gain and promote physical activity.

When making interventions in risk factors, it is important to take into account that some risk factors influence each other (e.g., positive and negative feedback). When decreasing the rate of a "negative" factor, the "side" effect of that intervention on human health may be either negative or positive. One example of risk factor "symbiosis" is apolipoprotein E allele 4 (apo epsilon4) and smoking. Each of them has been associated with an unfavorable lipid profile, while smoking may modulate the relation of apo epsilon4 to low-density lipoprotein (LDL) and total cholesterol levels, enhancing the genetic susceptibility to an unfavorable lipid profile among subjects with apo epsilon4 (Djousse et al., 2000). Smoking and obesity are an example of one of the most common "antagonistic" effects of the most widespread risk factors. Statistical control for cigarette smoking, as for factor influencing obesity-related mortality in smoking populations, requires careful consideration. The health benefits from quitting smoking are considerable, but may be partially negated by the weight gain following cessation, depending on the amount gained. Smoking prevalence started to decline modestly in some countries, including the United States, during the past 30 years, but the prevalence of overweight and obesity have risen steadily, and there is some evidence that these two trends might be partly related. Numerous studies have documented that cigarette smoking suppresses body weight and that cessation is commonly followed by weight gain: smoking cessation was estimated to be responsible for about one quarter of the increase in prevalence of overweight amongst men in the United States in the 1980s (Flegal et al., 1995; Wallenfeldt et al., 2001; Molarius et al., 1997). Smokers who quit tobacco entirely had an average weigh gain of 6.8 kg during the 9-year follow-up period (Rodu et al., 2004), and other recent studies have reported a range of weight gain amongst ex-smokers from 3 to 10 kg (Flegal et al., 1995; O'Hara et al., 1998), but the direct comparison between studies is difficult because of population inhomogeneity in age and duration of follow-up. It is still unclear if the smoking cessation weight gain of 6–10 kg is crucial in obesity "epidemy" in the United States and the health consequences on population level. For example, the reanalysis of the Framingham Heart Study data (Sempos et al., 1998) does not support the existence of correlation between smoking and measures of obesity: so, the estimated BMI was similar in both males and females, in smokers and nonsmokers, who were at the minimum risk of death. The similar result has been obtained in this study using the Metropolitan Relative Weight (MRW) criteria (other version of measuring overweight/obesity, calculated as the percentage of desirable weight; it is correlated with BMI), even after excluding persons with missing data, CVD at baseline, and persons who died within the first 4 years of follow-up.

While modeling/forecasting the future trends and population health effects of certain cancers (such as lung, esophageal, cervical), it is important to take into account the risk factors associated with cancer histotypes (e.g., two more common – adenocarcinoma and squamous cell carcinoma) because of the differences between the time trends of these histotypes (especially dramatic during the past 30 years), resulting in changed contributions of various risk factors in population cancer mortality and morbidity (see Chapter 7).

Clinical interventions usually involve complicated and costly research. In many respects, modeling approaches will be able to bear the burden of this investigation by detailed modeling of all steps involved in the intervention. Because of many attractive features, microsimulation approaches are good candidates as a methodology to support intervention design. Such features include (1) predicting required statistical power for these investigations, (2) analyzing relations between investigated quantities directed to potential predicting possible errors in design, (3) investigating contributions of possible biases, (4) analyzing effects of heterogeneity by considering various confounding variables, and (5) serving as a basis for model estimation procedure, e.g., by requirement to choose model parameters such that observed distributions will be statistically close to simulated. Microsimulation modeling may essentially enlarge horizons of intervention research, make them much cheaper, and provide comprehensive complement analyses and results.

## References


Akushevich I., Kravchenko J.S., Manton K.G., 2007. Health based population forecasting: effects of smoking on mortality and fertility. Risk Anal, 27(2):467–482.

Akushevich I., Kulminski A., Manton K., 2005. Life tables with covariates: life tables with covariates: dynamic model for nonlinear analysis of longitudinal data. Math Popul Stud 12(2):51–80.

Anderson R.N., 2002. Deaths: leading causes for 2000. Natl Vital Stat Rep 50(16):1–85.

Andreassi M.G., Botto N., Colombo M.G. et al., 2000. Genetic instability and atherosclerosis: can somatic mutations account for the development of cardiovascular disease? Environ Mol Mutagen 35:265–269.

Arabi M., Moshtaghi H., 2005. Influence of cigarette smoking on spermatozoa via seminal plasma. Andrologia 37(4):119–124.

Aragona M., Maisano R., Panetta S. et al., 2000. Telomere length maintenance in aging and carcinogenesis. Int J Oncol 17:981–989.

Ardestani S.K., Inserra P., Solkoff D., Watson R.R., 1999. The role of cytokines and chemokines on tumor progression: a review. Cancer Detect Prev 23:215–225.

Aviram M., 2000. Review of human studies on oxidative damage and antioxidant protection related to cardiovascular diseases. Free Radic Res 33(Suppl):S85–S97.

Bakker S.J., Gans R.O., 2000. About the role of shear stress in angiogenesis. Cardiovasc Res 45:270–272.

Banks D.A., Fossel M., 1997. Telomeres, cancer, and aging. Altering the human life span. JAMA 278:1345–1348.

Bartsch H., 2000. Studies on biomarkers in cancer etiology and prevention: a summary and challenge of 20 years of interdisciplinary research. Mutat Res 462:255–279.

Blobe G.C., Schiemann W.P., Lodish H.F., 2000. Role of transforming growth factor beta in human disease. N Engl J Med 342:1350–1358.

Brouilette S., Singh R.K., Thompson J.R., Goodall A.H., Samani N.J., 2003. White cell telomere length and risk of premature myocardial infarction. Arterioscler Thromb Vasc Biol 23:842–846.

Bruce W.R., Wolever T.M., Giacca A., 2000. Mechanisms linking diet and colorectal cancer: the possible role of insulin resistance. Nutr Cancer 37:19–26.

Cai H., Harrison D.G., 2000. Endothelial dysfunction in cardiovascular diseases: the role of oxidant stress. Circ Res 87:840–844.

CDC (Centers for Disease Control and Prevention), 1997. Smoking-attributable mortality and years of potential life lost – United States, 1984. MMWR Morb Mortal Wkly Rep 46(20):444–451.

CDC (Centers for Disease Control and Prevention), 2005. Annual smoking-attributable mortality, years of potential life lost, and productivity losses – United States, 1997–2001. MMWR Morb Mortal Wkly Rep 54(25):625–628.

Cerami C., Founds H., Nicholl I. et al., 1997. Tobacco smoke is a source of toxic reactive glycation products. Proc Natl Acad Sci USA 94:13915–13920.

Chambers J.C., Seddon M.D., Shah S., Kooner J.S., 2001. Homocysteine: a novel risk factor for vascular disease J R Soc Med 94:10–13.

Checkoway H., DiFernando G., Hulka B.S., Mickey D.D., 1987. Medical, lifestyle, and occupational risk factors for prostate cancer. Prostate 10:79–88.

Chen J.Z., Zhang F.R., Tao Q.M., Wang X.X., Zhu J.H., 2004. Number and activity of endothelial progenitor cells from peripheral blood in patients with hypercholesterolemia. Clin Sci (Lond) 107:273–280.

Chisolm G.M., Chai Y., 2000. Regulation of cell growth by oxidized LDL. Free Radic Biol Med 28:1697–1707.

Ciardiello F., 2000. Epidermal growth factor receptor tyrosine kinase inhibitors as anticancer agents. Drugs 60(Suppl 1):25–32.

Clemons M., Goss P., 2001. Estrogen and the risk of breast cancer. N Engl J Med 344: 276–285.

Coleman W.B., Tsongalis G.J., 1999. The role of genomic instability in human carcinogenesis. Anticancer Res 19:4645–4664.

Collins T., Cybulsky M.I., 2001. NF-kappaB: pivotal mediator or innocent bystander in atherogenesis? J Clin Invest 107:255–264.

Curran S., Murray G.I., 2000. Matrix metalloproteinases: molecular aspects of their roles in tumor invasion and metastasis. Eur J Cancer 36:1621–1630.

DHHS (U.S. Department of Health and Human Services), 2000. Healthy People 2010. With Understanding and Improving Health and Objectives for Improving Health. 2nd edition, Vol. 2. Washington, DC: U.S. Department of Health and Human Services.

DHHS (U.S. Department of Health and Human Services), 2004. The Health Consequences of Smoking. A Report of the Surgeon General U.S. Department of Health and Human Services. Rockville, MD: Public Health Service Office of the Surgeon General. http://www.cdc.gov/tobacco/sgr/sgr_2004/chapters.htm

Dimmeler S., Zeiher A.M., 2000. Endothelial cell apoptosis in angiogenesis and vessel regression. Circ Res 87:434–439.

Djousse L., Myers R.H., Coon H., Arnett D.K., Province M.A., Ellison R.C., 2000. Smoking influences the association between apolipoprotein E and lipids: the National Heart, Lung, and Blood Institute Family Heart Study. Lipids 35(8):827–831.

Doll R., Peto R., Boreham J. et al., 2005. Mortality from cancer in relation to smoking: 50 years observations on British doctors. Br J Cancer 92:426–429.

Dorfman S.F., 2008. Tobacco and fertility: our responsibilities. Fertil Steril 89(3):502–504.

Dreyer L., Olsen J.H., 1999. Risk for non-smoking-related cancer in atherosclerotic patients. Cancer Epidemiol, Biomarkers Prev, 8:915–918.

Edelberg J.M., 2002. Auto repair on the aging stem cell superhighway. Sci Aging Knowledge Environ 4 Sept:13. DOI: 10.1126/sageke.2002.35.pe13.

Enstrom J.E., 1984. Smoking and longevity studies. Science 225:878.

Evans W.K., Will B.P., Berthelot J.M., Wolfson M.C., 1995. Estimating the cost of lung cancer diagnosis and treatment in Canada: the POHEM model. Can J Oncol 5:408–419.

Feinleib M., Luoto J., 1984. Longevity of non-smoking men and women. Public Health Rep 99:223.

Fellows J.L., Trosclair A., Adams E.K., Rivera C.C., 2002. Annual smoking-attributable mortality, years of potential life lost, and economic costs – United States, 1995–1999. Morb Mortal Wkly Rep 51(14):300–303.

Feuer E.J., Etzioni R., Cronin K.A., Mariotto A., 2004. The use of modeling to understand the impact of screening on U.S. mortality: examples from mammography and PSA testing. Stat Methods Med Res 13:421–442.

Flegal K.M., Troiano R.P., Pamuk E.R., Kuczmarski R.J., Cambell S.M., 1995. The influence of smoking cessation on the prevalence of overweight in the United States. N Engl J Med 333:1165–1170.

Gajalakshmi G.K., Jha P., Ranson K., Nguyen S., 2000. Global patterns of smoking and smoking-attributable mortality. In: Jha P., Chalouka F. (eds.). Tobacco Control in Developing Countries. Oxford: Oxford Medical Publications, pp. 11–40.

George S.J., 2000. Therapeutic potential of matrix metalloproteinase inhibitors in atherosclerosis. Expert Opin Invest Drugs 9:993–1007.

Gerszten R.E., Mach F., Sauty A. et al., 2000. Chemokines, leukocytes, and atherosclerosis. J Lab Clin Med 136:87–92.
Gilbert N., Troitszch K.G., 1999. Simulation for the Social Scientist. Buckingham: Open University Press.
Giovino G.A., Schooley M.W., Zhu B.P., Chrismon J.H., Tomar S.L., Peddicord J.P., Merritt R.K., Husten C.G., Erikson M.P., 1994. Surveillance for selected tobacco use behavior – United States, 1900–1994. Morb Mortal Wkly Rep CDC Surveill Summ 43(3):1–43.
GMWKIII, 1998. Deaths from 282 Selected Causes by 5-year Age Groups, Race and Sex: Each State and the District of Columbia, 1995, 1996, 1997 and 1998. National Center for Health Statistics. http://cdc.gov/nchs/data/mortab
Goldschmidt-Clermont P., Peterson E.D., 2003. On the memory of a chronic illness. Sci Aging Knowledge Environ 12 Nov:8. DOI: 10.1126/sageke.2003.45.re8.
Harach H.R., Williams E.D., 1995. Childhood thyroid cancer in England and Wales. Br J Cancer 72:777–783.
Hatzistamou J., Kiaris H., Ergazaki M., Spandidos D.A., 1996. Loss of heterozygosity and microsatellite instability in human atherosclerotic plaques. Biochem Biophys Res Commun 225:186–190.
Health, United States, 2002. National Center for Health Statistics, http://www.cdc.gov/nchs/products/pubs/pubd/hus/02hustop.htm
Heart Disease and Stroke Statistics, 2005. Update. (American Heart Association, Dallas, TX), at http://www.americanheart.org/downloadable/heart/1105390918119HDSStats2005Update.pdf
Henderson B.E., Bogdanoff E., Gerkins V.R., SooHoo J., Arthur M., 1974. Evaluation of cancer risk factors in a retirement community. Cancer Res 34:1045–1048.
Hermann M., Berger P., 1999. Hormone replacement in the aging male? Exp Gerontol 34:923–933.
Hill J.M., Zalos G., Halcox J.P., Schenke W.H., Waclawiw M.A., Quyyumi A.A., Finkel T., 2003. Circulating endothelial progenitor cells, vascular function, and cardiovascular risk. N Engl J Med 348:593–600.
Holden C., 1983. Can smoking explain ultimate gender gap? Science 221:1034.
Hoyert D.L., Anderson R.N., 2001. Age-adjusted death rates: trend data based on the year 2000 standard population. Natl Vital Stat Rep 49(9):1–6.
Human Mortality Database (HMD), 2006. Internet address: http://www.mortality.org
IARC, 1990. Cancer: Causes, Occurrence, and Control. Vol. 100, Lyon, France: IARC Scientific Publications.
IARC, 2004. Tobacco Smoke and Involuntary Smoking. IARC Monographs on the Evaluation of Carcinogenic Risks to Humans. Vol. 83. Lyon: International Agency for Research on Cancer.
Irani K., 2000. Oxidant signaling in vascular cell growth, death, and survival: a review of the roles of reactive oxygen species in smooth muscle and endothelial cell mitogenic and apoptotic signaling. Circ Res 87:179–183.
Isner J.M., 1999. Cancer and atherosclerosis: the broad mandate of angiogenesis. Circulation 99:1653–1655.
Israels E.D., Israels L.G., 2000. The cell cycle. Oncologist 5:510–513.
Ivanov V.K., Manton K.G., Akushevich I., Gorsky A.I., Korelo A.M., Kravchenko J.S., Maksioutov M.A., Tsyb A.F., 2005. Risk of thyroid cancer after irradiation in children and adults. Curr Oncol 12(2):55–64.
JAMA, 2002. Annual smoking-attributable mortality, years of potential life-lost, and economic costs – United States, 1995–1999. JAMA 287:2355–2356.
Jensen T.K., Joffe M., Scheike T. et al., 2006. Early exposure to smoking and future fecundity among Danish twins. Int J Androl 29(6):603–613.
Jeremy J.Y., Rowe D., Emsley A.M., Newby A.C., 1999. Nitric oxide and the proliferation of vascular smooth muscle cells. Cardiovasc Res 43:580–594.

Joesoef M.R., Beral V., Aral S.O., Rolfs R.T., Cramer D.W., 1993. Fertility and use of cigarettes, alcohol, marijuana and cocaine. Ann Epidemiol, ISSN: 1047-2797, 3(6):592–594.

Jump D.B., Clarke S.D., 1999. Regulation of gene expression by dietary fat. Ann Rev Nutr 19:63–90.

Kerbel R.S., 2000. Tumor angiogenesis: past, present and the near future. Carcinogenesis 21:505–515.

Koohestani N., Chia M.C., Pham N.A., Tran T.T., Minkin S., Wolever T.M., Bruce W.R., 1998. Aberrant crypt focus promotion and glucose intolerance: correlation in the rat across diets differing in fat, n-3 fatty acids and energy. Carcinogenesis 19:1679–1684.

Koyi H., Branden E., Gnarpe J., Gnarpe H., Arnholm B., Hillerdal G., 1999. Chlamydia pneumonia may be associated with lung cancer. Preliminary report on a seroepidemiological study. APMIS 107(9):828–832.

Kravchenko J., Goldschmidt-Clermont P.J., Powell T., Stallard E., Akushevich I., Cuffe M.S., Manton K.G., 2005. Endothelial progenitor cell therapy for atherosclerosis: the philosopher's stone for an aging population? Sci Aging Knowledge Environ 2005(25):e18.

Kullo I.J., Gau G.T., Tajik A.J., 2000. Novel risk factors for atherosclerosis. Mayo Clin Proc 75(4):369–380.

Kune G.A., Kune S., Watson L-F., 1988. Colorectal cancer risk, chronic illness, operations, and medications: case control results from the Melbourne colorectal cancer study. Cancer Res 48:4399–4404.

Leinonen M., Saikku P., 2000. Infections and atherosclerosis. Scand Cardiovasc J 34:12–20.

Li C., Xu Q., 2000. Mechanical stress-initiated signal transductions in vascular smooth muscle cells. Cell Signal 12:435–445.

Lichtenstein P., Holm N.V., Vercasalo P.K., Iliadou A., Kaprio J., Koskenvuo M., Pukkala E., Skytthe A., Hemminki K., 2000. Environmental and heritable factors in the causation of cancer – analyses of cohorts of twins from Sweden, Denmark, and Finland. N Engl J Med 343(2):78–85.

Littman A.J., White E., Jackson L.A., Thornquist M.D., Gaydos C.A., Goodman G.E., Vaughan T.L., 2004. *Chlamydia pneumoniae* infection and risk of lung cancer. Cancer Epidemiol Biomarkers Prev 13:1624–1630.

Manole D., Schildknecht B., Gosnell B., Adams E., DerWahl M., 2001. Estrogen promotes growth of human thyroid tumor cells by different molecular mechanisms. J Clin Endocrinol Metab 86(3):1072–1077.

Manton K.G., Akushevich I., 2003. State variable methods for demographic analysis: a mathematical theory of physiological regeneration and aging. Nonlinear Phenom Complex Syst 6(2):1–15.

Manton K.G., Stallard E., 1988. Chronic disease risk modeling: measurement and evaluation of the risks of chronic disease processes. In the Griffin Series of the Biomathematics of Diseases. London, England: Charles Griffin Limited.

Manton K.G., Stallard E., Singer B.H., 1992. Projecting the future size and health status of the U.S. elderly population. Int J Forecast 8:433–458.

Manton K.G., Stallard E., Woodbury M.A., Dowd J.E., 1994. Time-varying covariates in models of human mortality and aging: Multidimensional generalizations of the Gompertz. J Gerontol 49:B169–B190.

Manton K.G., Yashin A.I., 2000. Mechanisms of Aging and Mortality: Searches for New Paradigms. Odense, Denmark: Odense University Press.

Markle H.V., 1997. Coronary artery disease associated with *Helicobacter pylori* infection is at least partially due to inadequate folate status. Med Hypotheses 49:282–292.

Marmot M., Elliott P., 2005. Coronary Heart Disease Epidemiology. From Aetiology to Public Health. 2nd edition. New York: Oxford University Press, p. 932.

Marnett L.J., 2000. Oxyradicals and DNA damage. Carcinogenesis 21:361–370.

Martin J.A., Hamilton B.E., Ventura S.J., 2001. Births: preliminary data for 2000. Natl Vital Stat Rep. 49(5):1–20.

Martin J.A., Hamilton B.E., Ventura S.J., Menacker F., Park M.M., Sutton P.D., 2002b. Births: final data for 2001. Natl Vital Stat Rep 51(2):1–102.
Martin J.A., Park M.M., Sutton P.D., 2002a. Births: preliminary data for 2001. Natl Vital Stat Rep 50(10):1–20.
Maxwell S.R., 1998. Women and heart disease. Basic Res Cardiol 93(Suppl 2):79–84.
Mayo M.W., Baldwin A.S., 2000. The transcription factor NF-kappaB: control of oncogenesis and cancer therapy resistance. Biochim Biophys Acta 1470:M55–M62.
McCully K.S., 1994. Chemical pathology of homocysteine. II. Carcinogenesis and homocysteine thiolactone metabolism. Ann Clin Lab Sci 24:27–59.
Miller G.H., Gerstein D.R., 1983. The life expectancy of non-smoking men and women. Public Health Rep 98:343–349.
Minino A.M., Arias E., Kochanek K.D., Murphy S.L., Smith B.L., 2002. Deaths: final data for 2000. Natl Vital Stat Rep 50(15):1–119.
Misra A., 2000. Risk factors for atherosclerosis in young individuals. J Cardiovasc Risk 7:215–229.
MMWR, 2005. Annual smoking-attributable mortality, years of potential life lost, and productivity losses – United States, 1997–2001. MMWR 54(25):625–628.
MMWR, 2006. Tobacco use among adults – United States, 2005. MMWR 55(42):1145–1148.
Mokdad A.H., Marks J.S., Stroup D.F., Gerberding J.I., 2004. Actual causes of death in the United States, 2000. JAMA 291(10):1238–1245.
Molarius A., Seidell J.C., Kuulasmaa K., Dobson A.J., Sans S., 1997. Smoking and relative body weight: an international perspective from the WHO MONICA project. J Epidemiol Community Health 51:252–260.
Moore M.A., Tsuda H., 1998. Chronically elevated proliferation as a risk factor for neoplasia. Eur J Cancer Prev 7:353–385.
Morre S.A., Stooker W., Lagrand W.K. et al., 2000. Microorganisms in the etiology of atherosclerosis. J Clin Pathol 53:647–654.
Mostafa T., Tawadrous G., Roaia M.M. et al., 2006. Effect of smoking on seminal plasma ascorbic acid in infertile and fertile males. Andrologia 38(6):221–224.
Neugut A.I., Jacobson J.S., Ghada S., Sherif G., Ahsan H., Garbowski G.C., Waye J., Forde K.A., Treat M.R., 1995. Coronary artery disease and colorectal neoplasia. Dis Colon Rectum 38:873–877.
Neugut A.I., Rosenberg D.J., Ahsan H., Jakobson J.S., Wahid N., Hagan M., Rahman M.I., Khan Z.R., Chen L., Pablos-Mendez A., Shea S., 1998. Association between coronary heart disease and cancers of the breast, prostate, and colon cancer. Cancer Epidemiol Biomarkers Prev 7:869–873.
Newby A.C., Zaltsman A.B., 2000. Molecular mechanisms in intimal hyperplasia. J Pathol 190:300–309.
Niho N., Takahashi M., Kitamura T., Shoji Y., Itoh M., Noda T., Sugimura T., Wakabayashi K., 2003a. Concomitant suppression of hyperlipidemia and intestinal polyp formation in Apc-deficient mice by peroxisome proliferators-activated receptor ligands. Cancer Res 63:6090–6095.
Niho N., Takahashi M., Shoji Y., Takeuchi Y., Matsubara S., Sugimura T., Wakabayashi K., 2003b. Dose-dependent suppression of hyperlipidemia and intestinal polyp formation in Min mice by pioglitazone, a PPAR gamma ligand. Cancer Sci 94:960–964.
Nugent M.A., Iozzo R.V., 2000. Fibroblast growth factor-2. Int J Biochem Cell Biol 32:115–120.
O'Brien E.R., Urieli-Shoval S., Garvin M.R. et al., 2000. Replication in restenotic atherectomy tissue. Atherosclerosis 152:117–126.
O'Hara P., Connett J.E., Lee W.W., Nides M., Murray R., Wise R., 1998. Early and late weight gain following smoking cessation in the lung health study. Am J Epidemiol 148:821–830.
Okura Y., Brink M., Itabe H. et al., 2000. Oxidized low-density lipoproteins is associated with apoptosis of vascular smooth muscle cells in human atherosclerotic plaques. Circulation 102:2680–2686.

Parsons R., Myeroff L.L., Liu B. et al., 1995. Microsatellite instability and mutations of the transforming growth factor beta type II receptor gene in colorectal cancer. Cancer Res 55:5548–5550.
Peate I., 2005. The effects of smoking on the reproductive health of men. Br J Nurs 14(7):362–366.
Pekkanen J., Manton K.G., Stallard E., Nissinen A., Karvonen M.J., 1992. Risk factor dynamics, mortality, and life expectancy differences between eastern and western Finland: the Finnish cohorts of the seven countries study. Int J Epidemiol 21:406–419.
Powles J.W., Sanz M.A., 1996. Vascular mortality not attributable to smoking trends in European populations since the 1950s. In: The XIV International Scientific Meeting of the International Epidemiological Association: global health in a changing environment. Program and abstracts. Nagoya: International Epidemiological Association, p. 189.
Ramlau-Hansen C.H., Thulstrup A.M., Aggerholm A.S. et al., 2007. Is smoking a risk factor for decreased semen quality? A cross-sectional analysis. Hum Reprod 22(1):188–196.
Ramsey S.D., McIntosh M., Etzioni R., Urban N., 2000. Simulation modeling of outcomes and cost effectiveness. Hematol Oncol Clin North Am 14(4):925–938.
Rauscher F.M., Goldschmidt-Clermont P.J., Davis B.H., Wang T., Gregg D., Ramaswami P., Pippen A.M., Annex B.H., Dong C., Taylor D.A., 2003. Aging, progenitor cell exhaustion, and atherosclerosis. Circulation 29, 457–463.
Reed J.C., 2000. Apoptosis and cancer: strategies for integrating programmed cell death. Semin Hematol 37(Suppl 7):9–16.
Robbins W.A., Elashoff D.A., Xun L., Jia J., Li N., Wu G., Wei F., 2005. Effect of lifestyle exposures on sperm aneuploidy. Cytogenet Genome Res 111(3–4):371–377.
Rodu B., Stegmayr B., Nasic S., Cole P., Asplund K., 2004. The influence of smoking and smokeless tobacco use on weight amongst men. J Intern Med 255:102–107.
Ron E., Kleinerman R.A., Boice J.D., LiVolsi V.A., Flannery J.T., Fraumen J.F., 1987. A population-based case–control study of thyroid cancer. J Natl Cancer Inst 79:1–12.
Ross J.S., Stagliano N.E., Donovan M.J., Breitbart R.E., Ginsburg G.S., 2001. Atherosclerosis and cancer. Common molecular pathways of disease development and progression. Ann NY Acad Sci 947:271–293.
Ruef J., Peter K., Nordt T.K. et al., 1999. Oxidative stress and atherosclerosis: its relationship to growth factors, thrombus formation and therapeutic approaches. Thromb Haemost 82:32–37.
Sasayama S., Okada M., Matsumori A., 2000. Chemokines and cardiovascular diseases. Cardiovasc Res 45:267–269.
Schwartz S.M., Murry C.W., 1998. Proliferation and the monoclonal origins of atherosclerotic lesions. Annu Rev Med 49:437–460.
Sempos C.T., Durazo-Arvizu R., McGee D.L., Cooper R.S., Prewitt T.E., 1998. The influence of cigarette smoking on the association between body weight and mortality. The Framingham Heart Study revisited. Ann Epidemiol 8(5):289–300.
Sepaniak S., Forges T., Monnier-Barbarino P., 2005. Consequences of cigarette smoking on male fertility. J Gynecol Obstet Biol Reprod (Paris) 34(1):3S102–3S111.
Sepaniak S., Forges T., Monnier-Barbarino P., 2006. Cigarette smoking and fertility in women and men. Gynecol Obstet Fertil 34(10):945–949.
Shackelford R.E., Kaufmann W.K., Paules R.S., 2000. Oxidative stress and cell cycle checkpoint function. Free Radic Biol Med 28:1387–1404.
Shi L., Macinko J., Starfield B., Politzer R., Xu J., 2005. Primary care, race, and mortality in U.S. States. Soc Sci Med 61(1):65–75.
Slater S., Oliver R.T., 2000. Testosterone: its role in development of prostate cancer and potential risk from use as hormone replacement therapy. Drugs Aging 17:431–439.
Smith R.A., Glynn T.J., 2000. Epidemiology of lung cancer. Radiol Clin North Am 38:453–470.
Smoking and Tobacco Control Monograph, 1997. No. 6, NIH Pub. No. 97-4213.

Strong J.P, Malcom G.T., McMahan C.A., Tracy R.E., Newman III W.P., Herderick E.E., Cornhill J.F., 1999. Prevalence and extent of atherosclerosis in adolescents and young adults: Implications for prevention from the Pathobiological Determinants of Atherosclerosis in Youth Study. JAMA 281:727–735.

Tabuchi M., Kitayama J., Nagawa H., 2006. Hypertriglyceridemia is positively correlated with the development of colorectal tubular adenoma in Japanese men. World J Gastroenterol 12(8):1261–1264.

Tardif J.C., 2000. Insights into oxidative stress and atherosclerosis. Can J Cardiol 16(Suppl):2D–4D.

Thambyrajah J., Townsend J.N., 2000. Homocysteine and atherothrombosis: mechanisms for injury. Eur Heart J 21:967–974.

Thompson M.M., Garland C., Barrett E., Khaw K., Friedlander N.J., Wingard D.L., 1988. Heart disease risk factors, diabetes, and prostatic cancer in an adult community. Am J Epidemiol 129:511–517.

TIPS (Tobacco Information and Prevention Source), November 2005. Smoking Prevalence Among U.S. Adults. URL: http//www.cdc.gov/tobacco/research_data/adults_prev/prevail.htm

Tunstall-Pedoe H. (ed.), 2003. Monica: Monograph and Multimedia Sourcebook. Geneva, Switzerland: World Health Organization.

Tzonou A., Hsich C.-C., Trichopoulos D., Aravandinos D., Kalandidi A., Margaris D., Goldman M., Toupadaki N., 1993. Induced abortions, miscarriages and tobacco smoking as risk factors for secondary infertility. J Epidemiol Community Health 47:36–39.

Vainio H., Bianchini F., 2002. IARC Handbooks of Cancer Prevention. Vol. 6: Weight Control and Physical Activity. Lyon, France: IARC Press.

Van den Akker-van Marle M.E., van Ballegooijen M., van Oortmarssen G.J., Boer R., Habbema J.D., 2002. Cost-effectiveness of cervical cancer screening: comparison of screening policies. J Natl Cancer Inst 94:193–204.

Vanness D.J., Tosteson A.N., Gabriel S.E., Melton L.J., 2005. The need for microsimulation to evaluate osteoporosis interventions. Osteoporos Int 16(4):353–358.

Ventura S.J., Hamilton B.E., Sutton P.D., 2003. Revised birth and fertility rates for the United States, 2000 and 2001. Natl Vital Stat Rep 51(4):1–18.

Waldron I., 1986. The contribution of smoking to sex differences in mortality. Public Health Rep 101(2):163–173.

Wallenfeldt K., Hulthe J., Bokemark L., Wikstrand J., Fagerberg B., 2001. Carotid and femoral atherosclerosis, cardiovascular risk factors and C-reactive protein in relation to smokeless tobacco use or smoking in 58 year old men. J Intern Med 250:492–501.

Wang X.L., 2000. Cigarette smoking. DNA variants in endothelial nitric oxide synthase gene and vascular disease. Contrib Nephrol 130:53–67.

Wang J.M., Deng X., Gong W., Su S., 1998. Chemokines and their role in tumor growth and metastasis. J Immunol Methods 220:1–17.

Witteman J.C.M., Grobbee D.E., Valkenburg H.A., van Hemert A.M., Stijnen T., Burger H., Hofman A., 1994. J-shaped relation between change in diastolic blood pressure and aortic atherosclerosis. Lancet 343:504–507.

Wolf D., 2001. The role of microsimulation in longitudinal data analysis. Can Stud Popul 28:165–179.

Wolfson M.C., 1994. POHEM – a framework for understanding and modeling the health of human populations. World Health Stat Q 47:157–176.

Wood D., 2001. Established and emerging cardiovascular risk factors. Am Heart J 141:49–57.

Woodbury M.A., Manton K.G., 1977. A random walk model of human mortality and aging. Theor Popul Biol 11:37–48.

Woodbury M.A., Manton K.G., 1983a. A mathematical model of the physiological dynamics of aging and correlated mortality selection. I. Theoretical development and critiques. J Gerontol 38:398–405.

Woodbury M.A., Manton K.G., 1983b. A theoretical model of the physiological dynamics of circulatory disease in human populations. Human Biol 55:417–441.
Woodbury M.A., Manton K.G., Stallard E., 1981. Longitudinal models for chronic disease risk: an evaluation of logistic multiple regression and alternatives. Int J Epidemiol 10:187–197.
Yamada K., Araki S., Tamura M., Sakai I., Takahashi Y., Kashihara H., Kono S., 1998. Relation of serum total cholesterol, serum triglycerides and fasting plasma glucose to colorectal carcinoma in situ. Int J Epidemiol 27:794–798.
Yancopoulos G.D., Davis S., Gale N.W. et al., 2000. Vascular-specific growth factors and blood vessel formation. Nature 407:242–248.
Zalba G., Beaumont J., San Jose G. et al., 2000. Vascular oxidant stress: molecular mechanisms and pathophysiological implications. J Physiol Biochem 56:57–64.
Zettler M.E., Pierce G.N., 2000a. Cell cycle proteins and atherosclerosis. Herz 25:100–107.
Zettler M.E., Pierce G.N., 2000b. Growth-promoting effects of oxidized low density lipoprotein. Can J Cardiol 17:73–79.

# Chapter 9
# Cancer Prevention

## 9.1 Brief Overview of Prevention History, Strategies, Conquests, and Uncertainty

Discussion of the possibility and effectiveness of cancer prevention started generations ago. The progress in cancer prevention is a story of evolution in both science and public perception of rapidly progressing research, coupled with changing societal attitudes and beliefs. In the first quarter of the twentieth century, the idea of cancer control was not widely promoted (Breslow et al., 1977). The earliest references to "cancer control" appeared in 1913, when the American Society for Control of Cancer was formed, that in 1945 became the American Cancer Society (ACS) (New York City Committee, 1994). In 1937, cancer control was officially recognized, when the National Cancer Institute (NCI) was formed. The Surgeon General was authorized to act through this Institute and the National Cancer Advisory Council to "cooperate with state health agencies in the prevention, control, and eradication of cancer" (Hiatt and Rimer, 2006). In 1973, the Division of Cancer Control and Rehabilitation was established at NCI as the first unit devoted to cancer control (Breslow et al., 1977). Later, in 1983, the Division of Cancer Control and Prevention was formed at the NCI to provide support for both scientific basis and practical interventions in cancer prevention. A five-phase program of research on cancer prevention and control was defined by Greenwald and Cullen in 1985: (1) Phase I – generation of hypothesis, (2) Phase II – methods development, (3) Phase III – controlled intervention trials, (4) Phase IV – studies in defined populations, (5) Phase V – demonstration projects, leading to health services programs and nationwide prevention strategies (Greenwald and Cullen, 1985; Greenwald, 1995). In 1999, the Division of Cancer Control and Population Science (DCCPS) and the Division of Cancer Prevention (DCP) were organized to succeed the old Division of Cancer Control and Rehabilitation, with the purpose of focusing on population/public health prospective studies integrated with behavioral factor research, as well as to incorporate recent biomedical discoveries in cancer prevention studies (Hiatt and Rimer, 1999). The

K.G. Manton et al., *Cancer Mortality and Morbidity Patterns in the U.S. Population*, Statistics for Biology and Health, DOI 10.1007/978-0-387-78193-8_9,
© Springer Science+Business Media, LLC 2009

importance of additional approaches to cancer prevention studies has been stressed, among them explicating the statement that basic research may significantly contribute to cancer control and population sciences and that progress in cancer prevention studies may not always occur in the linear fashion defined by the Greenwald–Cullen Five Phase Model. Contemporary studies require fundamental and applied scientists working together in transdisciplinary and interdisciplinary teams that allow developing a continuing and non-linear approach of reexamining the existing data/underlying knowledge base (Abrams, 1999; Turkkan et al., 2000; Hiatt and Rimer, 2006; Halfon and Hochstein, 2002).

A review of the progress made in cancer prevention and control in the latter half of the twentieth century reveals both successes and failures. Progress in early diagnosis has been made by inventing methods of primary and secondary cancer prevention, such as mammography, Pap smear test, colonoscopy, fecal occult blood testing (FOBT), efforts made to persuade the public to stop smoking, the development of hepatitis B vaccine and its impact on hepatocellular carcinoma morbidity, and the identification of chemopreventive agents for certain cancers.

One of the ways to assess the level of activity within a discipline is the evaluation of trends in research publications: data for trends in the number of citations in the National Library of Medicine's PubMed database for cancer prevention showed an increasing interest in this area of cancer research over a 30-year period (see Table 9.1). These trends reflect articles written in English, but not necessary by U.S. authors, therefore reflecting trends in the general medical literature. These trends must be interpreted with caution, because they may reflect changes in the way in which medical subject headings were applied to index the literature, rather than real increases in cancer-related research. Nevertheless, the number of articles on cancer prevention appears to increase markedly from 1985 to 2006, approximately 6.2-fold (see Table 9.1) in both human and animal studies. The number of articles on cancer prevention increased more substantially in the age group of people aged 65+, and the most dramatic – in the age group of 80+, reflecting the increasing interest in the old and oldest old. Although the number of research citations on cancer prevention increased during the past 30 years, they still represent only 7.2% of all cancer-related publications indexed in medical literature (compare with 3.2% in

**Table 9. 1** PubMed citations for cancer prevention, 1985–2006

| Year | Total number of cancer prevention publications | Human studies | Animal studies | Age groups, years old | | | |
|---|---|---|---|---|---|---|---|
| | | | | 0–18 | 19+ | 65+ | 80+ |
| 1985 | 958 | 823 | 214 | 179 | 360 | 181 | 0 |
| 1995 | 2,806 | 2,446 | 600 | 308 | 1,060 | 623 | 175 |
| 2000 | 3,890 | 3,338 | 952 | 370 | 1,381 | 811 | 203 |
| 2005 | 5,667 | 4,996 | 1,341 | 535 | 1,997 | 1,208 | 341 |
| 2006 | 5,976 | 5,266 | 1,358 | 586 | 2,101 | 1,222 | 380 |

1985). That might be evidence of the fact that researchers concentrate their efforts predominantly on studying existing cancer and its treatment, progression, and complications rather than on studying the possible ways cancer might be prevented, especially primarily.

It is possible that improvements in smoking behavior, some dietary modifications, and screening have played an important role in mortality decline of certain cancers during recent decades (see Chapter 1). However, there are still not often available to link cancer prevention and control interventions directly to declines in cancer mortality (World Cancer Research Fund, 1997; Devesa et al., 1989; Wingo et al., 1999; Hiatt and Rimer, 2006). According to the U.S. Surgeon General's report on the health consequences of smoking, the prevalence of smoking among adults decreased from 45% in 1964 to 23% in 2001 (Centers for Disease Control and Prevention, 2003). Now in the U.S. there are more former smokers than people who are current smokers (Hiatt and Rimer, 2006). Still controversial [supported with a systematic review (World Cancer Research Fund, 1997)] is the beneficial effects of increased fruit and vegetable consumption in preventing colon, mouth, esophagus, and stomach cancers. In a pooled analysis of 14 cohort studies, fruit and vegetable intakes were not strongly associated with decreased colon cancer risk overall, but might be associated with a lower risk of distal colon cancer (Koushik et al., 2007). Recently, the suggested beneficial amount of daily fruits and vegetables intake has been increased to nine or more servings (Hiatt and Rimer, 2006). Screening strategies seem to be effective in reducing cervical (from 7.7. in 1969 to 2.8 per 100,000 in 2000) and breast cancer mortality [trials reported mortality reductions ranging from no significant effects, as in the Canadian trial, to a 32% reduction in breast cancer mortality, with a summary RR of breast cancer death among women older than 50 of 0.77 (CI 95% 0.67–0.89), and for women aged 40–49 RR of 0.83 (CI 95% 0.64–1.04)] (SEER, 2003; US Preventive Services Task Force, 2002). The effectiveness of preventive measures has been estimated for colorectal cancer: mortality after 18 years of follow-up was 33% lower among persons who underwent FOBT than among controls who received usual care. The effectiveness of colonoscopy in a case–control study suggested decreased colon cancer incidence with OR = 0.47 (CI 95% 0.37–0.58) and lower mortality (OR = 0.43, CI 95% 0.30–0.63) (US Preventive Service Task Force, 2002; Mandel et al., 1993; Muller and Sonnenberg, 1995). Despite some success, there are not many positive effects of cancer prevention strategies for certain cancers. Substantial disparities in tobacco use between race and ethnic groups in the United States still exist, with over 25% of teenagers still smoking (Centers for Disease Control and Prevention, 2003). There are still many groups of females which are not appropriately screened for breast cancer (Swan et al., 2003). Obesity prevalence in the United States increased from 14.5 to 30.9% over the 1971–2000 period (MMWR, 2004). Then, there is still a need to seek optimal approaches to cancer prevention. That will take time and persistence.

New information about cancer risk factors obtained in some recent studies might provide additional strategic areas of future cancer prevention efforts. However, some actions should be taken with caution, taking into account *pro and contra* arguments, because of sometimes binary role of certain risk factors in cancer and noncancer diseases. Among recently published meta-analyses are the protective role of *Helicobacter pylori* infection in esophageal carcinoma and Barrett's esophagus (but not in squamous cell carcinoma) (Rokkas et al., 2007), the protective effect of recreational sun exposure for B-cell, but not T-cell, lymphomas (Kricker et al., 2007), and observed strong inverse relationships between atopic disease and glioma [unlikely to be explained by methodologic bias alone (Linos et al., 2007)], and etc. This "dualism" of some cancer risk factors (such as role of *H. pylori* in gastric cancer and esophageal adenocarcinoma, or the role of sun exposure in skin melanoma risk and non-Hodgkin's lymphoma) makes the task of estimating the contribution of certain risks to cancer incidence and/or mortality in populations complicated.

The estimates of cancer incidence support the claim that in an ideal world, more than 50% of cancers could be prevented, if what is already known about the etiology and the early cause of cancer were acted on and fully adopted (Wilson et al., 2002). To prevent cancer, an understanding of causality is a prerequisite for effective action. Causation in cancer risk factors (e.g., dietary patterns) is extremely difficult to establish: so, some campaigns, such as "5-a-day for better health" have been started during the past decade to encourage large-scale dietary changes, but their results are delayed because of a 20–30-year lag due to tumor latency.

At the time cancer is diagnosed, even with the advanced techniques now available, more than 90% of the biological lifetime of a tumor is over, with the best chance to control the malignant process having been missed (Meyskens, 2004). Because fewer than 50% of cancers, once established, are cured and because gains in treatment effectiveness have been increasingly incremental and expensive, early detection and prevention of cancer should be pursued aggressively as a means to reduce the burden of morbidity and mortality (Jemal et al., 2002). The role of prevention strategies in the overall management of cancer may be often neglected by clinical oncologists, although health-care planners and society are intensely interested in this topic (Young amd Wilson, 2002).

Cancer risk reduction may be divided into two strategies: (1) prevention (reduction of carcinogen exposure) and (2) protection (intervention that is aimed to stimulate mostly endogenous mechanisms of the organism to reduce the risk from exposure to carcinogens). Prevention, in its turn, has three levels: (1) primary – decrease the risk for normal asymptomatic individuals (e.g., screening programs, smoking cessation, diet modification), (2) secondary – decrease the progression of preneoplastic process (e.g., attempts to reverse preneoplasia such as oral leukoplakia or Barrett's esophagus) with chemoprevention and/or early detection, and (3) tertiary prevention – decrease the morbidity of established disease (e.g., chemoprevention of second malignancies).

Chemoprevention is a relatively new approach to cancer prevention with precedences in cardiology (i.e., cholesterol-lowering, antihypertensive, and antiplatelet agents in high-risk individuals). There are several important criteria that can improve the selection of molecular targets for cancer prevention, such as under-/overexpression of aberrant proteins in early neoplasias, biologic contribution from the aberrant protein (as it was demonstrated in genetically manipulated mouse models), pharmacologic accessibility, modulation that correlates with reduction in cancer incidence, and specificity of action within neoplasia rather than normal tissue (Viner et al., 2006). Chemopreventive strategy could potentially either prevent further DNA damage that might enhance carcinogenesis or suppress the appearance of the cancer phenotype, which is especially important for individuals at high risk for specific cancers. Chemoprevention is the use of either natural or synthetic substances, or their combination, to block, reverse, or retard carcinogenesis. Chemopreventive agents can be grouped into two general classes: (1) blocking agents (e.g., flavonoids, oltipraz, indoles, isothiocyanates) that prevent carcinogenic compounds from reaching or reacting with critical target sites by preventing the metabolic activation of carcinogens or tumor promoters by enhancing detoxification systems and by trapping reactive carcinogens, and (2) suppressing agents [e.g., vitamin D and related compounds, NSAIDs, vitamin A and retionoids, 2-difluoromethylornithine (DFMO), monoterpens, calcium] that prevent the evolution of the neoplastic process in cells that would otherwise become malignant (Keloff et al., 1994; Wattenberg, 1996). The idea of chemoprevention of human cancer has been widespread for more than 30 years. In 1976, Michael Sporn first used the term "chemoprevention" (Sporn et al., 1976) in his paper on vitamin A and retinoids and their effect in retarding chemical carcinogenesis.

The NCI's chemoprevention program in the early 1980s has developed into a major effort in which more than 400 potential chemopreventive agents have been studies, including more than 60 clinical trials. It is generally considered that large randomized Phase III clinical trials are the best for testing effectiveness of chemopreventive intervention. These studies may also provide opportunities to validate the potential biomarkers as surrogate end points for cancer. Among the large-scale interventional studies are (1) the Polyp Prevention Trial (began in 1991 and studied the effect of a low-fat, high-fiber, high-vegetable, and high-fruit dietary pattern on the recurrence of adenomatous colon polyps); (2) the Women's Health Initiative [began in 1993 and studied the effects of a low-fat dieting pattern (high in vegetables, fruits, and fiber), hormone replacement therapy, and calcium and vitamin D supplementation as potential preventive agents for cancer, CVD, and osteoporosis]; (3) the Linxian Trials (began in 1986 and conducted by the NCI in collaboration with the Chinese Academy of Medical Sciences, they studied the effect of daily intake of vitamin/mineral supplements on esophageal cancer in a high-risk population); (4) the Women's Health Study (began in 1992, it studied the risk and benefits of low-dose aspirin and the antioxidants beta-carotene and vitamin E in the primary prevention of CVD and cancer in healthy postmenopausal women); (5) the Breast Cancer Prevention Trial (began in 1992, it tested the ability of tamoxifen – a synthetic compound with antiestrogenic activity – to prevent the development of breast cancer in healthy women who were at an increased risk for disease as determined by age, number of first-degree relatives with breast cancer, age at first live birth, number of benign breast biopsies, age at menarche, and presence of atypical hyperplasia); (6) the Prostate Cancer Prevention Trial (studied the ability of finasteride to prevent the development of early-stage prostate cancer in men considered to be at increased risk based on age); (7) the Completed Beta-Carotene Trial (it included PHS, CARET, and ATBC studies; PHS was the general population study involved 22,000 U.S. physicians that evaluated the effects of aspirin and beta-carotene supplementation on the primary prevention of CVD and cancer; ATBC (in Finland) and CARET were both conducted in populations at high risk for lung cancer). Results of these studies attracted public and physician attention to the chemoprevention of cancer. In 1998, the tamoxifen breast cancer prevention trial was completed: it demonstrated that a drug could reduce the development of breast cancer in high-risk women by approximately 40% (Fisher et al., 1998). Other important studies of chemopreventive effects included NSAIDs in preventing recurrent colon polyps, and selenium and vitamin E as potential preventive agents in prostate cancer (Janne and Mayer, 2000; Thompson et al., 2001). However, some failures of chemoprevention studies (such as vitamin E and $\beta$-carotene in lung cancer prevention, 13cRA and 4-HPR in bronchial metaplasia reversion in smokers, low-fat diet in reducing recurrent colorectal adenomas, and sunscreen usage to prevent skin melanoma, etc.) caused skepticism about the success of preventive efforts ($\alpha$-Tocoferol and $\beta$-Carotene Cancer Prevention Study Group, 1994; Lee et al., 1994; Kurie et al., 2000; Schatzkin et al., 2000; Garland et al., 1993). One of the underlying causes of these failures might be in that carcinogenesis occurs over decades, whereas prevention trials last only around 10 years (development of melanoma, e.g., may be associated with very early damage, which is not altered by using sunscreens later in life) (Young and Wilson, 2002).

The prioritized risks that may be targets for cancer preventive strategies are shown in Fig. 9.1.

During the latter two decades, many chemopreventive agents showed certain efficacy for various organs/systems: upper aerodigestive tract – isotretinoin (Hong et al., 1986, 1990; Lippman et al., 1993); lung – isotretinoin (Lippman et al., 2001), anethole dithiolthione (Lam et al., 2002), 9-cis retinoic acid (Kurie et al., 2003); stomach – anti-H. pylori antibiotics (Correa et al., 2000); colon adenoma – lysine acetylsalicylate (Benamouzig et al., 2003), aspirin (Baron et al., 2003; Sandler et al., 2003), sulindac (Labayle et al., 1991; Giardiell et al., 2002), celecoxib (Steinbach et al., 2000); breast – fenretinide (Veronesi et al., 1999), tamoxifen (Fisher et al., 1998; Cuzick et al., 2002); prostate – finasteride (Thompson et al., 2003), cervix uteri – HPV-16 virus-like particle vaccine (Koutsky et al., 2002), etc. Some approaches that are beneficial in cancer prevention may be anticipated to be effective in other diseases (such as

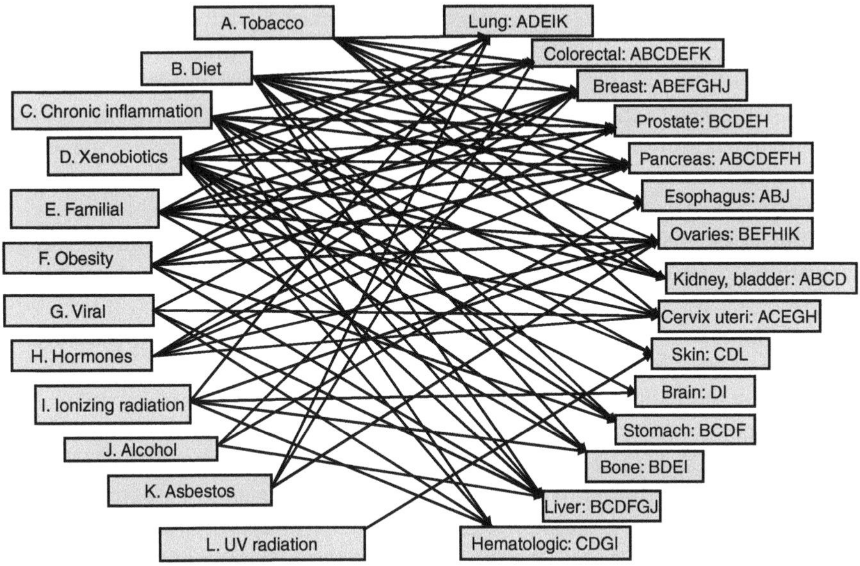


**Fig 9.1** Prioritized risks – potential targets for cancer prevention (the capital letters on the *right* show the associated with cancer risk factors, which are listed on the *left*) (Neiburgs and Vali, 2007)

neurodegenerative disease and atherosclerosis), sharing mechanisms of disease initiation or promotion (Lippman and Hong, 2002). Certain cancer chemopreventive agents may be effective in prevention or delaying of noncancer aging-related diseases: e.g., NSAID for colorectal cancer and rheumatoid arthritis, selective estrogen receptor modulators (SERMs) for osteoporosis and breast cancer, statins for CVD and colorectal cancer, peroxisome proliferator-activated receptor (PPAR) agonist for lipid homeostasis and cancer (Lundholm et al., 1994; Cummings et al., 1999). This cross-efficacy may suggest shared diseases mechanisms, and by using this potential, prevention may move beyond its classic "one disease-one drug" approach (Viner et al., 2006).

The development of multigenic models of cancer susceptibility will be an important future approach to predicting, preventing, and diagnosing some cancers. Even when there is evidence of genetic predisposition, future research efforts must focus on gene–environment interactions to fully develop effective cancer prevention and treatment strategies. Presently, genetic polymorphism monitoring might be used for the detection of patients who are more prone to specific types of cancer or to the adverse effects of specific pharmaceutical agents (Taningher et al., 1999). A systematic population screening of genetic polymorphisms to detect individuals predisposed to cancer seems to be an achievable goal only in the relatively distant future, once present uncertainties of metabolic pathways are resolved and techniques allowing the genotyping of large sets of genes are widely adopted (thus encouraging a profitable cost/effectiveness ratio). The possible exceptions could be cases of high occupational exposure to specific chemical agents or of high racial predisposition (analysis results should remain confidential to avoid risks of individual discrimination). Another exception could be exposure to tobacco smoke (though, in this case, antismoking campaigns might still prove to be the most effective preventive measures). Certain technologies could allow genotypes of relatively large sets of all the major families of Phase I and Phase II metabolism genes. It is possible to systematically explore not only DNA sequences but also mRNA levels in different target tissues. But it should be recognized that the crucial events of activation (Phase I) and detoxification (Phase II) take place not at the DNA or mRNA level, but rather at the protein enzymatic level (it is unclear if genotyping and analysis of gene expression will better reflect the overall phenotypic functions of Phase I and Phase II enzymes.)

There are many potential breakthroughs in cancer prevention and treatment, e.g., vaccines to prevent cancer, use of telomerase inhibitors for cancer treatment, use of cDNA microarrays to help individualize cancer treatment, use of p53 and similar inhibitors to spare normal tissue during cancer treatment, gene therapy, angiogenesis targeting treatment, improving quality of life for cancer patients (such as treating cachexia and anemia), and others. In the technical report "Health Status and Medical Treatment of the Future Elderly", the Technical Expert Panel estimated the likelihood of occurrence in 20 years (likelihood of occurrence means widespread use in clinical practice) of several potential medical breakthroughs in cancer treatment and prevention as following: telomerase inhibitors – 100% (effect on mortality: 50% will be cured, 50% will have a 25% prolongation of life), cancer vaccines – 10–20% (melanoma and renal cell carcinoma could be cured; all other cancers could have a 25% boost in survival), selective estrogen receptor modulators (SERMs) – 90% (breast cancer decrease of approximately 30%, decrease of osteoporosis with an increase of bone density in the spines of women with osteoporosis by 2%), and antiangiogenesis – 70–100% (cure for metastatic disease in 10–50%) (Goldman et al., 2004).

This paragraph briefly describes a new promising approach in cancer therapy – telomerase inhibitors. Telomerase inhibitors used for cancer treatment must be able to knock out the entire capacity of the cell to replace its telomeres, because the residual ability to replace telomeres may allow the cell to continue dividing, or the rise of cells that are resistant to the drugs through natural selection. Cancer cells have relatively shorter telomeres compared to normal cell types, thus making it possible to suggest that human cancers may be considerably more susceptible to being killed by agent inhibiting telomere replication than normal cells. One of the considerations is about the expected lag period between the time when telomerase is inhibited and the time cancer cells' telomeres shorten sufficiently to produce a deleterious effect on cellular proliferation (White et al., 2001). Cancer cells might become resistant to telomerase inhibitors or develop alternative mechanisms of telomere maintenance which is independent from telomerase, as this has been observed in experimentally immortalized human cell lines (Bryan et al., 1995). Inhibitors of telomerase also may potentially have effects on other human somatic cells that express telomerase (e.g., germline cells, hematopoietic stem cells, cells of the basal layer of the epidermis and intestinal crypts) (Wright et al., 1996). These effects might be minor because the stem cells of renewed tissues typically have much longer telomeres than cancer cells have and the deepest stem cells proliferate discontinuously. There is also concern that inhibiting telomerase might lead to an increase in malignancy by enhancing cells' genomic instability, as it has been shown in several animal models (Artandi and DePinho, 2000). However, presently, there is no proven evidence that this would be true in humans (White et al., 2001). Approaches for targeting telomerase in cancer therapy include inhibitors of the enzyme catalytic subunit and RNA component (hTERT), agents that target telomeres, telomerase vaccines, and drug targeting binding proteins (Phatak and Burger, 2007). With hTERT vaccines and antisense oligonucleotides (ODN) under clinical investigation, and several synthetic small molecules entering clinical trials, it is believed that telomerase therapy may soon become an integral part of cancer chemotherapy regimens. Several Phase I trials of hTERT immunotherapy have been completed in patients with breast, prostate, lung, and other cancers with promising results (Carpenter and Vonderheide, 2006; Hochreiter et al., 2006). When the U.S.-NCI trial concluded that the targeted hTERT peptide is not present on the surface of tumor cells and thus cannot be useful for immunotherapy of patients with cancer (Parkhurst et al., 2004), other studies showed obvious tumor responses (Brunsvig et al., 2006; Vonderheide et al., 2004; Phatak and Burger, 2007).

Cancer vaccines are another method of cancer therapy which was theorized to possibly cure certain cancers. It is usually combined with adjuvants and is active immunotherapy. Monoclonal antibodies, such as tumor cells, antiidiotype, ganglioside, and others were recently been proposed for melanoma treatment, and MAb CO17-1A – for treatment of colorectal carcinoma (Herlyn and Birebent, 1999). Tumor-associated antigens have been identified in melanoma, breast, colorectal, ovarian, lung, pancreatic, and other tumors. Some of these antigens, including carcinoembryonic antigen (CEA), MUC-1, prostate-specific antigen (PSA), Her-2/neu, gp72, gp75, gp100, tyrosinase, MAGE, CAGE, BAGE, and RAGE were targeted for immunotherapy of cancers of the bladder, breast, colon, lung, melanoma, pancreas, prostate, and renal cancers and sarcoma (Long et al., 1999; Scholm et al., 1998; Minev et al., 1999; Hwang et al., 1999).

## 9.2 Smoking

Tobacco was not known by Europeans prior to the voyages of Christopher Columbus – its written history began in 1492. Initially, the word "tabaco" was used by Tainos, the pre-Columbian natives of the West Indies, for naming the bifurcated tube used for inhaling "cohobba" (snuff). Historians of the exploration of the Western Hemisphere were confused and decided that "tabaco" was the name of the plant (Stewart, 1967). The main species of tobacco was named *Nicotiana tobaccum* after the French ambassador to Portugal Jean Nicot, who brought tobacco seeds in 1560 to his queen, Catherine de Medici, and has introduced snuff to the French court. Nicot experimented with tobacco trying to use it as medicine: he wrote about the case when he could "heal" a sore on one boy's cheek that was supposed to be cancerous, by tobacco application for a 10-day period. Tobacco was used in Europe for medical purposes over the almost 400 years for curing fistulas, abscesses, diarrhea, pain relief, wounds and burns, intestinal parasites, gout, tooth decay, persistent headaches, and even syphilis and madness (Doll, 1998; Stewart, 1967). Only since 1828, when chemists isolated nicotine, did researches start to gain certain knowledge of the poisoning components of tobacco. The disputes over tobacco medicine in the United States were continued until 1860, when tobacco was abandoned for medicinal use (Lock et al., 1998).

Approximately 5763 trillion cigarettes were manufactured in 2006 – an average of 2.4 per day for the whole 6,528,051,823 population of the world (US Census Bureau, 2006). Recently it has been projected that number of tobacco-attributable deaths for 2030 may range from 7.4 million in the optimistic scenario to 9.7 million in the pessimistic scenario (compared with 5.4 million deaths in 2005), declining by 9% in high-income countries and doubling in low- and middle-income countries (Mathers and Loncar, 2006; Jha et al., 2006). According to other forecasting, without preventing cancer through tobacco and infection control, the number of global cancer deaths may reach 11.5 million in 2030 (Mathers and Loncar, 2006) (tobacco smoking as a risk factor and as associated with smoking diseases are discussed in Chapters 3 and 7).

The first known public smoking ban was declared by Pope Urban VII in 1590, i.e., 30 years after tobacco seeds became known in Europe: he said that anyone who "took tobacco in the porchway of or inside a church, whether it be by chewing it, smoking it with a pipe or sniffing it in powdered form through the nose" would be excommunicated (Henningfield, 1985). The other earliest well-known antismoking measure has been taken by King James I, who issued in 1604 his "Counterblaste to Tobacco". Public smoking bans were enacted during the next 200 years in certain parts of Germany and Austria. Created in 1920s in the United States the Anti-Cigarette League succeeded in getting legal prohibition of cigarette smoking in many U.S. states. The nationwide tobacco ban was imposed in every German university, post office, and military hospital in the 1940s, under the supervision of Karl Astel's Institute for Tobacco Hazard Research (Proctor, 2001a, b). It was from the 1960s when multiple articles in medical journals and media, as well as the Surgeon General's Report started to cite smoking as a leading cause of lung cancer and heart disease. In 1975, Minnesota enacted the Minnesota Clean Indoor Air Act, making the first state to ban smoking in most public spaces, and in 1990 the city of San Luis Obispo, California, became the first city in the world to ban indoor smoking at all public places, including bars and restaurants. Bans on smoking in public places have been enacted in Ireland, Norway, and United Kingdom since 2004–2007.

There is strong evidence of the health benefits of smoking cessation for most of smoking-associated cancers and noncancer disease (U.S. Department of Health and Human Services, 1990; IARC, 2004). The health benefits are greatest when cessation occurs at an early age, but it is still obvious even when quitting smoking at age 60 (Peto et al., 2000). However, the absolute risk of developing cancer or other smoking-related diseases does not decrease after smoking cessation, but it increases with age at a slower rate in persons who stop smoking compared with those who continue. Estimating the effects of smoking cessation is more informative when using not the time since quitting, but age at cessation, because the benefits of cessation are not constant at every age (Thun and Henley, 2006).

The continuing decrease in male smoking prevalence in industrialized countries may be explained by cessation and death among smokers, as well as by the lower initiation rates of later cohorts. Smoking rates in women peaked about 20 years later than men's, it did not decline in next 15–20 years (women who stopped smoking or died are replaced by new smokers), and the rates of smoking-related diseases in most of industrialized countries continues to increase. Chronic obstructive pulmonary disease (COPD) needs at least 40 years after cessation to reduce its risk by more than half, so, the number of people with COPD is still rising in most countries among both, males and females (Lopez et al., 2006).

There are two complimentary approaches to reduce the tobacco effects on population health that were suggested by the U.S. Department of Health and Human Services (2000). One is a long-term intervention that includes the systematic application of primary prevention measures to reduce the initiation of tobacco use among young population. This approach has a long-term application, and the resulting significant reduction in tobacco consumption and decreases in smoking-associated diseases prevalence would be expected in the second half of the twenty-first century. The other, in the near term, approaches include counseling and treatment to facilitate cessation in current smokers.

The community-based interventions for reducing tobacco smoking prevalence include regulatory approaches (laws ensuring clean indoor air, restrictions on tobacco marketing, enforcement of laws restricting of persons who are under the legal age to gain access to tobacco), economic approaches (increasing the price of cigarettes through excise taxes), and to redefine social norms of tobacco use via countermarketing campaigns (Centers for Disease Control and Prevention, 1999). Several comprehensive tobacco control programs were effective among young people after applying in California, Massachusetts, and Florida (Bauer and Johnson, 2001). There are several recent reports about the sharp decrease of prevalence of smoking among U.S. high school students in

1997–2001 (Johnston, 2001; Centers for Disease Control and Prevention, 2002; Everett and Warren, 2001; Kopstein, 2001).

According to the Centers for Disease Control and Prevention (2000b), about 70% of current smokers in the United States report that they want to quit this habit. Various therapies are available now for a tobacco user who is going to quit, such as nicotine replacement (gum, patch, nasal spray, and inhaler), sustained-release bupropionhydrochloride, practical counseling (problem solving/skills training), social support, and help with securing social support outside of treatment, enabling almost 25% of attempting to quit to stay abstinent for 1 year after treatment (Anderson et al., 2001; Centers for Disease Control and Prevention, 2000a, b; U.S. Department of Health and Human Services, 2000).

Preventive strategies may be more effective when using the interregional heterogeneity to specify the interventions to local conditions and for different age groups, balancing between smoking prevention and cessation (Ezzati and Lopez, 2004).

## 9.3 Diet

Belief in the benefits of food for disease prevention and treatment dates back to ancient times. Hippocrates postulated: "Let food to be your medicine, and medicine be your food." Doll and Peto (1981) defined diet as a "chronic source of both frustration and excitement to epidemiologists", stressing that "for many years there has been strong but indirect evidence that most of the cancers that are currently common could be made less so by suitable modification of national dietary practices". In the 25 years which has passed since the Doll and Peto publication, there is still not enough precise and reliable evidences of what exact dietary changes would be of major importance in reducing various cancers risks in human population. Scientists repeatedly returned to attempts to estimate the percent of cancer deaths potentially avoidable by dietary changes. Doll and Peto suggested that the possible reduction of the cancer deaths in the United States by dietary modification could be around 35%, with the range of acceptable estimates from 10 to 70%, varying over cancer sites (almost 90% for the stomach and colorectal cancers, 50% for cancers of the endometrium, gallbladder, pancreas, and breast, and around 20% for the lung, larynx, bladder, cervix uteri, pharynx, and esophageal cancers). The method Doll and Peto used to make their estimates might be called an approach of "guesstimation". Later Doll narrowed the range of acceptable percentages of potentially avoidable cancer deaths by diet modification to 20–60% (Doll, 1992).

Traces of poor diet might be important throughout the several generations: it was shown that germ line damage to the sperm or egg is a likely cause of cancers of childhood, such as acute lymphocytic leukemia in children (Mayr et al., 1999). Poor diet in the father, or mother, or even in the grandmother when she was pregnant with the mother, interacting with genotype may be a contributor to this disease in offspring (Ames, 1999a; James et al., 1999).

In the United States, 80% of children and adolescents and 68% of adults did not meet the intake recommended by the National Cancer Institute and the National Research Council: five servings of fruits and vegetables per day (Krebs-Smith et al., 1995, 1996). In addition, almost 30% of the vegetables eaten by children and teenagers are potato chips or French fries, which are richer sources of starch and fat than vitamins (Ames, 1999b). The recommended dietary allowance of a micronutrient is mainly based on information on its acute deficiency, and the optimum amount for long-term health is generally not known. For many micronutrients, a sizable percentage of the population has an inadequate intake (according to Recommended Daily Allowance, Wilson et al., 1997). It has been shown that micronutrient deficiency, such as deficiency of folic acid, B12, B6, niacin, vitamins C and E, iron, and zinc can mimic radiation exposure in damaging DNA by causing single- and double-strand breaks, oxidative lesions, or both (Ames, 1998). The percentages of the population that consume less than half of the recommended daily allowance are 18% for zinc, 19% for iron (in menstruating women), 15% for vitamin C, and more than 20% for vitamin E. These deficiencies combined with approximately 10% of the population deficient for folate, more than 4% for vitamin B12, and about 10% for vitamin B6 include a considerable percentage of the U.S. population (Ames, 1998).

Table 9.2 presents some protective effects of dietary compounds resulted from studies in animal and human models.

It is difficult to analyze the association between specific food components and cancer risk. There are difficulties in collecting precise information about a person's dietary habits and complicated synergistic/antagonistic relationships

**Table 9. 2** Some protective effects of various dietary compounds against cancer in animal models and humans

| | Animals | Human |
|---|---|---|
| Cancer site | Anticancer | Anticancer |
| Brain | Curcumin*<br>Vitamin A* | Antioxidants* |
| Thyroid | – | Lemons* |
| Lung and bronchus | Silybinin (flavonone from milk thistle)*<br>Green tea | Fruits, vegetables<br>Black tea*<br>Selenium*<br>Apples*<br>Dairy products* |
| Oropharynx | Fish oil* | Bowman-bark (a soybean-derived compound)*<br>Vegetables*<br>Fruits*<br>Vitamin C*<br>Beta-carotene* |

**Table 9. 2** (continued)

| | Animals | Human |
|---|---|---|
| Cancer site | Anticancer | Anticancer |
| Esophagus | Zinc*<br>Fermented brown rice*<br>Vitamin E + selenium* | Fruits*<br>Vegetables* |
| Stomach | Catching* (from white tea)<br>Turmeric* | Fruits*<br>Vegetables* |
| Colon + rectum | Aliening (flavonoids from parsley, celery)*<br>Dietary flaxseed*<br>Casein from selenium-enriched milk*<br>Resveratrol (grapes)*<br>Chafuroside (flavone from oolong – Chinese tea)* | Vegetables*<br>Fruits*<br>Fiber intake*<br>Calcium*<br>Vitamin E*<br>Folic acid<br>Vitamin D*<br>Selenium* |
| Liver | Luteolin (flavonoid from Brussel sprouts, beets)* | Dairy products*<br>Fruits* |
| Pancreas | Curcumin + Cox-2 inhibitors*<br>Isothiocyanates (from cruciferous vegetables)*<br>Squalene (from shark liver)*<br>Fish oil* | Vitamin C*<br>Fruits*<br>Vegetables* |
| Kidney | No known protective dietary factors | Fatty fish* |
| Urinary bladder | Silymarin* (from milk thistle) | Drinking more fluids*<br>Vegetables* |
| Prostate | Dietary phytochemicals nobiletin and auraptene (from peel of citrus plants) | High intake of tomatoes<br>High intake of cruciferous vegetables<br>Vitamin E<br>Selenium<br>High intake of soy, beans, other legumes*<br>Fish*<br>Vitamin D* |
| Testis | No known protective dietary factors | High cheese consumption* |
| Uterus | Quercertine* (from onion, tomatoes)<br>Indole-3-carbinol* (from cruciferous vegetables) | Vegetables*<br>Fiber *<br>Folic acid* (for cervix uteri cancer)<br>Vitamin A* (for cervix uteri cancer) |
| Ovarian | Sulforaphane* (from cruciferous vegetables) | No known protective dietary factors |
| Breast | Caloric restriction<br>Flaxseed and its lignans | Moderate soy intake*<br>Calcium intake*<br>Vitamin D* |

* These data need further detailed studies.

between certain foods components, even inside the same food sample (i.e., vegetable, fruit). A subject of particular interest is how nutrient intakes could modify genetic susceptibility to diseases, and especially to cancer. That may provide a scientific basis for cancer preventative strategies through individual dietary modifications. One of the "side-effects" of the increasing interest to individualization of preventive and therapeutic approaches is very high activity of entrepreneurs selling dietary recommendations and supplements, which are claimed to be designed especially to individual's genetic susceptibility to certain disease (these products often include specifications as "nutrigenetic testing", "personalized supplements", "feed your genes right", "intelligent diet", and others). Recent testimony by the U.S. Government Accountability Office before the U.S. Senate Special Committee on Aging stated that the nutrigenetic tests, purchased from four Web sites, mislead consumers by providing dietary or lifestyle recommendations for disease prevention based on polymorphisms of some genes (Kutz, 2006). At present, our diet and its supplementation cannot be optimized according to genetic profiles to prevent cancer, until there is a strong scientific evidence on the interaction of genes and nutrients on cancer risk.

The question of the specific role of dietary components in cancer prevention remains largely unanswered. The overall recommendation to eat an abundant amount of fruits and vegetables has not changed for 25 years. The general recommendation to reduce total calories and fat consumption and to increase fiber is a good one with regard to CVD prophylaxis, but whether it is effective in cancer prevention remains unproven (Meyskens, 2004). Large prospective cohort studies and several large randomized trials recently indicated that fiber does not appear to be protective against colon cancer development (Schatzkin et al., 2000). An increasing amount of epidemiological data suggests that physical activity, basal metabolic index, and folate consumption may play critical roles in cancer prevention.

Approximately 20–30% of Americans consume multivitamin supplements daily, indicating strong public interest in the prevention of cancer and other chronic diseases through a nutrition-based approach. There are several large randomized clinical trials underway to clarify the effects of multivitamin supplements, including the Physicians' Health Study II, the Selenium and Vitamin E Cancer Prevention Trial, and a European study "Supplementation en Vitamins et Mineraux Antioxydants" (SU.VI.MAX). It has been shown that supplementation with antioxidant micronutrients vitamin C + vitamin E + $\beta$-carotene for 3 years was not an effective tool for gastric cancer control in a high-risk population (Plummer et al., 2007). Clinical studies did not show any reduction in cancer risk for $\beta$-carotene, lycopene, and vitamin E supplements. Some compounds presenting in whole foods (i.e., vegetables and fruits), and not included in supplements, may be of importance, but they are still unidentified. It is likely that, to reduce cancer risk, it is better to consume antioxidants mainly through food sources rather than supplements. However, supplements may be helpful for some people who are under endogenous and exogenous stress such as pregnant women, people with restricted dietary intake, athletes, and the elderly.

Nevertheless, it has been shown that certain supplements might be effective in prevention of certain cancers: vitamin A (25,000 IU per day) reduced the cutaneous squamous cell cancer risk in individuals with actinic keratoses, calcium intake decreased colorectal malignant polyps prevalence by 20% (Greenwald, 2002; Moon et al., 1997; Bonithon-Kopp et al., 2000), and taking vitamin D (from a balanced diet and supplementation together, with reasonable sun exposure) may be helpful in reducing risk of colon, prostate, and breast cancers.

It is unlikely that a single agent/dietary component acts as a "magic bullet" in cancer prevention. The best prospect is when multiple agents "hit" targets in the human organism prone to dysregulation for a given cancer. Optimization of the intake of specific foods and/or their bioactive components seems a noninvasive and cost-effective strategy for reducing the cancer burden, but this is not a simple process.

More than 5000 phytochemicals were identified in fruits, vegetables, and grains, with many still unknown. Brassica vegetables (such as broccoli) are rich sources of glucosinolate indole-3-carbinol (I3C) and the isothiocyanate sulforaphane, their predominately studied bioactive components. Heightened interest in these compounds as potential protective against cancer agents surfaced with the discovery of their potent ability to induce Phase II enzymes, participating in the detoxification of environmental carcinogens, and the ability to influence estrogen receptor-dependent and receptor-independent targets, involving cell proliferation and apoptosis. I3C is known to influence estrogen metabolism by mediating estrogen receptor binding and ultimately individual risk for the estrogen-mediated cancers of the breast and cervix (Kim and Milner, 2005). It influences cell proliferation by controlling the enzymes, regulating the cell cycle checkpoints, and proteins, and regulating DNA synthesis during mitosis (Cover et al., 1998).

Anticancer properties are also discovered in the isothiocyanates (ITCs), with sulfurophane receiving the most attention. It regulates gene expression and induces Phase II enzymes, including GSTM1 and glutathione-S-transferases, which conjugate carcinogens for urinary and bile excretion. Some studies showed that the high level of urinary ITCs was associated with a significant reduction in breast cancer risk in women, independent of their menopausal status (Fowke et al., 2003), and with reduction of lung cancer risk in male smokers (London et al., 2000). Several studies showed that sulforaphane induced apoptosis in leukemia, colon, and prostate cancers cells by interrupting the cell cycle transition to the second growth period (Choi and Singh, 2005).

Organosulfur compounds (e.g., contained in leek, onion, and garlic), such as diallyl sulfide, diallyl disulfide, and diallyl trisulfide, can induce apoptosis in cancer cells by triggering a series of genes activating apoptotic events, thus promoting apoptosis endogenously and exogenously (Martin, 2006), as well as stimulating Phase II enzymes for detoxification and inactivation of endogenous carcinogens. Diallyl sulfide showed its effectiveness as a potent inhibitor of colorectal and stomach cancer in chemically induced animal models of cancer (El-Bayoumy et al., 2006; Fleischauer et al., 2000), but its anticancer properties still remains a subject for a well-designed intervention studies in humans (Emenaker and Milner, 2006).

The interactions between the various components within a food, or in food combinations, may explain why isolated dietary components may not be as efficient for cancer prevention, as a whole food (see Table 9.3). What foods should be combined for maximum cancer prevention remain to be determined. For example, whole green tea is more effective than epigallocatechin gallate in inhibiting TNF-$\alpha$ release and increases the percentage of human lung cancer cells undergoing apoptosis. Combining soy phytochemicals and green tea extracts appeared to be more effective in inhibiting tumor angiogenesis, reducing estrogen receptor (ER)-alpha, and lowering serum insulin-like growth factor (IGF)-1 in estrogen-dependent human breast tumors, implanted into severely combined immunodeficient mice, than when either is provided alone. Consumption of tomato, but not only of its principal component, lycopene, inhibits N-methyl-N-nitrosourea and testosterone-induced prostate cancer. The dietary supplementation with cooked carrots increased the repair of 8-oxodG (as indicator of oxidative DNA damage) in white blood cells, whereas a similar amount of $\alpha$- and $\beta$-carotene provided as capsules had no effect. However, sometimes food may contain antagonistic components, e.g., soy, has a reduced ability to inhibit the aberrant crypt foci compared to isolated genistein in colon cancer in rats.

**Table 9.3** Some examples of combined chemopreventive effects of dietary components (studies on animal models with implanted human cancer cells)

| Combination | Cancer site | Effects | Reference |
|---|---|---|---|
| Soy + green tea | Breast | Inhibiting of tumor angiogenesis, reducing (ER[1])-alpha, lowering serum (IGF[2])-1 | Zhou et al. (2004) |
| | Prostate | Inhibit tumorogenicity and metastases | Zhou et al. (2003) |
| Orange + apple + grape + blueberry | N/A[3] | Increase in antioxidant activity: the median ED50[4] of each fruit in combination was five times lower than EC50 of each fruit alone | Liu (2004) |
| Vitamin D3 + 9-cis-retinoic acid | Breast | Possibly, regulation of genes involved with cell proliferation, differentiation and/or apoptosis. | Zu and Ip (2003) |
| Selenium + vitamin E | Prostate | Induction of apoptosis | Zu and Ip (2003) |

[1] Estrogen receptor.
[2] Insulin-like growth factor.
[3] Cancer site is not specified.
[4] Effective dose.

All of the major mechanism that have been examined as targets for cancer prevention respond to one or more dietary components. They include carcinogen metabolism, DNA repair, cell proliferation/apoptosis, inflammation, differentiation, oxidant/antioxidant balance, and angiogenesis (see Table 9.4). These data have been obtained from cell culture studies and, when interpreting the results from in vitro studies, it should be taken into account, that dose, cell type, culture conditions, and treatment time can affect the biological outcome.

**Table 9.4** Selected dietary components targeting carcinogenesis process (in vitro studies and in animals/human cell culture models)

| Targets | Dietary components | Dietary source |
|---|---|---|
| Carcinogen metabolism | Indoles | Cruciferous vegetables |
| | Isothiocyanates | Cruciferous vegetables |
| | Coumarins | Tonka bean, woodruff, plum |
| | Flavones | Parsley |
| | Allyl sulfides | Onion, garlic, leeks, chives |
| | Sinigrin | Cruciferous vegetables |
| | Phenolic acids | Eggplant, tomatoes, carrots, citruses, whole grains |
| DNA repair | Flavonols | Onion, leek, broccoli, green tea, tomato, red wine |
| | Vitamin E | Wheat germ oil, almonds |
| | Vitamin C | Citrus fruits, tomatoes |
| | Isothiocyanates | Cruciferous vegetables |
| Cell proliferation | Genistein | Miso, soybeans |
| | Epigallocatechin-3-gallate | Green tea |
| | Isothiocyanates | Cruciferous vegetables |
| | Vitamin D3 + 9-cis-retinoic acid | UVB rays + pork liver, carrots, sweet potato |
| Apoptosis | Selenium | Nuts, tuna |
| | Epigallocatechin-3-gallate | Green tea |
| | Phenylethyl isothiocyanate | Cruciferous vegetables |
| | Sulforaphane | Cruciferous vegetables |
| | Vitamin D3 + 9-cis-retinoic acid | UVB rays + pork liver, carrots, sweet potato |
| | Curcumin | Turmeric |
| | Apigenin | Parsley, celery, lettuce |
| | Quercetin | Onion, tomato |
| | Resveratrol | Grapes, red wine |
| | Lupeol | Mango, strawberry, grapes, olive |
| | Delphinidin | Pomegranate, strawberry |
| | Organosulfur compounds | Garlic, onion |
| | Lycopene | Tomato |
| | Genistein | Soybeans, miso |
| | Indol-3-carbinol | Cruciferous vegetables |
| | Luteolin | Celery, green pepper, peppermint |
| | Anthocyanins | Pomegranate |

**Table 9.4** (continued)

| Targets | Dietary components | Dietary source |
|---|---|---|
| | Caffeic acid phenethyl ester | Honey |
| | Gingerol | Ginger |
| | Capsaicin | Red pepper |
| Inflammation | Conjugated linoleic acid | Lamb, cheddar cheese, homogenized cow's milk |
| | Long-chain omega-3 fatty acids | Fish, beans, flaxseeds |
| | Butyrate | Butter, animal milk |
| | Epigallocatechin-3-gallate | Green tea |
| | Resveratrol | Grapes, red wine |
| | Genistein | Soybeans, miso |
| | Luteolin | Celery, green pepper, peppermint |
| | Quercetin | Onion, tomato |
| | Vitamin A | Pork liver, carrots, sweet potato, eggs |
| | Vitamin D3 | UVB rays |
| Differentiation | Vitamin D3 + 9-cis-retinoic acid | UVB rays + pork liver, carrots, sweet potato |
| Oxidant/ antioxidant balance | Flavonols | Onion, leek, broccoli, blueberry, tomato, red wine |
| | Vitamin E | Wheat germ oil, almonds |
| | Vitamin C | Citrus fruits, tomatoes |
| | Isothiocyanates | Cruciferous vegetables |
| Angiogenesis | Polyunsaturated fatty acids | Fish, soybean oil, corn oil, sunflower oil |
| | Epigallocatechin-3-gallate | Green tea |
| | Resveratrol | Grapes, red wine |
| | Curcumin | Turmeric |
| | Genistein | Soybeans, miso |

It has been shown in *in vitro* studies and in preclinical models that flavonoids (such as quercetin, rutin, genistein), phenols (such as curcumin, epigallocatin-3-gallate, resveratrol), isothiocyanates, diallyl sulfur compounds, indoles, and selenium may work as modulators of detoxification enzymes, playing a major role in the regulation of the mutagenic and neoplastic effects of chemical carcinogens (Phase II enzymes being mediated by the antioxidant response element located in the promoter region of specific genes). Cell proliferation might be modulated by genistein and epigallocathechin-3-gallate that cause cell-cycle arrest via the induction of CDK (cycline-dependent kinase) inhibitors p21 and p27 (they have tumor suppression activity, and their expression is controlled by the tumor suppression protein p53) and the inhibition of CDK4, CDK2, cyclin D1, and cyclin E (these kinases have a tumor stimulating activity). Isothiocyanates can induce p21 expression and inhibit cell proliferation at the G2-M checkpoint. Many of the dietary cancer-protective compounds target apoptosis: selenium, epigallocatechin-3-gallate, phenylethyl isothiocyanate, retinoic acid, sulforaphane, curcumin, apigenin, quercetin, and resveratrol. Dietary components such as conjugated linoleic acid, long-chain omega-3 fatty acids, butyrate, epigallocatechin-3-gallate, curcumin, resveratrol and others may reduce cancer risk by influencing the inflammatory process. It has also been suggested that dietary components might help inhibit tumor growth by preventing the expansion of new blood vessels in tumors, resulting in reduced tumor size and metastasis: polyunsaturated fatty acids, epigallocatechin-3-gallate, resveratrol, curcumin, and genistein may work as inhibitors of tumor angiogenesis. Some dietary components, such as epigallocatechin-3-gallate (in green tea), genistein (in soybeans), resveratrol (in red wine), and vitamin D3 (UVB rays) have cancer-protective effect targeting several stages of tumor progression (see Table 9.4).

Recent cell culture and animal studies also suggested that dietary compounds such as genistein, curcumin, epigallocathechin-3-gallate, resveratrol, indole-3-carbinol, proanthocyanadin, and vitamin D3 may enhance the efficacy of cancer chemotherapy and radiotherapy by modifying cell proliferation. When dietary components modify the same molecular targets as specific drugs, the dose of medicine used in cancer prevention may be reduced, avoiding or minimizing the potential adverse effects of the medicine.

Although in most cases there likely will be benefits from increased consumption of fruits and vegetables for cancer prevention, in a small subset of the population the opposite may occur. For example, there are data suggesting that some women who consume food or supplements with large amounts of natural DNA topoisomerase II inhibitors (e.g., flavonoids, including quercetin and genistein, also catechins – the basis for certain potent chemotherapy drugs) during pregnancy have a higher risk of infant acute myeloid leukemia, particularly associated with MLL+ gene (myeloid/lymphoid or mixed-lineage leukemia) (Ross, 2000; Greaves, 1997; Ross et al., 1996; Spector et al., 2005). These women appear to have a reduced ability to remove some of these substances from their organisms, which then accumulate and became toxic to the developing fetus. Cell cycling process is dynamic, and effects that appear to be beneficial for cell at one time could be harmful at another.

Many questions still remain unclear, including specificity of nutrient-cancer type responses, the minimum quantity, timing of exposure, and precise molecular target. No single nutrient or bioactive food component could be a "magic bullet" for cancer prevention, but they can act as modifiers of risk and/or tumor behavior. Many studies were published on specific foods, nutrients, lifestyle factors, and specific cancer risks, but until now no one study has provided clear results and recommendations on this subject, with single new reports sometimes causing contradictory or conflicting results to be overemphasized. However, the generally negative experience with $\beta$-carotene as a chemoprevention agent, particularly in lung cancer (see details in Chapter 3), mandates caution in extrapolating the potential benefit of a compound from epidemiologic observations without extensive supporting experimental data.

## 9.4 Obesity

In 1909, Moreschi observed that tumors transplanted into underfed mice did not grow as well as those transplanted into mice fed *ad libitum* (with free access to food and water). His finding stimulated a decade of research which showed that caloric restriction may negatively affected the growth of spontaneous tumors in animals. It has been supposed that the underlying mechanism might include enhanced DNA repair, moderation of oxidative damage to DNA, reduction of oncogenic expression, and affect insulin metabolism. Recent rodent studies demonstrated that caloric restrictions that supposed may inhibit or delay cancer progression are mediated by changes in energy balance, body mass, and body composition rather than calorie intake per se. Cancer risk may be increased not by consuming an excess number of calories, but by excess calorie retention. Limited caloric restriction studies in humans showed a possible association with reduction of heart disease and breast cancer incidence, but to date human studies on caloric restriction showed conflicting results (that may be due to the greater complexity of endocrinological control of energy production in humans).

At present, data provide convincing evidence of a positive association of overweight and obesity with cancers of the colon (in men), kidney, postmenopausal breast, endometrium, and probable evidence of a positive association with colon cancer (in women), esophageal adenocarcinoma and cancer of gastric cardia, and thyroid cancer (in women). Estimates of the population attributable risk (PAR) of cancer due to overweight and obesity have been summarized in the 2002 IARC report and are around 9% for colon cancer, 11% for postmenopausal breast cancer, 25% for renal cell cancer, 37% for esophageal cancer, and 39% for endometrial cancer (Ballard-Barbash et al., 2006). However, these estimates were based on international rates of overweight and obesity from the 1990s (IARC, 2002) and are higher in the United States today, given the continued increase in the prevalence of overweight and obesity. With the expectation that the prevalence of obesity in the United States is likely to continue, if not accelerate, overweight and obesity may become increasingly important contributors to cancer risk.

No controlled clinical trials have been performed on the effect of avoiding weight gain on cancer risk. However, many observational studies have shown that avoiding weight gain lowers the risk of cancers of the colon, breast (postmenopausal), endometrium, kidney, and esophagus, with the limited evidence for thyroid cancer and without substantial evidence for all other cancers (Vainio and Bianchini, 2002; Vainio et al., 2002).

Not enough evidence exists on how intentional weight loss can affect cancer risk (Obesity and Cancer, NCI). A very limited number of observational studies have examined the effect of weight loss, and a few have found some decreased risk for breast cancer among women who lost weight. However, most of these studies have not been able to evaluate whether the weight loss was intentional or related to other health problems (Trentham-Dietz et al., 2000; Kaaks et al., 1998; Ziegler et al., 1996). It has been shown, that intentional, but not unintentional, weight loss might be associated with reduced cancer risk (Parker and Folsom, 2003). There is an urgent need now to study the etiologic factors predisposing some people to overweight and obesity, and the relationship between overweight/obesity and cancer initiation (IARC, 2002).

## 9.5 Physical Activity

Presently there are a lot of epidemiologic data on the relation between physical activity and the risk of developing cancer. Although the direct evidence on this relation comes only from observational studies, randomized clinical trials have provided indirect evidence by examining the association of physical activity with other parameters associated with cancer risk, such as body weight and hormone levels. There have been no controlled clinical trials on the effects of regular physical activity on the risk of developing cancer in human (Obesity and Cancer, NCI).

Few data are available regarding the effects of physical activity on health outcomes among persons who already have cancer. Based on these limited data, the findings suggest that physical activity during cancer treatment is associated with better outcomes with regard to symptoms (such as nausea, fatigue, sleeping disorders), anxiety and depression, weight gain, and functional capacity of organs and systems. Physical activity also can improve short-term outcomes in patients who have completed medical treatment (Courneya, 2003). Presently, researchers are not certain if physical activity in cancer patients can improve the prognosis or progression of tumors (Lee and Oguma, 2006). One of the recent studies indicated that women with breast cancer who had the physical activity equal to approximately 1 hr/week of walking at 3 mph or more had a reduced mortality (Holmes et al., 2005).

A protective effect of physical activity on site-specific cancer risk with a dose–response association has been observed between physical activity and colon, and pre- and postmenopausal breast cancer (Thune and Furberg, 2001; Chen et al., 2003; Samad et al., 2005). The effect of physical activity on colon cancer risk has been examined in a limited number of studies including both lean and obese people. As a result, a protective effect of physical activity has been found in most of these studies in all groups of BMI (Vainio and Bianchini, 2002). Physical activity among postmenopausal women had a significant effect in reducing breast cancer risk in women who were of normal weight and no protective effect in overweight or obese women (McTiernan et al., 2003). The complicated nature of the physical activity variable, combined with a lack of knowledge regarding possible biological mechanisms operating between physical activity and cancer needs further studies, including controlled clinical randomized trials.

## 9.6 Alcohol

The association between alcohol consumption and cancer vary by site and type of alcoholic beverage. Clearly, alcohol consumption can be modified, and given evidence that alcohol consumption increases the risks of some cancers (e.g., cancers of oral cavity and pharynx, esophagus, liver, large bowel, larynx, breast), modification of alcohol consumption represents a ready means of decreasing cancer risk. In the United States in 2002, the incidence of cancer that appeared to be alcohol-related added up to almost 300,000 cases and 135,000 deaths (Marshall and Freudenheim, 2006). Control of alcohol intake would result in reduced incidence or mortality attributable to these cancer sites.

## 9.7 Nonsteroidal Antiinflammatory Drugs

Recent epidemiological studies suggest that nonsteroidal antiinflammatory drugs (NSAIDs) may reduce the risk of several cancers. An inverse association has been reported between use of aspirin and other NSAIDs and risk of colon cancer and adenomatous polyps (Baron and Sandler, 2000; Thùn et al., 2002). However, a large prevention trial found no reduction of colon cancer risk over 12 years of observation in patients taking aspirin (Habel and Friedman, 2006). Meta-analyses suggested that use of NSAIDs may be associated with a small decrease in risk of breast cancer (the combined RR for cohort studies was 0.78, with 95% CI = 0.62–0.99, for case–control studies it was 0.87, with 95% CI = 0.84–0.91) (Khuder and Mutgi, 2001).

Several mechanisms for the chemopreventive effects of NSAID have been proposed: restoration of apoptosis and angiogenesis inhibition, effecting apoptosis and tumor growth, and progression through prostaglandin-dependent [via inhibition of cyclooxygenase (COX) activity] and prostaglandin-independent pathways (Thun et al., 2002; Badawi, 2000; Michalowski, 2002). Unfortunately, the number of studies providing estimates of NSAIDs' effects for different cancer sites is still small. Several studies of esophageal, ovarian, and prostate cancers showed some inverse associations between NSAIDs and cancer risk. Those risks were only slightly less than 1.0 and not statistically significant. Most studies did not have detailed information about dose of NSAIDs and duration of treatment (Habel and Friedman, 2006). Additional epidemiologic studies may help to determine the optimal type, dose, and duration of NSAID to be tested in large interventional studies. Different population groups may have different preventive efficacy for NSAIDs, which may also be a subject of future studies.

## 9.8 Statistical and Modeling Approaches for Prognoses

The first hard efforts to prevent cancers like lung, breast, and cervix were made in the 1950s (Greenwald et al., 1987). In 1981, Doll and Peto (1981) examined the degree to which cancer incidence and mortality rates could be reduced in the

United States. They estimated the reduction theoretically by comparing the rates in the United States with those in other countries, based on the observation that the cancer incidence rate among migrants tends to be that found in the country to which they migrated, indicating that differential cancer incidence rates are partly due to environmental factors such as diet, exercise, occupational exposures, and smoking, and that cancer does not arise exclusively because of genetic factors. They included in their estimates only persons who were younger than 65, because the data on the incidence of cancer among older individuals were considered unreliable. The results of their analyses suggested that in 1970, from 75 to 80% of all cancers in the United States could be theoretically avoided if the population could be similar to the countries with the lowest incidences. The "environmental" factors that differed between the United States and low-risk populations included birth weight, age at puberty, life-long patterns of tobacco use, diet, physical activity, alcohol consumption, use of pharmacological agents, and reproduction. Although Doll–Peto's methods might seem rudimentary by contemporary methodological standards, they provided a foundation for later work (Curry et al., 2003).

In 1996, Willet and colleagues (1996) used an international comparison approach at the ecological level, similar to that used by Doll and Peto, to assess the degree to which cancer mortality could be reduced in the United States: they found that cancer mortality rate in the U.S. population could theoretically be reduced by 60%, with a more realistic reduction based on trends in risk behaviors would be 33%. They suggested that a 2/3 reduction in the number of smoking individuals would lead to a similar eventual reduction in smoking-associated cancer mortality. An almost 10% decrease in cancer mortality might be achieved by diet- and exercise-related changes, and alcohol-related cancer mortality rates could be reduced by a third, if percentage of those who consumed more than two drinks per day reduced their intake. Neither Willet and colleagues nor Doll and Peto provided a time frame for the reduction of the cancer mortality rate estimated in their analyses, and they did not specify the latency of effect (Curry et al., 2003).

In 1986, the NCI used another approach for estimation of the likely possibilities of reduction in specific cancer-related risk factors: they set a goal of a 50% reduction in the total cancer mortality from 1980 to 2000, as well as determined the goals for smoking, diet screening, and treatment to be achieved to 2000. In retrospect, most of these goals were overly optimistic, underestimating the latencies, including those of the social and political changes that would be needed to bring about large changes in the behavior of the population (Curry et al., 2003).

The American Cancer Society in 1996 set a "challenge goal" for a 25% reduction in cancer incidence, and a 50% reduction in the rate of mortality from cancer by 2015. Byers and colleagues (1999) examined the reasonability of reaching the ACS challenge goals by 2015 on the basis of possible reductions in selected major risk factors. They assumed at least a 10-year latency for the effects of tobacco and a 5-year latency for the effects of other factors. They estimated that if the reductions in the prevalence of risk factors accelerated more than it was in 1990, cancer incidence rate could decline by 19%, and the cancer mortality rate could decline by 29%. They found that past and future reductions in rates of tobacco use were the single largest contributor to the projected future declines in overall cancer incidence and mortality rates. Other risk factors for which declines in prevalence were projected to be important contributors to declines in cancer incidence and mortality rates were poor dietary pattern (low consumption of fruits and vegetables), especially for colorectal and lung cancers, alcohol consumption, and failure to be screened (especially for colorectal cancer and, to a lesser extent, for breast and prostate cancer). It is possible that the estimates of Bayers and colleagues are also overly optimistic. It is possible that people who have not adopted behavioral changes to date may be more resistant to changes, and over time "resistors" might become a larger proportion of the remaining "unchanged" population (Rogers, 1993). Also the projection by Bayers and colleagues did not take into account the possible future changes in obesity-related cancers, stating that obesity epidemic would be turned around in the coming years.

An important initiative is the NCI-sponsored Cancer Intervention and Surveillance Modeling Network, where investigators have evaluated the impacts of population-level changes in smoking, diet, physical activity, weight status, and the use of screening tests on the rates of cancer incidence and mortality.

At present, scientific estimates of the proportion of the cancer burden that can be eliminated, if the population distributions of major risk factors were to be shifted in a different direction, have focused on the important, but relatively narrow issues of strength of risk factor and disease associations and biological latency, while the issue of social and political latency to support behavioral change has gone underestimated (Curry et al., 2003). Another important area of research is how observed disparities in cancer incidence and mortality rates might be affected by divergent patterns of risk behaviors observed within certain racial, ethnic, and socioeconomic groups.

Most methods currently used to forecast population changes are based on the construction of hypotheses, which are expressed in a form of specific functions describing the age-dependence of mortality rates. These functions depend on a number of parameters estimated by fitting hypothetical curves to data. The functional forms for the rates often use empirical formulas (like Gompertz and Weibull formulas to describe the dependency of mortality on age), models derived from biologically motivated approaches [such as evolutionary explanation of changes in mortality, models based on repair capacity (Yashin et al., 2000)], models describing population heterogeneity [like frailty models, acquired heterogeneity models and etc. (Vaupel et al., 1979; Manton et al., 1986)], models considering the stochasticity of underlying processes [as Phase-type distribution models, unobserved life-history processes, quadratic hazard models, stochastic process models with Gaussian assumption, etc. (Manton and Stallard, 1988; Yashin and Manton, 1997; Akushevich et al., 2005)], and variants and composites of these approaches. The main feature of these methods is that they do not directly consider individual changes in health over life, rather, integrated characteristics (e.g., distribution of latent frailty, etc.) are used to describe a model, estimate model parameters, and forecast changes. This does not diminish the value and the importance of the above studies for specific tasks, but it leaves the domain of individual-level simulations incompletely explored.

The situation is similar in the field of forecasting health insurance in general, and Medicare expenditures in particular (Ash and Byrne-Logan, 1998; Lee and Tuljapurkar, 1998; Review of Assumptions and Methods of the Medicare Trustees' Financial Projections, 2000; Miller, 2001; Lee and Miller, 2002; Foster, 2003; Bhattacharya et al., 2004). Forecasts are made on the basis of time series of integrated characteristics. Most research addressing this domain is actuarial and not biomedical. Thus, the most significant imperfection of existing forecasting methods is that they take into account only changes in the size and structure of the population and do not directly account for changes within individual health structures.

In contrast, models incorporating individual changes possess flexibility and have the potential for performing specific simulations. They allow researchers to analyze the statistical uncertainty of their forecasts by conducting multiple projections and by calculating standard deviation over the sets of projections. These models, however, require many process parameters, knowledge of which is formulated as prior information. One solution is to use the microsimulation approaches introduced in Chapter 5 and applied for population projections in Chapter 8. The advantageous features of this approach are demonstrated in these chapters. The most important is the flexibility of the underlying model to allow the construction of new models, using previous models of biological subsystems as building blocks. In addition, microsimulation provides a natural basis for the investigation of systematic bias due to a variety of sources of uncertainty, e.g., in the initial population distribution, in rates of life events, in rates of investigated processes (e.g., rates of enigmatic processes), etc.

An important feature of such forecasting models is the ability to examine different "what-if" scenarios. Individual changes in health status over life are constructed as the result of processes of interest such as aging, interactions between an individual and society, or individual habits potentially impacting life event rates (e.g., smoking). Several features of a microsimulation approach (e.g., the notion of an individual state, use of microsimulation for the construction of forecasts, using experimental results for rates of life events) are similar to those implemented in a demographic-economic model – the Future Elderly Model (FEM), which actually combined three models: a model of health care costs, a model of health status transitions, and a model to predict characteristics of future, newly entering Medicare enrollees (Goldman et al., 2004). Important differences lie in the age range model (all ages in our model and ages 65+ in FEM) and in the approach to the design of interventions. In our approach, experimental measurements are applied to analyses of interventions. Our model is based on known (observed) semiparametric relative risks of selected disease(s) (attributable to a specific risk factor), infertility based on prevalence of specific risk factor(s), and the relative contribution of those selected diseases to mortality.

The microsimulation approach used in Chapter 8 for analysis of the health effects of smoking can be used for the analysis of each risk factor discussed in this Chapter. The series of data, such as relative risks of specific cancers incidences for the population group with and without a certain risk factor, has to be combined to perform such studies. These data can be obtained from auxiliary studies, or, if such studies are not available, specific analyses can be performed using data sets that include both types of data, i.e., cancer incidence/ survival and risk factors [one such data set is the NLTCS linked with daily clinical and diagnostic Medicare records (Manton et al. 2006, 2007b)].

Further generalization of the microsimulation model is to consider the contribution of each factor (smoking, obesity, diet, and other) to cancer incidence. Most diseases and injuries have multiple causes, including potentially preventable and nonpreventable factors and conditions that may contribute to a specific cancer onset, its progression, and death caused by this cancer. Microsimulation, as a method describing the effects of several risk factors on population health, is "flexible" enough to represent the complex impact when risk of one factor influences risk(s) of another risk factor(s). All techniques in this book can be generalized to the case of multiple dynamic risk factors. The only obstacle is availability of measurements of rates and relative risks.

Similarly, interventions to analyze financial outcomes can be constructed. For that one needs to relate (through experimental results or hypotheses about the efficacy of new medical technology) changes in certain rates (e.g., decreasing mortality from cancer) with expenditures on the development of the technology and financial benefits from increases in the life span.

The simple version of microsimulation used to identify the pure effect of a risk factor (e.g., smoking interventions, as discussed in Chapter 8) can be generalized to introduce new life events and interventions that change the rates (e.g., new medical technologies influencing the incidence of smoking-attributable diseases and fertility, socioeconomic changes). Generalization of this model to include risk factors other than smoking will produce results of practical use in health, social and actuarial sciences, as well as allow estimation of the socioeconomic effects of new medical technologies. Another way to generalize this model is to analyze more comprehensive phenomena. In this case, data are often not available to cover all relevant life events, e.g., infertility/ fertility rates. In such cases, a cohort, instead of an entire population, can be used in projections.

The population is always heterogeneous with respect to carcinogenesis risk. The basis of this heterogeneity is individually specific and is often unobservable. Microsimulation performs projections by simulating individual trajectories using the distribution of corresponding characteristics estimated at population level. Then individual trajectories are used to obtain predictions at a population level. Evidently, this approach possesses high efficiency for predictors of the risk of carcinogenesis for a specific individual. The trajectory should be simulated starting from specific, individually measured risk factors, and the population distributions used for the simulation have to be conditional on demographic and/or health characteristics of a group to which the individual belongs. A set of such simulated trajectories would provide individual prognoses (means over trajectories) and estimation for the accuracy of the prognoses (standard deviations over trajectories). Such individual prognoses also require analyses of uncertainties, which can be also performed by the methods of microsimulation (Akushevich et al., 2006, 2007; Manton et al., 2007a).

Thus, a model for analysis of the demographic, economic, and biomedical consequences of implementation of a certain prevention strategy has to have the potential to predict respective characteristics (e.g., economic efficiency, population losses) and to analyze them for different scenarios to make conclusions about the efficiency of the given prevention strategy. Components of the model usually include the following four models. The first is a probabilistic model for age (sex and race) patterns of demographic events such as death, onset of diseases, or recovery from disease. These probabilities are modeled based on the decision of expert panels (as in Goldman et al., 2004) or based on data from respective meta-analyses (as in Chapter 8 for analysis of smoking impacts). The second is a model for probabilities of birth or a model for the distribution of investigated characteristics at an initial age. Often additional data sets can be required to construct this distribution. The third is a model for medical costs associated with base demographic events, e.g., cost of a specific cancer treatment. The source of information for such a model is usually in the Medicare Service use files. The fourth is the projection models allowing for the simulation of individual life typically on a 'year-by-year' basis. Such simulation reproduces a "simplified" life of an individual, and population characteristics are calculated as means over simulated individual "trajectories." The specific choice of parameters of the model defines a certain scenario of the future. Conclusions concerning the efficiency of a prevention strategy are made by comparing the results of projections based on scenarios with and without implementing this strategy.

Moreover, this approach allows us to hope that in the near future individual prognoses can be realistically based on modeling of biological approaches to carcinogenesis (as discussed in Chapter 2). So far, biological models are oversimplified in respect to underlying biological approaches incorporated in a specific model because of methodological difficulties. However, using microsimulation for the joint modeling which will allow linked risk factors, its impact for triggering and promoting of the carcinogenesis process, and simulation of this process, from individual to population level, is promising for achieving progress in this area.

The task of analyzing effects of a specific technology is connected with a problem of labor force participation and human capital. The largest benefit of the intervention would be an increase in human capital as a result of the creation of additional healthy life years for each person in the population (Kravchenko et al., 2005). The most valuable new technologies will both improve the elderly's standard of living and increase per capita labor productivity by extending work life and enhancing age-specific physical and cognitive capabilities. Murphy and Topel (2006) evaluated the economic consequences of increases in quantity and quality of life. They suggest that the value of health increases for (1) larger population, (2) higher incomes, (3) higher existing levels of health, and (4) a population age distribution close to that of the age-at-onset of prevalent chronic diseases. They estimated that declines in U.S. mortality from 1970 to 2000 were worth \$95 trillion (\$60 trillion after removing health expenditures). They have found that half of the economic benefits from the health improvements 1970–2000 were due to reduced CVD mortality. That increase the future value of interventions for other causes of death and the value of a person-year in 2000 raised by 18%. In 1990, cancer mortality rates started to decline, and by 2004 they dropped for more than 10%. Murphy and Topel estimated that a 1% reduction in cancer mortality has a value of \$500 billion; thus, observed cancer mortality declines are worth \$5 trillion.

One important point which has to be taken into account in such interventional analyses is the description of health and active conditions of individuals living for additional years provided by a medical technology. Important questions here are as follows (Cairncross, 2007): (1) Will their health be good enough for them to work for longer? (2) And if it is, will they be as productive as younger folk, or will an older workforce in industrialized countries lose its competitive edge against industrializing countries that still have youthful employees? These questions return us to the discussion of whether modern medicine is able to cure or prevent the chronic diseases of old age, and what specific trends of chronic disability in elderly (65 + years old) are declining, how disability declines, and how improved health may increase human capital at later ages, stimulating the growth of gross domestic product and national wealth. Comprehensive analyses of these and related questions were recently performed by Manton et al., (2007b). Using longitudinally linked NLTCS and Medicare expenditure data to make projections of human capital growth on Medicare and Medicaid expenditures, they evaluated the optimal level of investment in research in the U.S. economy. They assumed human capital preservation at older ages stimulates growth of both the U.S. population and labor force ages (with growing proportions of the elderly population surviving in healthier states). Specifically, they demonstrated that health changes, as included by NLTCS data, can reduce total Medicare and Medicaid changes over the long term by 40–50% (from 24% of the growth domestic product (GDP) in 2080 to 12–15% ).

Reviewing the models of potential reductions of cancer incidence and mortality rates, it is important to keep in mind that the benefits of cancer's primary prevention strategies (as distinguished from, i.e., screening) will also improve other, noncancer aspects of public health, and these improvements may occur over the shorter period than changes in cancer incidence and mortality rates (Colditz and Gortmaker, 1995). A person who quit smoking or was prevented from starting smoking reduces not only his/her risk for lung cancer (e.g.) decades later but also the risk of respiratory and cardiovascular diseases earlier (US DHHS, 1990). If a person adopted a healthier diet pattern and increased physical activity, it will reduce his/her risk of CVD, stroke, and diabetes, as well as some other diseases with a shorter latency period than cancer (Knowler et al., 2002).

## 9.9 The Present and Future of Microsimulation Models in Tasks of Cancer Prevention

The microsimulation model started to be used in analyses of health effects relatively recently. Basic goals of these studies were directed to investigation of the cost-effectiveness analyses of certain prevention strategies and to the prediction of future changes, also motivated by problems of health economics. One important example of this kind of study mentioned above is the comprehensive demographic-economic microsimulation model (i.e., FEM) which was conceived at the RAND Corporation in 1997. The purpose of its designing was to help policy analysts and private firms understand future trends in health and health spending, longevity, and medical technologies. It tracks elderly, Medicare-eligible individuals over time to project their health conditions, their functional status, and their Medicare and total health-care expenditures (Goldman et al., 2004). Another example would be an analysis of efficacy of screening strategy presented by Myers et al (2000). They used a Markov model to attempt to describe a natural history of HPV infection, and the lifetime costs and life expectancy of a hypothetical cohort of women screened for cervical cancer in the United States (Kulasingam and Myers, 2003). The most recent studies included (1) estimation of the long-term effectiveness of screening in Mayo study participants for individual-level data provided from the single-arm study of helical computed tomographic (CT) screening: the Lung Cancer Policy Model was applied to this study – a comprehensive microsimulation model of lung cancer development, screening findings, treatment results, and long-term outcomes, which in the study of McMahon et al. (2008) predicted changes in lung cancer-specific and all-cause mortality as functions of follow-up time after simulated enrollment and randomization, (2) a microsimulation model, applied to combine baseline mortality with the mortality and health-related quality of life (HRQoL) associated with mitral valve replacement (MVR – operation of choice for mitral valve disease) in order to estimate the long-term survival and quality-adjusted life years (QALE) resulting from two procedures – conventional MVR and MVR with subvalvular apparatus[1] preservation (SVP) (Rao et al., 2008), (3) using a microsimulation model to compare three

[1] Subvalvular apparatus is an integral part of the heart mitral valve structural complex that includes the left ventricular free wall, two papillary muscles, and chordae tendinae. Preservation of the subvalvular apparatus maintains function of the cardiac left ventricular and thus improves survival.

screening strategies (i.e., mammography, magnetic resonance imaging, and both methods combined) versus clinical surveillance for breast cancer in women with BRCA1 gene mutations (Lee et al., 2008).

The direction to develop the microsimulation strategies to the tasks of health economics and public health has clear perspectives and will essentially contribute to future scientific knowledge. However, we would like to stress, another set of tasks is also of a great importance. The elaborating microsimulation strategies allowing investigation of the mechanisms of carcinogenesis at all stages of process, relating with the effects of risk factors, and analyzing the way capable of prevention or reduction of cancer risks at both individual and population levels. Several examples of how it could be possible to solve these problems were discussed earlier (see Chapter 8). However, the spectrum of future applications is much broader. The important questions which have to be investigated and that can contribute to cancer prevention strategies include: (1) analysis of risk factor effects, estimation of the frames of its applicability and quantifying its effects on cancer development, (2) analysis of population effects, i.e., specific effects of certain risk factors on population, (3) deeper analysis of underlying mechanisms by which the specific risk factors act, and role of exposure to risk factor in initiation and promoting carcinogenesis, and (4) studying the effects of interactions of different risk factors, and clarifying the ways these effects can be modulated. In all these and related analyses, one has to investigate the limits of applicability of development strategies, their specific impacts on certain population groups, and also answer the traditional questions about the cost-effectiveness and cost-benefits.

An example of a comprehensive approach to the analysis of a specific problem was recently presented by the Cancer Intervention and Surveillance Modeling Network (CISNET), e.g., studying the impact of mammography and adjuvant therapy on U.S. breast cancer mortality in 1975–2000 [collective results from the CISNET are presented in the Journal National Cancer Institute Monographs (2006)]. In the broad spectrum of approaches, the microsimulation models were at the one end of the spectrum, and mechanistic or analytic models were at the other. In microsimulation models, population effects were investigated through simulating individual life histories with and without interventions, while in mechanistic models analytically derived equations were used, which described the relationships between key health states and/or tumor growth and metastasis. Specific modeling approaches combined the best methodological features available in each of the basic models.

## 9.10 Summary

Cancer prevention will be a significant focus of research and intervention during the next decades. It will be based on the ability to identify the individual susceptibility to specific cancers and the molecular targets that can alter or stop the process of carcinogenesis. Public and governmental action toward medical research on prevention will increase, especially as baby boomers are ageing and tending to use their political acumen to push for medical progress against the diseases that threaten them most, with cancer and CVD as the leading causes of mortality at late ages (Young and Wilson, 2002).

The efforts on decreasing widely recognized risk factors of certain cancers (like smoking – for lung cancer, HPV – for cervical cancer, and others) will continue to be effective in cancer prevention, as it was during the past 30 years. However, this strategy was effective mostly in decreasing the incidence of certain cancer histotypes: e.g., of squamous cell carcinomas, but not of adenocarcinomas [which incidence rates in several cancer sites began rising during the last two decades (see Chapter 7)]. That makes it very important to study risk factors associated with both specific histotypes and to develop specific prevention strategies.

Pharmacological and genetic researchers are joining efforts to develop chemoprevention agents designed to affect molecular targets linked to specific premalignant conditions. Technical developments, including DNA chips, make population screening easier; however, such screening should be cost-effective and ethically acceptable. The individualization of preventive strategies is the ideal scenario. Cancer researchers and practitioners have two positions related to this subject. The first is so called "the head-in-clouds" position, when it is believed that medicine of the twenty-first century will be focused on individual's susceptibility factors. Identifying them will allow us to move from a "diagnose and treat" to a "predict and prevent" health-care system. Susceptibility would be defined in population screenings, and then specifically "designed" chemoprevention or lifestyle changes will be used to prevent disease. The second position is so called "the head-in-sand", based on believe that most diseases do not have Mendelian[2] subgroups, and most susceptibility factors are not strong enough to effectively predict the disease risk of an individual. It may be unrealistic, for many people to sufficiently modify their lifestyle to avoid modest genetic risks (e.g., based on experience with smoking, obesity and diet) (Strachan and Read, 1999). Both positions are overgeneralized. In practice, changing one person's lifestyle provides a health-conscious minority of individuals with a useful option to reduce cancer risk though the overall impact on population health may be small. The simple changes capable reducing disease risk may have more chances to improve population health.

Recently it has been established that many (if not a majority) of risk factors feature an "U-shaped" population risk distribution both for certain cancers and many of noncancer diseases (e.g., CVD): i.e., risk factor values deviate to both extremes – the highest and the lowest values, are associated with increased risk of disease incidence or mortality (e.g., cholesterol level, body weight, body weight at birth, alcohol consumption, supplementation with, or environmental

---

[2] Mendelian disorder caused by a single mutation in the DNA, which causes a single basic defect with pathologic consequences.

exposure to, certain vitamins and microelements, such as vitamin A, folic acid, vitamin C, iodine, etc.). Therefore, it is important to find the optimal values of risk factors when making health forecasts, e.g., risk factor intervention as preventive strategies. Computer microsimulation may be a useful instrument, flexible enough to include multiple characteristics that would be employed in defining a "golden ratio" value for every risk factor, which then will be implemented in specific prevention strategies.

A detailed analysis of the efficiency of newly developed preventive medical technologies and their effects on cost and health status described by disability, comorbidity, cancer recovery, and cognitive ability, as well as their effects on various financial and health risks will be of growing importance for both clinicians and health policymakers.

## References

Abrams D.B., 1999. Nicotine addiction: paradigms for research in the 21st century. Nicotine Tob Res 1:S211–S215.

Akushevich I., Kulminski A., Manton K., 2005. Life tables with covariates: life tables with covariates: dynamic model for nonlinear analysis of longitudinal data. Math Popul Stud 12(2):51–80.

Akushevich I., Manton K.G., Kulminski A., Kovtun M., Kravchenko J., Yashin A. (2006) Population models for the health effects of ionizing radiation. Radiats Biol Radioecol 46(6):663–674.

Akushevich I., Kravchenko, J.S., Manton K.G., 2007. Health based population forecasting: effects of smoking on mortality and fertility. Risk Anal 27(2):467–482.

Ames B.N., 1998. Micronutrients prevent cancer and delay ageing. Toxicol Lett 102–103:5–18.

Ames B.N., 1999a. Cancer prevention and diet: help from single nucleotide polymorphisms. PNAS 96(22):12216–12218.

Ames B.N., 1999b. Micronutrient deficiencies. A major cause of DNA damage. Ann N Y Acad Sci.889:87–106.

Anderson C.M., Burns D.M., Major J.M. et al., 2001. Changes in adolescent smoking behaviors in sequential birth cohorts. In: Burns D., Amacher R., Ruppert W. (Eds.). Changing Adolescent Smoking Prevalence. Smoking and Tobacco Control Monograph No. 14. NIH Publ. No. 02-5086. Bethesda, MD: U.S. Department of Health and Human Services, National Institutes of Health, National Cancer Institute, pp. 141–155.

Artandi S.E., DePinho R.A., 2000. Mice without telomerase: what can they teach us about human cancer? Nat Med 6:852–855.

Ash A.S., Byrne-Logan S., 1998. How well do models work? Predicting health care costs. In Proceedings of the Section on Statistics in Epidemiology, American Statistical Association, August 1998, pp. 42–49.

Badawi A.F., 2000. The role of prostaglandin synthesis in prostate cancer. BJU Int 85(4):451–462.

Ballard-Barbash R., Friedenreich C., Slattery M., Thune I., 2006. Obesity and body composition. In: Cancer Epidemiology and Prevention. Schottenfeld D., Fraumeni J.F. Jr. (Eds.). 3rd edition. Oxford: Oxford, University Press.

Baron J.A., Sandler R.S., 2000. Nonsteroidal anti-inflammatory drugs and cancer prevention. Annu Rev Med 51:511–523.

Baron J.A., Cole B.F., Sandler R.S. et al., 2003. A randomized trial of aspirin to prevent colorectal adenomas. N Engl J Med 348:891–899.

Bauer U., Johnson T., 2001. Predictors of tobacco use among adolescents in Florida, 1998–1999. In: Burns D., Amacher R., Ruppert W. (Eds.) Changing Adolescent Smoking Prevalence. Smoking and Tobacco Control Monograph No. 14. NIH Publ. No. 02-5086. Bethesda, MD: U.S. Department of Health and Human Services, National Institutes of Health, National Cancer Institute, pp. 121–140.

Benamouzig R., Deyra J., Martin A. et al., 2003. Daily soluble aspirin and prevention of colorectal adenoma recurrence: one-year results of the APACC trial. Gastroenterology 125:328–336.

Bhattacharya J., Cutler D.M., Goldman D.P., Hurd M.D., Joyce G.F., Lakdawalla D.N., Panis C.W.A., Shang B., 2004. Disability forecasts and future Medicare costs. Front Health Policy Res, 7, September 2004.

Bonithon-Kopp, C., Kronberg, O., Giacosa, A et al, 2000. Calcium and fiber supplementation in presentation of colorectal adenoma recurrence: a randomized intervention trial. Lancet 14:1300–1306.

Breslow L., Agran L., Breslow D.M. et al., 1977. Cancer control: implications from its history. J Natl Cancer Inst 59:671–686.

Brunsvig P.E., Aamdal S., Gjertsen M.K. et al., 2006. Telomerase peptide vaccination: a phase I/II study in patients with non-small cell lung cancer. Cancer Immunol Immunother 55:1553–1564.

Bryan T.M., Englezou A., Gupta J. et al., 1995. Telomere elongation in immortal human cells without detectable telomerase activity. EMBO J 14:4240–4248.

Byers T.E., Mouchawar J., Marks J., Cady B., Lins N., Swanson G.M., Bal D.G., Eyre H., 1999. The American Cancer Society challenge goals. How far can cancer rates decline in the U.S. by the year 2015? Cancer 86(4):715–727.

Cairncross F., (2007) Economics: age, health and wealth. Nature 448:875–876 (23 August 2007) | doi:10.1038/448875a; Published online 22 August 2007

Carpenter E.L., Vonderheide R.H., 2006. Telomerase-based immunotherapy of cancer. Expert Opin Biol Ther 6:1031–1039.

Centers for Disease Control and Prevention, 1999. Best Practices for Comprehensive Tobacco Control Programs. Atlanta: CDC< National Center for Chronic Disease Prevention and Health Promotion, Office on Smoking and Health.

Centers for Disease Control and Prevention, 2000a. Cigarette smoking among adults – United States, 2002. MMWR 51:642–645.

Centers for Disease Control and Prevention, 2000b. Use of FDA-approved pharmacologic treatments for tobacco dependence – United States, 1984–1998. MMWR 49:665–668.

Centers for Disease Control and Prevention, 2002. Youth risk behavior surveillance - United States, 2001. MMWR Surveill Summ 51:SS4.

Centers for Disease Control and Prevention, 2003. Cigarette smoking among adults – United States, 2001. MMWR 52:953–966.

Chen K., Qiu J-L., Zhang Y., Zhao Y-W., 2003. Meta analysis of risk factors for colorectal cancer. World J Gastroenterol (7):1598–1600.

Choi S., Singh S.V., 2005. Bax and Bak are required for apoptosis induction by sulforaphane, a cruciferous vegetable-derived cancer preventive agent. Cancer Res 65(5): 2035–2043.

Colditz G.A., Gortmaker S.L., 1995. Cancer prevention strategies for the future: risk identification and preventive intervention. Milbank Q 73(4):621–651.

Correa P., Fontham E.T., Bravo J.C. et al., 2000. Chemoprevention of gastric dysplasia: randomized trial of antioxidant supplements and anti-helicobacter pylori therapy. J Natl Cancer Inst 92:1881–1888.

Courneya K.S., 2003. Exercise in cancer survivors: biopsychosocial outcomes. Med Sci Sports Exerc 35:1846–1852.

Cover C.M. et al., 1998. Indole-3-carbinol inhibits the expression of cyclin-dependent kinase-6 and induces a G1 cell cycle arrest of human breast cancer cells independent of estrogen receptor signaling. J Biol Chem 273:3838–3847.
Cummings S.R., Eckert S., Krueger K.A. et al., 1999. The effect of raloxifene on risk of breast cancer in postmenopausal women: results from the MORE randomized trial. Multiple Outcomes of Raloxifene Evaluation. JAMA 281:2189–2197.
Curry S.J., Byers T., Hewitt M. (Eds), 2003. Fulfilling the Potential of Cancer Prevention and Early Detection. Institute of Medicine, National Research Council of the National Academies. Washington, D.C.: The National Academies Press, 542 pages.
Cuzick J., Forbes J., Edwards R. et al., 2002. First results from the International Breast Cancer Intervention Study (IBIS-I): a randomized prevention trial. Lancet 360:817–824.
Devesa S.S., Young J.L. Jr., Brinton L.A. et al., 1989. Recent trends in cervix uteri cancer. Cancer 64:2184–2190.
Doll R. 1998. Uncovering the effects of smoking: historical perspective. Statist Meth Med Res 7:87–117.
Doll R., 1992. The lessons of life: keynote address to the Nutrition and Cancer Conference. Cancer Research, Suppl. 52:2024S–2029S.
Doll R., Peto,R., 1981. The causes of cancer: quantitative estimates of avoidable risks of cancer in the United States today. J Natl Cancer Inst 66(6):1191–1308.
El-Bayoumy K. et al., 2006. Cancer chemoprevention by garlic and garlic-containing sulfur and selenium compounds. J Nutr 136(3):864S–869S.
Emenaker N.J., Milner J.A., 2006. Eating for cancer prevention: a molecular approach. AgroFOOD Industry Hi-Tech 17(3):24–26.
Everett S., Warren C., 2001. Trends and subgroup differences in tobacco use among high school students in the United States, 1991–1997. Changing Adolescent Smoking Prevalence. Smoking and Tobacco Control Monograph No. 14. NIH Publ. No. 02-5086. Bethesda, MD:U.S. Department of Health and Human Services, National Institutes of Health, National Cancer Institute, pp. 35–50.
Ezzati M., Lopez A.D., 2004. Regional, disease specific patterns of smoking-attributable mortality in 2000. Tob Control 13:388–395.
Fisher B., Costantino J.P., Wickerham D.L., et al., 1998. Tamoxifen for prevention of breast cancer: Report of the National Surgical adjuvant Breast and Bowel Project P-1 Study. J Natl Cancer Inst 90:1371–1388.
Fleischauer A.T. et al., 2000. Garlic consumption and cancer prevention: meta-analyses of colorectal and stomach cancers. Am J Clin Nutr 72(4):1047–1052.
Foster R.S., 2003. The Financial Outlook for Medicare. *Testimony before the House Committee on Energy and Commerce Subcommittee on Health.*
Fowke J.H. et al., 2003. Urinary isothiocyanate levels, brassica and human breast cancer. Cancer Res 63(14):3980–3986.
Garland C.F., Garland F.C., Gorham E.D., 1993. Rising trends in melanoma: an hypothesis concerning sunscreen effectiveness. Ann Epidemiol 3:103–110.
Giardiell F.M., Yang V.W., Hylind L.M. et al., 2002. Primary chemoprevention of familial adenomatous polyposis with sulindac. N Engl J Med 346:1054–1109.
Goldman D.P., Shekelle P.G., Bhattacharya J. et al., 2004. Health status and medical treatment of the future elderly. Final Report. TR-169-CMS. Rand Health.
Greaves M.F., 1997. Etiology of acute leukemia. Lancet 349:344–349.
Greenwald P., 1995. Introduction: history of cancer prevention and control. In: Greenwald P., Kramer B.S., Weed D.L. (Eds): Cancer Prevention and Control. New-York: Marcel Dekker, Inc.
Greenwald, P., 2002. Cancer prevention clinical trials. J Clin Oncol 20:14S–22S.
Greenwald P., Cullen J.W., 1985. The new emphasis in cancer control. J Natl Cancer Inst 74:543–551.

Greenwald P. Cullen J.W., McKenna J.W., 1987. Cancer prevention and control: from research through applications. J Natl Cancer Inst 79:389–400.

Habel L.A., Friedman G.D., 2006. Pharmaceticals other than hormons. In: Schottenfeld D., Fraumeni J.F. Jr. (Eds). Cancer Epidemiology and Prevention. 3rd edition. Oxford: Oxford University Press. pp. 489–506.

Halfon N., Hochstein M., 2002. Life course health development: an integrated framework for developing health, policy, and research. Milbank Q 80:433–479.

Henningfield J.E., 1985. Nicotine: An Old-Fashioned Addiction. New York, NY: Chelsea House Publishers, pp. 96–98.

Herlyn D., Birebent B., 1999. Advances in cancer vaccine development. *Annals of Medicine* 31(1):66–78.

Hiatt R.A., Rimer B.K., 1999. A new strategy for cancer control research. Cancer Epidemiol Biomarkers Prev 8:957–964.

Hiatt R.A., Rimer B.K., 2006. Principles and applications of cancer prevention and control interventions. In: Schottenfeld D., Fraumeni J.F. Jr. (Eds). Cancer Epidemiology and Prevention. 3rd edition. Oxford: Oxford University Press. pp. 1283–1291.

Hochreiter A.E., Xiao H., Goldblatt E.M. et al., 2006. Telomerase template antagonist GRN163L disrupts telomere maintenance, tumor growth, and metastasis of breast cancer. Clin Cancer Res 12(10):3184–3192.

Holmes M.D., Chen W.J., Feskanich D. et al., 2005. Physical activity and survival after breast cancer diagnosis. JAMA 293:2479–2486.

Hong W.K., Endicott J., Itri L.M. et al., 1986. 13-cis-retinoic acid in the treatment of oral leukoplakia. N Engl J Med 315:1501–1505.

Hong W.K., Lippman S.M., Itri L.M. et al., 1990. Prevention of second primary tumors with isotretinoin in squamous-cell carcinoma of the head and neck. N Engl J Med 323:795–801.

Hwang L.C., Fein S., Levitsky H., Nelson W.G., 1999. Prostate cancer vaccines: current status. *Seminars in Oncology* 26(2):192–201.

IARC Working Group, 2002. IARC Handbooks of Cancer Prevention, Vol. 6. Weight control and physical activity. Lyon, France: IARC Press, pp. 1–315.

IARC, 2002. Weight, Control and Physical Activity. Lyon: IARC.

IARC, 2004. Tobacco Smoke and Involuntary Smoking. IARC Monographs on the Evaluation of the Carcinogenic Risk of Chemicals to Humans, Vol. 83. Lyon: IARC.

James S.J., Progribna M., Progribny I.I. et al., 1999. Abnormal folate metabolism and mutation in the methylenetetrahydrofolate reductase gene may be maternal risk factors for Down syndrome. Am J Clin Nutr 70:495–501.

Janne P.A., Mayer R.J., 2000. Chemoprevention of colorectal cancer. N Engl J MedNEJM 342:1960–1968.

Jemal A., Thomas A., Murray T et al., 2002. Cancer statistics. CA Cancer J Clin 52:23–47.

Jha P., Chaloupka F.J., Corrao M. et al., 2006. Reducing the burden of smoking world-wide: effectiveness of interventions and their coverage. Drug Alcohol Rev 25(6):597–609.

Johnston L., 2001. Changing Demographic Patterns of Adolescent Smoking over the Past 23 years: National Trends from the Monitoring the Future Study. Changing Adolescent Smoking Prevalence. Smoking and Tobacco Control Monograph No. 14. NIH Publ. No. 02-5086. Bethesda, MD: U.S. Department of Health and Human Services, National Institutes of Health, National Cancer Institute, pp. 9–34.

Kaaks R., Van Noord P.A.H., Den Tonkelaar I., et al., 1998. Breast cancer incidence in relation to height, weight and body-fat distribution in the Dutch "DOM" cohort. Int J Cancer 76(5):647–651.

Keloff G.J., Boone C.W., Crowell J.A. et al., 1994. Chemopreventive drug development: perspectives and progress. Cancer Epidemiol Biomarkers Prev 3:85–98.

Khuder S.A., Mutgi A.B., 2001. Breast cancer and NSAID use: a meta-analyses. Br J Cancer 84(9):1188–1192.

Kim Y.S., Milner J.A., 2005. Targets for indole-3-carbinol in cancer prevention. J Nutr Biochem 16(2):65–73.
Knowler W.C., Barrett-Connor E., Fowler S.E., Hamman R.F., Lachin J.M., Walker E.A., Nathan D.M., 2002. Reduction in the incidence of type 2 diabetes with lifestyle intervention or metformin. N Engl J Med 346(6):393–403.
Kopstein A., 2001. Trends in adolescent smoking in the United States: data from the national household survey on drug abuse 1994 through 1998. Changing Adolescent Smoking Prevalence. Smoking and Tobacco Control Monograph No. 14. NIH Publ. No. 02-5086. Bethesda, MD: U.S. Department of Health and Human Services, National Institutes of Health, National Cancer Institute, pp. 51–68.
Koutsky L.A., Ault K.A., Wheeler C.M. et al., 2002. A controlled trial of a human papillomavirus type 16 vaccine. N Engl J Med 347:1645–1651.
Koushik A., Hunter D.J., Spiegelman D. et al., 2007. Fruits, vegetables, and colon cancer risk in a pooled analysis of 14 cohort studies. J Natl Cancer Inst 99(19):1471–1483.
Kravchenko J., Goldschmidt-Clermont P.J., Powell T., Stallard E., Akushevich I., Cuffe M.S., Manton K.G., 2005. Endothelial progenitor cell therapy for atherosclerosis: the philosopher's stone for an aging population? Sci Aging Knowledge Environ Jun 22; 2005 (25):18.
Krebs-Smith S.M., Cook A., Subar A.F. et al., 1995. U.S adults' fruit and vegetable intakes, 1989 to 1991: a revised baseline for the Health People2000 objective. Am J Public Health 85:1623–1629.
Krebs-Smith S.M., Cook A., Subar A.F. et al., 1996. Fruit and vegetable intakes in children and adolescents in the United States. *Arch. Pediatr Adolesc Med* 150:81–86.
Kricker A., Armstrong B.K., Hughes A.M. et al., 2007. Personal sun exposure and risk of non-Hodgkin lymphoma: a pooled analysis from the Interlymph Consortium. Int J Cancer 122(1):144–154.
Kulasingam S.I., Myers E.R., 2003. Potential health and economic impact of adding a human papillomavirus vaccine to screening programs. JAMA 290(6):781–789.
Kurie J.M., Lee J.S., Khuri F.R. et al., 2000. N(4-hydroxphenyl) retinamide in the chemoprevention of squamous metaplasia and dysplasia of the bronchial epithelium. Clin Cancer Res 6:2973–2979.
Kurie J.M., Lotan R., Lee J.J. et al., 2003. Treatment of former smokers with 9-cis-retinoic acid reverses loss of retinoic acid receptor-beta expression in the bronchial epithelium: results from a randomized placebo-controlled trial. J Natl Cancer Inst 95:206–214.
Kutz G. (2006). Nutrigenic testing: test purchased from four web sites misled consumers. At: www.gao.gov/new.items/d06977t.pdf
Labayle D., Fischer D., Vielh P. et al., 1991. Sulindac causes regression of rectal polyps in familial adenomatous polyposis. Gastroenterology 101:635–639.
Lam S., MacAulay C., Le Riche J.C. et al., 2002. A randomized phase IIb trial of anethole dithiolethione in smokers with bronchial dysplasia. J Natl Cancer Inst 94:1001–1009.
Lee R., Tuljapurkar S., 1998. Population Forecasting for Fiscal Planning: Issues and Innovations. Burch Working Paper No. B98-05, University of California, Berkeley, December.
Lee R., Miller T., 2002. An approach to forecasting health expenditures, with application to the U.S. Medicare System. Health Serv Res, **37**(5), October 2002, pp. 1365–1386
Lee I-M., Oguma Y., 2006. Physical activity. In: Cancer Epidemiology and Prevention. Schottenfeld D., Fraumeni J.F .Jr. (Eds). 3rd edition. Oxford: Oxford University Press
Lee J.S., Lippman S.M., Benner S.E. et al., 1994. Randomized placebo-controlled trial of isotretinoin in chemoprevention of bronchial squamous metaplasia. J Clin Oncol 12:937–945.
Lee J.M., Kopans D.B., McMahon P.M. et al., 2008. Breast cancer screening in BRCA1 mutation carriers: effectiveness of MR imaging – Markov Monte Carlo decision analysis. Radiology Mar; 246(3):763–771
Linos E., Raine T., Alonso A. et al., 2007. Atopy and risk of brain tumors: a meta-analysis. J Natl Cancer Inst 99(20):1544–1550.

Lippman S.M., Hong W.K., 2002. Cancer prevention science and practice. Cancer Res 62:5119–5125.
Lippman S.M., Batsakis J.G., Toth B.B. et al., 1993. Comparison of low-dose isotretinoin with beta-carotene to prevent oral carcinogenesis. N Engl J Med 328:15–20.
Lippman S.M., Lee J.J., Karp D.D. et al., 2001. Randomized phase III intergroup trial of isotretinoin to prevent second primary tumors in stage I non-small-cell lung cancer. J Natl Cancer Inst 93:605–618.
Liu R.H., 2004. Potential synergy of phytochemicals in cancer prevention: mechanism of action. J Nutr 134:3479S–3485S.
Lock S., Reynolds L.A., Tansey E.M. (Eds), 1998. Ashes to Ashes: The History of Smoking and Health. 2nd edition, Amsterdam: Rodopi. 256 pages.
London S.J. et al., 2000. Isothiocyanates, glutathione-S-transferase M1 and T1 polymorphisms, and lung cancer risk: a prospective study of men in Shanghai, China. Lancet 356(9231):724–729.
Long L., Glover R.T., Kaufman H.L., 1999. The next generation of vaccines for the treatment of cancer. Curr Opin Mol Ther 1:57–63.
Lopez A.D., Shibuya K., Rao C. et al., 2006. Chronic obstructive pulmonary disease: current burden and future projections. Eur Respir J 27:397–412.
Lundholm K., Gelin J., Hyltander A. et al., 1994. Anti-inflammatory treatment may prolong survival in undernourished patients with metastatic solid tumors. Cancer Res 54:5602–5606.
Mandel J., Bond J., Church T. et al., 1993. Reducing mortality from colorectal cancer by screening for fecal occult blood. N Engl J Med 328:1365–1371.
Manton K.G., Stallard E., 1988. Chronic Disease Risk Modeling: Measurement and Evaluation of the Risks of Chronic Disease Processes.//In the Griffin Series of the Biomathematics of Diseases. London, England: Charles Griffin Limited.
Manton K.G., Stallard E., Vaupel J.W., 1986. Alternative models for the heterogeneity of mortality risks among the Aged. J Am Stat Assoc 81(395):635–644
Manton K.G., Gu X., Lamb V.L., 2006. Change in chronic disability from 1982 to 2004/2005 as measured by long-term changes in function and health in the U.S. elderly population. Proc Natl Acad Sci U S A. 103(48):18374–18379.
Manton K.G., Akushevich I., Kulminski A., 2008. Human mortality at extreme ages: data from the national long term care survey and linked Medicare records, Math Popul Stud 15:137–159.
Manton K.G., Lowrimore G.R., Ullian A.D., Gu X., Tolley H.D., 2007b. Labor force participation and human capital increases in an aging population and implications for U.S. research investment. Proc Natl Acad Sci U S A. 104(26):10802–10807.
Marshall J.R., Freudenheim J.O., 2006. Alcohol. In: Cancer Epidemiology and Prevention. Schottenfeld D., Fraumeni J.F. Jr. (Eds). 3rd edition. Oxford: Oxford University.
Martin K.R., 2006. Targeting apoptosis with dietary bioactive agents. Exp Biol Med 231(2):117–129.
Mathers C.D., Loncar D., 2006. Projections of global mortality and burden of disease from 2002 to 2030. PLoS Med 3(11):e442.
Mayr C.A., Woodall A.A., Ames B.N., 1999. In Preventive Nutrition: The Comprehensive Guide for Health Professionals. Eds Bendich A., Deckelbaum R.J. Totowa, NJ: Humana.
McMahon P.M., Kong C.Y., Johnson B.E. et al., 2008. Estimating long-term effectiveness of lung cancer screening in the Mayo CT screening study. Radiology 248:278–287.
McTiernan A., Kooperberg C., White E. et al., 2003. Recreational physical activity and the risk of breast cancer in postmenopausal women: The Women's Health Initiative Cohort Study. JAMA 290(10):1331–1336.
Meyskens, F.L. (2004). Cancer prevention, screening, and early detection. In: Abeloff, M.D., Armitage, J.O., Niederhuber, J.E., Kastan, M.B., McKenna, W.G. (eds.) Clin Oncol. 3rd edition, pp. 425–472. Amsterdam: Elsevier.

Michalowski J., 2002. COX-2 inhibitors: cancer trials test new uses for pain drug. J Natl Cancer Inst 94(4):248–249.

Miller T., 2001. Increasing longevity and Medicare expenditures. *Demography*, **38**(2), May 2001, pp. 215–226

Minev B.R., Chavex F.L., Mitchell M.S., 1999. Cancer vaccines: novel approaches and new promise. Pharmacol Ther 81(2):121–139.

MMWR, 2004. Trends in intake of energy and macronutrients – United States, 1971–2000. Mor Mortal Wkly Rep 53:80–82.

Moon, T.E., Levine, N., Cartmel, B. et al, 1997. Effect of retinol in preventing squamous cell skin cancer in moderate-risk subjects: a randomized, double-blind, controlled trial. Southwest Skin Cancer Prevention Study Group. Cancer Epi Biom Prev 6:949–956.

Muller A.D., Sonnenberg A., 1995. Protection by endoscopy against death from colorectal cancer. A case–control study among veterans. Arch Intern Med 155:1741–1748.

Murphy K.M., Topel R.H. (2006) The value of health and longevity. J Political Econ 114:871–904.

Myers E.R., McCrory D.C., Nanda K. et al., 2000. Mathematical model for the natural history of human papillomavirus infection and cervical carcinogenesis. Am J Epidemiol 151(12):1158–1171.

National Cancer Institute, Acrylamide in Foods. Fact Sheet. www.nci.nih.gov/cancertopics/factsheet/acrylamideinfoods

National Cancer Institute, Artificial Sweeteners and Cancer. Questions and Answers. www.nci.nih.gov/cancertopics/factsheet/Risk/artificial-sweeteners

National Cancer Institute, Heterocyclic Amines in Cooked Meats www.nci.nih.gov/cancer topics/factsheet/Risk/heterocyclic-amines

Nieburgs H.E., Vali V.E., 2007. Oncogenic cofactors that may act in synergistic, interactive or sequential manner on various target sites. Cancer Detection and Prevention 31(6): back cover.

New York City Cancer Committee, 1994. New York City Cancer Committee: History of the America society for the control of cancer, 1913–1943. New York: New York City Cancer Committee.

Nutrition and Physical Activity Guidelines 2006. American Cancer Society. At: www.cancer.org/docroot/PED/content/PED_3_2X_Diet_and_Activity_Factors_That_Affect_Risks.asp

Obesity and Cancer: questions and answers. National Cancer Institute. At: www.cancer.gov/cancertopics/factsheet/Risk/obesity/

Parker E.D., Folsom A.R., 2003. Intentional weight loss and incidence of obesity-related cancers. The Iowa Women's Health Study. Int J Obes Relat Metab Disord 27(12):1447–1452.

Parkhurst M.R., Riley J.P., Igarashi T. et al., 2004. Immunization of patients with the hTERT:540-548 peptide induces peptide-reactive T lumphocytes that do not recognize tumors endogenously expressing telomerase. Clin Cancer Res 10:4688–4698.

Peto R., Darby S., Deo H. et al., 2000. Smoking, smoking cessation and lung cancer in the UK since 1950: combination of national statistics with two case–control studies. BMJ 321: 323–329.

Phatak P., Burger A.M., 2007. Telomerase and its potential for therapeutic intervention. Br J Pharmacol 152:1003–1011.

Plummer M., Vivas J., Lopez G. et al., 2007. Chemoprevention of precancerous gastric lesions with antioxidant vitamin supplementation: a randomized trial in a high-risk population. J Natl Cancer Inst 99(2):137–146.

Proctor R.N., 2001a. Commentary: Schairer and Schoninger's forgotten tobacco epidemiology and the Nazi quest for racial purity. Int J Epidemiol 30:31–34.

Proctor R.N., 2001b. Commentary: pioneering research into smoking and health in Nazi Germanu – the "Wissenschaftliches Institute zur Erforschung der Tabakgefahren" in Jena. Int J Epidemiol 30:35–37.

Rao C., Hart J., Chow A. et al., 2008. Does preservation of the sub-valvular apparatus during mitral valve replacement affect long-term survival and quality of life? A Microsimulation Study. J Cardiothorac Surg 3(1):17.

Review of Assumptions and Methods of the Medicare Trustees' Financial Projections. December 2000.

Rogers E., 1993. Diffusion of Innovation. London: The Free Press.

Rokkas T., Pistolas D., Sechopoulos P. et al., 2007. Relationship between Helicobacter pylori infection and esophageal neoplasia: a meta-analysis. Clin Gastroenterol Hepatol 5(12):1413–1417.

Ross J.A., 2000. Dietary flavonoids and the MLL gene: a pathway to infant leukemia? PNAS 97(9):4411–4413.

Ross J.A., Potter J.D., Reaman G.H., Pendergrass T., Robinson L.L., 1996. Maternal exposure to potential inhibitors of DNA topoisomerase II and infant leukemia (United States): a report from the Children's Cancer Group. Cancer Causes Control 7:581–590.

Samad A.K., Taylor R.S., Marshall T., Chapman M.A., 2005. A meta-analysis of the association of physical activity with reduced risk of colorectal cancer. Coloraectal Dis 7(3):204–213.

Sandler R.S., Halabi S., Baron J.A. et al., 2003. A randomized trial of aspirin to prevent colorectal adenomas in patients with previous colorectal cancer. N Engl J Med 348:883–890.

Schatzkin A., Lanza E., Corle D., et al. (2000). Lack of effect of a low-fat, high-fiber diet on the recurrence of colorectal adenomas. Polyp Prevention Trial Study Group. N Engl J Med 342:1149–1155.

Scholm J., Tsang K.Y., Kantor J.A., Abrams S.I., 1998. Cancer vaccine development. Expert Opin Investig Drugs 7(9):1439–1452.

Spector L.G.. Xie Y., Robinson L.L. et al., 2005. Maternal diet and infant leukemia: the DNA topoisomerase II inhibitor hypothesis: a report from the Children's Oncology Group. Cancer Epidemiol Biomarkers Prev 14(3):651–655.

Sporn M.B., Dunlop N.M., Newton D.L. et al., 1976. Prevention of chemical carcinogenesis by vitamin A and its synthetic analogs (retinoids). Fed Proc 35:1332–1338.

Steinbach G., Lynch P.M., Phillips R.K. et al., 2000. The effect of celecoxib, a cyclooxygenase-2 inhibitor, in familial adenomatous polyposis. N Engl K Med 342:1946–1952.

Stewart G.G., 1967. A history of the medicinal use of tobacco, 1492–1860. Med Hist 11(3):228–268.

Strachan T., Read A.P., 1999. Human Molecular Genetics 2. 2nd edition. New York, NY: John Willey & Son, Publishers. BIOS Scientific Publishers, LTD. 576 pages.

Surveillance, Epidemiology and End Results (SEER) Program, 2003. SEER Stat Database: Mortality – All COD, Public Use with State, Total U.S. (1969–2000). Underlying mortality data provided by NCHS (www.cdc.gov/nchs), National Cancer Institute, DCCPS, Surveillance Research Program, Cancer Statistics Branch.

Swan J., Breen N., Coates R.J. et al., 2003. Progress in cancer screening practices in the United States: results from the 2000 National Health Interview Survey. Cancer 97:1528–1540.

Taningher M., Malacarne D., Izzotti A. et al, 1999. Drug metabolism polymorphisms as modulator of cancer susceptibility. Mutat Res 436:227–261.

Thompson I.M., Kouril M., Klein E.A. et al., 2001. The Prostate Cancer Prevention Trial: current status and lessons learned. Urology 57(Suppl. 1):230–234.

Thompson I.M., Goodman P.J., Tangen C.M. et al., 2003. The influence of finasteride on the development of prostate cancer. N Engl J Med 349:215–224.

Thune I., Furberg A.S., 2001. Physical activity and cancer risk: dose-response and cancer, all sites and site-specific. Med. Sci. Sports Exerc. 33(6), Suppl., pp. S530–S550.

Thun M.J., Henley S.J., 2006. Tobacco. In: Schottenfeld D., Fraumeni J.F. Jr. (eds.). Cancer Epidemiology and Prevention. 3rd edition. Oxford: Oxford University Press, pp. 217–242.

Thun M.J., Henley S.J., Patrono C., 2002. Nonsteroidal anti-inflammatory drugs as anticancer agents: mechanistic, pharmacologic, and clinical issues. J Natl Cancer Inst 94(4):252–266.

Trentham-Dietz A., Newcomb P.A., Egan K.M. et al., 2000. Weight change and risk of postmenopausal breast cancer (United States). Cancer Causes Control 11(6):533–542.

Turkkan J.S., Kaufman N.J., Rimer B.K., 2000. Transdisciplinary tobacco use research centers: a model collaboration between public and private sectors. Nicotine Tob Res 2:9-13.U.S. Department of Health and Human Services, 1990. The Health Benefits of Smoking Cessation: A Report of the Surgeon General. Rockville, MD. U.S. Department of Health and Human Services, Public Health Service, centers for Disease Control, Center for Chronic Disease Prevention and Health Promotion, Office on Smoking and Health.

US DHHS, 1990. The Health Benefits of Smoking Cessation. A Report of the Surgeon General. Rockville, MD: U.S. Department of Health and Human Services, Public Health Service, Centers for Disease Control, Center for Chronic Disease Prevention and Health Promotion, Office on Smoking and Health.

U.S. Department of Health and Human Services, 2000. Reducing Tobacco Use: A Report of the Surgeon General. Atlanta: U.S. Department of Health and Human Services, Centers for Disease Control and Prevention, National Center for Chronic Disease Prevention and Health Promotion, Office on Smoking and Health.

U.S. Preventive Services Task Force, 2002. Guide to Clinical Preventive Services. 3rd edition. Recommendations. Washington, D.C.: Office of Disease Prevention and Health Promotion.

US Census Bureau, 2006. World Information. Washington, D.C., USA: US Census Bureau. www.census.gov/cgi-bin/ipc/idbagg

Vainio H., Bianchini F., 2002. IARC Handbooks of Cancer Prevention. Volume 6: Weight Control and Physical Activity. Lyon, France: IARC Press.

Vainio H., Kaaks R., Bianchini F, 2002. Weight control and physical activity in cancer prevention: international evaluation of the evidence. Eur J Cancer Prev 11(Suppl2): S94–S100.

Vaupel J.W., Manton K.G., Stallard E., 1979. The impact of heterogeneity in individual frailty on the dynamics of mortality. Demography 9:439–454

Veronesi U., De Palo G., Marubini E. et al., 1999. Randomized trial of fenretinide to prevent second breast malignancy in women with early breast cancer. J Natl Cancer Inst 91:1847–1856.

Viner J.L., Hawk E., Lippman S.M., 2006. Cancer chemoprevention. In: Schottenfeld D., Fraumeni J.F. Jr. (Eds.). Cancer Epidemiology and Prevention. 3rd edition. Oxford: Oxford University Press. pp. 1318–1340.

Vonderheide R.H., Domchek S.M., Schultze J.L. et al., 2004. Vaccination of cancer patients against telomerase induces functional antitumor CD8 T lymphocytes. Clin Cancer Res 10:828–839.

Wattenberg L.W., 1996. Chemoprevention of cancer. Prev Med 25:4445.

White L.K., Wright W.E., Shay J.W., 2001. Telomerase inhibitors. Trends Biotechnol 19(3):114–120.

Willet W.C., Colditz G.A., Mueller N.E., 1996. Strategies for minimizing cancer risk. Sci Am 275(3):88–91, 94–95.

Wilson J.W., Enns C.W., Goldman J.D. et al, 1997. Data tables: combined results from USDA's 1994 and 1995 Continuing Survey of Food Intakes by Individuals and 1994 and 1995 Diet and Health Knowledge Survey. Beltsville Human Nutrition Research Center, Riverdale, MD.

Wilson S., Jones L., Coussens C., Hanna K. (Eds), 2002. Cancer and the Environment. Gene-environment Interaction. Washington, D.C.: Institute of Medicine. National Academy Press.

Wingo P.A., Ries L.A.G., Giovino G.A. et al., 1999. Annual report to the nation on the status of cancer, 1973–1996, with a special section on lung cancer and tobacco smoking. J Natl Cancer Inst 91:675–690.

World Cancer Research Fund, American Institute for Cancer Research. Food, Nutrition and the Prevention of Cancer: A Global Perspective. Washington D.C.: American Institute for Cancer Research, 1997. At: www.wcrf.org/research/fnatpoc.lasso

Wright W.E., Platyszek M.A., Rainey W.E. et al., 1996. Telomerase activity in human germ-line and embryonic tissues and cells. Dev Genet 18:173–179.

Yashin A.I., Manton K.G., 1997. Effects of unobserved and partially observed covariate processes on system failure: a review of models and estimation strategies. Stat Sci 12(1):20–34.

Yashin A.I., Iachine I.A., Begun A.Z., 2000. Mortality modeling: a review. Math Popul Stud 8(4):305–332

Young R.C., Wilson C.M., 2002. Cancer prevention: past, present and future. Clin Can Res 8:11–16.

Zhou J.R., Yu L., Zhong Y. et al., 2003. Soy phytochemicals and tea bioactive components synergistically inhibit androgen-sensitive human prostate tumors in mice. J Nutr 133:516–521.

Zhou J.R., Yu L., Mai Z. et al., 2004. Combined inhibition of estrogen-dependent human breast carcinoma by soy and tea bioactive components in mice. Int J Cancer 108:8–14.

Ziegler R.G., Hoover R.N., Nomura A.M. et al., 1996. Relative weight, weight change, height, and breast cancer risk in Asian American women. J Natl Cancer Inst 88(10):650–660.

Zu K., Ip C., 2003. Synergy between selenium and vitamin E in apoptosis induction is associated with activation of distinctive initiator caspases in human prostate cancer cells. Cancer Res 63:6988–6995.

$\alpha$-Tocoferol, $\beta$-Carotene Cancer Prevention Study Group, 1994. The effect of vitamin E and $\beta$-carotene on the incidence of lung cancer and other cancers in male smokers. N Engl J Med 330:1029–1035.

# Conclusion and Outlook

Significant breakthroughs in the scientific confrontation of human knowledge and cancer continue to be postponed. Though some success in the decreasing incidence of specific cancers has been achieved, it is premature to speak about general results. While for certain cancers screening or risk factor intervention strategies are effective, there are cancers and cancer histotypes for which trends have not changed significantly over the last half century. In this book, we advocate the point of view that interdisciplinary teams will be necessary for the ultimate breakthroughs necessary in cancer research.

To make faster progress against cancer, we need to not only discover the specific mechanisms of cancer but also find general laws about the relation of different cancer features. In this way, we will develop innovative systematic approaches to allow the integration of theoretical and experimental information from multiple sources. Often, progress has been limited by the absence of integrative analysis, as well as by the fact that such knowledge has been collected in different fields of sciences, which cannot be simply combined because of differences in scientific languages (e.g., linguistic gaps between mathematicians and medical doctors).

Therefore, interdisciplinary approaches performed by groups of scientists, who are not only the experts in their specific areas but able to understand the language of colleagues' other areas, would be beneficial. An example of such an interdisciplinary team are the authors of this book. In this book, we summarized information on (1) biological and medical aspects of cancer initiation and progression; (2) development of biological concepts of carcinogenesis; (3) mathematical modeling of concepts, comparing these models with available datasets and prediction of future trends and tendencies; and (4) epidemiologic, population, and demographic approaches to analysis of datasets. Based on that, we suggest new methods of analysis (including those modified from related fields) and applied them to data hypothesizing necessary new concepts in tasks of cancer risks and prevention.

We expect that future breakthroughs will result from progress in (1) systematic approaches for finding shared features controlling initiation, promotion, and progression of specific cancers and cancer histotypes; (2) development of new models of carcinogenesis combining broad classes of parameters, from long-term epidemiologic cohort studies to individual measurements of cancer-related factors and characteristics; (3) development of approaches to individual cancer-risk modeling, capable of better predicting risks for individuals or homogeneous groups of individuals; and (4) microsimulation approaches to predict short- and long-term consequences, aka effectiveness, of various medical/therapeutic interventions and preventive strategies. Taking together, efforts in these directions may add the necessary interdisciplinary background for future breakthroughs in human cancer research.

# Keywords

Cancer morbidity
Cancer mortality
Carcinogenesis
Interdisciplinary approach
Mathematical modeling

# Abbreviations

| | |
|---|---|
| **AC** | adenocarcinoma |
| **ACS** | American Cancer Society |
| **ADH** | alcohol dehydrogenase |
| **AIC** | Akaike information criterion |
| **ALDH** | aldehyde dehydrogenase |
| **ALE** | active life expectancy |
| **APC** | adenomatous polyposis coli (gene) |
| **APC** | annual percent change |
| **Apo epsilon4** | apolipoprotein E allele 4 |
| **APOE** | apolipoprotein E |
| **ARCIC** | Atherosclerosis Risk in Communities (surveillance study) |
| **BAGE** | B melanoma antigen |
| **Bax** | Bcl-2-associated X (protein) |
| **Bcl-2** | B-cell leukemia 2 (gene) |
| **BiP** | binding immunoglobulin protein |
| **BRCA1** | breast cancer 1 (gene) |
| **C/EBP** | CAAT/enhancer-binding protein (purified C/EBP selectively recognizes CCAAT homologies) |
| **CAGE** | cancer/testis antigen cancer-associated gene |
| **c-AMP** | cyclic adenosine monophosphate (or 3′-5′-cyclic adenosine monophosphate) |
| **CDC** | Centers for Disease Control and Prevention |
| **CDK** | cycline-dependent kinase |
| **cDNA** | complementary DNA |
| **CDR** | crude death rate |
| **CEA** | carcinoembryonic antigen |
| **CHD** | coronary heart disease |
| **CHOP** | C/EBP homologous protein |
| **CIN** | cervical intraepithelial neoplasia |
| **CISNET** | Cancer Intervention and Surveillance Modeling Network |
| **CJIC** | gap junctional intercellular communication |
| **CLIA** | Clinical Laboratory Improvement Act |
| **CNS** | central nervous system |

| | |
|---|---|
| **COPD** | chronic obstructive pulmonary disease |
| **CVD** | cardiovascular disease |
| **DASDR** | direct age standardized age-specific death rate |
| **DCCPS** | Division of Cancer Control and Population Science |
| **DCIS** | ductal carcinoma *in situ* |
| **DCP** | Division of Cancer Prevention |
| **DMPA** | depot medroxyprogesterone acetate |
| **DNA** | deoxyribonucleic acid |
| **DR** | death receptor (e.g., DR5) |
| **DSB** | double-strand break |
| **DZ** | dizygotic |
| **ED50** | median effective dose (produces desired effect in 50% of population) |
| **EGFR** | epidermal growth factor receptor |
| **EPA** | Environmental Protection Agency |
| **EPC** | endothelial progenitor cell |
| **ERR** | excess relative risk |
| **ETRC** | Extended Techa River Cohort |
| **EYL** | extra years of life |
| **FDA** | U.S. Food and Drug Administration |
| **FEM** | Future Elderly Model |
| **FGF-2** | fibroblast growth factor 2 |
| **FOBT** | fecal occult blood testing |
| **GBD** | global burden of disease |
| **GMCSF** | granulocyte macrophage colony-stimulating factor |
| **GoM** | Grade of Membership |
| **GST** | glutathione-*S*-transferase (e.g., GSTP1) |
| **HBV** | Hepatitis B virus |
| **HCA** | heterocyclic amine |
| **HCC** | hepatocellular carcinoma |
| **HCV** | Hepatitis C virus |
| **HIF** | hypoxia-inducible factor |
| **HIV** | human immunodeficiency virus |
| **HMD** | human mortality database |
| **hMLH1** | human mutL homolog 1 |
| **hMSH2** | human mutS homolog 2 |
| **HNPCC** | hereditary nonpolyposis colorectal cancer |
| **Hp** | *Helicobacter pylori* |
| **HPP1/TPEF** | hyperplastic polyposis protein 1/transmembrane protein containing epidermal growth factor (and follistatin domains) |
| **HPV** | *Human papillomavirus* |
| **HSP** | heat-shock protein |
| **HSV** | *Herpes simplex virus* |
| **HSV-TK** | *Herpes simplex virus* thymidine kinase |

| | |
|---|---|
| **hTERT** | human telomerase reverse transcriptase |
| **I3C** | glucosinolate indole-3-carbinol |
| **IAP** | inhibitor of apoptosis protein |
| **IARC** | International Agency for Research on Cancer |
| **ICAM** | intercellular adhesion molecule (e.g., ICAM1) |
| **ICD-9-CM** | International Classification of Diseases, Ninth Revision, Clinical Modification |
| **IDP** | intraductal proliferation |
| **IGF** | insulin growth factor |
| **IGF2** | insulin-like growth factor 2 |
| **IL-1** | interleukin-1 |
| **IR** | ionizing radiation |
| **ITC** | isothiocyanate |
| **LDL** | low-density lipoprotein |
| **LE** | life expectancy |
| **LEEP** | loop electrosurgical excision procedure |
| **LLS** | Linear Latent Structure (analysis) |
| **LNT** | linear no-threshold |
| **LSS** | Life Span Study |
| **MAb CO17-1A** | monoclonal antibody against the 17-1A antigen |
| **MAGE** | melanoma antigen family (gene) |
| **MASDR** | marginal age standardized age-specific death rates |
| **MCMC** | Markov chain Monte Carlo method |
| **MDR1** | human multidrug-resistance 1 (gene) |
| **MGMT** | ($O^6$-methylguanine–DNA methyltransferase) DNA-repair gene |
| **MINT2** | MSX2-interacting nuclear target protein |
| **MMWR** | Morbidity and Mortality Weekly Report |
| **MOE** | margin of exposure |
| **MONICA** (study) | Monitoring Cardiovascular Disease (study) |
| **MRW** (criteria) | Metropolitan Relative Weight (criteria) |
| **MSI+** | microsatellite instability-positive (colon cancer) |
| **mtDNA** | mitochondrial DNA |
| **MTHFR** | Methylenetetrahydrofolate reductase |
| **MUC-1** | mucin 1, transmembrane (protein) |
| **MVK** (model) | Moolgavkar-Venzon-Knudson (model) |
| **MZ** | monozygotic |
| **NAT** | *N*-acetyltransferase |
| **NCHS** | National Center for Health Statistics |
| **NCI** | National Cancer Institute |
| **NFkB** | nuclear factor kappa B |
| **NHANES** | National Health and Nutrition Examination Survey |
| **NHEFS** | National Health Epidemiologic Follow-up Study (NHANES I) |
| **NHLBI** | National Heart, Lung, and Blood Institute |

| | |
|---|---|
| **NIH** | National Institutes of Health |
| **NLTCS** | National Long-Term Care Survey |
| **NNK** | (tobacco-specific) nitrosamines 4-(methylnitrosamino)-1-(3-pyridyl)-1-butanone |
| **NNK** | 4-(methyl-nitrosamino)-1-(3-pyridyl)-1-butanone |
| **NO** | nitrogen oxide |
| **Nrf2** | NF-E2-related factor-2 |
| **NSAID** | nonsteroidal antiinflammatory drug |
| **NVSR** | National Vital Statistics Reports |
| **Oct4** | octamer-4 |
| **ODN** | oligonucleotide |
| **8-oxodG** | 8-oxodeoxyguanosine |
| **OXPHOS** | oxidative phosphorylation |
| **PAH** | polynuclear aromatic hydrocarbon |
| **PAR** | population attributable risk |
| **PDGF** | platelet-derived growth factor |
| **pH** | potential of hydrogen (a measure of the acidity or alkalinity of a solution) |
| **PPAR α** (agonist) | peroxisome proliferator-activated receptor α (agonist) |
| **PSA** | prostate-specific antigen |
| **RAGE** | renal tumor antigen |
| **Rb** (gene) | retinoblastoma (gene) |
| **RF** | radiofrequency |
| **RNA** | ribonucleic acid |
| **rRNA** | ribosomal ribonucleic acid |
| **S/HMO** | Social Health Maintenance Organization |
| **SCC** | squamous cell carcinoma |
| **SEA** | state economic area |
| **SEER** | Surveillance, Epidemiology, and End Results (Register) |
| **SERM** | selective estrogen receptor modulators |
| **SMR** | standardized mortality ratio |
| **SOA** | Society of Actuaries |
| **SSA** | Social Security Administration |
| **SSA life tables** | life tables for the U. S. Social Security Area |
| **SU.VI.MAX** | Supplementation en Vitamins et Mineraux Antioxydants |
| **Sv** | Sievert |
| **TGF-β** | transforming growth factor beta |
| **TGFβIIr** | transforming growth factor β type II receptor |
| **TIMP3** | TIMP metallopeptidase inhibitor 3 |
| **TNF** | tumor necrosis factor |
| **TRAIL** | TNF-related apoptosis-inducing ligand |
| **tRNA** | transfer RNA |
| **TSCE** | two-stage clonal expansion model |
| **TSH** | thyroid-stimulating hormone |

| | |
|---|---|
| **TURP** | transurethral resection of the prostate |
| **UNSCEAR** | United Nations Scientific Committee on the Effects of Atomic Radiation |
| **UV** (radiation) | ultraviolet (radiation) |
| **UVB** (rays) | ultraviolet B (rays) |
| **VEGF** | vascular endothelial growth factor |
| **WHI** | Women's Health Initiative |

# Glossary

This glossary provides definitions of background terms from interdisciplinary perspectives: i.e., biomedical terms are defined for specialists on mathematical modeling, and, in contrast, statistical/mathematical terms are written in a form understandable for biologists and medical doctors. If the reader would like to get deeper insights to these termins, he/she could use specialized sources (including references from this book related to the mentioned terms).

To define a specific term, we used various sources, such as the Dictionary of Cancer Terms of the NCI (at http://www.cancer.gov/dictionary/), the NCI Drug Dictionary (at http://www.cancer.gov/drugdictionary/), the Glossary of Statistical Terms of NCI (at http://www.cancer.gov/statistics/glossary), the NCI Dictionary of Genetics Terms (at http://www.cancer.gov/cancertopics/genetics terms-alphalist/), the definitions of SEER Cancer Statistics Review 1973–1991 at NCI (at http://www.meds.com/lung/seer.html), Wikipedia (a: http://en.wikipedia.org/wiki/Main_Page), the Medical Dictionary at MedicineNet.com (at http://www.medterms.com/script/main/), the Medical Dictionary at the Free Dictionary by Farlex (at http://medical-dictionary.thefreedictionary.com/), the Health and the Science sections at Answers.com (at http://www.answers.com/topic/), the Merriam–Webster online dictionary (at http://www.merriam-webster.com/dictionary/), the Online Medical Dictionary (at http://cancerweb.ncl.ac.uk/cgi-bin/), The Immune System Glossary (at http://www.lymphomation.org/glossary.html), the Talking Glossary of Genetic Terms at the National Human Genome Research Institute, NIH (at http://www.genome.gov/glossary.cfm), Atlas of Genetics and Cytogenetics in Oncology and Heamotology (at http://atlasgeneticsoncology.org/Genes/), the Dictionary of Genetic Terms: Genomics and Its Impact on Medicine & Society, 2001, Primer, Dept Energy (DOE) (at: http://www.ornl.gov/sci/techresources/Human_Genome/publicat/primer2001/glossary.shtml), as well as our own publications, and publications of other authors to which we refered in the chapters where the term was used.

## A

**α-ketoglutarate** – a salt or ester of α-ketoglutaric acid, which is formed as an intermediate compound in the Krebs cycle – the energy-producing process.

Alpha-ketoglutarate is used by cells during their growth and when healing from injuries. Alpha-ketoglutarate plays an important role as nitrogen transporter in various metabolic pathways by which the amino groups of amino acids are attached to it by transamination and carried to the liver (for utilization in the urea cycle).

**Accessory cancer gene** – genes which relate to cancer indirectly, i.e., by increasing mutation rates of oncogenes and antioncogenes, and/or by facilitating cell proliferation of intermediate cells and/or cancer progression (e.g., xeroderma pigmentosum gene, ataxia telangiectasia gene, etc.).

**Adenocarcinoma (AC)** – a cancer histotype that originates in glandular tissue – the part of an epithelial tissue, which includes skin, glands, and other tissues that lines the organ/body's cavities.

**Adjuvant therapy** – an additional treatment that is usually given after the primary treatment (e.g., surgery) to increase the chance of curing the tumor (e.g., when all detectable tumor has been removed, but when there remains a risk of relapse). It may include radiation therapy, chemotherapy, hormone therapy, or biological therapy (immunotherapy).

**Age-adjusted rate** – a weighted average of the age-specific rates that would have existed if the population under study had been distributed over age the same way as in the "standard" population.

**Age–period–cohort (APC) analysis** – an approach that aims to simultaneously determine the effects of age groups, time periods, and birth cohorts. This methodological problem is important for demography, epidemiology, and statistics.

**Age-related maculopathy/age-related macular degeneration** – a progressive loss of central vision with its onset at age 60 and older, when the macula area of the eye's retina (i.e., the center of the inner lining of the eye) becomes thinner due to atrophy. This is the leading cause of central vision loss in the United States at ages 60+.

**Akaike information criterion (AIC)** – a criterion introduced by Hirotsugu Akaike as a measure of the goodness of fit of an estimated statistical model (i.e., as a tool for selection of an optimal mode from within a set of proposed models). Thus, models may be ranked according to their AIC, with the one having the lowest AIC being the best.

**Alcohol dehydrogenases (ADH)** – a group of seven enzymes participating in the conversion from alcohol to acetaldehyde (a toxic substance). If acetaldehyde would not be further converted to acetate by ALDH, it causes the cells' injuries.

**Aldehyde dehydrogenase (ALDH)** – a group of enzymes that catalyse the oxidation (dehydrogenation) of acetaldehyde to acetate (following the step of alcohol conversion to acetaldehyde), which then could be metabolized to less toxic substances. These enzymes are found in many tissues, but are at the highest concentration in the liver.

**Anchorage-independent growth** – oncogene-transformed cells typically do not require exogenous growth factors or cell anchorage for proliferation, e.g., they do not require a surface on which to flatten out and divide, thus having an "anchorage-independent growth". Anchorage-independence strongly correlates with tumorigenicity and invasiveness (e.g., in a small-cell lung carcinoma).

**Angiogenesis** – a process characterized by the growth of new blood vessels from preexisting vessels. Being a physiological process in growth and development, as well as in wound healing, it can, however, become a major step in tumor progression.

**Antioncogene** (or suppressor gene) – a gene which suppresses the expression of an oncogene (see **Oncogene**) of other gene(s) so that its inactivation or deletion could lead to the carcinogenesis cascade.

**Antisense oligonucleotides (ODN)** – low-molecular weight macromolecules (circa 8000 mW) that bind via hydrogen bonding to the appropriate complementary strand of a macromolecular nucleic acid target (RNA or less often, DNA) thus blocking the macromolecules activity. ODNs may be used either as therapeutic agents or as tools to study gene function (e.g., the antisense technology is an important approach for the sequence-specific knockdown of gene expression, when the sequence, complementary by virtue of Watson–Crick bp hybridization, is applied to a specific mRNA, thus inhibiting its expression and inducing a blockade in the transfer of genetic information from DNA to protein). Recently, the Food and Drug Administration has approved the first antisense oligonucleotide, Vitravene (for cytomegalovirus retinitis).

**APC regulatory pathway** (adenomatous polyposis coli gene) – a tumor suppressor gene whose mutations result in uncontrolled proliferation of intestinal epithelial cells and are associated with a familial adenomatous polyposis (an inherited type of colon cancer characterized by the development of thousands of polyps in the colon) and with the early stages of colorectal carcinogenesis.

**Apoptosis** – programmed cell death that involves a series of biochemical events leading to morphological changes (e.g., changes to the cell membrane such as loss of membrane asymmetry and attachment, cell shrinkage, nuclear fragmentation, chromatin condensation, and chromosomal DNA fragmentation). Regulated processes of disposal of cellular debris which does not damage the organism differentiate apoptosis from **necrosis.**

**Armitage–Doll model of carcinogenesis** – a statistical model of carcinogenesis, proposed in 1954 by Peter Armitage and Richard Doll, which suggested that a sequence of distinct genetic events preceded cancer onset. It assumes that cancer develops from a single cell by going through a series of irreversible, heritable, mutation-like events (stages). It predicts cancer incidence as $I(\text{age}) = \text{c}\cdot\text{age}^{m-1}$, where $m$ is the number of stages of carcinogenesis.

**Ataxia-teleangiectasia syndrome (AT)** – a progressive neurological disorder with the first symptoms occurring during the first years of life, but which sometimes remains undiagnosed until the second decade of life. AT patients are significantly predisposed to cancer, particularly lymphomas and leukemia, which present in 10–20% of patients and may be the first manifested symptoms, as well as an increased sensitivity to ionizing radiation, premature aging, and hypogonadism.

**ATP (adenosine-5′-triphosphate)** – a nucleotide that participates in energy transfer within cells for using this energy in various processes, such as biosynthetic reactions, motility, cell division, and signal transduction pathways. ATP is produced in photosynthesis and cellular respiration.

**Autocrine signaling** – a form of signaling in which a cell secretes a hormone, or chemical messenger (called the autocrine agent) that binds to autocrine receptors on the same cell, leading to changes in the cell (e.g., cytokine IL-1 in monocytes). It differs from **Paracrine signaling** (see below)

**Average years of life lost (AYLL)** – the average years lost to a particular cancer among all persons who died of that cancer. It is calculated by dividing the person-years of life lost (PYLL) for a particular cancer by the number of deaths from that cancer (see also **Person-years of life lost**).

## B

**β-catenin** – a subunit of the cadherin protein complex (a transmembrane protein) that plays an important role in cell adhesion, ensuring that cells within tissues are bound together. Beta-catenin has been implicated as an integral component in the Wnt signaling pathway, which is well known for its role in embryogenesis and cancer (e.g., an increase in $\beta$-catenin production has been noted in patients with basal cell carcinoma).

**BAGE** (B melanoma antigen) – a gene expressed in melanoma. It belongs to the family of genes that contains 15 nearly identical sequences at chromosomes 9, 13, 18, and 21. *BAGE* is also expressed in other cancers, such as breast, gastric, and hepatocellular cancers (see also **MAGE** and **CAGE**).

**Bax** – protein of the Bcl-2 gene family; promotes apoptosis (see also **Bcl-2**).

**Bcl-2 genes** – B-cell leukemia/lymphoma-2 genes responsible for blocking apoptosis in normal cells. This family of genes produces the proteins which can be either pro-apoptotic (e.g., Bax, BAD, Bak, Bok) or antiapoptotic (e.g., Bcl-2 proper, Bcl-xL, Bcl-w). The Bcl-2 genes play the role in carcinogenesis of melanoma, prostate, breast, and lung cancers, and lymphoma. It is involved in resistance to certain cancer treatments. It also participates in pathogenesis of some noncancer diseases, such as schizophrenia and autoimmune diseases.

**Bernoulli life table** – a procedure of analysis of mortality in population developed by Daniel Bernoulli (1700–1782), a Swiss mathematician, for analyzing the smallpox morbidity and mortality based on censored data to demonstrate the efficacy of vaccination. In those life tables, he attempted to estimate the heterogeneity of the exposed population with respect to its susceptibility in death from specific infections.

**Beta-distribution** – a general type of statistical distribution which is related to the gamma distribution, and which provides a good fit for the age-specific incidence of many of adult tumors: $I(t) = (\alpha t)^{k-1}\ (1-\beta t)$. It can be viewed as the superposition, at each age, of two types of cell dynamics: (1) cancer creation, which is most simply modeled with the multistage assumption as caused by somatic mutation and promotion steps from genetic and environmental risks/exposures, and (2) cancer extinction, which is modeled as a cumulative probability linearly increasing to age 100. An apoptosis and cell senescence (e.g., loss of proliferative ability due to senescence) are candidate mechanisms of extinction. If the rate of telomere shortening were uniform over tissue type and time, this mechanism could be modeled as causing cell senescence with the age-dependent probability $\beta t$ and thus could become the $(1-\beta t)$ cancer extinction age factor in the beta-distribution model.

**Biodemography** – a new branch of demography that involved both the complementary biological and demographic determinants studying the interactions between the birth and death processes in individuals, cohorts, and populations. The biological component brings the theoretical background of evolution theory in demography, and the demographic component provides an analytical foundation for many of the principles upon which evolutionary theory rests.

**Biomarker** – a biological, physiological, behavioral, or molecular indicator of a process, disease, or system. Biomarkers are objectively measurable parameters (e.g., enzyme concentration, hormonal concentration, specific gene phenotype, presence of biological substance) characterizing an organism's state of health or disease, or response to a therapeutic intervention.

**BiP (binding immunoglobulin protein)** – a protein of an endoplasmic reticulum (ER) that has a function of a stress sensor, triggering the so-called unfolded protein response, by using ATP/ADP cycling to regulate other proteins' folding.

**Boron neutron capture therapy (BNCT)** – a type of radiation therapy that brings together two components that, when kept separate, have only minor effects on cells. At first, a stable isotope of boron (boron-10) is used – it can be concentrated in tumor cells by attaching it to tumor-seeking compounds. Second, patient receives a beam of low-energy neutrons. Boron-10 collected by tumor disintegrates after capturing a neutron beam, and the high-energy heavy-charged particles destroy cancer cells in close proximity, leaving adjacent normal cells largely unaffected. For example, BNCT is used for the treatment of gliomas and melanomas.

**BRCA1** (BReast CAncer 1) – gene on chromosome 17 that normally helps to suppress cell growth. A person who inherits a mutated BRCA1 gene has a higher risk of getting breast, ovarian, or prostate cancer (see also **BRCA2**).

**BRCA2** (Breast CAncer 2) – gene on chromosome 13 that normally helps to suppress cell growth. A person who inherits a mutated BRCA2 gene has a higher risk of getting breast, ovarian, or prostate cancer (see also **BRCA1**).

**Breakage-fusion-bridge cycle** – damage that happens to a dicentric chromosome during each cell cycle. This process has been described by Barbara McClintock (1902–1992) – the American cytogeneticists and the Nobel Prize Laureate in Physiology or Medicine – as "if chromosomes are broken by various means, the broken ends appear to be adhesive and tend to fuse with one another... As the two centromeres of the terminally united chromosomes pass to opposite poles in this mitotic anaphase, a chromatid bridge is produced". That creates a mechanism for generating genetic heterogeneity in ceratin human cancers (e.g., osteosarcoma, breast cancer).

## C

**CAGE (cancer-associated gene)** – a cancer/testis antigene family, which is associated with tumor progression. Its expression is testis restricted in normal tissues. It is expressed in lung, gastric, cervical, and other cancers (see also **BAGE** and **MAGE**).

**CAGE (cap analysis gene expression)** – method of gene expression analysis and the profiling of transcriptional start points (TSPs). This method can be used to synthesize primers to clone mRNAs, and for single nucleotide polymorphisms (SNPs) analysis in promoter regions.

**Cancer Intervention and Surveillance Modeling Network (CISNET)** – a consortium of investigators whose focus is to use modeling to improve an understanding of the impact of cancer control interventions (e.g., prevention, screening treatment) on population trends in incidence and mortality. This network is also used to project future trends and to help determine optimal cancer-control strategies.

**Cancer precursor** – a lesion that precedes the appearance of invasive cancers (from the Latin "*praecursor*," composed of "*prae-*" = before and "*curro*" = to run). In order of increasing cancer outcome, these lesions are atypia, dysplasia, and carcinoma *in situ*. Cancer precursor can be defined by the following criteria: (1) evidence must exist that it is associated with an increased risk of cancer; (2) when it progresses to cancer, the resulting cancer arises from cells within the precancer; (3) it differs from the normal tissue from which it arises; (4) it differs from the cancer into which it develops, although it has some, but not all, of the molecular and phenotypic properties that characterize the cancer; (5) there is a method by which the precancer/cancer precursor can be diagnosed.

**Cancer progression** – increase in the size of a tumor and/or spread of cancer in the body. This phase involves development of metastatic tumor cells, formation of groups of tumor cells of various sizes, migration of these groups through the circulatory system, their arrest at distant sites and development of metastatic foci at distant sites.

**Carcinogenesis** – a basic multistage process by which normal cells are transformed into cancer cells.

**Carcinoma** – a type of cancer that rises form epithelial cells, such as skin cell or the cells of lining/covering of internal organs. Based on histopathological characteristics, carcinomas are divided into adenocarcinomas (see **Adenocarcinoma**) and squamous cell carcinomas (see **Squamous cell carcinoma**).

**Carcinoma *in situ* (CIS)** – a group of abnormal cells that have a potential to become a cancer and spread into nearby normal tissue (from Latin "*in situ*" = in its place). CIS is considered a cancer precursor that may, if left untreated long enough, transform into a malignant neoplasm.

**Case-control study** – a study that compares two groups of people: those with the disease or condition under study (cases) and a similar group who do not have the disease or condition at the beginning of the study (controls). Medical and lifestyle histories of the people in each group are studied to analyze what factors may be associated with the disease/condition.

**Caspases** – a family of cysteine proteases, which play essential roles in apoptosis, necrosis, and inflammation (e.g., some caspases are required for cytokine maturation). Caspases may be potential therapeutic targets (e.g., for Alzheimer's disease).

**cDNA (complementary DNA)** – a DNA in which the sequence on one strand of the double-stranded structure chemically matches the sequence on the other strand as " key and a lock" that operates at the molecular level (so, the two strands are complementary to one another). The cDNA is used in certain research techniques (e.g., a polymerase chain reaction – PCR). Also cDNA is produced by retroviruses (e.g., HIV) being this way integrated into the host's genome.

**Cell-cycle effectors** – a key component of the cell-cycle progression machinery and cell-cycle checkpoint system. Genetic instability is considered to be a major driving force of malignancy of cancer cells, and some cancer-associated genetic instability is known to be caused by defects in the cell-cycle checkpoint control. Cell-cycle effectors together with sensors and checkpoint signal transducers are three components of the cell checkpoint control system.

**Cell proliferation** – an increase in the number of cells as a result of cell growth and cell division.

**Cervical intraepithelial neoplasia (CIN)** – an abnormal growth of potentially precancerous cells in cervix. The major cause of CIN is human papillomavirus (HPV) infection, which is sexually transmitted. The high-risk HPV types more often associated with cervical carcinoma are 16 and 18. CIN is classified in grades: (1). CIN1 (Grade I) is a mild dysplasia, or abnormal cell growth, corresponds to a low-grade squamous intraepithelial lesion (LGSIL) and confined to the basal 1/3 of the epithelium; (2) CIN2 (Grade II) is a moderate dysplasia confined to the basal 2/3 of the epithelium; and (3) CIN 3 (Grade III) is a severe dysplasia that affects more than 2/3 of the epithelium (it may be referred as cervical carcinoma *in situ*). Both CIN2 and CIN3 correspond to high-grade squamous intraepithelial lesions (HSIL).

**Chaos theory** – a theory that describes the behavior of complex natural systems that are so sensitive that small initial changes can cause unexpected final results, thus giving an impression of randomness. This happens even though these systems are deterministic, meaning that their future dynamics are fully defined by their initial conditions, with no random elements involved. This behavior is known as deterministic chaos. Chaotic behavior is observed in most of natural systems.

**Chaperon** – protein that assist the noncovalent folding/unfolding and the assembly/disassembly of other macromolecular structures (it does not occur in these structures when the latter are performing their normal biological functions). Many chaperones are heat-shock proteins (proteins expressed in response to elevated temperatures or other cellular stresses).

**Chemoprevention** – the strategy of using either natural or synthetic substances or their combination, to block, reverse, or retard carcinogenesis. Chemopreventive strategy could potentially either prevent further DNA damage that might enhance carcinogenesis, or suppress the appearance of the cancer phenotype, which is especially important for individuals at high risk for specific cancers.

**Chemopreventive agents** – agents or drugs that inhibit, reverse, or retard cancer development. Chemopreventive agents can be grouped into two general classes: (1) blocking agents (they prevent carcinogenic compounds from reaching or reacting with critical target sites by preventing the metabolic activation of carcinogens or tumor promoters by enhancing detoxification systems and by trapping reactive carcinogens, e.g., flavonoids, oltipraz, indoles, isothiocyanates) and (2) suppressing agents (they prevent the evolution of the neoplastic process in cells that would otherwise become malignant, e.g., vitamin D and related compounds, NSAIDs).

**Chemotherapy** – a method of disease treatment by chemical compounds that kill certain cells (i.e., microorganisms or cancer cells). This term usually refers to anticancer drugs or to the combination of these drugs into a standardized treatment regimen. In its nononcological use, the term may also refer to antibiotics (i.e., antibacterial chemotherapy). The chemotherapeutic drugs can be divided into alkylating agents, antimetabolites, anthracyclines, plant alkaloids, topoisomerase inhibitors, and other antitumor agents (all of these drugs affect cell division or DNA synthesis and function in some way). Some newer agents don't directly interfere with DNA, such as monoclonal antibodies and the new tyrosine kinase inhibitors (e.g., imatinib mesylate), which directly targets a molecular abnormality in certain types of cancer (chronic myelogenous leukemia, gastrointestinal stromal tumors). Also some drugs may modulate tumor cell behavior without directly attacking those cells (e.g., hormone therapy).

**Chromophobe renal cell carcinoma (ChRCC)** – a distinct subtype of renal cell carcinoma, possibly originating from the collecting renal tubules. This is a rare type of kidney cancer: it represents approximately 5% of all renal tumors.

**Chromosome instability** – a state of continuous formation of novel chromosome mutations, at a rate higher than in normal cells. The increased probability of acquiring chromosomal aberrations affects the DNA repair, replication, or chromosome segregation. Chromosomal instability is a common finding in malignant tumors, however, its precise pathogenetic role remains to be established.

**Chronic obstructive pulmonary disease (COPD)** – a group of diseases characterized by the pathological limitation of airflow in the airway that is not fully reversible (i.e., because of obstruction of airflow, an air is trapped in the lungs). COPD is the umbrella term for chronic bronchitis, emphysema, and a range of other lung disorders (most often due to tobacco smoking, but can be due to other airborne irritants such as solvents, as well as congenital conditions such as alpha-1-antitrypsin deficiency). COPD is the 4th leading cause of death in the United States.

**Clemmensen's hook** – the phenomenon of age-specific effect of breast cancer exponential growth with the subsequent decrease around the age of 50 years before continuing to rise again. It has been observed for both incidence and mortality and is interpreted as the overlapping of two curves corresponding to pre- and postmenopausal breast cancers, respectively.

**Clinical Laboratory Improvement Act (CLIA)** – an act that purposes to set minimum standards for all laboratories to follow and to determine whether laboratories are achieving those standards. The Congress passed the CLIA in 1988 to ensure the accuracy, reliability, and timeliness of patient test results regardless of where in the United States the test was performed.

**Clonal expansion** – a production of daughter cells all arising originally from a single cell (e.g., in a clonal expansion of lymphocytes, all progeny share the same antigen specificity).

**c-myc** – a gene that regulates other genes by coding a protein that binds to the DNA of other genes. When *myc* is mutated, or overexpressed, the protein doesn't bind correctly and thus may cause a cancer.

**Cohort prevalence** – a prevalence of a specific disease or characteristics/factor among individuals of a studied cohort, i.e., who share a demographic, clinical, or other statistical characteristic (e.g., age, study site).

**Colony-stimulating factors (CSFs)** – secreted glycoproteins which bind to receptor proteins on the surfaces of hemopoietic stem cells, thus activating the intracellular signaling pathways, which can cause the cells to proliferate and differentiate into a specific kind of blood cell (e.g., white blood cells). They may be synthesized and administered exogenously. The name "colony-stimulating factors" comes from the method by which they were discovered.

**Complexity theory** – a theory which is closely linked to chaos theory (see **Chaos theory**), attempting to explain the complex phenomenon which is not explainable by traditional (mechanistic) theories. It integrates ideas derived from chaos theory, cognitive psychology, computer science, evolutionary biology, general systems theory, fuzzy logic, information theory, and other related fields to deal with the natural and artificial systems as they are, and not by simplifying them (breaking them down into their constituent parts). It recognizes that complex behavior emerges from a few simple rules, and that all complex systems are networks of many interdependent parts which interact according to those rules.

**Connexins,** or gap junction proteins – a family of structurally related transmembrane proteins that assemble to form the cells gap junctions. Each gap junction is composed of two hemichannels (connexons) each is constructed out of six connexin molecules. Gap junctions are essential for many physiological processes, such as the coordinated depolarization of cardiac muscle, and proper embryonic development. For this reason, mutations in connexin-encoding genes can lead to functional and developmental abnormalities.

**Cost–benefit analysis** – analysis that converts effects into monetary terms (i.e., the cost) and compares them.

**Cost-effectiveness analysis (CEA)** – a type of analysis that converts effects into health terms and describes the costs for some additional health gain (e.g., cost per additional cancer case prevented). In the context of pharmacoeconomics, the cost effectiveness of a therapeutic or preventive intervention is the ratio of the cost of the intervention to a relevant measure of its effect. Cost effectiveness is typically expressed as an incremental cost-effectiveness ratio (ICER), the ratio of change in costs to the change in effects. A special case of CEA is cost–utility analysis, where the effects are measured in terms of years of full health lived, using a measure such as quality-adjusted life years or disability-adjusted life years.

**Covariate** – is a variable that is possibly predictive of the outcome under study. It is used in a regression analysis or in more sophisticated approaches like the quadratic hazard model as explanatory variable (independent variable or predictor). It can be time-or age-dependent.

**Cowden syndrome** – an inherited disorder characterized by multiple tumor-like growths called hamartomas (these small, noncancerous growths are most commonly found on the skin and mucous membranes, such as the lining of the mouth and nose, but can also occur in the intestinal tract and other parts of the body) and an increased risk of certain forms of cancer, such as cancers of the breast, thyroid, and uterus.

**Cox regression**, or Cox proportional-hazards regression – a regression method for modeling survival times which uses the maximum likelihood method (it is also called proportional hazards model because it characterize the ratio of the risks, i.e., a hazard ratio). This model includes the predictor variables (i.e., prognostic variables) and the outcome variable (e.g., whether the patients survived or died during follow-up). Cox regression is described by $\lambda(t, \mathbf{x} = \lambda_0(t)$

$\exp(\beta \mathbf{x}(t))$, where $\lambda_0(t)$ is an unspecified baseline hazard function, $x(t)$ is a vector of covariate values, possibly time-dependent, and $\beta$ is a vector of unknown regression parameters.

**Cross-sectional study** – a study in which a study population is ascertained at one point in time (e.g., when all individuals in the study population were asked about their current disease status and their current or past exposure status).

**Crude death rate (CDR)** – a number of deaths occurring within the year divided by the mid-year population size, expressed per 1000 population. It is "crude" in the sense that all ages are represented in the rate (i.e., it does not take into account the variations in risks of dying at particular ages).

**Crude rate** – a rate based on the frequency of disease/disorder in the entire population, ignoring demographic subdivisions such as age (although rates are usually given separately for males and females because of the different disease patterns by sex). The measure can be useful for summarizing the extent of the disease burden, but its utility for comparing risk is limited because of the different demographic structures in populations worldwide, both geographically (between populations) and within a given population over time.

**Cyclin E** – a member of cyclines family (see definition above) that involved in regulation of cell cycle (i.e., phase $G_1$) participating in the transition from cell cycle phase $G_1$ (period in the cell cycle of the major cell growth during its lifespan, when new organelles are synthesized) to phase S (short for "synthesis phase", when DNA synthesis or replication occurs).

**Cyclin-dependent kinase (CDK)** – an enzyme that belongs to a group of protein kinases originally discovered as being involved in the regulation of the cell cycle. CDKs are also involved in the regulation of transcription and mRNA processing. A CDKs are activated by association with a cyclins (see definition below), forming a cyclin-dependent kinase complex. CDKs are considered a potential target for anticancer medication.

**Cyclins** – a family of proteins involved in the progression of cells through the cell cycle. Cyclins form a complex with its partner cyclin-dependent kinases (see definition above), activating kinases' function. Cyclins are so named because their concentration varies in a cyclical fashion during the cell cycle: i.e., they are produced or degraded as needed in order to drive the cell through the different stages of the cell cycle.

**Cyclins D** – a family of three closely related proteins termed cyclin **D1**, **D2,** and **D3** that are expressed in an overlapping redundant fashion in all proliferating cell types and collectively control the progression of cells through the cell cycle (see definition of **Cyclins** above). Since cyclins D are essential to cell division, they may also be involved in cancer (e.g., amplification or overexpression of **cyclin D1** is important in the development of many cancers including parathyroid adenoma, breast, prostate and colon cancers, lymphoma, and melanoma).

**Cytochrome c** – a small heme protein found loosely associated with the inner membrane of the mitochondrion. It is found in plants, animals, and many unicellular organisms. Cytochrome *c* can catalyze several reactions such as hydroxylation and aromatic oxidation and shows peroxidase activity by oxidation of various electron donors. Cytochrome *c* is also an intermediate in apoptosis.

**Cytochrome P450 (P450, CYP450)** – a very large family of hemoproteins. The most common reaction catalyzed by cytochrome P450 is a monooxygenase reaction, e.g., insertion of one atom of oxygen into an organic substrate, while the other oxygen atom is reduced to water. More than 7700 distinct CYP sequences are known (as of September 2007). The name cytochrome P450 is derived from the fact that these are colored ("chrome") cellular ("cyto") proteins, with a "pigment at 450 nm", so named for the characteristic peak formed by absorbance of light at wavelengths near 450 nm, when the heme iron is reduced and complexed to carbon monoxide. Most CYPs can metabolize multiple substrates, and many can catalyze multiple reactions. The cytochrome P450 is a major system involved in oxidative metabolism, i.e., chemical modification or degradation of chemicals including drugs and endogenous compounds.

**Cytokines** – a category of small signaling proteins and glycoproteins (in the range of 5–30 kD) that are released by cells and have specific effects on cell–cell interaction, communication and behavior of other cells. They are produced by a wide variety of hematopoietic and nonhematopoietic cell types and can have effects on both, nearby cells or throughout the organism, sometimes strongly dependent on the presence of other chemicals. There are several types of cytokines, such as lymphokines, interleukins, and chemokines, which differ by their functions and target of action. Apart from their role in the development and functioning of the immune system, as well as their aberrant modes of secretion in a variety of immunological, inflammatory, and infectious diseases, cytokines are also involved in embryogenesis.

## D

**Death receptor (DR)** – the cell surface receptors that can detect the presence of extracellular death signals (initiated by specific ligands such as Fas ligand, TNF alpha and TRAIL) and, in response, very rapidly activate an apoptosis (e.g., by activating a caspase cascade).

**Degenerative disease** – a disease in which the function or structure of the affected tissues or organs will progressively deteriorate over time (e.g., Alzheimer's disease, Parkinson's disease, osteoarthritis, osteoporosis).

**Depot medroxyprogesterone acetate (DMPA)** – a progestin-only hormonal contraceptive birth control depot injection drug which is injected every 3 months.

**Deterministic function**, or nonstochastic function, – a function which does not contain stochastic parameters or noise.

**Disease marker** – a specific molecular signature of disease, physiological measurement, genotype structural or functional characteristic, metabolic changes, or other determinants that may simplify the diagnostic process, make diagnoses more accurate, distinguish diagnoses before symptoms appear and help track disease progression.

**Dizygotic (DZ) twins** – siblings that are due to independent fertilization, i.e., fertilization of two different ovas by different sperms (when two fertilized eggs are implanted in the uterine wall at the same time, thus forming two zygotes). Also they are known as fraternal twins, or nonidentical twins, or biovular twins.

**D-loop (displacement loop)** – a DNA structure where the two strands of a double-stranded DNA molecule are separated for a stretch and held apart by a third strand of DNA. The third strand has a base sequence which is complementary to one of the main strands and pairs with it, thus displacing the other main strand in the region. Within that region the structure is thus a form of triple-stranded DNA (with a shape resembling a capital "D", where the displaced strand formed the loop of the "D"). D-loops occur in a number of situations, including DNA repair, telomeres, and as a semistable structure in mitochondrial circular DNA molecules.

**DNA (deoxyribonucleic acid)** – a nucleic acid (macromolecule composed of chains of monomeric nucleotides, i.e., organic compounds consisting of a nitrogen base, a sugar, and a phosphate group) that contains the genetic information used as instructions to develop an organism and regulate its functions. Nucleotides inside DNA are arranged in a specific sequence, according to which specific proteins with various functions are produced. DNA is usually double stranded.

**Double-strand breaks (DSBs)** – breaks, in which both strands in the double helix are severed, which are particularly hazardous to the cell because they can lead to genome rearrangements. There are two mechanisms for DSBs repair: nonhomologous end joining (NHEJ) and recombinational repair (also known as template-assisted repair or homologous recombination repair).

## E

**Ectocervix** – a portion of cervix projecting into the vagina (referred to as the *portio vaginalis*), 3 cm long and 2.5 cm wide, on an average.

**Ectopia (ectopy)** – a displacement or malposition of an organ in the body. Most ectopias are congenital, but some may happen later in life and may be physiological (e.g., cervical ectopia during puberty or pregnancy).

**ED50 (median effective dose)** – a dose that produces the desired effect in 50% of population (i.e., study group).

**Effector cells** – cells that perform a specific function as response to a various stimuli. For example, to kill a pathogen or a cancer cell, immune system cells (e.g., natural killers or cytotoxic T-cells) are activated.

**Empirical Bayes approaches** – a class of statistical methods which uses data collected by observations in "real life" situations (in contrast to theory) to evaluate the conditional probability distributions arising from Bayes' theorem. These methods allow one to estimate quantities (probabilities, averages, etc.) of individuals by combining information from measurements on the individual and on the entire population.

**Endocervical canal** – a canal between ectocervix's (see **Ectocervix** above) opening to vagina and the uterine cavity (it is about 7–8 mm at its widest in reproductive-aged women).

**Endopeptidases (or endoproteinases)** – proteolytic peptidases that break peptide bonds of nonterminal amino acids (i.e., within the molecule). They are usually very specific for certain amino acids (e.g., trypsin, chymotrypsin, elastase, pepsin, endopeptidase, etc.).

**Endoplasmic reticulum (ER)** – an extensive network of fine tubules, vesicles, and cisternae interspersed throughout the cytoplasm of the cell, used for the transport of substances inside of a cell (e.g., proteins) to be used in the cell membrane or to be secreted (exocytosed) from the cell (e.g. digestive enzymes). Also ER participates in sequestration of calcium, and production and storage of glycogen, steroids, and other macromolecules.

**Endothelial progenitor cells (EPCs)** – multipotent stem cells, i.e., they have an ability to develop into more than one cell type of the body. They are one of the three types of stem cells to be found in bone marrow.

**Epidermal growth factor (EGF)** – naturally occurring proteins that can stimulate cellular proliferation, and cellular differentiation and growth. The EGF receptor (EGFR) is the cell-surface receptor for members of EGF-family proteins. Mutations leading to EGFR overexpression (known as upregulation) or overactivity are associated with a number of cancers, such as lung cancer and glioblastoma multiforme.

**Estrogen receptor alpha (ER-alpha)** – a nuclear receptor which is activated by the sex hormone estrogen.

**Eukaryotes** – organisms whose cells are organized into complex structures enclosed within membranes (e.g., animals, plants, fungi, and protists). The nucleus (from the Greek $\epsilon\upsilon$, meaning "good/true", and $\kappa\upsilon\rho\upsilon o\nu$, "nut") is the cell structure that differs eukaryotic from prokaryotic cells. Also eukaryotic cells contain other organelles, such as mitochondria, chloroplasts, and Golgi bodies.

**Exocervix** – the lower third of the cervix (see also **Ectocervix**)

**Extracellular matrix (ECM)** – the extracellular part of tissue that provides a structural support to the cell, segregates tissues from one another, and participates in regulation of intercellular communications. ECM plays an important role in growth, wound healing, and fibrosis. Cancer metastasis often involves the destruction of ECM.

**Extra years of life (EYL)** – number of years that life expectancy is increased as a result of different interventions.

# F

***Fas*** – a gene that encodes one of several proteins important to apoptosis. The *Fas* gene is a member of the tumor necrosis factor (TNF) receptor superfamily (it has also been known as APO-1, APT1, and CD95). Defective *Fas*-mediated apoptosis may lead to oncogenesis, as well as to developing of a drug resistance in existing tumors. Germ line mutation of *Fas* is associated with autoimmune lymphoproliferative syndrome (ALPS), which is a childhood disorder of apoptosis.

**Fecal occult blood test (FOBT)** – a test used to detect the presence of microscopic blood in stool ("occult" means that the blood is hidden from view). This is a screening test for colorectal cancer that is usually based on the detection of peroxidase activity in stool. Hemoccult II is the most popular test kit that uses a guaiac-impregnated paper. Other tests (such as Hem- Select and FlexSure) directly detect human hemoglobin in the stool by using antihuman hemoglobin antibodies.

**Fibroblast growth factor (FGF)** – a group of growth factors that act on the fibroblasts (i.e., "building blocks" of fibrous tissue) in many organs, such as blood vessels, skin, kidney, heart, bone, etc. EGF participates in wound healing of normal tissues, and in tumor development, regulating the formation of new blood vessels (angiogenesis processs). Also FGF is a critical component of human embryonic stem cell culture, where it is necessary for the cells to remain in an undifferentiated state.

**Fokker–Plank equation** – an equation that describes the time evolution of the probability density function of the health state that represented by a set of covariates. It is named after Adriaan Fokker and Max Planck (it is also known as the Kolmogorov forward equation).

**Future Elderly Model (FEM)** – a demographic-economic model framework of health spending projections. It enables the user to answer "what-if" questions about the effects of changes in health status and disease treatment on future health care costs. It includes a multidimensional characterization of health status allowing the user to include a richer set of demographic parameters and comorbid conditions. This model was designed as a tool to help policy analysts and private firms better understand future trends in health, health spending, medical technology, and longevity.

**Fuzzy set logic** – a form of logic derived from fuzzy set theory (see **Fuzzy set theory**) for cases when approximation is more important than precision. The degree of truth of a statement can range between 0 and 1 and is not constrained to the two truth values {true, false} as in classic predicate logic. This approach can be used in various studies, from control theory to artificial intelligence.

**Fuzzy set theory** – a theory, that is a generalization of a classical set theory (in which the membership of elements in a set is described by a bivalent probability – an element either belongs or does not belong to the set) and permits a gradual assessment of the membership of elements in a set. Fuzzy set theory was first introduced in 1965 by Lotfi Zadeh at the University of California, Berkeley and fuzzy logic was derived from it for use in applications (see **Fuzzy set logic**)

## G

**G2-M checkpoint** – a control checkpoint that determines whether the cell is ready to proceed to enter mitosis and divide (checkpoint is a surveillance system responsible for monitoring the proper completion of an event within a cell). It is located at the end of the $G_2$ phase of the cell cycle (i.e., the phase between DNA synthesis and mitosis, when cell continues growing and producing new proteins) and triggers the start of the M phase (i.e., mitosis). If this checkpoint is passed, the cell initiates the beginning of mitosis. When DNA was damaged prior to mitosis, then, to prevent a transmission of this damage to daughter cells, the cell cycle is arrested.

**Gamma distribution** – in probability theory and statistics, a gamma distribution is a two-parameter continuous probability distribution. It has a scale parameter $\theta$ and a shape parameter $k$. This is a general type of statistical distribution that is related to the beta-distribution and arises naturally in processes for which the waiting times between Poisson distributed events are relevant. It is not often used to model life data by itself, however, its ability to behave like other more commonly used life distributions may be used to determine which of those life distributions should be used to model a particular set of data.

**Gap junction** – a junction between cells that allows the different molecules and ions, mostly small intracellular signaling molecules (intracellular mediators), to pass freely between cells. One gap junction is composed of two connexons (or hemichannels). Gap junctions are expressed in almost all tissues of the body (excluding sperm and erythrocytes). Mutations in gap junction gene are associated with some human genetic disorders (e.g., celiac disease).

**Gaussian distribution/ standard normal distribution** – a common probability distribution displayed by population data, named after Carl Friedrich Gauss. It may be used in many fields (e.g., in natural and behavioral sciences). It is characterized by two parameters (i.e., *location* and *scale*), such as the mean ("average") $\mu$, and variance (standard deviation squared) $\sigma^2$. If the values of the distribution are plotted on a graph's horizontal axis and their frequency on the vertical axis, then it is displayed as symmetric bell-shaped, with the central value or mean representing the most frequently occurring value.

**Gene amplification** – a process of making multiple copies of a gene. It plays a role in carcinogenesis, when cancer cell amplifies, or copies, DNA segments as a result of cell signals and sometimes environmental events.

**Gene chip technology** – a development of cDNA (see **cDNA**) microarrays from a large number of genes. Used to monitor and measure the changes in gene expression (see **Gene expression**) for each gene represented on the chip.

**Gene expression** – a process by which coded by genes information is converted into the structures present and operating in the cell. Expressed genes include those that are transcribed into messenger RNA (see m**RNA**) and then translated into protein, and those that are transcribed into m**RNA** but not translated into protein (e.g., transfer and ribosomal RNAs).

**Gene mapping** – the mapping of genes to specific locations on chromosomes, what is important to understand their association with diseases. There are two types of gene mapping: (1) genetic mapping that uses a linkage analysis to determine the relative position between two genes on a chromosome (two loci, i.e., locations of genes, are linked when they are inherited together), and (2) physical mapping that determines the absolute position of a gene on a chromosome.

**Genetic code** – a sequence of nucleotides coded in triplets (codons) along the messenger RNA (mRNA), which determines the sequence of amino acids in protein synthesis. The genetic code can be used to predict the amino acid sequence.

**Genetic polymorphism** – a condition in which a genetic character occurs in more than one form, resulting in the coexistence of more than one morphological type in a given population, such as an occurrence of differences in DNA sequences inside the population.

**Genome** – all the genetic material in the chromosomes and extrachromosomal genes of a particular organism, with its size given as its total number of base pairs (i.e., a complete DNA component of an organism).

**Genotype** – the genetic constitution of an organism (the complete set of genes both dominant and recessive) as distinguished from its physical appearance (its phenotype).

**Glioma** – type of primary central nervous system (CNS) tumor that arises from glial cells. The most common site of gliomas occurrence is the brain, but gliomas can also affect the spinal cord or any other part of the CNS, such as the optic nerves.

**Glutathione-S-transferase (GST)** – a family of enzymes regulating cytosolic, mitochondrial, and microsomal proteins that are capable of multiple reactions with multiple substrates, both endogenous and xenobiotic. GST participates in the phase II of biotransformation of xenobiotics, through which various drugs, poisons, and other compounds are modified to be able to be excreted from the body: GST conjugates these compounds to make them dissolvable and thus excretable (out of the body).

**Golgi (Goldgi) apparatus** – an organelle found in most eukaryotic cells, which was identified in 1898 by the Italian physician Camillo Golgi. Its main function is to process and package the macromolecules such as proteins and lipids, which are synthesized by the cell (that is particularly important in the processing of proteins for their secretion).

**Gompertz law of mortality** – a law proposed by Benjamin Gompertz in 1825 that assumes the exponential increase in death rates with age.

**Goodness of fit** – a characteristic of a statistical model that describes how well a model fits a set of observations. It is used in statistical hypothesis testing.

**Grade of Membership (GoM) model** – a model for analyzing high dimensional discrete response data by estimating (using maximum likelihood principles) two types of parameters: (1) the first describes the probability that a person who is exactly like one of the *K* analytically defined types has a particular response on a given variable, and (2) the second describes each individual's degree of membership (in each of the *K* types).

**Granulocyte macrophage colony stimulating factor (GMCSF)** – a protein secreted by macrophages, T cells, mast cells, endothelial cells, and fibroblasts. It stimulates stem cells to produce granulocytes (neutrophils, eosinophils, and basophils) and monocytes.

## H

**Heat shock proteins (HSPs)** – a group of proteins whose expression increases when the cells are exposed to elevated temperatures or other stress. The HSPs are named according to their molecular weights (e.g., Hsp60, Hsp70, and Hsp90 refer to families of heat shock proteins on the order of 60, 70, and 90 kDa, respectively).

**Hematopoietic stem cell therapy** – a method of treating patients with cancers (as well as with other disease of blood and immune systems). A hematopoietic stem cell is a cell isolated from the blood or bone marrow that can renew itself, can differentiate to a variety of specialized cells, can mobilize out of the bone marrow into circulating blood, and can undergo apoptosis. In animal experiments, hematopoietic stem cells are able to form muscle, blood vessel, and bone cells. Applied to human cells, they potentially may replace a wide array of cells in human body.

**Her-2/neu** – a protein associated with higher breast cancer aggressiveness. It is a member of the ErbB protein family (known also as the epidermal growth factor receptor family). Because of its prognostic role as well as its ability to predict response to treatment with trastuzumab, breast tumors are routinely checked for overexpression of HER2/neu. Overexpression also occurs in other cancers, such as ovarian or stomach.

***Herpes simplex* virus thymidine kinase (HSV-TK)** – a key enzyme in the pyrimidine (one of the two classes of bases in DNA and RNA) salvage pathway with a very broad substrate specificity. It is a target of antiviral medication. Also it can be used in gene therapy of human cancers.

**Heterocyclic amines (HCAs)** – organic compounds with a ring structure containing nitrogen atoms in addition to carbon. These carcinogenic chemicals formed from the cooking of muscle meats of beef, pork, fowl, and fish. The carcinogenes are formed when amino acids (i.e., the building blocks of proteins) and creatine (i.e., a chemical waste molecule that is generated from muscle metabolism) react at high cooking temperatures. There are at least 17 different HCAs that may increase cancer risk in human.

**Histone** – a chief protein component of chromatin. It acts as a spool around which the DNA winds. It participates in gene regulation. Histones are necessary for DNA compaction to make the genome fit inside the cell nuclei (i.e., compacted molecule is 30,000 times shorter than an unpacked molecule).

**hMLH1 (human mutL homolog 1)** – a protein involved in the mismatch repair process after DNA replication. Mutation of hMLH1 gene is associated with hereditary nonpolyposis colorectal cancer (HNPCC), as well it predisposes to cancers of endometrium, ovary, urinary tract, stomach, small bowel, biliary tract, and brain.

**hMSH2 (human mutS homolog 2)** – a protein involved in the mismatch repair process after DNA replication. Mutation in hMSH2 gene is associated with hereditary nonpolyposis colorectal cancer (HNPCC). It is also associated with increased risk of endometrial, urinary, gastric, intestinal, biliary, and brain cancers.

**Hodgkin lymphoma** – a type of lymphoid tissue tumor (described by Thomas Hodgkin in 1832) characterized by involvement of lymph nodes and development of systemic symptoms with advanced disease. It occurs most frequently at age 15–35 and at age 55+. Hodgkin's lymphoma was one of the first cancers to be cured by radiation and later one of the first to be cured by combination chemotherapy.

**Hormone replacement therapy (HRT)** – a treatment for after-surgery menopausal, perimenopausal and postmenopausal women that helps to decrease a discomfort (e.g., relief from menopausal symptoms, such as hot flashes, irregular menstruation, fat redistribution, etc.) and associated health problems (e.g., risk of osteopenia that leads to osteoporosis) caused by imbalance in estrogen and progesterone levels. This treatment artificially boosts hormone levels by using estrogens, progesterone or progestins (and sometimes testosterone).

**hTERT (human telomerase reverse transcriptase)** – an enzyme involved in DNA replication which is a catalytic subunit of telomerase (see also **Telomerase**).

**Hypermethylation** – a process of increasing the epigenetic methylation that changes genes' activity, thus causing oncogenes to produce proteins that cause cells malignization (see also **Methylation**).

**Hyperthermia** (also known as thermal therapy, or thermotherapy) – a type of cancer treatment when body tissue is exposed to high temperatures (up to

113°F) thus damaging or killing cancer cells (usually with minimal injury to surrounded normal tissues). Local hyperthermia is applied to tumor using various techniques such as external, intraluminal/endocavitary, and interstitial hyperthermia. Regional hyperthermia is applied to a large areas of tissue, such as a body cavity, organ, or limb (e.g., deep-tissue, regional perfusion, hyperthermic peritoneal perfusion). Whole body hyperthermia is used to treat metastatic cancer that has spread throughout the body.

**Hypomethylation** – a decrease in the epigenetic methylation that changes genes activity, thus quieting genes that under the normal conditions suppress cancer (tumor suppressor genes) (see also **Methylation**).

**Hypoxia inducible factor (HIF)** – a transcription factor responding to changes in available oxygen in the cellular environment, particularly induced by hypoxia (i.e., decrease in oxygen). HIFs are vital to organism development (e.g., mutation in mammals HIF-1 gene results in perinatal death).

## I

**IAPs (inhibitors of apoptosis proteins)** – a family of functionally and structurally related proteins which serve as endogenous inhibitors of programmed cell death, i.e., apoptosis. The human IAP family consists of at least six members.

**ICAM1 (intercellular adhesion molecule)** – a type of molecule continuously presents in low concentrations in the membranes of leukocytes and endothelial cells. It can be induced by interleukin-1 (IL-1) and tumor necrosis factor alpha (TNFα) and is expressed by the vascular endothelium, macrophages, and lymphocytes.

**IGF2 (insulin-like growth factor 2)** – one of the insulin-like growth factors (see **IGFs**) which promote a cells growth during gestation (i.e., period from conception to birth). It is sometimes produced in nonislet cell tumors (usually of mesenchymal or epithelial cell types) causing hypoglycemia.

**Immunomodulator** – a drug used for its effect on the immune system. Based on their effects they are divided on immunosuppressants and immunostimulators.

**Incidence** – the number of newly diagnosed every year cases of disease. **Incidence rate** is the number of new disease cases diagnosed annually per 100,000 of population. Crude incidence rate is the total incidence rate at a given time, specific incidence rate is the incidence rate divided by categories (age specific, disease specific, and mortality), and adjusted incidence rate is the rate adjusted to a standard population based on characteristic influencing the outcome of analysis (e.g., age-adjusted).

**Insulin-like growth factors (IGFs)** – polypeptides which share the structural similarity to insulin, and by which cells communicate with their environment. This is a complex system that plays an important role in aging, cancer, and diabetes and consists of two cell-surface receptors (IGF1R and IGF2R), two ligands (IGF1 and IGF2), six high-affinity IGF binding proteins (IGFBP 1-6), and associated IGFBP degrading enzymes (proteases).

**Interferons (IFNs)** – natural proteins produced by the cells of the immune system in response to challenges by foreign agents such as viruses, parasites, and tumor cells. IFNs belong to the large class of glycoproteins – cytokines (see **Cytokines**). IFNs inhibit viral replication within host cells, activate natural killer cells and macrophages, increase antigen presentation to lymphocytes, and induce the resistance of host cells to viral infection.

**Interleukins (ILs)** – a group of cytokines (see **Cytokines**) first seen to be expressed by white blood cells (leukocytes, hence the *-leukin*) with "communicative" purpose (*inter-*). ILs are produced by many types of cells and play an important role in regulation of the immune system function, including participation in carcinogenesis.

**Intraoperative radiotherapy (IORT)** – a form of external irradiation given during the surgery to treat localized cancers that cannot be completely removed or that have a high risk of recurring in nearby tissues (e.g., for treating of thyroid, colorectal, gynecological, intestinal, and pancreatic cancers, as well as it may be used for treating of certain types of brain tumors and pelvic sarcomas). A high-energy dose of radiation is used aiming directly at the tumor site during surgery (nearby healthy tissue is protected with special shields).

**Invasive cancer** – a cancer that spreads beyond the layer of tissue in which it originally developed and grows into surrounding healthy tissues (also called infiltrating cancer).

***In vitro* study** – a biological study which is carried out in isolation from a living organism.

***In vivo* study** – a biological study which takes places within a living biological organism.

**Isothiocyanates (ITCs)** – a group of naturally occurring compounds, such as thioglucoside conjugates (glucosinolates), in certain plants and cruciferous vegetables (e.g., Brussels sprouts, broccoli, cabbage, horseradish, radish, turnip, etc.). In experiment, they inhibit cancer development, thus may be used as chemopreventive agents. One of the main mechanisms by which ITCs inhibit carcinogenesis is through the inhibition of cytochrome P450 enzymes (see **Cytochrome P450**) that participate in production of carcinogenic compounds (e.g., benzo[a]pyrene and other polycyclic aromatic hydrocarbons).

# K

**Kalman filter model (state space model)** – a useful technique for estimating of the state of a dynamic system from a series of incomplete and noisy measurements by (1) using indirect measurements of the state variables and (2) by using the covariance information of both the state variables and the indirect measurements. This method is used to describe how measurements of a particular aspect of a system are correlated to the actual state of the system (as well as how the various measurements are correlated to one another). So, Kalman filtering is a technique for using the correlation information to derive better estimates of unknown quantities.

**Kolmogorov equation** – see **Fokker–Planck equation.**

# L

**Latent/latency period** – an interval between the stimulus (exposure to factor) and the response. In medical studies, this is an interval between an exposure to infectious agent or a carcinogen and the consequent clinical manifestation of disease. In carcinogenesis modeling, as it has been defined by Armitage–Doll, this is an interval between an exposure to carcinogen and the appearance of the first cancer cell.

**Latent structure analysis** – a statistical method for finding unobserved structure which is responsible for generating observed categorical data. Examples of specific models include latent class model, Rasch (or latent trait) model, Grade of Membership, and linear latent structure analysis.

**Leiomyoma** – a benign smooth muscle neoplasm which is not premalignant. It occurs in any organ, but more commonly in uterus, intestine, and esophagus.

**Levy type distribution** – one of the few distributions that are stable and that have probability density functions which are analytically expressible (named after the French mathematician Paul Pierre Lévy). It is used as a model of heterogeneity when studying nonhomogeneous data (e.g., population).

**Li-Fraumeni syndrome** – a rare autosomal dominant hereditary disorder named after Frederick Pei Li and Joseph F. Fraumeni, Jr., the American physicians, who first recognized and described it. Patients with this syndrome have an increased susceptibility to cancer due to a mutation in the p53 tumor suppressor gene, which normally helps control cell growth.

**Linear Latent Structure (LLS) analysis** – a mixture distribution model constrained to satisfy the local independence assumption. A feature of LLS analysis is the existence of a high-performance numerical algorithm, which reduces parameter estimation to a sequence of linear algebra problems.

**Linear no-threshold function** – a linear function of dose response that assumes event risk has no threshold for dose of exposure associated with this event (e.g., assuming that cancer risk is proportional to the dose of radiation even at low doses).

**Liposomal therapy** – a new strategy for anticancer drug delivery by using liposomes (vesicles composed of a phospholipid bilayer surrounding an aqueous milieu that can be used as transporters of various substances into the cell). Liposomal drugs accumulate within neoplastic tissues, thus increasing concentration of cytotoxic agents in tumor, whereas healthy tissues are spared from toxicity. Liposomal therapy is used for treatment of metastatic breast and ovarian cancers, and Kaposi's sarcoma.

**Longitudinal study** – research that involves repeated observations of the same items over a long period of time, often many decades. Types of longitudinal studies include cohort studies and panel studies. Cohort study samples a cohort (group experiencing some event – typically birth – in a selected time period) and studies it at intervals through time. Panel study uses a cross-section sample surveying it at specific intervals. A retrospective study is a longitudinal study that looks back in time.

**Loop electrosurgical excision procedure (LEEP)** – one of the most commonly used approaches to treat a high-grade cervical dysplasia (discovered by colposcopic examination). Various shapes and sizes of loop can be used depending on the size and orientation of the lesion. The cervical transformation zone and lesion are excised to an adequate depth, which in most cases is at least 8 mm, and extending 4–5 mm beyond the lesion. A second pass with a more narrow loop can also be done to obtain an endocervical specimen for further histologic evaluation.

**Lotka–Volterra equations** – also known as the *predator–prey equations*, are a pair of first order, nonlinear, differential equations frequently used to describe the dynamics of biological systems in which two species interact, i.e., one is a predator and other is a prey. This method was proposed independently, by the U.S. mathematician, physical chemist, and statistician Alfred J. Lotka in 1925, and Italian mathematician and physicist Vito Volterra in 1926.

**Low-density lipoprotein (LDL)** – a lipoprotein that transports cholesterol and triglycerides from liver to peripheral tissues and regulates cholesterol synthesis. LDLs can be retained in arteries by arterial proteoglycans and start the formation of plaques. Increased levels of LDLs are associated with atherosclerosis and its complications (e.g., heart attack, stroke, and peripheral vascular disease).

**Lumpectomy** – a surgical procedure designed to remove a lesion (benign or malignant) and a small amount of normal tissue around it from an affected man or woman's breast. This is a relatively noninvasive procedure compared to a mastectomy, with the preservation of the essential anatomy of the breast.

# M

**MAb CO17-1A** (monoclonal antibody against the 17-1A antigen) – an antibody to the cell-surface glycoprotein 17-1A expressed on epithelial tissues and on various carcinomas. It has been used in developing of anticancer drug (i.e., edrecolomab) for treating of colon and breast carcinomas with metastasis to the lymph nodes (however, it did not demonstrate any benefit during a phase III study, compared to conventional chemotherapeutic agents).

**MAGE** (melanoma antigen family genes) – these genes encode proteins with 50–80% sequence identity to each other. They have been implicated in some hereditary disorders, such as dyskeratosis congenita.

**Markov chain Monte Carlo method (MCMC)** – a class of algorithms for sampling from probability distributions based on constructing a Markov chain. It originated in physics as a tool for exploring equilibrium distributions of interacting molecules. Monte Carlo methods tend to be used when it is infeasible or impossible to compute an exact result with a deterministic algorithm. The term "Monte Carlo" was coined in the 1940s by physicists working on nuclear weapon projects in the Los Alamos National Laboratory.

**Markov process/chains** – a sequence of random variables in which the distribution of each element depends only on the value of the previous one.

**Mastectomy** – the surgical removal of one or both breasts, partially or completely, more often done to treat breast cancer. Total mastectomy – removal of breast tissue and nipple. Modified radical mastectomy – removal of the breast, most of the lymph nodes under the arm and often the lining over the chest muscles. Radical mastectomy (removal of the breast, lymph nodes and chest muscles) is no longer common. (See also **Lumpectomy**).

**Meta-analysis** – an analysis that combines the results of several studies addressing a set of related research hypotheses. The first meta-analysis was performed by Karl Pearson in 1904, in an attempt to overcome the problem of reduced statistical power in studies with small sample sizes. This approach shifts an emphasis from a single to multiple studies, stressing the practical importance of the size effect instead of the statistical significance of individual studies. The results of a meta-analysis are often shown in forest plots.

**Metalloproteinases** – a family of enzymes from the group of proteinases, named by the nature of the most prominent functional group in their active site. They play an important role in embryonic development, morphogenesis, reproduction, tissue remodeling, arthritis, cancer, and cardiovascular disease.

**Methotrexate** – an antimetabolite and antifolate drug used in cancer and autoimmune diseases treatment. It acts by inhibiting the metabolism of folic acid. It was originally used as part of combination chemotherapy regimens. More recently it has been used for treatment of some autoimmune diseases (e.g., ankylosing spondylitis, Crohn's disease, psoriasis, psoriatic arthritis, and rheumatoid arthritis).

**Methylation** – the attachment or substitution of a methyl group on various substrates. In biochemistry, methylation more specifically refers to the replacement of a hydrogen atom with the methyl group. Methylation is involved in modification of heavy metals, regulation of gene expression and protein function.

**Methylenetetrahydrofolate reductase (MTHFR)** – an enzyme of the cells' cytoplasm which is a key regulatory enzyme in the metabolism of folate (i.e., folic acid that is a key factor in the synthesis of DNA and RNA). Its polymorphism is associated with neural tube defects in offspring, arterial and venous thrombosis, cardiovascular disease, and a decreased risk for certain leukemias and colon cancer (only when the dietary intake of folate is high).

**Metropolitan Relative Weight (MRW) criteria** – version of measuring of overweight/obesity. It calculated as the percentage of desirable weight (it correlates with BMI).

**MGMT** ($O^6$-methylguanine–DNA methyltransferase) – a human DNA-repair gene which plays a significant role in carcinogenesis (e.g., in patients with glioblastoma multiforme, a type of brain tumor, the methylation state of the *MGMT* gene determines the response to treatment with temozolomide).

**Microarray (DNA microarray, chip)** – a technology used in molecular biology and in medicine that consists of series of arrays of thousands of microscopic spots of DNA oligonucleotides, each containing picomoles of a specific DNA sequence. In standard microarrays, the probes are bound to a solid surface by covalent attachment to a chemical matrix (via epoxy-silane, amino-silane, lysine, polyacrylamide or other substrates). The solid surface can be either a glass or a silicon chip, in which case they are commonly known as *gene chip* (or *Affy Chip* when an Affymetrix chip is used).

**Microsatellite instability** – a condition manifested by DNA damage due to defects in the normal DNA repair process. Sections of DNA called microsatellites (which consist of a sequence of repeating units of 1–6 bp in length) become unstable and can shorten or lengthen. This is a key factor in carcinogenesis of colorectal, endometrial, ovarian, gastric cancers, and others.

**Microsimulation** (a.k.a. microanalytic simulation) – a research area in applied econometrics that simulates the behavior of individuals or other objects over time. Microsimulation can be either dynamic or static. If it is dynamic the behavior of people changes over time, whereas in the static a constant behavior is assumed.

**MINT** (MSX2-interacting nuclear target protein) – a large 400-kDa nuclear matrix protein first identified by expression cloning using radio-labeled MSX2 gene (it regulates production of a protein that is necessary for proper development throughout the body). MINT is a nuclear DNA- and RNA-binding protein highly expressed in central nervous system, lymphoid tissue, cardiac tissue, and osteoblasts.

**Mitochondrial DNA (mtDNA)** – the DNA located in organelles called mitochondrias (structures located in the cell's cytoplasm and responsible for energy production in cells). Due to higher mutation rate of mtDNA than of nuclear DNA (nDNA), mtDNA is a powerful tool for tracking ancestry through females (matrilineage). It has been used to track the ancestry of many species back hundreds of generations. Also human mtDNA can be used to identify individuals.

**Mitotic index** – a measure for the proliferation status of a cell population. It is defined as the ratio between the number of cells in mitosis and the total number of cells. This index can be calculated from a slide, as well as by light microscopy, as the number of cells containing visible chromosomes divided by the total number of cells in the field of view.

**Mixing distribution** (in latent structure analysis) – a distribution of the latent variables that contains information regarding the phenomenon under study. Specific models of latent structure analysis vary by assumptions, regarding properties of the mixing distribution.

**Monoclonal antibody therapy** – a use of monoclonal antibodies (the highly specific antibodies produced in large quantity by the clones of a single hybrid cell formed in the laboratory by the fusion of a B cell with a tumor cell) to specifically targeted cells. The main objective of this therapy is stimulating the patient's immune system to attack the malignant tumor cells, as well as the prevention of tumor growth by blocking specific cell receptors. Variations exist within this treatment, e.g., radioimmunotherapy, when a radioactive material is delivered to specific cells by using a protein that binds to the surface of the target cells.

**Monoclonal cells** – a group of cells derived from a single ancestral cell by repeated cellular replication, thus forming a "clone" (e.g., monoclonal population of tumor cells).

**Monozygotic (MZ) twins** – identical twins occur when a single egg is fertilized to form one zygote (the cell that results from fertilization – i.e., the union of a spermatozoon and an ovum) which then divides into two separate embryos.

**Morbidity** – a total number of disease cases in a population at a particular point in time (from Latin "*morbidus*" = sick, unhealthy) (compare to incidence of a disease – i.e., the number of *new* cases in a particular population during a particular time interval).

**Mortality/mortality rate** – the number of deaths caused by a disease during a period of time (usually, per year). Mortality rate is the number of deaths "per standard unit of population" during the time period (usually per 100,000 of population per year). Mortality rates can be calculated as crude, specific, or adjusted (see also **Incidence/incidence rate**).

**mRNA (messenger RNA)** – a type of RNA that contains information (a complimentary DNA copy) to specify the amino acid sequence of proteins and carries this exact nucleoside sequence ("message") of DNA to the protein-producing units in the cell called ribosomes (i.e., proteins are synthesized in accordance to mRNA's "message"). The sequence of mRNA that encodes a protein is oriented in only one direction, which is known as the "sense" orientation.

**MUC-1** (mucin 1, cell surface associated) – a human gene, a member of the mucin family, which encodes a membrane bound glycosylated phosphoprotein. This protein has a protective function by binding to pathogens. It also participates in a cell signaling pathways. Overexpression, aberrant intracellular localization, and changes in glycosylation of this protein are associated with carcinomas.

**Multidrug resistance** – a condition enabling a disease-causing organism (bacteria, virus) or cancer cells to resist distinct drugs or chemicals of a wide variety of structure and function targeted at eradicating the organism/cancer cell.

**Myeloid/lymphoid or mixed-lineage leukemia (MLL) gene** – a regulator of genes transcription that belongs to the group of histone-modifying enzymes and is involved in the epigenetic maintenance of transcriptional memory and the pathogenesis of human leukemias.

# N

**N-acetyltransferase (NAT)** – a conjugating enzyme that catalyzes the transfer of acetyl groups from acetyl-CoA to arylamines. It is expressed from the genes NAT1 and NAT2. It plays an important role in metabolic activation and detoxification of certain human precarcinogens, such as homo- and heterocyclic arylamines. Polymorphisms of NAT1 and NAT2 genes are associated with increased risks of head and neck, lung, breast, larynx, urinary bladder, and colorectal cancers.

**National Long Term Care Survey (NLTCS)** – a longitudinal survey designed to study changes in the health and functional status of older Americans (aged 65+). It also tracks health expenditures, Medicare service use, and the availability of personal, family, and community resources for caregiving. The survey began in 1982, and follow-up surveys were conducted in 1984, 1989, 1994, 1999, and 2004.

**Necrosis** – (from the Greek "nεkρός" = dead) – an accidental death of cells, that begins with cell swelling, chromatin digestion, and disruption of the plasma and organelle membranes, and at later stages is characterized by extensive DNA hydrolysis, vacuolation of the endoplasmic reticulum, organelle breakdown, and cell lysis. In contrast to apoptosis, cleanup of cell debris due to necrosis by phagocytes of the immune system is difficult. It is hard for the immune system to locate and recycle dead cells which have died through necrosis than if the cell had undergone apoptosis. There are many causes of necrosis, including injury, infection, cancer, infarction, poisons, and inflammation.

**NF-kB (nuclear factor kappa B)** – a transcription factor involved in cellular responses to stimuli such as stress, cytokines, free radicals, ultraviolet irradiation, bacterial or viral antigens, and others. It plays a key role in regulating the immune response to infection. NF-*k*B family members share a structural homology with the retroviral oncoprotein v-Rel, thus being classified as NF-*k*B/Rel proteins.

**Non-Hodgkin lymphoma** – a hematological malignancy developing in organs associated with the lymphatic system (e.g., spleen, lymph nodes, tonsils). The etiology of most lymphomas is not known. Some types of lymphomas are associated with viruses: i.e., Burkitt's lymphoma, extranodal NK/T cell lymphoma, Hodgkin's disease and most of AIDS-related lymphomas are associated with *Epstein–Barr virus*; an adult T-cell lymphoma is caused by the *HTLV-1 virus*; gastric lymphoma is associated with *Helicobacter pylori* infection.

**Nonsteroidal antiinflammatory drugs (NSAIDs)** – a group of drugs with analgesic, antipyretic and, in higher doses, antiinflammatory effects. The term "nonsteroidal" is used to distinguish these drugs from steroids, which (among a broad range of other effects) have a similar eicosanoid-depressing, antiinflammatory action. Most NSAIDs act as nonselective inhibitors of the enzyme cyclooxygenase, inhibiting both the cyclooxygenase-1 (COX-1) and the cyclooxygenase-2 (COX-2) isoenzymes. Cyclooxygenase catalyzes the formation of prostaglandins and thromboxane from arachidonic acid. Prostaglandins act as messenger molecules in the process of inflammation.

**Nuclear DNA (nDNA)** – a DNA contained within a nucleus of eukaryotic organisms. In most cases it encodes more of the genome than the mitochondrial DNA (mtDNA). Nuclear DNA is the most common DNA used in forensic examinations.

## O

**Oncocytoma** – kidney, salivary or endocrine gland tumor made up of oncocytes, large cells with small irregular nuclei and dense acidophilic granules due to the presence of abundant mitochondria. Renal oncocytoma represents 5–15% of surgically resected neoplasms of kidney. This tumor is usually asymptomatic and discovered incidentally on a tomography or ultrasound examination.

**Oncogene** – mutated and/or overexpressed version of normal gene that affect cell growth, cell differentiation, and could prevent a cell from initiating apoptosis. Genetic mutations resulting in the activation of oncogenes increase the chance that a normal cell will develop into a tumor cell. Alterations can be inherited or caused by an environmental exposure to carcinogens.

**Oropharynx** – oral part of the pharynx from the soft palate to the level of the hyoid bone. It includes the soft palate, the base of the tongue, and the tonsils.

**Ovulation** – the phase of the menstrual cycle during which a mature ovarian follicle ruptures and discharges an ovum (also known as an oocyte, female gamete, or casually, an egg) that participates in reproduction if fertilized.

**Oxidative phosphorylation (OXPHOS)** – a metabolic pathway that uses energy released by the oxidation of nutrients to produce adenosine triphosphate (ATP). During oxidative phosphorylation, electrons are transferred from electron donors to electron acceptors, such as oxygen, in a redox reaction which release energy to form ATP. Oxidative phosphorylation produces reactive oxygen species, such as superoxide and hydrogen peroxide that, by forming free-radicals, damage cells and contribute to aging and disease.

**8-oxodG (8-oxodeoxyguanosine)** – a potential biomarker of oxidative DNA damage. It has implications for the study of mutagenesis, carcinogenesis, and free radical toxicity.

**Octamer-4 (oct-4)** – a protein critically involved in the self-renewal of undifferentiated embryonic stem cells. It is involved in tumorigenesis of adult germ cells (progenitor of gamete – cell that fuses with another gamete during fertilization in organisms). In animals, an ectopic expression of its gene causes dysplastic lesions of skin and intestine.

# P

**p16** – a tumor suppressor gene important in regulating the cell cycle. Mutations in *p16* increase the risk of developing a variety of cancers (e.g., melanoma).

**p53** – a transcription factor that regulates the cell cycle and acts as a tumor suppressor. It is important in multicellular organisms as it helps to suppress cancer cells.

**Paget's disease** – a chronic disorder results in enlarged and deformed bones (named after Sir James Paget, the British surgeon who first described this disease). The excessive breakdown and formation of bone tissue that occurs with Paget's disease can cause bone to weaken, resulting in bone pain, arthritis, deformities, and fractures. **Paget's disease of the breast,** also known as **Paget's disease of the nipple**, is a condition that outwardly may have the appearance of eczema – with skin changes involving the nipple of the breast. This condition may be fatal.

**Palliative care** – medical or comfort care that reduces the severity of a disease or slows its progress rather than providing a cure. Usually is used in cases of incurable diseases, when active treatment is not recommended due to coexistent severe health conditions, and when the patient does not want to be treated.

**Pap smear test** (**Papanikolaou test** or **Papanicolaou test**) – a medical screening method of detection premalignant and malignant processes in the ectocervix (see also **Ectocervix**).

**Paracrine signaling** – a form of cell signaling when the target cell is close to (i.e., "para") the signal-releasing cell (e.g., growth factor as paracrine signaling agent). Overproduction of some paracrine growth factors is associated with cancer development. It differs from autocrine signaling (see **Autocrine signaling**): the paracrine signaling affects cells of a different type than the cell performing the secretion, while autocrine signaling affects cells of the same type.

**Paraglioma** – a neoplasm derived from the chromoreceptor tissue of a paraganglion (groups of chromaffin cells distributed in extraadrenal sites, such as around aorta, in kidney, liver, heart, and gonads).

**PCR (polymerase chain reaction)** – a technique in molecular genetics that permits the analysis of any short sequence of DNA (or RNA), without having to clone it, to reproduce (amplify) selected sections of DNA. PCR is used to diagnose genetic diseases, do a DNA fingerprinting, detect bacteria and viruses, study human evolution, etc.

**Perimenopause** – a period prior to menopause during which estrogen levels begin to drop (usually about 3–5 years).

**Peroxisome proliferator-activated receptor (PPAR)** – a group of nuclear receptor proteins functioning as transcription factors which regulates gene expression. They participate in regulation of cells differentiation and metabolism. There are three subtypes of PPAR: PPAR-alpha, PPAR-beta/delta, and PPAR-gamma.

**Person-years of life lost (PYLL)** – the sum of years of life lost by all persons in a population who died of a particular disease. Actuarial (life-expectancy) tables are used to determine the years of life remaining if the persons had not died of disease at that particular age.

**Phenotype** – biochemical or/and physical characteristics of an individual resulting from the interaction between individual's genotype and environment.

**Pheochromocytoma** – a neuroendocrine tumor of the medulla of the adrenal glands (originating in the chromaffin cells) which secretes excessive amounts of catecholamines (e.g., epinephrine and norepinephrine). Patient with pheochromocytoma has severe headache, palpitations, rapid heart rate, sweating, flushing, abdominal pain, increased appetite with the loss of weight, etc.

**Platelet-derived growth factors (PDGF)** – growth factors regulating cell growth and division that plays an important role in angiogenesis (including angiogenesis in malignant tumors).

**Poisson regression** – a form of regression analysis used to model count data and contingency tables. It assumes that a response variable $Y$ has a Poisson distribution and assumes that the logarithm of its expected value can be modeled by a linear combination of unknown parameters. A Poisson regression model is sometimes known as a *log-linear model*, especially when used to model contingency tables.

**Polynuclear aromatic hydrocarbons (PAHs)** – hydrocarbon compounds with multiple benzene rings. They are formed usually from incomplete burning of carbon-containing materials like oil, wood, garbage, or coal. PAHs are components of asphalts, fuels, and greases. Also they may be formed during some cooking processes.

**Population attributable risk (PAR)** – a reduction in incidence that would be observed if the population was entirely unexposed compared with its current (actual) exposured status.

**Premature death** – a death that has occurred earlier than would be statistically expected.

**Primary prevention** – prevention of disease in population through promotion of health/healthy lifestyle and specific protection such as immunization.

**Progestogens (progestagens)** – hormones producing effects similar to those of progesterone (the only natural progestogen, all other progestogens are synthetic and are often referred to as progestins). Progestogens differ by their potency (affinity for progesterone receptors) and side-effects.

**Prokaryotes** – a group of organisms that lack a cell nucleus and other membrane-bound organelles (from the Old Greek "*pro*" = before and "*karyon*" = nut or kernel, referring to the cell nucleus), thus differing from the eukaryotes which have a cell nucleus. Most of prokaryotes are unicellular, but some are multicellular organisms.

**Propagation of uncertainty** (or propagation of error) – the effect of variables' uncertainties (or errors) on the uncertainty of a function based on them. When the variables are the values of experimental measurements, they have uncertainties due to measurement limitations (e.g. instrument's precision), which propagate to the combination of variables in the function.

**Prostate specific antigen (PSA)** – a protein produced by the cells of the prostate gland. PSA is present in small quantities in the serum of normal men and is often elevated in prostate cancer (and several other prostate diseases). A blood test to measure PSA is the most effective test currently available for the early detection of prostate cancer.

**Puberty** – a process of physical changes by which a child's body becomes an adult body capable of reproduction. Puberty is initiated by hormonal signals coming from the brain to the gonads (i.e., ovaries and testes), and gonads produce a variety of hormones that stimulate the growth and participate in regulation of function of reproductive organs, breasts, bones, muscle, skin, and brain.

## Q

**Quadratic hazard model** – a model that assumes health state is described by a set of risk factors or covariates and that the hazard function (e.g., mortality or cancer incidence) is a quadratic function of the covariates.

**Queuing theory** – a theory that deals with stochastic models describing transformations of random flows. It enables mathematical analysis of several related processes, including arriving at the (back of the) queue, waiting in the queue (essentially a storage process), and being served by the server(s) at the front of the queue. The theory permits the derivation and calculation of several performance measures including the average waiting time in the queue or the system, the expected number waiting or receiving service and the probability of encountering the system in certain states, such as empty, full, having an available server or having to wait a certain time to be served.

## R

**Radiosensitizer** – a drug that makes tumor cells more sensitive to radiation therapy.

**RAGE** – a transmembrane receptor from immunoglobulin superfamily that is linked to some chronic diseases resulted from vascular damage (e.g., Alzheimer's disease, atherosclerosis, myocardial infarction, congestive heart failure, diabetic mellitus, and some cancers).

**Random variable** – a mathematical entity describing chance and probability in a mathematical way which was developed for analysis of games of chance, stochastic events, and the results of scientific experiments. There are two types of random variables: discrete and continuous.

**Randomized controlled trial (RCT)** – a study in which people are allocated at random (by chance) to receive clinical interventions. The RCT is used in testing health-care services and health technologies.

**Ras** – a signal transduction protein which communicates signals to other cells. When a DNA mutation turns this signal permanently on, an unlimited cell growth is triggered, leading to development of cancer. Mutations in the *Ras* family (e.g., H-*Ras*, N-*Ras* and K-*Ras*) are found in about 30% of all human tumors.

**Respiratory distress syndrome (in infants)** – a syndrome occurred in prematurely born infants due to insufficiency of surfactant (a substance that coats and lubricates the microscopic air spaces – alveoli – in lungs, preventing lungs from collapsing between breaths) production and structural immaturity of their lungs. It may also result from a genetic disorder of surfactant production. This syndrome affects about 1% of newborns and is the leading cause of death in preterm infants.

**Retinoblastoma gene (Rb)** – a tumor suppressor gene that is dysfunctional in a number of cancers. It is named by retinoblastoma cancer for which it was described first.

**Retinoid compounds/retinoids** – a class of chemical compounds related to vitamin A. Retinoids functions in human organism involve regulation of cell proliferation and differentiation, growth of bone tissue, vision, immune function, and activation of tumor suppressor genes.

**Retrovirus** – a virus belonging to the *Retroviridae* family. Retroviruses use an enzyme reverse transcriptase to perform the reverse transcription of their genomes from RNA into DNA, which then can be integrated into the host's genome. The virus then replicates as part of host's DNA.

**Risk factor** – an aspect of personal behavior or lifestyle, environmental exposure, or inborn or inherited characteristic, which, on the basis of epidemiological evidence, is known to be associated with a certain disease or health condition.

**RNA (Ribonucleic acid)** – a nucleic acid (see also **DNA**) transcribed from DNA by enzymes called RNA polymerases. RNA plays a crucial role in protein synthesis. Also it carries the viruses' genetic information. RNA is usually single-stranded.

**rRNA (ribosomal RNA)** – a central component of the ribosome (i.e., protein manufacturing machinery of cell). Its function is to provide a mechanism for decoding messenger RNA (see also **mRNA**) into amino acids and to interact with the transfer RNA (see also **tRNA**) during translation by providing peptidyl transferase activity.

## S

**SAGE** – a sequence-based highly sensitive technology for gene identification and quantitation in which short (10–14 bp) sequences, called *tags,* are extracted from specific positions within a transcript. The expression profile is then computed to identify the gene corresponding to each tag.

**Screening** – a strategy used in a population to detect a disease in asymptomatic individuals. Its purpose is to identify disease at early stage, thus enabling earlier treatment.

**Secondary prevention** – an identification and detection of disease in its earliest stages, before the development of noticeable symptoms to be successfully cured, its progression could be slowed, its complications minimized or prevented, and disability reduced. Another goal of secondary prevention is to prevent the spread of communicable diseases (i.e., diseases that can be transmitted from one person to another).

**Selective estrogen receptor modulators (SERMs)** – a class of medications acting at the estrogen receptor. A characteristic that distinguishes these substances from pure receptor agonists and antagonists is that their action is different in various tissues, thereby granting the possibility of selectively inhibiting or stimulating estrogen-like action in various tissues.

**Senescence** – the biological processes of a living organism approaching an advanced age (from the Latin word "*senex*" = old man, or old age, or advanced in age). *Organismal senescence* is the aging of whole organisms. *Cellular senescence* is when isolated cells demonstrate a limited ability to divide (also known as replicative senescence, the Hayflick phenomenon, or the Hayflick limit – named after Leonard Hayflick who first published description of this phenomena in 1965).

**Signal transduction** – a process by which a cell converts one kind of signal or stimulus into another. It involves ordered sequences of biochemical reactions inside the cell, which are carried out by enzymes, activated by second messengers, thus creating a *signal transduction pathway*. The number of proteins and other molecules participating in the events involving signal transduction increases resulting in a signal cascade (i.e., amplification of the signal).

**SMAD** – a class of proteins that modulates activity of transforming growth factor beta ligands. The SMAD proteins are homologs of both, the drosophila protein (mothers against decapentaplegic – MAD) and the *C. elegans* protein – SMA (the abbreviation is SMAD, i.e., in *Drosophilas* mutation in mother's *MAD* repressed the gene in the embryo– the phrase "Mothers against" was added since mothers often form organizations opposing various issues).

**Somatic mutation** – a mutation acquired by chance or resulting from exposure to toxins, as opposite to a **germ line mutation,** which is inherited.

**Squamous cell carcinoma (SCC)** – a malignant tumor of squamous epithelium (thin, flat cells that look under the microscope like fish scales, from Latin *"squama"* = scale of fish or serpent). Squamous cells are found in the tissue that forms the surface of the skin, the lining of hollow organs, and the passages of the respiratory and digestive tracts. SCCs may occur in many different organs, such as skin, lips, mouth, esophagus, urinary bladder, prostate, lungs, vagina, and cervix.

**Standardized mortality ratio (SMR)** – a ratio of the actual number of deaths in a population to the number of deaths one would expect if the population had the same death rate as the standard population. It is express as a ratio of observed to expected deaths, multiplied by 100.

**State space model** – see also **Kalman filter model.**

**Stereotactic radiosurgery** – a therapy that uses a large dose of radiation to destroy tumor tissue without involving actual surgery. It is used in the treatment of small benign and malignant brain tumors (including meningiomas, acoustic neuromas, and pituitary cancer), metastatic brain tumors (cancer that has spread to the brain from another part of the body). It can also be used to treat other diseases, such as Parkinson's disease or epilepsy. Stereotactic radiosurgery can be performed by using a linear accelerator to administer high-energy photon radiation to the tumor (i.e., LINAC-based stereotactic radiosurgery), using a gamma knife with cobalt 60 and by using heavy charged particle beams (such as protons and helium ions) to deliver stereotactic radiation to the tumor.

**Stereotactic radiotherapy** – a radiotherapy that uses multiple small fractions of radiation as opposed to one large dose. Giving multiple smaller doses may improve outcomes and minimize side effects (e.g., stereotactic radiotherapy is used to treat brain tumors).

**Stroma** – from Greek "στρώμα" = bed – a connective supportive framework of a biological cell, tissue, or organ. It is contrasted with parenchyma and is synonymous with the interstitial space.

**Sudden infant death syndrome (SIDS)** – a syndrome marked by the symptoms of sudden and unexplained death of an apparently healthy infant aged one month to one year.

**Survival rate** – a percentage of patients who are alive for a given period of time after diagnosis or treatment (commonly referred to cancer). This is often measured 5 years after diagnosis or treatment and called the 5-year survival rate. Five-year survival rates are often used to compare the effectiveness of various therapies.

**Sv (Sievert)** – a unit of exposure to ionizing radiation (IR) that reflects biological effects of radiation (opposed to the physical aspects, which are characterized by the absorbed dose, measured in grays, Gy). It equals to the absorbed dose multiplied by a "radiation weighting factor" (this factor depends on the type of IR and energy range).

## T

**Telomerase** – expression plays a role in cellular senescence, as it is normally repressed in postnatal somatic cells resulting in progressive shortening of telomeres. Deregulation of telomerase expression in somatic cells may be involved in oncogenesis. Studies in mice suggest that telomerase also participates in chromosomal repair, since *de novo* synthesis of telomere repeats may occur at double-stranded breaks.

**Telomere** – a region of repetitive DNA at the end of chromosomes, which protects the end of the chromosome from destruction. Derived from the Greek word *"telos"* = end and "meres" = part.

**Tertiary prevention** – an improvement of life quality in sick people by limiting their complications and disabilities, reducing the severity and progression of disease, and providing them with rehabilitation. Unlike primary and secondary prevention, tertiary prevention involves the disease treatment conducted by health-care practitioners, rather than public health agencies.

**TGFβr** (transforming growth factor $\beta$ receptors) – the serine/threonine kinase receptors which can be found in brain, heart, kidney, liver, testes, etc. Over-expression of TGF can induce fibrosis in organs, e.g., in kidneys that causes kidney failure with diabetes. Recent developments have found that using certain types of protein antagonists against TGF$\beta$ receptors can halt and sometimes even reverse the effects of renal fibrosis.

**Time trend** – a change in a disease incidence or mortality rate over the time of observation.

**TIMP3** (TIMP metallopeptidase inhibitor 3) – a human gene that belongs to the tissue inhibitors of metalloproteinases, which encodes proteins – inhibitors of the matrix **metalloproteinases** (see also **Metalloproteinases**) that involved in degradation of the **extracellular matrix** (see also **Extracellular matrix**). Mutations in this gene are associated with the autosomal dominant disorder – Sorsby's fundus dystrophy (a rare autosomal dominant degenerative disease of the macula which manifests by symptoms of night blindness or sudden loss of vision acuity in the third–fourth decades of life).

**Transforming growth factor beta (TGF-β)** – a secreted protein that exists in three isoforms (i.e., TGF-$\beta$1, TGF-$\beta$2, and TGF-$\beta$3). The TGF-$\beta$ family is part of a superfamily of proteins known as the transforming growth factor beta superfamily, which includes inhibins, activin, antimüllerian hormone, bone morphogenetic protein, decapentaplegic and Vg-1. TGF beta controls proliferation, cellular differentiation, and other functions in most cell types. It can also act as a negative autocrine growth factor. Specific receptors for TGF-$\beta$ activation trigger apoptosis when activated.

**Transurethral resection of the prostate (TURP)** – an urological operation used to treat a benign prostatic hyperplasia. It is performed by visualizing the prostate through the urethra and removing prostate tissue by electrocautery or sharp dissection.

**tRNA (transfer RNA)** – a small ribonucleic acid (see also **RNA**) that usually consists of about 74–95 nucleotides and that transfers a specific amino acid to a growing polypeptide chain at the ribosomal site of protein synthesis (during the translation phase). Each type of tRNA molecule can be attached to only one type of amino acid. However due to the genetic code contains multiple codons that specify the same amino acid, tRNA molecules that bearing different anti-codons may also carry the same amino acid.

**Tsallis entropy** – a theory that gives an insight into the chaos of physical systems that could be applied to subjects from the locomotion of microorganisms to the motions of stars. Its definition has been proposed by Conctantino Tsallis in 1988. This is a generalization of a standard Boltzmann–Gibbs entropy that described by equation $S = k \ln W$ and which provides the mathematical definition of entropy and serves as the cornerstone of "statistical mechanics".

**Tumor initiation** – a process in which normal cells are changed thus becoming able to form cancer. Substances that cause that effect are tumor initiators.

**Tumor necrosis factor-alpha (TNFα)** – a cytokine involved in systemic inflammation which stimulates the acute phase reaction. It participates in regulation of apoptotis, cellular proliferation, differentiation, inflammation, tumorigenesis, and viral replication. Dysregulation and, in particular, overproduction of TNF have been implicated in a variety of diseases, including cancer.

**Tumor promotion** – a process in which an existing tumor is stimulated to grow. Tumor promoters themselves can not cause cancer.

**Tumor suppressor gene** (see also **Antioncogene**) – a gene that protects a cell from become a cancer cell. When mutated this gene losses or has reduced its function, thus letting the cell to progress to cancer (usually other genetic changes occur resulting in the transformation into cancer).

**Two-disease model** – the model of survival analysis which assumes that study population is represented as a mixture of two subpopulations with different risk factors susceptibilities and, probably, with different mechanisms of tumorigenesis.

**Two-stage clonal expansion model (TSCE)** – the most popular version of the two-stage MKV model of carcinogenesis. It assumes that the number of normal cells is either constant or described by a deterministic function and that all rates are time-independent.

**Tyrosinase** – a copper-containing enzyme that oxidizes the amino acid tyrosine and other phenolic compounds, thus forming brown and black pigments (i.e., melanin). Mutation of gene controlling this enzyme in human causes albinism.

## V

**Vascular dementia (senility)** – a group of syndromes characterized by declining in individual's mental capacity and intellectual ability. It is caused by different mechanisms resulting in vascular brain lesion (e.g., cerebrovascular disease). This is the second most common form of dementia in elderly (following Alzheimer's disease).

**Vascular endothelial growth factor (VEGF)** – a subfamily of growth factors which are important signaling proteins involved in both vasculogenesis (the *de novo* formation of the embryonic circulatory system) and angiogenesis (the growth of blood vessels from preexisting vasculature).

## W

**Weibull hazard function** – a continuous probability distribution used in biomedical studies for survival analysis. It is characterized by high flexibility, i.e., it could describe the behavior of other statistical distributions (e.g., normal or exponential distributions).

## X

**Xenobiotic** – a chemical compound that is foreign to organism (e.g., antibiotics, pesticides, etc.).

# Index


**A**
Accessory cancer gene, 52
Acrylamide, 120, 135, 138–139
Active life expectancy, 157
Adenocarcinoma, 45, 91, 218, 237, 240, 248, 257, 283, 295–298, 342, 354, 371, 382
Age–period–cohort, 63
Age
  breast cancer and, 163, 235, 259, 279–283
  cancer and, 115–118
  cervical cancer and, 283–290
  *Helicobacter pylori* infection and, 30, 288
  hepatocellular carcinoma and, 238–239, 258–259
  lung cancer and, 232–235, 248–256, 338–339
  melanoma and, 18, 20, 62, 288, 341
  ovarian cancer and, 260
  prostate cancer and, 293–295
  thyroid cancer and, 131–134, 226
  uterine cancer and, 239–240
Aging, cancer, 283–290
Airline flight crews, cancer risk, 125
Akaike information criterion, 263
Alcohol
  cancer risk factor, 105, 113–114, 122, 124, 291, 299, 315, 332, 373, 375
  hepatocellular carcinoma, 238, 259, 290–291
  liver cancer and, 22, 258, 290–291
  squamous cell carcinoma and, 256–257, 297–298, 357
Anchorage-independent growth, 38
Androgen(s), 5, 94, 120, 238, 259, 294, 295
  prostate cancer and, 5, 294
Angiogenesis, 6, 38, 45, 331, 358, 369
Antioncogene, 52, 56, 60
APC pathway, 50
Apoptosis, 44, 330, 367, 368
  alterations in, 37, 330
  mitochondria and, 44
Armitage–Doll model
  basics, 53
  applications, 54
  generalizations, 58–59
Artificial sweeteners, 139–140
Asbestos, 94, 98, 99, 100, 120, 206, 340, 357
Aspartame, 139
Ataxia telangiectasia, 117
Atherosclerosis, 327–333
Average years of life lost, 12, 13

**B**
Barrett's esophagus, 98, 354
  Esophageal carcinoma and, 354
Bateman's solution, 52
Bernoulli life table, 152, 191
Beta-carotene, 355, 356, 363
Beta-distribution model, 62
Billroth, Christian Theodore, 4
Biomarker
  of effect, 90
  of exposure, 89–90
  of susceptibility, 90
Body mass index
  breast cancer and, 288
  covariate in life table, 174
  cancer risk and, 30, 107, 108,
  esophageal cancer and, 237, 257
Boveri, Theodor, 47
BRCA1
  breast cancer and, 118, 281, 282
  screening for, 380–381
BRCA2
  breast cancer and, 281, 282
  screening for, 281

Breast cancer
adenocarcinoma of, 248, 280
age and, 163, 235, 259, 279–283
body mass index and, 288
BRCA genes and, 282
chemoprevention of diet, 45
and familial, 60, 282
duct breast carcinoma, 259, 276, 277
familial relative risk, 60, 93
fat intake and, 98, 110, 288, 329
genetic factors and, 279–283
height and, 288
high-penetrance genes in, 282
hormone replacement therapy and, 159–160, 235
*in situ*, 248, 259
incidence, age pattern, 63, 220, 221, 249, 252, 255
lobular breast carcinoma, 276
low-penetrance genes in, 282
mammography of, 235, 259, 281, 352, 381
Manton and Stallard model (two-disease model), 60, 151, 161–164, 265, 281
mortality from, age pattern, 33, 220
obesity and, 235
oral contraceptives and, 98, 103, 281
Parmigiani model, 281
Pike model, 280–281
postmenopause and, 280–281, 286
precursors, 299–300
premenopausal, 280, 288
progesterone and, 239
progestine and, 159
puberty and, 288
race and, 156, 164, 217, 226, 231
Rosner and Colditz model
survival for, 17, 18, 20, 156
time trends, 156
U.S., 7, 14, 161–164, 219, 220, 221, 235

C
Cancer, 1–33, 37–76, 89–140, 151–186, 191, 192, 193, 195, 196, 197, 199, 205–212, 217–242, 247–300, 313–327, 327–333, 335, 338, 339, 340, 341, 342, 351–383, 393
Cancer age pattern, 285
Cancer histotype, 218, 237, 242, 247–300, 342, 382, 393
Cancer initiation, 55, 75, 117, 265, 290, 372, 393
Cancer morbidity, 6–31, 32, 94, 151–186, 196, 217–242, 247–300
Cancer mortality, 7, 18, 21–31, 53, 54, 55, 61, 69, 118, 151, 158, 160, 161, 163, 164, 173, 185, 186, 205–212, 217–223, 232, 233, 234, 237, 240, 256, 281, 338, 340, 342, 353, 374, 375, 379, 381
Cancer progression, 50, 51, 52, 53, 331, 371
Cancer promotion, 43, 49, 51, 65, 135, 282, 290, 293, 295
Cancer risk, 3, 29, 30, 40, 54, 63, 70, 89–140, 151, 163, 164–174, 185, 195, 205, 206, 208, 218, 235, 239, 240, 241, 247, 258, 259, 281, 282, 283–290, 294, 295, 331, 332, 333, 341, 353, 357, 363, 365, 370, 372, 373, 381, 393
Carcinogenesis
cell cycle defects in, 38, 42
cellular growth defects in, 38–46, 48
epigenetic alterations in, 47, 48–49, 51, 89, 115
genetic events in, 48, 50, 57, 67, 295
multistage hypothesis, 264
Carcinogens, 3, 5, 6, 32, 37–39, 41, 43, 44, 46–74, 89–94, 117, 119–123, 125, 130, 133–135, 137–139, 151, 161–164, 175, 191, 193, 199, 247, 256, 262, 264, 265, 275–277, 281, 282, 284–292, 298, 329, 331, 354–356, 366, 368, 369, 377, 378, 381, 382, 393
Carcinoma *in situ*, 12, 45, 130, 259, 260, 280, 284
*Caspases, in apoptosis*, 44
Cervical cancer
*Human papillomavirus* and, 29, 120, 121, 284, 289
Adenocarcinoma, 240
squamous cell carcinoma, 240, 283, 295–298
age and, 96, 226, 240, 277, 283–290
incidence of, 3, 226, 240, 260, 283–290, 297, 382
*in situ*, 240, 260, 277, 284, 297
screening for, 240, 248, 260, 261, 283–290, 293
oral contraceptives and, ,98, 99, 103, 106, 121, 239
*Human immunodeficiency virus infection* and, 30, 121
time trends in, 295
race and, 231, 240
U.S., 285
Invasive, 20, 223, 240, 261, 284, 289
mortality from, 21–31
parity and, 239

pregnancy and, 3, 163, 289, 314, 370,
puberty and, 284–289
perimenopause and, 284, 285, 287, 288
postmenopause and, 286, 329
prevention of, 240, 289, 292, 294, 300, 326, 341
sexual risk factors for, 3, 261, 286–289
survival for, 7, 13–16, 152–161, 167, 172, 182, 184, 192, 193, 197, 198, 212, 217, 232, 237, 290, 298, 358, 377, 380
Cervix
anatomy of, 217, 220, 223, 224–226, 240, 248, 260–261
cancer precursors, 286, 299
CIN, 289, 290
screening for, 260–261
squamous dysplasia, 95, 104, 105, 218, 237, 250, 256, 277, 283, 288, 295–298
transformation zone, 289
Chemoprevention
definition of, 355, 356, 382
trials for, 355, 356
Chernobyl nuclear power plant accident
thyroid cancer and, 127, 130
Childhood cancer, 132
Cholecystectomy, 96, 112, 195, 258
Cholesterol
atherosclerosis and, 199, 333, 334, 336
cancer risk and, 110, 135, 173, 333–339
Cigarette smoking
age at initiation, 173, 316, 337
cervical cancer and, 54, 104
cessation, 234, 317, 328, 342, 354, 361, 362
esophageal cancer and, 137, 237
historical perspective on, 21
lung cancer and, 30, 52, 54, 60, 94, 114, 169, 315, 338–340, 360, 366, 382
mortality burden of, 12–20, 21–31, 163, 165, 169, 173, 198, 232, 234, 235, 237, 238, 241, 256, 299
nasopharyngeal cancer and, 91
oropharyngeal cancer and, 17, 91
prevalence of, 197–198, 234–235, 237–238, 252, 256–258, 276, 297, 317
squamous cell carcinoma and, 104, 105, 114, 218, 237, 240, 250, 283, 295–298, 342, 354
U.S trends in, 6–31, 205–212, 217–241, 248–261, 325–326
CIN (cervical intraepithelial neoplasia), 260, 289, 290
CISNET (Cancer Intervention and Surveillance Modeling Network), 375, 381
Coal tar
occupational exposure to, 123, 340, 374
Clonal expansion, 51, 52, 54–56, 62, 116, 262, 264, 327, 330
Clonal expansion model, 56, 62, 264
Colon, cancer
age and, 237
body mass index and, 112, 288
diet and, 75, 110, 135–137, 257, 327, 329, 331, 333, 353, 355, 356, 362, 364–370
familial, 55, 96
fat and, 356
fiber and, 364
incidence of, 55
mortality from, 15, 18, 20, 21–31, 54, 75, 134, 169, 217, 219, 220–241
NSAIDS and, 355, 356, 373
obesity and, 237, 241, 257, 297, 331, 340, 342, 353, 371–372, 375, 377, 382
physical activity and, 30, 258, 295, 297, 331, 337, 365, 372, 374, 375, 380
prevention of, 236, 237, 331, 353, 365, 371, 373
race and, 236–237
sex and, 27, 29, 99, 108, 267, 270, 271
smoking and, 54, 103, 104, 169, 237, 331, 353
*See also* Colorectal cancer
Colorectal cancer
age and, 95, 96, 106, 107, 110, 111, 112, 134, 236–237
body mass index and, 112, 173, 288
diet and, 99, 103, 104, 106, 107, 110, 134, 137, 257
distal, 91, 236–237, 258, 353
familial adenomatous polyposis and, 55, 95, 96
fat and, 99, 110, 332
fecal occult blood test, 352
fiber and, 99, 364
fruits and vegetables, 363, 365, 370
incidence of, 236, 257
mortality from, 237
NSAIDS and, 355, 356, 373
obesity and, 237, 257
physical activity and, 258
prevention of, 353, 357, 362, 375
proximal, 236, 258
race and, 236, 257
screening, 236–237
sex and, 236, 257
smoking and, 257
time trends of, 236–237
*See also* Colon cancer

Complexity and chaos theory, 64
Connexins, 37, 43
Corpus uteri, cancer, 14, 15, 17, 27, 54, 62, 96, 99, 217, 219–226, 229, 231, 239–240, 248, 250, 252, 254, 259, 269, 272, 274, 275, 332
Cox
  proportional hazard regression, 159, 290
  regression, 64, 159, 161
Crude death rate, 208–209
Cyclamate, 101, 139
Cytochrome P450
  CYP1A1, 91
  CYP1A2, 91, 137
  CYP2C19, 91
  CYP2D6, 91
  CYP2E1, 91
Cytokines, 42, 45, 72, 73, 133, 288, 329, 331

**D**
Dairy products
  prostate cancer and, 98, 103, 110, 364
Death certificate, 154, 155, 156, 218, 220
Deoxyribonucleic acid (DNA)
  repair, 38, 39, 43, 89, 90, 92, 93, 124, 125, 328, 329, 368, 371
  mitochondrial DNA, 39–41
  nuclear DNA, 38–39, 40, 53, 75, 94, 125
Diabetes mellitus, 96, 154, 177, 331, 332, 334
Diet
  cancer prevention and, 362–370
  carcinogens and, 366, 369
  chemoprevention and, 370
  epidemiological studies of, 136
  fat in, 363, 365
  fiber in, 364, 365
  folic acid in, 363, 364, 383
  fruits and vegetables in, 257, 353, 363, 365, 370, 375
  macronutrients in, 137
  meat in, 98, 99, 100, 103, 111, 112, 134, 135, 137, 138
  meta-analysis and, 103, 110, 112, 118
  micronutrients in, 137
Direct age standardized age-specific death rate (DASDR), 209, 210, 211

**E**
Electromagnetic fields
  brain cancer and, 115, 333
  leukemia and, 118
Empirical Bayesian approach, 33
Endometrial cancer, *see* Uterus/corpus uteri cancer
Endoplasmic reticulum, 41–42, 126
Energy intake, 40
Environmental risk factor, 33, 52, 53, 54, 118–119, 194, 205, 313–326
Epidermal growth factor receptor, 327
Epithelial progenitor cells, 337
Esophageal cancer
  alcohol and, 237
  Barrett's esophagus and, 98, 354
  chemoprevention of, 353–370
  cigarette smoking and, 237, 257
  diet and, 137
  *Helicobacter pylori* infection and, 354
  incidence of, 237, 256
  mortality from, 237, 256
  age and, 237, 256
  gender and, 237, 256
  time trends in, 237, 257
  race and, 237, 256
  U.S., 237, 256–257
  International, 294, 371
  Adenocarcinoma, 237
  squamous cell carcinoma, 237
  obesity and, 237, 257
  survival for, 237
Estrogen, 5, 94, 95, 99, 100, 106, 114, 115, 120, 135, 136, 159, 163, 237, 239, 280–282, 328, 337, 356–358, 366, 367
Excess relative risk, 56, 126, 287
Extended Techa River Cohort, 195
Extra years of life (EYL), 338, 339

**F**
Familial adenomatous polyposis, 55, 95, 96
Fat, dietary, 329
Fecal occult blood test, 352
Fiber, dietary, 98, 355, 364
Fibroblast growth factor 2, 327
Fokker-Planck equation, 33, 64, 73, 74, 164, 165, 167
Folic acid, 5, 363, 364, 383
Food
  additives, regulation of, 124, 138, 139
  cooking methods, 139
Framingham Heart Study
  life table, 159, 173
  quadratic hazard model application for, 164, 300, 313

Fruits and vegetables, 257, 353, 363, 365, 370, 375
Future Elderly Model, 376
Fuzzy
 set logic, 33
 state model, 66

## G

Galen, Claudius, 2
Gap junctional intercellular communication, 42, 47
Gastric carcinoma, *see* Stomach cancer
Gaussian
 process/distribution, 68, 165, 166, 171, 172, 182, 183
Generalized frailty model
 Predictions, 184
Genetic polymorphism, 91, 93, 137, 357
Geo-epidemiological/mapping study, 205–212
GLOBOCAN, 7
Glutathione-S-transferase, 117, 282, 294, 366
Goldgi apparatus, 41–42, 126
Gompertz, Benjamin, 191
Gompertz tumor growth function, 58, 72, 73
Goodness of fit test, 65
Grade of Membership (GoM), 199–205
Granulocyte macrophage colony stimulating factor, 327
Greenwald-Cullen Five Phase Model, 352

## H

Halsted, William Stewart, 4
Heat shock protein, 41
Height and cancer risk, 95, 96, 173, 285, 288
*Helicobacter pylori*, 30, 101, 121, 134, 329, 354
*Hepatitis B virus* (HBV)
 age and, 290–291
 drug users and, 292
 hepatocellular carcinoma and, 290–293
 prevalence and, 291
 race and, 291
 time trends in, 291
*Hepatitis C virus* (HCV)
 age and, 290–291
 drug users and, 292
 hepatocellular carcinoma and, 290–293
 prevalence and, 291
 race and, 291
 time trends in, 291
Hepatocellular carcinoma
 age and, 291
 alcohol and, 290
 cirrhosis and, 291
 diabetes mellitus and, 291
 diet and, 294
 HBV infection and, 291
 HCV infection and, 291
 incidence of, 291
 iron stores and, 238
 mortality from, 332
 nonalcoholic steatohepatitis and, 291
 race and, 293
 sex and, 258
 survival for, 290
Heterocyclic amines, 138
Heterogeneity
 breast cancer of, 63
 cancer histotype and, 218, 247, 261, 264, 265
 cervical cancer of, 103, 264
 hepatocellular carcinoma of, 299
 bias, 102
 prostate cancer of, 261, 265
 white population of, 211, 212, 317, 319
Hippocrates, 2, 362
Histon, 39, 137
Hormone replacement therapy, 159, 235, 355
*Human immunodeficiency virus* (HIV), 12, 30, 44, 121, 287, 292, 368
Human Mortality Database, 219, 325
*Human papillomavirus* (HPV), 29, 120, 121, 284, 289
Hypoxia inducible factor, 41
Hysterectomy
 cervical cancer and, 283
 ovarian cancer and, 277
 uterine cancer and, 277

## I

Immune response
 host of, 284, 286, 287, 288
Immunomodulation, 45–46
Incidence
 adenocarcinomas of, 295
 age and, 13, 17, 53, 61, 62, 75, 126, 178, 182, 191, 226
 age-specific, 57, 62, 177, 178, 182, 224, 226, 231, 248, 255, 262, 264, 278, 280, 285, 290, 295, 300, 340
 data sources for, 26, 314, 317
 international, 90, 138, 219, 294, 371, 374

Incidence (*cont.*)
modeling of, 37–74
non-homogeneity of, 235, 239, 281, 282, 292–295, 299
race and, 94, 95, 97, 117, 156, 164, 180, 182, 206, 211, 217, 218, 226, 231, 236, 237, 238, 240 SEER data on, 62, 164, 248, 283
sex and, 25–29, 47, 94, 95–100, 130–133, 156, 164, 180, 205, 206, 211, 218–226, 236–238, 241, 252, 257, 258–278, 284, 286–289, 296–297, 316–317
squamous cell carcinomas of, 218, 295–298, 382
time trends in, 7, 63, 218, 219, 220–226, 232, 233, 248, 252, 257, 262, 295, 296, 323, 325, 342
U.S., 13, 17, 18, 19, 30, 139, 161, 164, 177–179
Infertility
Primary, 316
Secondary, 198, 316, 319
smoking and, 198, 316
microsimulation in, 377
Insulin growth factor, 327
Interdisciplinary
approach, 31–33, 140, 393
studies, 31, 32
team, 31, 32, 290, 352, 393
Interferon, 45
Intermediate cells (initiated cells), 51, 52, 54, 55, 57, 58, 117
International Agency for Research on Cancer (IARC), 7, 120, 138, 139, 290, 291, 315, 328, 341, 361, 371, 372
International Classification of Diseases for Oncology, 219
Iodine-131
Chernobyl nuclear power plant accident, 123, 126
thyroid cancer and, 134
Ionizing radiation
atomic bomb, 124
cerebrovascular disease and, 124
Chernobyl nuclear power plant and, 123, 126
children exposure of, 131
chronic exposure, 195
dirty bomb and, 126
dose of exposure, 124
external irradiation, 126, 130
Hiroshima and Nagasaki, 124
internal irradiation, 126
low-dose exposure, 124, 126
peri-menopause and, 133
puberty and, 133
thyroid cancer and, 127–134
Isothiocyanates, 364, 368, 369

**K**

Kalman filter model (state space model), 60, 66
Kaplan–Meier, or the log rank test, 159
Kidney cancer, *see* Renal cancer
Knudson's two-hit hypothesis, 55–58
Kolmogorov equation, 33, 59, 165

**L**

Latent period/latency, 56, 64, 69, 72, 117, 130, 132, 136, 161, 163, 164, 265, 281, 285, 291, 340, 354, 374, 375, 380
Latent structure analysis, 201, 202, 203, 204
Leukemia, 5, 6, 14, 16, 21, 26, 28, 59, 97, 103, 104, 106, 107, 108, 111, 118, 119, 126, 315, 328, 332, 362, 366, 370
Levy type distribution, 170
Life table models
extensions of, 185
covariates in, 159
SSA, 153
Linear Latent Structure (LLS) analysis, 200, 201, 203–205, 212
Linear no-threshold function, 65
Liver cancer, 7, 22, 30, 107, 220, 224, 238, 241, 258–259, 290, 291, 292
age and, 238
alcohol and, 238
cirrhosis and, 239
diet and, 257
HBV infection and, 238, 258
HCV infection and, 238, 258
incidence of, 238
international, 374
iron stores and, 238
mortality from, 232–242
nonalcoholic steatohepatitis and, 291
race and, 238, 258–259
sex and, 238
survival for, 238
U.S., 291
*See also* Hepatocellular carcinoma
Logistic regression, 160, 192, 335
Longevity, 32, 40, 63, 195, 330, 338, 380
Lotka–Volterra equations, 64
Low-density lipoprotein (LDL), 330, 337, 341

Lung cancer
 Adenocarcinoma, 295
 age cohort studies of, 232
 asbestos and, 340
 beta-carotene and, 355
 birth cohort studies of, 232
 chemoprevention of, 354
 cigarette smoking and, 232
 incidence of, 232
 international, 374
 mortality from, 232
 N-acetyltransferase and, 91
 prevention of, 294
 race and, 233
 screening for, 248–256
 sex and, 252
 small cell, 250, 252
 smoking cessation and, 361
 squamous cell carcinoma, 250
 survival for, 232
 time trends of, 232
 tobacco smoking and, 360
 U.S., 252
 vitamin A and, 355, 363
 vitamin E and, 355, 356, 363

**M**

Mammography, 235, 259, 281, 352, 381
Manton and Stallard model, breast cancer, 53, 54, 60, 61, 62, 72, 74, 132, 154, 155, 156, 157, 161, 162, 163, 165, 192, 207, 265, 281, 313, 314, 335, 375
 *See also* Breast cancer
Manton–Woodbury–Yashin approach, 193
Margin of exposure, 139
Marginal age standardized age-specific death rates (MASDR), 209, 210, 211
Markov
 Markov chain Monte Carlo method, 60
 Markov process, 58, 66
Mathematical modeling, 31, 32, 64, 74, 134, 175, 289, 393
Mayak nuclear facility, 124, 126
Meat, cancer risk, 98, 99, 100, 103, 111, 112, 134, 135, 137, 138
Mechanism of carcinogenesis, 38, 56
Melanoma (cutaneous), 1, 15, 17, 18, 20, 22, 27, 62, 96, 97, 101, 109, 112, 288, 315, 341, 354, 356, 358, 359
Menarche, 95, 96, 133, 135, 279, 280, 287, 288, 356
Menopause, 95, 96, 97, 98, 133, 160, 162, 163, 235, 280, 284, 285, 286, 287, 288, 329
Meta-analysis, 102, 104, 105, 106, 107, 108, 109, 110, 111, 112, 113
Metalloproteinases/matrixins, 46, 51
Methotrexate, 5
Metropolitan Relative Weight criteria, 342
Microarray chip, 6
Microsatellite instability-positive colon cancer (MSI+), 237
Microsimulation
 cancer prevention and, 197
 dynamic and, 197
 forecasting and, 199
 in life table model, 197
 interventions and, 196–199
 screening effect and, 197
 sensitivity analysis and, 314
 smoking-attributable health effects and, 315
 static, 197–198
Milk/diary products, cancer risk, 89, 99, 110, 130, 138, 363, 364, 369
Mitochondrial DNA (mtDNA), 39–41, 125
Mixed models of carcinogenesis, 60
Mixing distribution, 68, 69, 72, 183, 200
Mobile phones (cell phones), cancer risk, 114–115
Model of carcinogenesis
 Armitage-Doll model, 54, 58–59, 62, 65, 74
 barrier breaking mechanism model, 57, 58
 beta-distribution model, 62
 gamma-frailty model, 298
 Moolgavkar-Venzon-Knudson two-stage model (MVK), 55–58
 multievent model, 58
 multiple pathway model, 59–60
 multi-stage model, 48–49
 stochastic multivariate model, 73–74
 two-stage clonal expansion model, 56
Mortality
 adenocarcinomas of, 218
 age and, 217
 age-specific, 220, 222, 224, 226, 231, 232, 241
 data sources for, 314
 international, 218, 219, 374
 modeling of, 242
 non-homogeneity of, 242
 race and, 217, 218
 sex and, 217
 squamous cell carcinomas of, 218, 237, 240
 time trends in, 218, 219, 223, 226, 232, 241
 U.S., 217, 219, 222, 223, 227, 239, 241
Multidisciplinary studies, 31
Multistage model, 52, 54, 58, 67, 75, 117, 162, 164, 191

N

National Long Term Care Survey (NLTCS)
- age pattern of mortality rate, 182
- age pattern of cancer incidence, 184
- sensitivity analysis

Necrosis, 41, 45, 67, 291
Negative Binomial regression, 160, 207, 208
Nicotine, 52, 252, 256, 360, 362
Nitroso-compounds, 92, 112, 134
Non-Hodgkin lymphoma, 95, 111
Non-steroidal anti-inflammatory drugs (NSAIDs), 42, 355, 356, 373
Nuclear factor kappa B (NFkB), 327, 331
Nunney model, breast cancer, 281
Nutrients, and cancer risk
- Macronutrients, 137
- Micronutrients, 137

O

Obesity
- adenocarcinoma and, 371
- breast cancer and, 371
- colorectal adenoma and, 356
- esophageal cancer and, 237
- prevalence of, 371

Oncogene, 6, 44, 47, 48, 52, 56, 60, 93, 286
Oral contraceptives
- breast cancer and, 98
- cervical cancer and, 99
- colorectal cancer and, 236
- endometrial cancer and, 239
- hepatocellular carcinoma and, 238
- ovarian cancer and, 239

Organosulfur compounds, 366, 368
Ovarian cancer
- age and, 260
- age at menarche and, 135
- age at menopause and, 96
- body mass index and, 112
- cigarette smoking and, 316
- hysterectomy and, 277
- incidence of, 260
- mortality from, 281
- oral contraceptives and, 99
- pregnancy and, 289
- race and, 260
- time trends in, 260
- survival for, 260

Oxidative phosphorylation (OXPHOS), 39

P

p53, 41, 42, 44, 50, 53, 63, 66, 104, 137, 282, 358, 369
Paget's disease, 4
Pancreatic cancer
- alcohol and, 238
- cigarette smoking and, 238
- dairy products and, 364
- diabetes mellitus and, 331
- fruits and vegetables and, 353
- incidence of, 238
- mortality from, 238
- race and, 238
- sex and, 238
- survival for, 238
- time trends in, 238

Pap smear test, 352
Parmigiani model, breast cancer, 281
- *See also* Breast cancer

Perimenopause, 284, 285, 287, 288
Person-years of life lost, 12, 13
Phenotype, 39, 45, 51, 52, 91, 93, 115, 116, 137, 203, 293, 355
Physical activity, 30, 235, 258, 295, 297, 331, 337, 341, 365, 372, 374, 375, 380
Pike model, breast cancer, 280
- *See also* Breast cancer

Platelet-derived growth factor, 327
Poisson regression, 160
Polymorphism, 39, 45, 91, 93, 137, 357, 365
Polynuclear aromatic hydrocarbons, 92, 138
Population attributable risk, 371
Postmenopausal hormone therapy, 98, 99, 239
Precursors, cancer, 135, 286, 294, 295, 299, 329, 334
Prevention of cancer
- Primary, 17, 30, 355, 356, 361, 379
- Secondary, 22, 197
- Tertiary, 355

Progesterone
- breast cancer and, 239
- endometrial cancer and, 239
- ovarian cancer and, 239

Progestine, 159
Progestogens, 239
Prostate cancer
- Adenocarcinoma, 295
- age and, 261
- androgens and, 294, 295
- calcium and, 355
- chemoprevention of, 355
- dairy products and, 98
- diet and, 294

early detection and, 293
heterogeneity of, 261
incidence of, 235
international, 294
latency period of, 340
mortality from, 236
prostate-specific antigen (PSA) test, 42, 61, 236, 261, 294, 359, 369
race and, 235
screening, 236
survival for, 236–237
testosterone and, 295
time trend in, 295
two-disease model of, 295
U.S., 294
Prostate gland
Prostate-specific antigen (PSA) test, 42, 61, 236, 261, 294, 359, 369
*See also* Prostate cancer
Puberty, 133, 284, 285, 286, 287, 288, 289, 374

**Q**
Quadratic hazard model
life table prediction, 164
microsimulation, 167
stochastic process model, 164–174
U-shaped hazard, 382
Queuing theory

**R**
Race
adenocarcinoma and, 237
breast cancer and, 235
cervical cancer and, 240
colorectal cancer and, 236
endometrial cancer and, 239
esophageal cancer and, 237
kidney cancer and, 240
liver cancer and, 258
lung cancer and, 248
ovarian cancer and, 260
pancreatic cancer and, 238
prostate cancer and, 293–295
squamous cell carcinoma and, 295–298
stomach cancer and, 236–237
Radiation
Carcinogenesis, 119–134
ionizing, *see* ionizing radiation
thyroid cancer, 127–134, 287, 333
Radionuclides, 121, 122, 126, 130, 195
Radiotherapy, 6, 195, 370
Radon, 57, 94, 98, 114, 119, 122, 123, 340
*Ras* genes, 50
Rectal cancer
age and, 237
body mass index and, 288
diet and, 355
distal, 236, 237
familial adenomatous polyposis and, 258
fat and, 355
fecal occult blood test, 352
fiber and, 355
fruits and vegetables, 355
incidence of, 236
mortality from, 237
NSAIDS and, 355
obesity and, 353
physical activity and, 258
prevention of
proximal, 236, 237, 258
race and, 236
screening, 236, 237
sex and, 236
smoking and, 257
time trends of, 236
*See also* Colorectal cancer
Relative risk, 56, 93, 102, 112, 126, 130, 132, 196, 198, 256, 286, 287, 291, 316, 317, 318, 319, 322, 323, 326, 340, 341, 376, 377
Renal cancer
age and, 240
cigarette smoking and, 241
diet and, 241
hypertention and, 241
incidence of, 240
mortality from, 240
race and, 240
sex and, 240
survival for, 240
time trends in, 240–241
Retinoblastoma
Knudson's two-hit hypothesis, 55–58
Retinoblastoma gene, 6
Retinoids, 43, 67, 355
Risk factor
age pattern of, 151
controllable, 94–115
dynamic model of, 117, 124, 127, 130
environmental, 118–140
intervention, 94
microsimulation, 119
noncontrollable, 94–115
Risk intervention, 313–343
Roentgen, Wilhelm Conrad, 5, 6

Rosner and Colditz model, breast cancer, 280–281
See also Breast cancer
rRNA, 39

S
Salt
gastric cancer and, 257
Screening
Biases, 247
breast cancer, 259
cancer prevention, 351–383
cervical cancer, 240
colorectal cancer, 257–258
lung cancer, 248–256
modeling, 294
prostate cancer, 261, 293–295
Senescence, 38, 40, 46, 62, 63, 75, 161, 286, 334
Smoking, *see* Cigarette smoking
Solar radiation, 122
Soot, occupational, 3, 123
Specific viral infections, 7, 46
Squamous cell carcinoma, 104, 105, 114, 218, 237, 240, 250, 283, 295–298, 342, 354, 382
Stage at diagnosis, 17, 20
Standardized mortality ratio, 208
State economic area, 206
Stochastic process model
radiation dose and, 194, 195, 196
application to the Extended Techa River Cohort, 195
parameter estimation, 194, 195
likelihood, 194, 195
microsimulation and, 196–199
Stomach, cancer
Adenocarcinoma, 257
age and, 257
alcohol and, 257
chemoprevention of, 324
diet and, 237
fruits and vegetables and, 257
*Helicobacter pylori* infection and, 237, 257, 329
incidence of, 257
international, 257
mortality from, 237
nitrate and, 135
prevention of, 257
race and, 237, 257
salt and, 257
sex and, 257
survival for, 257
time trends in, 257
U.S., 257
Surveillance, Epidemiology, and End Results ( SEER) program, 12, 16, 19, 20, 32, 62, 130, 162, 164, 217, 218, 219, 223, 225, 248, 251, 252, 260, 265, 275, 277, 283, 284, 290, 297, 300, 353
Survival rate, 5-year, 13, 14, 15, 17, 18, 20, 237
Susceptibility
age and, 283–290
genetics and, 282
host immune response and, 284
sex hormones "shift" and, 286

T
Tamoxifen, 99, 122, 235, 356
Techa river cohort, 195
Telomeres, 38, 43–44, 49, 61, 115, 116, 282, 330, 335, 358, 359
Testicular cancer
age and, 224–226
incidence of, 224
mortality from, 223
race and, 224
survival for, 223
time trend in, 224
Testosterone, 43, 292, 295, 328, 367
Thyroid cancer
age and, 131
Chernobyl nuclear power plant accident and, 132
cytokine "storm" and, 133
follicular, 130, 134
hormonal "shifts" and, 133
host factor and, 133
incidence of, 133, 241
iodine-131 and, 134
ionizing radiation and, 127–134, 194
mortality from, 134, 241
papillary, 134, 241
peri-menopause and, 133
puberty and, 133
race and, 241
sex and, 241
survival for, 241
time trend in, 241
thyroid-stimulating hormone and, 241
Tumor necrosis factor alpha, 41, 45
Tobacco, *see* Cigarette smoking
Tomatoes, 364, 368, 369

Transforming growth factor beta (TGF-β), 327, 330
Transurethral resection of the prostate, 236
tRNA, 39
Tsallis entropy, 33, 170
Tumor growth rate, 66–74, 205
Two-disease model
  age pattern of cancer incidence and, 285
  application, breast cancer mortality, 288
  breast cancer and, 277
  cervix cancer and, 283
  liver cancer and, 290–292
  prostate cancer and, 277, 293–295
Two-parameter Weibull function
  applications, 298–299
  generalizations, 192
Two-stage clonal expansion model
  applications, 57–58
  identifiability of parameters, 57
  specific case of two-stage MKV model, 56

U
Ultraviolet radiation, 121, 124
Uterine/corpus uteri cancer
  age and, 239–240
  hormone replacement therapy and, 239
  hysterectomy and, 240
  incidence of, 239
  menopause and, 239
  mortality from, 239
  obesity and, 239
  oral contraceptive and, 239
  parity and, 239
  race and, 239
  survival for, 240
  time trends in, 240

V
Vitamin A, 100, 135, 237, 355, 363, 364, 366, 369, 383
Vitamin C, 363, 364, 365, 368, 369, 383
Vitamin D, 355, 364, 411
Vitamin E, 355, 356, 363, 364, 365, 367, 368, 369
Viral infection
  cervical cancer, 37, 12, 75, 226, 240, 248, 260, 261, 277, 283–290, 297, 328, 380, 382
  *Hepatitis B virus*, 100, 121, 290
  *Hepatitis C virus*, 100, 121, 290
  hepatocellular carcinoma, 91, 104, 106, 226, 238, 248, 250, 258, 277, 288, 290–293, 297, 332, 352
  *Human immunodeficiency virus*, 30, 121
  *Human papillomavirus*, 29, 101, 120, 121, 284, 289
  liver cancer, 7, 22, 30, 107, 220, 224, 238, 241, 258–259, 290, 291, 292
  *See also* specific viral infections
Virchow, Rudolf, 4, 46, 47

W
Weibull hazard function, 54, 67, 71, 161
Weibull, E.H. Waloddi, 54, 62, 67, 68, 69, 70, 71, 72, 74, 161, 162, 163, 182, 183, 184, 191, 192, 262, 298, 375

X
X-rays, 3, 5, 119, 195

Z
Zinc, 46, 99, 100, 101, 237, 363, 364

Springer
the language of science

springer.com

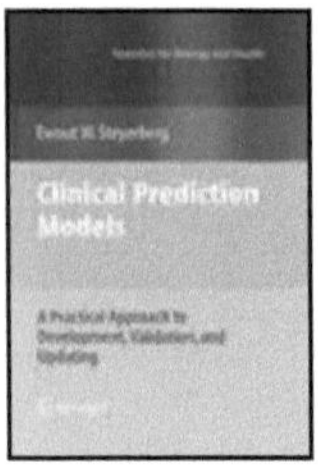

## Clinical Prediction Models

### A Practical Approach to Development, Validation, and Updating

**Ewout W. Steyerberg**

This book provides insight and practical illustrations on how modern statistical concepts and regression methods can be applied in medical prediction problems, including diagnostic and prognostic outcomes. Many advances have been made in statistical approaches towards outcome prediction, but these innovations are insufficiently applied in medical research. Old-fashioned, data hungry methods are often used in data sets of limited size, validation of predictions is not done or done simplistically, and updating of previously developed models is not considered. A sensible strategy is needed for model development, validation, and updating, such that prediction models can better support medical practice.

2008. 462 pp. (Statistics for Biology and Health) Hardcover
ISBN 978-0-387-77243-1

## Statistical Methods for Environmental Epidemiology with R

### A Case Study in Air Pollution and Health

**Roger D. Peng and Francesca Dominici**

This book provides an overview of the methods used for investigating the health effects of air pollution and gives examples and case studies in R which demonstrate the application of those methods to real data. The book will be useful to statisticians, epidemiologists, and graduate students working in the area of air pollution and health and others analyzing similar data.

2008. 150 pp. (Use R!) Softcover
ISBN 978-0-387-78166-2

## The Statistical Analysis of Functional MRI Data

**Nicole A. Lazar**

This book offers researchers who are interested in the analysis of fMRI data a detailed discussion from a statistical perspective that covers the entire process from data collection to the graphical presentation of results. The book is a valuable resource for statisticians who want to learn more about this growing field, and for neuroscientists who want to learn more about how their data can be analyzed.

2008. 289 pp. (Statistics for Biology and Health) Hardcover
ISBN 978-0-387-78190-7

**Easy Ways to Order▸** Call: Toll-Free 1-800-SPRINGER • E-mail: orders-ny@springer.com • Write: Springer, Dept. S8113, PO Box 2485, Secaucus, NJ 07096-2485 • Visit: Your local scientific bookstore or urge your librarian to order.

MIX
Papier aus verantwortungsvollen Quellen
Paper from responsible sources
FSC® C105338

If you have any concerns about our products,
you can contact us on
**ProductSafety@springernature.com**

In case Publisher is established outside the EU,
the EU authorized representative is:
**Springer Nature Customer Service Center GmbH**
**Europaplatz 3, 69115 Heidelberg, Germany**

Printed by Libri Plureos GmbH
in Hamburg, Germany